Quantum Transport in Semiconductor Devices

Simulation using particles

Online at: https://doi.org/10.1088/978-0-7503-5237-6

Quantum Transport in Semiconductor Devices

Simulation using particles

David K Ferry
School of Electrical, Computer, and Energy Engineering, Arizona State University, Tempe 85287-6206, AZ, USA

Xavier Oriols
Departament d'Enginyeria Electrònica, Universitat Autònoma de Barcelona, Escola d'Enginyeria, 08193, Cerdanyola del Vallès, Spain

Josef Weinbub
Institute for Microelectronics, Technische Universität Wien, Austria

IOP Publishing, Bristol, UK

ISBN 978-0-7503-5237-6 (ebook)
ISBN 978-0-7503-5234-5 (print)
ISBN 978-0-7503-5238-3 (myPrint)
ISBN 978-0-7503-5236-9 (mobi)

DOI 10.1088/978-0-7503-5237-6

Version: 20231101

IOP ebooks

British Library Cataloguing-in-Publication Data: A catalogue record for this book is available from the British Library.

Published by IOP Publishing, wholly owned by The Institute of Physics, London

IOP Publishing, No.2 The Distillery, Glassfields, Avon Street, Bristol, BS2 0GR, UK

US Office: IOP Publishing, Inc., 190 North Independence Mall West, Suite 601, Philadelphia, PA 19106, USA

To our daughters and sons.

Contents

Preface

The use of particles in the simulation of transport or in the study of device physics is actually older than quantum mechanics. Newton's laws of dynamics were described in the late 17th century, and these yield equations of motion that are certainly relevant for particles. In classical electronics, a particle description was available at the time of the first vacuum tubes, as it was the essence of the Langmuir–Child law for the vacuum diode. Even with the arrival of quantum mechanics, the use of particles was a central part of de Broglie's wave *and* particle picture, and it was clear from others' discussions, such as those of Kennard, that quantum mechanics primarily presented particle transport with additional (quantum) potentials. The rapid acceleration of science in the mid-20th century was accomplished by the use of particle simulations in many areas of physics, notably neutron transport, which was studied by von Neumann using the newly available computing machines. The use of statistical approaches appeared at the same time in the form of the Monte Carlo algorithms suited to these new computing machines.

Monte Carlo procedures were introduced to semiconductors in the mid-1960s, and their use blossomed following this introduction. Complicated physical systems could be described by their appropriate equations and the difficult solutions left to the computers to solve in a statistical manner. In the late 1980s, this approach was described in intimate detail for a wide range of scientific fields by Hockney and Eastwood's new book, *Computer Simulation with Particles*. This book represented a major teaching tool for new generations of students in engineering and science. Indeed, the current authors came to use it as a suitable starting point for new students in the field of electronics. In some sense, this book is the inspiration for this new work, which seeks to apply these tools to the more complex field of transport in quantum devices.

The arrival of quantum mechanics, and especially its applications in the fields of telecommunications, computing, and semiconductor devices, has certainly expanded the realm of simulation and modeling. Indeed, algorithms that account for quantization and an effective quantum potential have become central tenets of technical computer-aided design (TCAD), an approach that is fundamental in changing from the physics of devices to those of integrated circuits. However, in many approaches to quantum TCAD, the emphasis has shifted away from particle-based approaches to, for example, the use of nonequilibrium Green's functions (NEGFs), even in full-band approaches. As our colleagues have pointed out, this latter approach is computer intensive and does not possess any of the simplicity inherent in the detailed microscopic picture provided by Monte Carlo particle simulations. As a result, the absolute need for TCAD now suffers from the economic cost of needing ever-larger computing systems with which to implement NEGF simulations. In addition, simulation tools are more than just mathematical solutions for differential equations. They are expected to provide new views that illustrate our understanding of electron devices. The aforementioned Monte Carlo solution, apart from being a mathematical solution of the Boltzmann equation, also offers an appealing and intuitive view of the manner in which electrons traverse the active region. This intuitive view is far from the view that emerges from NEGF. In fact, this intuitive view clarifies discussions as to which types of perspectives are offered by most modern quantum simulators and avoids the usual endpoint of mere 'shut up and calculate' arguments that deny the need for physical intuition.

This brings us to the subject of this book, which the authors feel is necessary to bring the community back to the view that even quantum simulations can be accomplished with particle-based approaches. These approaches provide computational efficiency along with a view of the detailed physics but without the burdens imposed by NEGF. In other words, contrary to the directions some are taking, *particles are still important and efficient*! And, in a great many of these particle approaches to quantum transport and simulation, the Monte Carlo approach (no longer merely for the Boltzmann equation but also for new quantum equations of motion) remains a viable and integral method with which to proceed.

It has to be admitted that the introduction of quantum mechanics into the world of TCAD brings with it an enormous amount of baggage, which lies in the philosophical interpretation and/or understanding of quantum mechanics. From the beginning, there was debate among the chief advocates—Bohr and his acolytes on one side (known today as the orthodox approach) and on the other side, Einstein, de Broglie, Schrödinger, and their acolytes. The difference between these groups led to the initial 'debates,' and the arguments have not gone away even today. One well-known quantum scientist has claimed that Bohr brainwashed his acolytes into believing that the philosophical questions had been solved decades ago. Another stated that the result of Bohr's philosophy led to a situation in which no one could understand quantum mechanics. Thus, the philosophical questions and the debates continue. Actually, since the early days, several philosophical approaches have arisen that differ significantly from Bohr's orthodox viewpoint and yet still give a full account of quantization. Most importantly, many of these different philosophical approaches blend nicely into physics as it evolved from classical terms. The debate is discussed at various points within this book, not to highlight the debate, but to explain the philosophical points which enable one to use particles in quantum transport and quantum devices. This book provides an important relief for those readers who view electrons as particles traversing the active region from contact to contact. We hold that this view is not a naïve one when it comes to quantum devices. On the contrary, it is so rigorous that quantum transport models can be built upon it, as we endeavor to demonstrate in this book.

Often, the debate on how the quantum world needs to be understood leads to the question of just what a measurement entails and how it may be defined in quantum terms. This led to at least one early (and continuing, but orthodox) view in which measurement is said to 'collapse' the wave function, in essence destroying the previous quantum nature of the system. It is the case that an extremely wide range of physical measurements that can be made today and measurement after measurement after measurement tells us that the quantum system still exists. These facts suggest that perhaps this notion that the quantum nature disappears in response to the wave function collapse is an attempt to avoid a proper description of quantum reality.

Indeed, it may be assumed that modern measurements lay the 'complete collapse' assertion into its grave. But this is an argument for elsewhere, not for this book. For quantum transport simulators, perhaps the worst inheritance one receives from the orthodox fathers is the belief that electrons inside the active device cannot have microscopic properties. How can such an orthodox view be described as intuitive?

As quantum approaches to semiconductor transport have evolved in their application, it is clear that particle approaches have remained a goal worthy of achieving. The research literature describes several efforts to use particles to statistically treat the wave function itself, which is considered the basis of the statistical properties of quantum mechanics. Particles have naturally been used to follow the trajectories defined by the quantum potential in the hydrodynamical evaluation of Schrödinger's theory, an approach that began immediately after Schrödinger's own work in 1926. It is significant that recent studies in quantum optics have demonstrated that photons, used in a two-slit experiment, actually follow these quantum trajectories along with the quantum potential. These optical approaches have, in fact, verified this hydrodynamic approach to quantum transport. In addition, Monte Carlo approaches to evaluating transport using the Wigner function began to appear almost three decades ago, and these approaches have shown to be fully compatible with analytic approaches in predicting phase interference and correlations with real space potentials. Feynman himself held the view that the paths that formed the basis of his famous path integral represented complete particle trajectories and were to be interpreted in precisely this manner. Finally, it has recently been shown that even NEGF can be evaluated quickly and efficiently by particle Monte Carlo methods in certain situations and using a certain choice of basis set to formulate the problem.

A major question is: just how does quantum transport differ from classical transport? This is a topic that is discussed throughout this book. Briefly, it is not only the constraining dimensions of small systems that lead to quantization, but one should not ignore the importance of the environment, i.e. the world in which the device is embedded. It has been known for years, particularly in the area of integrated circuits, that the circuit affects the device performance and vice versa. Nowhere is this more true than in microwave circuits. But classically, it is felt that the circuit environment merely changes the boundary conditions that are seen by the internal physics of the device. This is dramatically different when one considers quantum systems. Here, the environment can physically change the entire nature of the quantization within the device. The coupling between the device and its environment can lead to some quantum states disappearing or becoming mere classical states but it also can lead to new quantum states that do not exist on their own in either the device or the environment. These ideas are the purpose of the book, as suggested by the title. In summary, it is evident that the laws of quantum mechanics can vary significantly from those of classical mechanics. However, physical theories encompass more than just laws; they also discuss the ontology, which are the elements used to understand the world. In this sense, there is no inherent reason to preclude quantum ontology from aligning with classical ontology. Consequently, quantum devices can be meticulously modeled and simulated using quantum laws, while their comprehension can align with a quantum ontology that mirrors the classical one dealing with particles. Different laws, yet an identical ontology to understand classical and quantum devices.

The range of devices in which quantum effects are significant is quite wide and ranges from the small devices of modern logic circuits to large, high-power devices in the microwave realm. Certainly, most would recognize the onset of quantization in the small devices, if for no other reason than the size quantization arising from the physical smallness of, and the resulting confinement of particles within, the device. In high-power devices, quantum effects actually become important because of the frequency at which these devices are operated. In the push to terahertz frequencies, the timescale over which transport occurs means that quantum states become quite important in the device physics. This means that quantum effects impact *all* devices, and most importantly, the transport within these devices. In some cases, it may be difficult to determine this, as the quantum effects may be masked by other observable effects. Nevertheless, in the microscopic world of transport, these effects become important and must be considered in order to obtain accurate TCAD representations. The authors hope that this book will enable a wider audience to more fully address the world of quantum transport in quantum semiconductor devices by considering a particle picture. Finally, additional resources may be found on the website: https://quantum-transport-in-semiconductor-devices-simulation-using-particles.uab.es, or by following the qr-code illustrated below.

David K Ferry
Xavier Oriols
Josef Weinbub

Acknowledgements

This book stands as a culmination of years of dedicated research in the field of quantum electron transport.

We are deeply grateful to the many colleagues and friends who engaged in valuable discussions that motivated us in writing this book. The list of those deserving recognition is too extensive to write it explicitly without omitting relevant names.

Our heartfelt appreciation extends to our families for their unwavering encouragement and understanding. Their patience and support sustained us through long hours of work.

We wish to acknowledge the support from numerous research entities in USA and Europe that enabled us to devote our research activity to these topics.

In particular, in this last year, Xavier Oriols would like to express gratitude to Spain's Ministerio de Ciencia, Innovación y Universidades for their support under Grant PID2021-127840NB-I00 (MICINN/AEI/FEDER, UE) and to the European Union's Horizon 2020 research and innovation programme under Grant 881603 GrapheneCore3 and Grant DIGITAL-101084035-DIGIQ: Digitally Enhanced European Quantum Technology Master. Josef Weinbub acknowledges support by the Austrian Science Fund (FWF) projects P33609-N and P37080-N.

Finally, our sincere thanks go to the IOP team whose belief in the potential of this project and guidance led it to fruition.

August 28, 2023

David K Ferry
Xavier Oriols
Josef Weinbub

Author biographies

David K Ferry

David K Ferry is Regents' Professor Emeritus in the School of Electrical, Computer, and Energy Engineering at Arizona State University (ASU). He moved to ASU in 1983, following shorter stints at Texas Tech University, the Office of Naval Research, and Colorado State University. He received his doctorate from the University of Texas, Austin and spent a postdoctoral period at the University of Vienna, Austria. His research is focused on semiconductors, particularly as they apply to nanotechnology and integrated circuits, as well as quantum effects in devices and materials. In 1999, he received the Cledo Brunetti Award from the Institute of Electrical and Electronics Engineers and is a fellow of this group as well as the American Physical Society and the Institute of Physics (UK). He is the author of some 40 books and book chapters and more than 900 refereed contributions.

Xavier Oriols

Xavier Oriols received BS and MS degrees in physics and a PhD degree in electronic engineering with an extraordinary doctoral award from the Universitat Autònoma de Barcelona (UAB), Barcelona, Spain in 1993, 1994, and 1999, respectively. In 1997, he worked at the Institut d'Électronique de Microélectronique et de Nanotechnologie, Lille, France. In 2001 and 2002, he was a visiting professor at The State University of New York, USA. Dr Oriols received the prize for young researchers in the framework of the Spanish I3 Program and several other prizes for research excellence. He is a full professor at UAB. He has authored more than 150 contributions to scientific journals and conferences. His research activity combines both his practical interest in quantum nanodevices and his interest in the foundations of quantum mechanics, particularly Bohmian mechanics. His research covers a wide spectrum from fundamental issues of physics to practical engineering of nanodevices.

Josef Weinbub

Josef Weinbub is an associate professor of high-performance simulation in micro- and nanoelectronics, an IEEE senior member, and the current vice-chair of the IEEE Nanotechnology Council's Modeling and Simulation Technical Committee. He obtained a doctoral degree in computational microelectronics and the *venia docendi* (habilitation) in the field of micro- and nanoelectronics from TU Wien. He was a visiting researcher at the EPCC, University of Edinburgh and at the Device Modelling Group, University of Glasgow, Scotland, UK, as well as at SILVACO Inc., Santa Clara, CA, USA. He founded and now chairs the master's program in Computational Science and Engineering at TU Wien and is involved with several international scientific conferences in various management and scientific roles. He is an associate editor of the Journal of Computational Electronics, a principal investigator of various research projects, and published over 200 contributions in scientific journals and conference proceedings. Together with his team, he investigates cutting-edge research problems in computational micro- and nanoelectronics with a particular focus on developing and utilizing particle Wigner-based quantum transport simulations.

Part I

Introduction

Chapter 1

Introduction

Even classical transport has many parts. People living in the Stone Age certainly understood the role of forces and the reactions that those forces caused. Even wind forces have been controlled by sailing ships for more than 4000 years. However, it was Newton's *Principia* [1] and laws of motion that established classical mechanics as a scientific field. Newton's laws described how a force would act upon a body and how the body would move in response to that force. Naturally, in the world of small things such as electrons, these laws were still held to apply. But electrons generally come in bunches, especially in things like semiconductor devices, which led to the role of statistical physics. As early as 1868 [2], Boltzmann had developed the idea of a distribution function in which the role of a single electron could be envisioned even though it depended upon all the other electrons. He showed that this distribution could be described by a phase space point that would move along a trajectory and in a sense also established the possibility of using particles to simulate the motion of the whole. The introduction of the *Stosszahlansatz* (molecular chaos hypothesis) to uncouple the electron from the rest of the group, which led to the Boltzmann transport equation, came just a few years later [3]. The important point here is that the introduction of the single-particle distribution function allows an ensemble of particles to be treated in a simplified but analytical manner. The use of particles in computational schemes to simulate the dynamics of the distribution function came later, yet led to the various techniques that have arisen for this purpose.

Generally, quantum mechanics is thought to have arrived with Planck and his need to assume that light was composed of particles in order to bring his new theory of black-radiation into agreement with experiments [4]; the light particle had an energy $E = hf, \quad \hbar = h/2\pi$, where h is Planck's constant and f is the light frequency. However, it is fair to point out that light had been assumed to be composed of particles since the time of the ancient Greeks and is described in this way in Newton's work [5], where he discussed the corpuscular nature of light. Even so, light became known as a wave in the eyes of many scientists beginning in the 17th

doi:10.1088/978-0-7503-5237-6ch1

century, a view supported by the diffraction studies of Young [6]. This was all cemented by Maxwell's wave theory of electromagnetic radiation [7]. The nature of light became a conundrum after Planck and particularly after the experiments of Taylor [8], in which weak light was shown to diffract around a needle but to arrive at a photographic plate as a particle.

Thus, light was shown to have a dual personality. It was both a particle and a wave, and the relevant description depended upon the nature of the measurement being made. On the other hand, the electron was clearly a particle, at least for the time being. This would soon change when Bohr presented his theory of the atom, in which the existence of Planck's quantization would lead to the quantization of the orbiting electron's angular momentum [9]. This would certainly be a wavelike property, in that the orbit had to contain a fixed number of wavelengths or fractions thereof.

Before proceeding too deeply into our topic, it will be necessary to make some comments on the fundamental understanding(s) of quantum mechanics in the following sections. In principle, it might seem strange that a book devoted to engineers and scientists interested in nanoscale semiconductor devices would say something about the foundations of quantum mechanics. After all, the role of engineers is usually thought of as just using mature physical theories to provide practical applications that improve society. In the present case, this would take the form of understanding faster and smaller semiconductor devices. But the key question in this regard is this: is the quantum theory really a mature discipline that is free from ambiguities? It was developed more than a century ago, so it seems old enough to be a mature theory that engineers can use without any concerns. But it is still true today that both philosophers and physicists continue to have a lively debate on what the quantum theory really means. In fact, there are strong disagreements among the scientific community about how the quantum world should be understood. Therefore, this chapter introduces a discussion about how this quantum world is to be understood. We will return to this below. But first, some discussion of the forces and behaviors in classical systems is in order.

1.1 Particles in classical transport

Newton gave the world three laws [1], in which the force acting upon a body is intrinsic to the second and third laws. Here, he established that (second law)

$$\frac{d(mv)}{dt} = \boldsymbol{F}, \tag{1.1}$$

where v is the velocity of the particle, m is its mass, and $\boldsymbol{F}$ is the applied force. In Newton's day, the law was used for macroscopic things, for which the mass was always a constant and could be brought outside the derivative. (Here, the bold characters are indicative of vectors in three-dimensional Cartesian space.) In semiconductors, this is no longer the case, as will be detailed later. The third law showed that two bodies which interacted did so in a manner that depended upon their relative distance apart, and the resultant forces were equal and opposite for the two particles. This led to a form found for both coulomb interactions and gravitational interactions, namely

$$\boldsymbol{F}_1 = -\frac{K}{\boldsymbol{x}_1^2 - \boldsymbol{x}_2^2}, \tag{1.2}$$

where K is a constant that depends upon the nature of the force field and the $\boldsymbol{x}_i$ are the two positions of the particles in three dimensions.

Equation (1.2) allows us to discuss the situation in which there are a great many particles. If there are N particles, one needs to have $6N$ coordinates in configuration space as well as time, which represents the progress of the system of particles. This is a lot of coordinates to deal with, and one can see why Boltzmann sought ways to treat a one-particle distribution, which we deal with below. A century after Newton, Lagrange published his view of classical mechanics [10], in which the trajectory of each particle would define a path in which the action was minimal. The principal quantity here is the Lagrangian

$$L = T - V, \tag{1.3}$$

in which T is the kinetic energy and V is the potential energy, and the latter arises from an external potential or the interaction potential between the particles. The kinetic energy arises from

$$T = \sum_i \frac{m_i}{2} v_i^2 = \sum_i \frac{p_i^2}{2m_i}, \tag{1.4}$$

where p_i is the momentum of particle i. The action mentioned above is the integral of the Lagrangian over time, as follows:

$$S = \int_{t_1}^{t_2} \sum_i L(\boldsymbol{x}_i, \boldsymbol{p}_i, t)dt. \tag{1.5}$$

Here, the Lagrangian L exists in the phase space of position and momentum. The position primarily arises from the potential energy V, and the momentum primarily arises from the kinetic energy of (1.4). Minimizing this function gives rise to the principal equation of motion, which is analogous to Newton's laws:

$$\frac{d}{dt}\left(\frac{\partial L}{\partial v_i}\right) - \frac{\partial L}{\partial \boldsymbol{x}_i} = 0, \tag{1.6}$$

which, of course, reduces to (1.1). Later, Hamilton reformulated the approach (using a function that has come to be called the Hamiltonian [11]) with

$$H = T + U, \tag{1.7}$$

(here, we use U for potential energy and V for electric potential to avoid confusion) for which the relevant equations become

$$\begin{aligned}\frac{d\boldsymbol{x}_i}{dt} &= \nabla_p H \\ \frac{d\boldsymbol{p}_i}{dt} &= -\nabla H\end{aligned}, \tag{1.8}$$

where the subscript on the right of the first equation denotes differentiation with respect to momentum, while the unsubscripted ∇ in the second line denotes differentiation with respect to position. The first of these gives the rate of change of position, i.e. the velocity, and the second gives us the generalized form of Newton's second law. We now have the dynamics of the entire set of N particles.

The dynamics may be connected to an N-particle distribution function by defining the latter as $f_N = f_N(\boldsymbol{x}_1 \ldots \boldsymbol{x}_N; \boldsymbol{p}_1 \ldots \boldsymbol{p}_N; t)$, which may be considered to be the probability of finding the ensemble of particles in a configuration in which each particle has one of the coordinates and momentum defined in the description. The reduction of this N-particle probability distribution to a one-particle distribution is normally referred to as the Bogoliubov–Born–Green–Kirkwood–Yvon (BBGKY) hierarchy [12–15]. To begin with, the total time derivative is given by the chain rule of differentiation, as follows:

$$\frac{df_N}{dt} = \frac{\partial f_N}{\partial t} + \sum_{i=1}^{N} \frac{\partial \boldsymbol{x}_i}{\partial t}\frac{\partial f_N}{\partial \boldsymbol{x}_i} + \sum_{i=1}^{N} \frac{\partial \boldsymbol{p}_i}{\partial t}\frac{\partial f_N}{\partial \boldsymbol{p}_i} = 0, \tag{1.9}$$

which is required in the absence of any scattering in order to conserve phase space during the evolution of the ensemble. If the equations of motion from (1.6) and (1.8) are used, this can be written as

$$\frac{\partial f_N}{\partial t} + \sum_{i=1}^{N} \frac{\boldsymbol{p}_i}{m_i}\frac{\partial f_N}{\partial \boldsymbol{x}_i} + \sum_{i=1}^{N} \boldsymbol{F}_i \frac{\partial f_N}{\partial \boldsymbol{p}_i} = 0. \tag{1.10}$$

Here, the force is the sum of the interparticle interaction as well as the external potential that may be applied to the system

$$\boldsymbol{F}_i = -\frac{\partial V_{\text{ext}}}{\partial \boldsymbol{x}_i} - \sum_{j=1, \neq i}^{N} \frac{\partial \varphi_{ij}}{\partial \boldsymbol{x}_i}, \tag{1.11}$$

while φ is the interparticle potential leading to the force (1.2).

The normalization of f_N is normally accomplished by integrating over all the position variables in three dimensions in order to get the unity value. This requires normalization by appropriate volumes. But this can all be simplified merely by defining the reduced distribution, an s-particle distribution, as follows:

$$f_s = \int f_N(\boldsymbol{x}_1 \ldots \boldsymbol{x}_N; \boldsymbol{p}_1 \ldots \boldsymbol{p}_N; t) d\boldsymbol{x}_{s+1} \ldots d\boldsymbol{x}_N d\boldsymbol{p}_{s+1} \ldots d\boldsymbol{p}_N. \tag{1.12}$$

In quantum mechanics, the distribution function would be replaced by the density matrix and the integral in (1.12) by a trace over the important parts of the matrix [16]. Here, we will continue to follow the classical route. Using this definition, nearly all of the terms in (1.10) and (1.11) follow in a straightforward manner. The exception is the last term of (1.11). This term involves two particles. If both particles are in the first set $\{1, s\}$, the choice is easy. If both particles are in the set $\{s+1, N\}$, they merely provide a constant shift in the potential. The actual term that is important is the term that occurs when one particle is in the first set and the second lies in the second set. In this latter case, the variables can be interchanged among themselves (for the second set), so that the important potential is that of a particle in the first set and the $s+1$th particle. This now leads to the equation

$$\frac{\partial f_s}{\partial t} + \sum_{i=1}^{s} \frac{\boldsymbol{p}_i}{m_i} \frac{\partial f_s}{\partial \boldsymbol{x}_i} - \sum_{i=1}^{s} \left(\frac{\partial V_{\text{ext}}}{\partial \boldsymbol{x}_i} - \sum_{j=1,\neq i}^{s} \frac{\partial \varphi_{ij}}{\partial \boldsymbol{x}_i} \right) \frac{\partial f_s}{\partial \boldsymbol{p}_i} = (N-s) \sum_{i=1}^{s} \int \frac{\partial \varphi_{i,\,s+1}}{\partial \boldsymbol{x}_i} \frac{\partial f_{s+1}}{\partial \boldsymbol{p}_i} d\boldsymbol{x}_{s+1} d\boldsymbol{p}_{s+1} \tag{1.13}$$

The Boltzmann equation is achieved by letting $s = 1$ in (1.13) and then invoking the *Stosszahlansatz* (molecular chaos hypothesis). First, using reduction to a single particle gives

$$\frac{\partial f_1}{\partial t} + \frac{\boldsymbol{p}_1}{m_1} \frac{\partial f_1}{\partial \boldsymbol{x}_1} - \left(\frac{\partial V_{ext}}{\partial \boldsymbol{x}_1} \right) \frac{\partial f_1}{\partial \boldsymbol{p}_1} = (N-1) \int \frac{\partial \varphi_{12}}{\partial \boldsymbol{x}_1} \frac{\partial f_2}{\partial \boldsymbol{p}_1} d\boldsymbol{x}_2 d\boldsymbol{p}_2 . \tag{1.14}$$

In the last term (on the right-hand side), the two-particle distribution appears and gives rise to the correlation between the various particles. One often sees the two-particle function written as

$$f_2(\boldsymbol{x}_1, \boldsymbol{x}_2) = f_1(\boldsymbol{x}_1) f_1(\boldsymbol{x}_2) - g(\boldsymbol{x}_1, \boldsymbol{x}_2), \tag{1.15}$$

where g is the explicit two-particle correlation. The *Stosszahlansatz* is the explicit invalidation of this last correlation term. The last term on the right-hand side of (1.14) then simply becomes the Hartree potential, which gives the interaction between an electron and the smooth background of charge. This potential is usually found from Poisson's equation and contributes to the self-consistent potentials used in semiconductor device simulations [17].

In general, one now can determine the average velocity of the particles by integrating the velocity over this distribution function rather than by taking an average over the ensemble of particles. However, there is still the problem of the normalization of this distribution function. There are various manners in which this can be accomplished. One approach is to say that this one-particle distribution function (1.14) gives the probability of finding a particle in boxes of volume $\boldsymbol{\Delta x}$ (centered around $\boldsymbol{x}$) and $\boldsymbol{\Delta p}$ (centered around $\boldsymbol{p}$). In this sense, the distribution function is described in a six-dimensional phase space, and the entities $\boldsymbol{x}$ and $\boldsymbol{p}$ do not refer to the actual position of any single carrier but to the position in phase space. The requirement that the probability must sum to unity then requires

$$\int \int f(\boldsymbol{x}, \boldsymbol{p}, t) d\boldsymbol{x} d\boldsymbol{p} = 1. \tag{1.16}$$

However, this is not the only definition that can be made, nor is it the most convenient one to use.

An alternative method, and the one actually used in (1.13), is to define the distribution function as the average number of particles in a box of volume $\boldsymbol{\Delta x}$ (centered around $\boldsymbol{x}$) and a box $\boldsymbol{\Delta p}$ (centered around $\boldsymbol{p}$). The integration over the distribution then yields

$$\int\int f(\boldsymbol{x}, \boldsymbol{p}, t) d\boldsymbol{x} d\boldsymbol{p} = N(t), \tag{1.17}$$

where $N(t)$ is the total number of particles in the entire system at time t. It is important to note that the Fermi–Dirac distribution does not satisfy either of these two definitions, since this latter distribution has a maximum amplitude of one and the integral always yields the Fermi energy E_F. In condensed matter, such as semiconductors, the differentials have correction factors that have to be included for accuracy. First, it may be recalled that the momentum wave vector k is not treated as a continuous variable in condensed matter. Rather, k is quantized into discrete values that are separated by numerical factors, so that a factor of $1/(2\pi)^3$ has to added to the differential of momentum [18]. In addition, the actual momentum $\boldsymbol{p}$ differs from the wave vector k by a factor of $\hbar$ in each physical dimension, so this adds a factor of $1/\hbar^3$ to the differential. Finally, in quantum mechanics, each state determined by a single wave number value may hold two electrons of opposite spin. Therefore, a final factor of two must be added to the differential to fully count all the available states. Hence, we can already see that quantum mechanics has complicated the interpretation of the distribution function. It will lead to other complications.

However, one must remember that this distribution function is a representation of an ensemble of particles. Before computers became available, it was easier to do computations by hand using the distribution function and the Boltzmann equation (1.14). But this changed around the middle of the twentieth century. In fact, the concept of using random techniques dates to an even earlier time, that of Leclerc [19], who tried to estimate the value of π by dropping needles on the floor. Closer to the former time, Fermi apparently experimented with the use of a Monte Carlo process to estimate neutron transport, but Ulam is generally credited with developing the modern Markov chain process, which was named Metropolis by his colleague [20]. This approach was employed by von Neumann to study neutron transport during World War II, using the new digital computer at the University of Pennsylvania [21]. This approach, referred to as kinetic Monte Carlo in recent times, was apparently first used for electron transport in semiconductors by Kurosawa [22]. Subsequently, the use of particles in computer simulation was discussed for a wide range of fields [23]. This latter work opened the door to a world of applications for simulations based on particles, especially in semiconductor devices.

1.2 The quantum mechanical view of particles

It was remarked at the beginning of this chapter that the first discussion of quantization in the atom was the 1913 paper by Bohr [9]. This is not quite true.

Rutherford first proposed a model of the atom in which the atom consisted of a central, small heavy core concentrated almost at a point and surrounded by the various electrons [24]. Earlier, Nagaoka had developed a planar, almost two-dimensional model of electrons orbiting a central core [25], which was subsequently discovered by Rutherford and discussed in his group, which included Bohr. Bohr then developed his model of the atom, assuming Planck's quantization of the photon, and found that this led to quantization of the electron's angular momentum [9]. On discovering this, he reiterated his work with the initial assumption that the angular momentum was quantized and as a result found Planck's quantization [9]. But it would remain for Sommerfeld to work out the full three-dimensional, spherically symmetric model of the atom with the electrons in shells around the central core [26].

Subsequently, the quantization of the angular momentum was expressed by Einstein through the integral [27]

$$\Gamma_C = \oint \boldsymbol{p} \cdot d\boldsymbol{l} = 2\pi n\hbar. \tag{1.18}$$

This was improved by Brillouin [28] and Keller [29] to account for other phase contributions, as follows:

$$\Gamma_C = \oint \boldsymbol{p} \cdot d\boldsymbol{l} = 2\pi\hbar\left(n + \frac{\mu}{4} + \frac{b}{2}\right), \tag{1.19}$$

where μ and b are Maslov indices, and (1.19) is more commonly known as Einstein–Brillouin–Keller (EBK) quantization. Here, μ normally counts the number of turning points in a classical trajectory, and b counts the number of hard wall reflections in this trajectory. Equation (1.19) is also associated with the later development of Berry's phase in classical dynamics and chaos [30] and the Gutzwiller trace formula for computing the spectra of periodic orbits in chaotic systems [31]. While (1.18) and (1.19) are generally described as semiclassical quantization, they were to appear again in the work of Heisenberg and again in the works of Madelung, Kennard, and Bohm, the latter of which are discussed in later sections of this chapter.

This is all useful, but the title of this section indicates that there may be a question about how one is to understand the word 'particle.' However, we will see that there is not even a consensus on the reality of the wave function itself. In this case, what is an electron? What is a particle? For example, Bohr was adamant in later years that a particle (in particular, the electron) had no meaning in his description of the atom. This was a result of his view of the role of measurement in defining the quantum system. On the other hand, other interpretations of quantum mechanics had a more realistic view of the world and accepted that particles could play a role. Nevertheless, there is certainly a lack of consensus in these definitions. Each interpretation of quantum mechanics and the reality of the wave function introduces its own concept of a particle. These various concepts and interpretations will be discussed in the following sections, and this also creates differences in just what is meant by a particle.

1.2.1 Heisenberg and quantization

In early 1925, Heisenberg spent time on a remote island north of Germany, seeking to relieve his allergies. While there, he worked out his view of how quantum mechanics should be described. While he was a student of Max Born in Göttingen, he had spent considerable time in Copenhagen with Bohr, so he was familiar with the positivist or logical empiricism viewpoints Bohr espoused that derived from the work of Ernst Mach in the previous century. These views held that reality existed only when a measurement was actually made (these views will be discussed below in section 1.3). Heisenberg's new theory of quantum mechanics was published later that year [32], and here he begins with the statement: 'The present paper seeks to establish a basis for theoretical quantum mechanics founded exclusively upon relationships between quantities which in principle are observable.' Observability is thus a central concept, and Bohr would extend this in his orthodox theory to mean that any concept of a particle has no meaning until such an observation is made. In this orthodox view, the electron is nothing (or at least, it is nothing with a well-defined position) before the measurement takes place. When a positional measurement takes place, the wave function that describes the quantum system is both formed and then collapsed into some eigenstate of the position, and the electron 'appears' as an entity with a well-defined position. Notice that the properties of an electron (a wave or a particle) strongly depends on the details of the measurement.

In his original work, Heisenberg spent considerable effort comparing the new quantum mechanics with classical mechanics, pointing out that they should begin to merge in their predictions for large quantum numbers, yet he acknowledges that the Einstein–Bohr approaches given by and large by (1.18) are already a significant departure from classical mechanics, so much so that even the simplest quantum problem is failed by classical approaches. In his development, he depends upon condition (1.18) for periodic motion, and this allows him to focus upon the frequencies of transition between different atomic states (each of which has its own periodic motion), finding that these frequency differences were naturally quantized, as would be the energy differences, due to Planck's relation. He found that these atomic energy levels were given by

$$E_n \sim -\frac{1}{n^2}, \quad n = k + l + 1, \tag{1.20}$$

where n is the principal quantum number, k is the radial quantum number, and l is the angular momentum quantum number. The number of constants here arises from the different turning points for motion in the different coordinate directions (r, φ, and θ in spherical coordinates). These energy levels have different values than those proscribed by Bohr in his early theory, but the transitions remain unchanged. The problem lay in the interpretations, for Heisenberg is said to have claimed that quantities which could not be measured [32] '…had no sense…,' and this has strong results for the orthodox approach. While this may be viewed as saying that these unmeasurable properties do not have reality, as would be claimed by Bohr, this view is not supported either in classical mechanics or in some modern quantum mechanics.

Born recognized that the mathematics described in Heisenberg's paper was akin to matrix mechanics and induced Jordan to work on the topic. Jordan first assured himself that the operations described by Heisenberg were consistent with matrix operations, that they also satisfied Hamilton's canonical equations of motion (1.8), and that this approach would lead to classical results in the appropriate limit. Here, Heisenberg gives us a conundrum, in that the particle needs to exist if it is to respond to Hamilton's canonical equations, while still making '…no sense….' It can hardly be described as having both characteristics.

Heisenberg's work also led to the important commutator relationship between position and momentum [33]

$$xp - px = i\hbar, \tag{1.21}$$

where x and p are the position and momentum *operators*. While this is written as a scalar, it is true for vectors, but only when these two operators act in the same direction, e.g. both are $\hat{x}$-components (where $\hat{x}$ is here a Cartesian axis). An additional result that he obtained was that the operators would attain their own time dependence from their interaction with the Hamiltonian, even if they had no intrinsic time dependence, through the relationship

$$\frac{dO}{dt} = \frac{1}{\hbar}(OH - HO), \tag{1.22}$$

where O is an arbitrary operator and H is the Hamiltonian (1.7). Once more, we see the conundrum. Are position and momentum real properties, or does the uncertainty relation (1.21) apply only to observables arising from measurements? Uncertainly exists classically between Fourier transform pairs, as anyone who has watched a train pass understands. How, then, does (1.21) differ from classical values of this uncertainty?

Proofs of Heisenberg's paper reached England quickly, and these were passed to another student, Paul Dirac. Dirac recognized the matrix mechanics and quickly developed a mathematical formalism using this result [34]. In this effort, Dirac replaced Heisenberg's dependence upon experimental observables with the importance of a proper mathematical theory. Reality now naturally flowed from the basis upon which the theory was produced. That is, variables within the theory had reality whether or not they were observed, or even observable. More importantly, Dirac developed the theory further and introduced a many-body wave function in the form of a determinant [35].

Heisenberg was to continue in his view of the necessity of measurement and developed the uncertainty principle mentioned above [36]. In this, he goes further in arguing that, if one wishes to know the position of an object, then one first has to describe the experimental method by which that position is to be determined. Of course, some quite similar arguments have to be made classically if one is interested in the position of an object; one often forgets that simply looking at the object is, in fact, a form of measurement. The argument continues to be whether or not that object exists prior to the act of measurement. Nevertheless, Heisenberg asserts that the quantum uncertainty goes beyond that naturally found in classical systems.

In general, it may be said that the Heisenberg approach to quantum mechanics is one based upon the Hamiltonian, with the resulting form of dynamics and kinematics. But his first paper on the topic is largely a long discussion of the differences between various atomic states. Nevertheless, the continued development of this approach has been termed matrix mechanics to differentiate it from other forms of the new quantum mechanics. An important point is that Heisenberg said nothing about causality in this first paper. If Jordan was correct, and the approach is equivalent to Hamilton's canonical equations of motion (1.8), then it must be assumed that these forces lead to causal behavior. This was to be disputed later, as discussed below.

1.2.2 De Broglie and Schrödinger

Even earlier than Heisenberg, Louis de Broglie had worried about the duality of photons as particles and waves. Following the experiments of Taylor [8], he began to describe his 'atoms of light' as riding on a nonmaterial light wave [37]. The frequency of this wave was determined by Planck's relation. This was an early version of his pilot wave theory, in which the wave guided the particle in its motion. In this approach, it was said that he created the idea of wave *and* particle [38]. Bohr's atomic theory led de Broglie to conclude that if the electron motion around the atom (the angular momentum) was quantized, then the electron must also be a wave, as this quantization arose from the finite wavelength contained in a closed orbit [39]. This contrasted dramatically with Bohr's ideas of reality, as there was now both a wave and a particle, both of which had a reality from the beginning and did not depend upon a measurement later in time. But with the need to design as well as to measure, Bohr's philosophical views are both difficult to understand and difficult to accept, given the need to understand a reality when dealing with semiconductor devices.

Einstein had already shown that the photon had a momentum hf/c [40], where f is the frequency and c is the speed of light. Hence, it was relatively easy for de Broglie to connect this momentum to a wavelength ($\lambda = c/f$) as follows [39]:

$$\lambda = \frac{h}{p}. \tag{1.23}$$

Here, however, de Broglie pointed out that p was not the momentum of the light wave, but of the particle itself that accompanied the light wave. This is in direct contrast to Bohr's view, and the particle now clearly exists from the beginning. This momentum is the electron's momentum when applied to electron particles and waves, and (1.18) would assure quantization of the momentum. This connection between the particle electron and the wave was a new process in understanding how waves and particles represented a duality of meaning and interpretation. It might be an understatement to say that Bohr was not pleased with this viewpoint. The later discovery of electron diffraction in x-ray scattering cemented the view that these particles had wavelike properties as well.

Einstein was intrigued with these new ideas. He felt that de Broglie had made a direct link between the waves that represented matter, the particles of that matter, and the atom itself [41]. The simple formula (1.23) suggested that all physical matter, such as particles, could also be viewed as a wave in the quantum world. But this raised a new problem, as these waves did not have an appropriate equation. Light waves or radio waves had Maxwell's equations but matter waves had nothing.

Erwin Schrödinger had seen de Broglie's work and took it to heart. Over a winter holiday at the end of 1925, he developed his famous equation [42], used it to determine the frequencies of Planck's oscillators [43], and showed how the matrix mechanics followed from his equation [44]. His famous equation, in one dimension, is given as follows:

$$i\hbar\frac{\partial\psi(x, t)}{\partial t} = -\frac{\hbar^2}{2m}\frac{\partial^2\psi(x, t)}{\partial x^2} + U(x)\psi(x, t), \tag{1.24}$$

where $\psi(x, t)$ is the wave function of matter. Note that this wave function is a complex quantity. $U(x)$ is the potential, just as it appears in (1.3) or (1.7). One might say that this result crashed upon the physics community like a tidal wave. It produced all the correct results from Heisenberg and/or Bohr, and it was easy to understand. Classical physicists understood waves; they could get a good grasp of this new form of mechanics. These waves were continuous, even if their frequencies were not. Moreover, perturbation theory showed that particles would move smoothly, over time, from one state to another, appearing to jump only when viewed on a long-term basis. (This was much like old movies in which the small number of frames per second made the moving actors appear to move from one place to another in a series of small jumps.)

What made the Schrödinger equation different was the complex nature of the wave functions and the explicit imaginary coefficient of the time derivative term. This was new territory in science. However, in the stationary situation, in which there was no time variation, the remaining equation was well known to the community. It was merely a particular form of the Sturm–Liouville equation, as found in almost any bound-state eigenvalue problem from classical mathematics [45]. These problems are commonly referred to as boundary-value problems. And nearly all mathematicians and physicists were taught how to solve these problems. The unique importance of such boundary-value problems is that their solutions form the basis for the mathematics of linear algebra; that is, for various sets whose members are the solutions to the particular problem. In such linear algebra, a field is a function with a specific value at each point space. A vector field also has a specific direction in space. These series are familiar to engineers trained in electromagnetic fields, modern circuit theory, or system theory. The clarity of the Schrödinger picture is thus easily accepted by engineers and scientists with these backgrounds.

What was new was the idea of differential operators, which had already been used by Heisenberg. In this new wave approach, the momentum operator was expressed as

$$p = -i\hbar\frac{\partial}{\partial x} \tag{1.25}$$

in one dimension. The right-hand side of (1.24) could then be recognized as

$$\left(\frac{p^2}{2m} + U\right)\psi = H\psi, \tag{1.26}$$

since the term in parentheses on the left is clearly the Hamiltonian (1.7) with (1.4). The difference from classical mechanics lies solely in the fact that the momentum is now an operator. Since the right-hand side of (1.24) or (1.26) is a simple eigenvalue problem, there exists a set of solutions $\phi_i(x)$, each of which has its own eigenvalue E_i, whether of classical or quantum origin. This can be used by defining the wave function in terms of this set, as follows:

$$\psi(x) = \sum_i a_i\phi_i(x), \tag{1.27}$$

where the constants a_i are found from the solutions and boundary values applied to the equation. However, (1.27) can be used in (1.24) to find the time-dependent solutions, as follows:

$$\psi(x, t) = \sum_i a_i\phi_i(x)e^{-iE_i/\hbar}. \tag{1.28}$$

The familiarity with classical waves was now an advantage in learning the new quantum mechanics. It was no coincidence, then, that Schrödinger's new wave mechanics was favored over matrix mechanics as the use of quantum mechanics expanded in the community. The problem is that this is a wave theory, and the role of the particles had been lost. They were to be recovered sooner rather than later.

1.2.3 The hydrodynamic or Bohmian view

Almost as soon as Schrödinger's papers appeared, it was noted by Madelung that the wave equation could be cast in a hydrodynamic form [46]. Noting that Schrödinger's equation already involved complex variables, Madelung suggested writing the wave function in the form (now in the full three-dimensional space)

$$\psi(\mathbf{x}, t) = R(\mathbf{x}, t)\exp[iS(\mathbf{x}, t)/\hbar], \tag{1.29}$$

where R is an amplitude and S is the action, although this latter quantity appears in a slightly different form than (1.5). Both of these two quantities are taken to be real functions. If (1.29) is used in the three-dimensional form of (1.24), a long, complex equation is found, after the exponential is factored from the equation, as follows:

$$\begin{aligned} -R\frac{\partial S}{\partial t} + i\hbar\frac{\partial R}{\partial t} &= \frac{R}{2m}(\nabla S)^2 - \frac{i\hbar R}{2m}\nabla^2 S \\ &- \frac{i\hbar}{m}\nabla \mathrm{S}\cdot\nabla R - \frac{\hbar^2}{2m}\nabla^2 R + VR \end{aligned}. \tag{1.30}$$

For this equation to balance with the equality, it is required that both the real and imaginary terms balance separately. Consider the real parts, which lead to the equation

$$-R\frac{\partial S}{\partial t} = \frac{R}{2m}(\nabla S)^2 - \frac{\hbar^2}{2m}\nabla^2 R + UR. \tag{1.31}$$

Here, one can divide by the factor R and then rearrange the terms to yield

$$-\frac{\partial S}{\partial t} = \frac{(\nabla S)^2}{2m} + U - \frac{\hbar^2}{2mR}\nabla^2 R. \tag{1.32}$$

Madelung noted that the negative time derivative of the action is the energy, so that the left-hand side and the first two terms on the right-hand side are just (1.7) if the momentum is defined as

$$p = \nabla S. \tag{1.33}$$

Such a definition is consistent with Hamilton's equations of motion, particularly with the second equation of (1.8), especially if the time derivative of S is the classical Hamiltonian. But (1.32) contains another term, specifically

$$Q = -\frac{\hbar^2}{2mR}\nabla^2 R. \tag{1.34}$$

This term is often called the *quantum potential.* While the hydrodynamic connection was not particularly highlighted in the early years, it was later rediscovered by Bohm [47], so that (1.34) is often termed the Bohm potential.

The quantum potential was investigated for particles moving through a pair of slits (this is the particle version of Young's two-slit experiment [48]) and then used to compute the trajectories of particles following the hydrodynamic view. This clearly established that the hydrodynamic view would exhibit the results of the two-slit experiment. This simulation has been done more recently, as shown in figure 1.1 [49]. Here, two sources are used to create the particle and wave beams entering the region of interest. Clearly these two sources can be considered to be the outputs of a two-slit experiment. The wave and the particles have a velocity initially set as part of the initial conditions. Wave propagation and particle propagation, the latter of which is guided by the quantum potential, are then followed in time. In panel (a), the quantum wave function is plotted in position and time and the amplitude is color coded. Red represents the lowest values and blue the highest values. In panel (b) of this figure, the actual particle trajectories flowing from the two sources are shown, and one can see how the interference fringes are actually formed in the resulting pattern. Finally, in panel (c), the trajectories of (b) are overlaid onto the wave function of (a). In essence, the quantum potential gives rise to nonlocal forces that lead to the interference pattern that appears on any screen placed to intercept these particles.

The imaginary terms in (1.30) may be collected to give yet another equation that must be satisfied, and this is

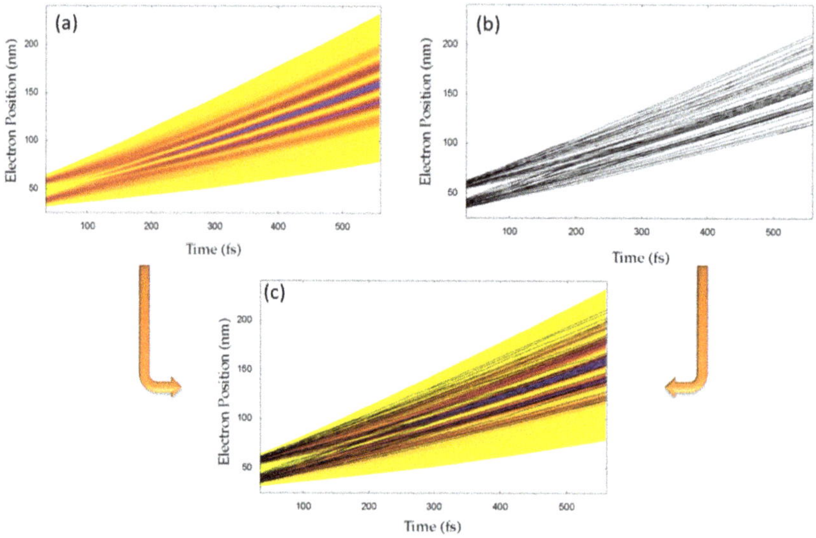

Figure 1.1. (a) The quantum potential found by solving the Schrödinger equation for waves passing through two slits. (b) Trajectories for particles initialized in the two slits as they progress under the action of the quantum potential. (c) The two are now superimposed on each other. One can easily see how the interference pattern forms from the nonlocal effects of this potential.

$$i\hbar\frac{\partial R}{\partial t} = -\frac{i\hbar R}{2m}\nabla^2 S - \frac{i\hbar}{m}\nabla S \cdot \nabla R, \tag{1.35}$$

which may be rearranged to give

$$\frac{\partial R}{\partial t} + \frac{R}{2m}\nabla^2 S + \frac{1}{m}\nabla S \cdot \nabla R. \tag{1.36}$$

If each term is premultiplied by a factor of $2R$, then this last equation may be rewritten as

$$\frac{\partial R^2}{\partial t} + \frac{1}{m}\nabla \cdot (R^2 \nabla S) = 0. \tag{1.37}$$

It may immediately be recognized that $R^2 = |\psi|^2$ from (1.29). This quantity is recognized as the probability that the 'particle' may be found at any point $\boldsymbol{x}$ at time t. In devices, it is often taken to be the local particle density following a proper normalization of the wave function with the total number of particles in the system. Strictly speaking, since the wave function mentioned in (1.24) and (1.29) is a one-particle wave function dealing with one degree of freedom, it would describe just one electron in the device. However, as we did in the classical case when reducing the N-particle probability distribution to a one-particle distribution, this one-particle wave function can also be comprehended as a statistical description of the total electrons within the device. This dual role of the one-particle wave function will only become problematic when discussing correlations among electrons, where an N-particle wave function becomes the appropriate tool. The intricacies of the many-body problem in quantum mechanics will emerge notably, for instance, in section 3.1 of chapter 3. The resulting (1.37) may be recognized as a continuity equation that conserves the probability within the system volume. Consistent with this continuity equation is a connection between the momentum and the velocity

$$v = \frac{1}{m}\nabla S. \tag{1.38}$$

It is important to note that the velocity, as well as the momentum itself, is buried in the exponent of (1.29). Bohm referred to these variables as 'hidden' variables, as it is quite difficult to make measurements of the phase of a wave function [47]. This led to considerable acrimony from the proponents of Bohr's philosophy (to be discussed below further), since such variables could not be measured and were therefore considered to be unreal.

Kennard was to bring us back to particles by showing that the hydrodynamic form was applicable to a system of particles [50]. His purpose was to show that the entire quantum theory of mechanical systems, particularly those composed of particles, could be explained simply and without loss of generality by the use of Schrödinger's equation. His development followed the hydrodynamic form of Madelung, although he was apparently unaware of this earlier work. Nevertheless, it was Kennard who expressly pointed out that the use of particles was suitable for this approach. In the end, he clearly noted that the particles would follow their classical trajectories from Hamilton's equations as modified by the additional presence of a quantum potential (1.34). This quantum potential may be seen to serve the role of de Broglie's pilot wave in guiding the particle, as demonstrated using the Bohm approach [48].

Another important point to notice from (1.29) is that the action is only defined modulo 2π. Since the momentum is the gradient of the action, this immediately requires the momentum to be quantized through (1.18) or (1.19). In classical hydrodynamics, the integral in these equations is the vorticity of the fluid, so that the quantization leads to quantization of the vorticity. That is, the flow of the wave function in this viewpoint can exhibit vortices, or whorls. In devices, these vortices can arise due to complex inhomogeneities in the density and or nodes in the space of energy and velocity [51]. In many cases, the full charge density within the device can thermally average some of these vortices, except at low temperatures. In figure 1.2,

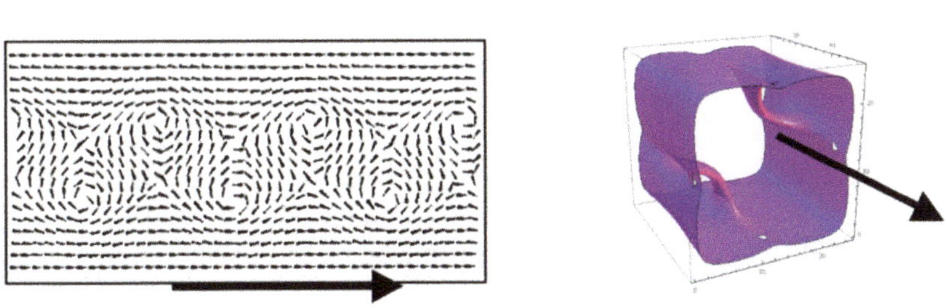

Figure 1.2. (a) The velocity field for a periodic array of vortices forming in a nanowire device. These vortices are time dependent and their movement leads to a *vortex tube*, illustrated in panel (b). Reprinted from Barker [51], with the permission of AIP Publishing.

these vortices are shown for a semiconductor nanowire [51]. In panel (a), the velocity field is plotted and several vortices are shown by the current arrows. It appears that in this case, the vortices are forming a periodic array. Actually, these are time-dependent structures which form a vortex tube that moves through the device. Such a tube is shown in panel (b). Typically, these vortices have a lifetime in the presence of scattering. If the angular frequency of rotation is given by $\omega = v/r$, where v is the angular velocity and r is the size of the vortex, then one needs to have $\omega\tau > 1$, where τ is the dephasing time (typically the optical phonon scattering time), in order for the vortex to remain stable [52]. As the vortex is quantized in the present case, the condition on the vortex size is $r < \sqrt{N\hbar\tau/m}$ and the frequency is $\omega = N\hbar/mr^2$. The formation of such vortices is connected with the classical concept of turbulence in hydrodynamic flows, such as those found in whirlpools and even the gravity-induced vortex in the bath drain.

At the computational level, and based on all that has been explained in this section 1.2.3, the hydrodynamic and Bohmian views are essentially equivalent. However, there are distinctions at a more fundamental level. In the hydrodynamic view, particles are considered as a continuous fluid behavior (reinterpreting the behaviour of the wave function solution of the Schrödinger equation). In contrast, the Bohmian view assigns each particle a well-defined position and trajectory. Notably, the hydrodynamic perspective is less preoccupied with addressing the quantum measurement problem. On the other hand, Bohmian trajectories, applicable to both unitary and non-unitary evolutions, provide an additional framework for explaining single- and multi-time quantum measurements without invoking the orthodox concept of collapse of the wave function, as will be further discussed in section 2.5 of chapter 2 [47, 49].

1.2.4 Consistent histories

One of the strange problems in the orthodox theory has us believe that reality, and thus the wave function, is created with the act of measurement. Yet, the wave is destroyed (it collapses) with this measurement. Certainly, if we have, for example, a nanowire transistor, measuring the current passing through the wire does not destroy the quantum effects, no matter how many times this current is measured. The world is full of such quantum systems in which measurements are made and the quantum effects are still there. Alongside other theories, such as the Bohmian theory, which does not necessitate collapse as a fundamental concept, as will be clarified in section 2.5 of chapter 2, there is another interpretation of quantum theory in which measurement is a natural part of the evolution of the quantum system. This is the *consistent histories* approach, in which measurement involves a mathematical procedure by which conditional probabilities can be calculated during the evolution of the system [53]. The probabilities are introduced from the beginning as part of the axiomatic basis of the quantum theory, irrespective of measurements. One may think of a history as a sequence of events or alternatives at specific instances of time [54]. These alternatives may be simple statements such as yes or no, but each opens the pathway for probabilities in each temporal state. In a strict interpretation, this alternative is the result of a measurement that formally projects the entire Hilbert space onto a set of allowed states [55], and in a sense the history is a trajectory moving through the state space. By nature, these projection operators compress the phase space and thus introduce decoherence at each specific time. (While this approach is often called decoherence theory, this name will be left to a later section.)

In some sense this approach arose from the obvious problems that exist in Bohr's world. Griffiths suggested that the orthodox world of Bohr was difficult for beginners to understand, both because the mathematics might be new to them and because the measurements seem to have a magical quality and suddenly appear as the primary aspect of the theory [56]. In some sense, this approach is required if one insists upon too much accuracy. Suppose the measurements are made at a set of specific points or positions in real space. This, by its nature, now creates problems due to the uncertainty principle (whether quantum or classical). The propagation of the wave function and its important phase from one point to another creates the quantum evolution. Yet the phase involves the momentum through the dependence of the action in (1.29) on this quantity. Since the positions are defined, the momentum must be almost uncertain, and this means that the time of arrival of the evolution at each point cannot be determined with any accuracy. While time is not an operator in nonrelativistic quantum mechanics, it does have an uncertainty related to the existence of measurements. Therefore, the uncertainty with which the evolution actually arrives at a particular point reinforces the statistical nature of this evolution. Thus, it is natural that the consistent histories approach has become a statistical theory, fully in keeping with Schrödinger's views but in direct disagreement with Bohr, whose Machian beliefs did not allow for statistical theories.

In the hydrodynamic (or Bohmian) approach, probabilities stem from the initial conditions of particles (similar to classical mechanics). Therefore, the hydrodynamic (or Bohmian) approach can be regarded as a mechanistic interpretation of the statistical aspects of quantum mechanics, much like classical mechanics aids in comprehending statistical classical mechanics [57]. An important point in the consistent histories approach is that the measurements do not lead to a complete collapse of the quantum system. In the consistent histories approach, measurement only leads to decoherence at particular times but does not decohere the entire quantum system.

1.3 Electronic devices as complex systems

As was pointed out above, Heisenberg thought that quantum mechanics should describe the observables that arose from measurements. Bohr went further. He took the view that in quantum mechanics, nothing was real unless it had been measured by an experiment. He is purported to have said that anything we regard as real is actually made up from things that cannot be regarded as real. It is clear that Bohr adopted the ideas of logical positivism, as put forward by Ernst Mach in the last half of the 19th century. Mach neither accepted Boltzmann's statistical mechanics nor even the idea of the atom. It is said that he announced that he didn't believe in atoms because he could not see them. Mach was said to hold that the best theories were those that dealt only with the relationships among observable quantities [58]. In regard to Bohr's new model of the atom [9], perhaps he felt that it was not sufficiently radical to only show quantization of the angular momentum. He had to assert that even thinking about concepts on the atomic level required an entire new view of science, which could only exist within Mach's philosophy. It is certainly true that some thought that the propagation of this view to an entire generation of young physicists was the result of their brainwashing by Bohr [59]. This led to a general view among physicists that the Bohr interpretation was the correct view of quantum mechanics. But it was not the only possibility. Nevertheless, a major debate began to form between the Bohr party and the Schrödinger–de Broglie–Einstein party. In the view of Bohr, if reality only set in with the measurement, then there could be no question of causality and determinism. In the wave function view, both causality and determinism followed directly from the equations. It is often said that the positivist view followed from Aristotle, who was said to have observed that experience is the source of all real knowledge, but it has to be said that he also remarked that experience could tell you that a fire was hot, but it could not tell you why!

As early as Kennard [50], the view was expressed that quantum mechanics evolved relatively smoothly from classical mechanics, and the latter was a limit of the former. Moreover, this approach reinforces the use of particles in science along with waves. In this situation, one would have to ask what Mach and Bohr would have said about the entire world of subatomic particles and more specifically their wide range of quantum properties (color, charm, and position, as seen, for example in the bottom quark, etc). In Einstein's world, reality already exists when a theory, such as general relativity, describes reality and also predicts what measurements will tell us. Even the theory associated with subatomic particles told the world that the Higgs boson existed long before any confirmation was obtained through measurement. If everyone had believed Bohr, the world would not have spent the vast fortune needed to verify the Higgs boson.

In spite of these arguments it was observed that even at the end of the twentieth century, the 'leadership' in physics still adhered to the positivist viewpoint. In his oral history, Selleri [60] asked how this could be so and whether it was the result of '...repression, and control of positions and publications?' Such actions would be even more evident in science later in the 21st century. Even Feynman made the comment '...nobody understands quantum mechanics...' [61]. Here, however, he was referring to the orthodox view of quantum mechanics (according to Bohr) and not his own version of quantum mechanics. The early developments in quantum mechanics discussed above were Heisenberg's matrix mechanics and Schrödinger's wave mechanics. Feynman introduced a third version based upon his path integral approach using Lagrangian mechanics [62]. This approach depended heavily upon particles and their paths. But here, the probability amplitude of a path is associated with the entire motion of the particle rather than an evaluation at a given point in time. In Lagrangian mechanics, the classical path is determined by the minimum of the action (1.5). In Feynman's quantum mechanics, a single path is not specifically determined, but in the ensemble of possible paths, each path is weighted by the exponential phase determined by integrating the action along the path; this phase involves the action, just as in (1.29). Thus, particles and causality/reality are directly involved by extension from classical mechanics. Dyson went further in identifying these particles as asymptotic parts of the wave function [63].

When the wave function and the quantum system are time dependent, then the wave function certainly changes as the system evolves. As a result, (1.34) tells us that the quantum potential also evolves and must be re-evaluated at each instant of time. This is a part of the larger importance of time in the real world and is especially true for semiconductor devices. Devices are nonequilibrium systems, which means they are irreversible: time has a distinct direction known as the arrow of time. Of course, the system normally involves dissipation and decoherence, especially in quantum systems. Moreover, the quantum system and Schrödinger's equation depend upon probabilities. In this sense, the response of the device system has to be considered to be probabilistic, but this was known from the earliest of classical theories. Quantum mechanics is a probabilistic theory and should be compared with classical statistical mechanics, although this does not forbid the use of particles in the simulation of any probability distribution as classical mechanics can be employed to elucidate statistical classical mechanics. However, this view differs from the orthodox interpretation of quantum mechanics, in which the measurement creates the probabilities and the presence of the wave function. The Schrödinger–de Broglie–Einstein picture differs in that the probabilities and wave function exist from the beginning. But the difference has a deeper contrast. We can summarize by stating that in the orthodox view, there are two strange beliefs. First, wave function destruction (collapse) is considered to be the byproduct of measurements. Second, there is a belief that if the quantum system is opened to the environment, all the quantum effects tend to disappear, and the system appears to be classical in nature. This latter point will be treated in the next section, and we now turn to the former.

The continuation of belief in Bohr's positivist viewpoints grew almost like a religion. His true believers came to occupy leadership positions in physics, and this became a self-fulfilling growth typical of new religions. For our purposes here, it may not matter, because an approach based upon particles has the support of a great fraction of the actual founders of quantum mechanics, and the results obtained using this approach tend to agree with theory.

In Heisenberg's early views of the quantum world, the only variables that the theory addressed were those capable of being observed [32]. In any connection with classical theory, there is a definitive problem with this approach. Modern electronic systems, whether they are small devices, or continent-wide power distribution systems, have an enormous number of dynamical variables. It is never easy to decide whether all of these variables are either observable or controllable. There are two different problems here: (i) can a distinct dynamical variable actually be observed in the measurable output of the system, and (ii) can this distinct dynamical variable actually be controlled by the inputs available to the system? The answers to these two questions determine to a large extent just how well the system can be operated stably. Consider, for example, a modern fighter aircraft in which acrobatic or unusual modes of aircraft orientation are often required. The pilot has a 'joystick' and some controls with which to exercise global control but does not truly control the aircraft in detail. The actual control of the exceedingly large number of dynamical variables is handled by the computers that make billions of decisions per second. If these computers fail, the pilot is in trouble. In fact, it is often said that even if the computers fail during landing, the best option is to point the aircraft toward an uninhabited area and activate the ejection seat. That is, the aircraft is to all intents and purposes uncontrollable if the computers fail. This is no different than a modern car, in which the transmission clutch(s) and gear selection may be operated under computer control. If this computer fails, the car is not going anywhere, and the best option is to bail out and seek another source of transportation. The point here is that the operator does not have control of, or the ability to observe, a large majority of the dynamical variables of the system. This is also true of any semiconductor device. Whether it is a bipolar transistor or a field-effect transistor, there is no real ability to measure the detailed spatial variation of the carrier density. For these examples, there is an environment in which the actual transistor or system exists. For the transistors, the current measurements can be made on the device. However, the entire set of dynamical variables, which correspond to the available states within the system, cannot be fully measured. Therefore, a useful quantum theory has to go beyond just that determined by measurements. A pertinent example concerning the performance of a typical electron device, where only the current is measured using an external ammeter, becomes relevant in this context. Can one adequately explain the operation of an electron device by solely discussing the measured output current, without referencing unmeasured quantities such as energy, momentum, or the position of electrons within the device? Any textbook on electron devices shows that the answer is negative.

Why, then, are we to believe that quantum systems are so primitive that all possible dynamical variables may be measured as a way to define reality and the existence or collapse of an appropriate wave function? Are we now able to really observe subatomic particles such as quarks and their values of color and charm? In reality, high-energy physics revolves around the cloud chamber and the trajectories that these particles exhibit in their passage, along with some interactions that may be observed. Here, the truth of the cloud chamber, and the quantum physics it is used to observe, lies in the trajectories. As Feynman remarked, these trajectories represent the entire motion of a particle and not merely its position at a particular instant of time. Perhaps it is more appropriate to point out that the trajectory represents an almost continuous measurement that describes a consistent history for that particle and any interactions that occur. This continuous measurement does not seem to destroy the quantum mechanics here. Yet a particular set of trajectories revealed in a given cloud chamber trace represents a cross-section of the entire universe of subatomic particles and their quantum physics. There remain a great many unobservable dynamical variables (and particles) in a given set of trajectories. As an example, consider how long it took to finally observe just the single Higgs boson. It too was unobservable for a long period of history.

It seems there is a lot more reality in the world beyond merely what we observe, or even what we can seem to measure. And this must be true even with quantum physics. A viable quantum theory that describes reality and has the ability to predict the future that measurements may provide is far more useful to the person who has to apply this theory [64] than that provided by the orthodox views, and this forces us to look to those theories that can provide this capability.

As we have discussed already, there are quantum theories or interpretations that do not rely upon the explicit role of the observer and thereby avoid the objections noted in the preceding sections. These interpretations began already with de Broglie [37], Madelung [46], Kennard [50], and Bohm [47] till the consistent histories [53]. The point is that in these approaches, properties such as position are intrinsic to the theory, and the evolution of the system can be discussed in terms of trajectories. Even so, less observable quantities such as momentum and velocity can still be considered to be intrinsic properties of the quantum system.

1.3.1 Electronic devices as open systems

The situation is further complicated because electronic devices are open systems, and the quantum mechanics of such open systems is different than that of closed systems. The active region is actually restricted to a small, nanometer-scale size in one or more directions, which is embedded in a large body comprising the device's source and drain regions. This, in turn, is embedded within a larger environment. Depending on the time and energy scales involved, a variety of quantum effects can be present, including well-known effects from the semiclassical world but also including phase coherence, etc. These make the carrier transport quite complex. This complexity arises because these devices interact strongly with their environment. This is in no small part a result of the requirement that the measurements naturally occur in the environment. This connection to the environment may be through a set of contacts or through interactions with the various scattering mechanisms. The environment can change the quantum effects within the device. The active region must have an effect upon the environment in order to allow measurements of the region's properties to take place.

Consider, for example, one of the electrons within a semiconductor device that may pass through one of the contacts and thus into the environment. The measurement of a property of this electron, such as its position, requires an interaction between the electron and the measurement system. This means that new degrees of freedom are introduced into the quantum system. Sometimes, these new degrees of freedom can be considered to be among those of the environment. The process of measurement can create a back-reaction upon the electron itself as well as responding to the effect of the electron on the measurement's degrees of freedom. Even though the measurement may produce a modification of the electron's trajectory, it has no special role in defining the reality of this electron. In this sense, quantum reality may be very close to classical reality.

The complications that arise are focused upon the fact that the device itself exists within an environment, such as that provided by the measuring process. The entire system then becomes quite a complex system. This means that there are multiple parts to the overall system, and these parts may all interact with one another. More importantly, the system is far from equilibrium when the device is operating.

The device system is extremely irreversible, since excitation follows a different path through the phase space of variables than the path followed in the relaxation of the system. There may often be hysteresis in the operation. While we have discussed much of the quantum nature and the properties desired for the quantum theory of interest, there is much more to be said, especially about the connections between the device and its environment. But this will be delayed until chapter 3.

1.3.2 Electronic devices as nonequilibrium systems

In real semiconductor devices, the applied voltages (fields) drive the system well out of equilibrium. Moreover, the response of the device to its external voltages does not evolve smoothly from the equilibrium state in which no voltages are applied. Rather, it is driven rapidly out of equilibrium while seeking a steady state in which a balance is obtained between the driving voltages (fields) and the dissipation within the device and to its environment [65]. There are a variety of equations that are then used to determine this steady-state condition, if it exists. Some devices, such as Gunn diodes, often do not reach a steady state but rather obtain a controlled temporal oscillation that is strongly impacted by their environmental circuits [66]. The classical approach is dominated by the Boltzmann equation and, in modern approaches, is normally evaluated using the particle-based Monte Carlo method. A hierarchy of quantum equations is also used, such as the density matrix, the Wigner function, and Green's functions. Some of these also utilize particle-based methods, which are the theme of this book.

Another important point is that the relaxation of the device once the voltages are removed does not follow a simple reversal of the excitation trajectory through the current–voltage phase space. Rather, a great many devices exhibit hysteresis in their excitation/relaxation responses when viewed in the current–voltage phase space. This naturally follows from the fact that most devices involve capacitance and nearly all carrier responses have an inductive nature. These two effects contribute significantly to the hysteresis, but there is also a nonlinear contribution that is most evident in devices such as the Gunn diode and the resonant-tunneling diode. Devices generate entropy due to the dissipation within them during the switch-on (excitation) process as well as during the period of time that they stay on. They also generate even more entropy during the switch-off (relaxation) process, since dissipation continues in this regime as well.

As an example of the hysteretic effect of switch-on and switch-off, consider the simple Gunn diode [67]. While the prototypical device is based upon GaAs, it can be fabricated on any direct-gap semiconductor for which the satellite X and L valleys can be reached by excitation in an electric field and subsequent intervalley transfer. The basic effect depends upon carriers (typically electrons) that are excited from the central, high mobility Γ valley to one of the satellite, low mobility X or L valleys. This transfer produces a negative differential mobility in the velocity–field relationship and a negative differential conductance in the current–voltage relationship [68]. This latter curve is plotted in figure 1.3. It may be seen that there are regions of the curve where a given current density may exist for two different values of the electric field. The region between these two field values exhibits negative differential

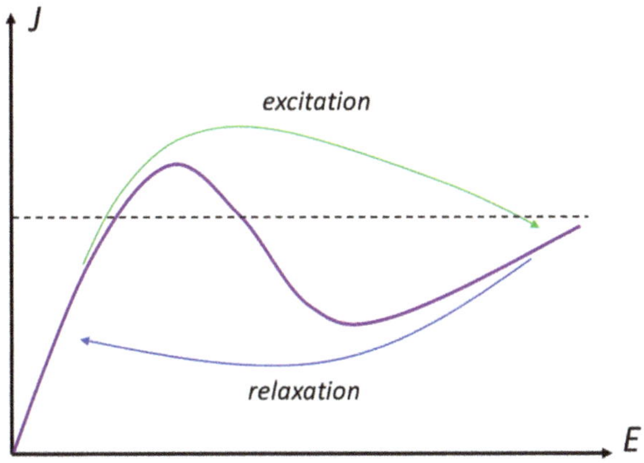

Figure 1.3. The main curve (purple) represents the current–field relationship for a GaAs Gunn diode. The negative differential conductance (where the current decreases with increasing field strength) leads to instability. Two different electric field values result in domains which form, and the carrier density is inhomogeneous in order to support these domains. The excitation (green) occurs without the presence of domains, while the relaxation (blue) has fully formed domains. This leads to the hysteresis shown.

conductance, which means that the current actually decreases with an increase in electric field. This region is unstable, and the electric field breaks up into domains corresponding to the high value and the low value of the field. Since the domains are supported by an inhomogeneous electron density, they often propagate with the carrier velocity. Thus, the switch-on traverses the current–voltage phase space *before* the domains are formed. On the other hand, the switch-off occurs in the presence of fully formed domains, so that the trajectories are quite different, as shown in the figure. The hysteresis is usually heavily affected by the environment in which the device is embedded.

In modern small devices, the evaluation of these processes becomes more complicated. This is because the timescales upon which excitation (and perhaps relaxation) occurs may be dominated by the transient response characteristics of the velocity and distribution function within the device. The relaxation behavior of the device is usually on a different timescale than the excitation, in part because of this internal timescale. In addition, the small device size necessarily leads to the presence of quantization within the device. This quantization is seen in solutions of the Schrödinger equation within the device. Thus, normal wave functions are used to describe this spatial quantization. However, studies of the quantum transport through a device often convert these wave functions into field operators, such as those discussed in the previous section. In the next section, we turn to the various approaches to quantum transport.

1.4 Probability and particles

It is perhaps time to stop and address an important point. In the de Broglie perspective, there is both a particle and a wave. This carries over to the hydrodynamics developments of Madelung, Kennard, and Bohm. Hence, a quantum electron can be defined as something with a well-defined position in ordinary physical space. In this sense, a quantum electron is then quite similar to what we discussed before in the section on classical transport. As we will see later, this point of view was even defended by Einstein. Of course, in this view, the electron is a real thing regardless of whether or not a measurement is made. Despite its conceptual simplicity, it is well known that this point of view provides predictions that completely agree with all known quantum phenomena. The wave function thus describes the propagating support for the particle motion.

On the other hand, in the Schrödinger perspective, the wave function is the entirety of the physical particle, and a localized particle is considered to be described by a wave packet. Such a wave packet is a group of waves with very similar momenta and energy, so that they sum to a sizable magnitude at a single point in space. The reader is likely familiar with the concept of a delta function in space, whose Fourier transform is uniform throughout momentum space. Thus, a localized wave packet requires a very broad spectrum in momentum space. This is very different from the *particle and the wave* view of the previous paragraph. Yet both of these approaches give the particle a reality beyond that defined by any measurement.

With photons, the particle is once more considered to be the localized wave packet. Diffraction then occurs because of the presence of the waves, but the photon impinging upon a photographic plate or a screen is detected because of the localized packet, which appears to be a particle. However, if one wants to study the diffraction of electrons, either the use of a localized wave packet or the coupled particle and wave viewpoint will suffice to illustrate the effects. Nevertheless, there is a difference between the interpretation that follows from de Broglie and that which follows from Schrödinger. This distinction is seldom discussed in applications that, by and large, follow from use of the Schrödinger equation.

When simulations utilize particles, it is not clear whether or not this distinction between the two interpretations is important. For example, with particle simulation using the Wigner function, the particles are often thought of exactly as in the case of the classical Monte Carlo method—they are merely a representation of the distribution function, i.e. the Boltzmann distribution in a classical context or the Wigner distribution in a quantum mechanical context. In methods based on the hydrodynamic approach, one has to consider just what the simulation particles are meant to represent. Are they the real particles that accompany the wave function, or are they representations of the wave function itself? If the former, then the wave function evolution must also be found.

In most simulations, the particles may ultimately be considered to be representations of the distribution function, whatever it may be, and these particles respond to all forces, classical and quantum. In this sense, regions of space in which the number of particles is large correspond to regions where the carrier density is also large. This means that the classical distribution or the quantum distribution is also large in these regions. In the normal interpretation of the wave function, it is usually asserted that the probability for the occupancy of a state or of a region of space is given by

$$P(x) \sim |\psi(x)|^2. \tag{1.39}$$

Thus, quantum probabilities depend upon the details of the wave function and its variation in position. In this sense, there are important differences in probability between classical and quantum situations. In classical mechanics, individual trajectories can be employed to construct ensemble probabilities. Such classical probability exerts no influence on the behavior of individual trajectories. In contrast, in quantum mechanics, the ensemble probability has a dramatic influence on the behavior of individual systems. For example, each individual Bohmian trajectory is influenced by the ensemble probability, as seen in the quantum potential (1.34), which relies on the probability (1.39). This quantum probability is also connected to the local density of carriers when one deals with devices. The normalization is then such that the local density is the product of the probability and the total number of carriers in the structure.

1.4.1 Probability

As discussed earlier, probability in the Feynman picture is related to the entire motion of a particle from an initial time to a final time [62]. In this case, the probability is connected to the summation over the paths to find the total path integral. Nevertheless, it is necessary to discuss the summation of probabilities in quantum mechanics and how it differs from classical mechanics. In this, the discussion of Feynman is followed [62].

It is reasonable to think about the probability of an event being P_{ab} if a measurement of A gives the result a and a subsequent measurement of B gives b. Note that this assumes that measurement B occurs after that of A. Similarly, P_{ac} is the probability that a measurement of A gives the result a and a subsequent measurement of C gives c. Similarly, the probability of an event being P_{abc} results from a measurement of A that gives the result a, a subsequent measurement of B that gives b, and a subsequent measurement of C that gives c. If the events that occur between A and B are independent of those occurring between B and C and a measurement of B gives a complete description of the state at that time, then one may say that

$$P_{abc} = P_{ab}P_{bc}. \tag{1.40}$$

Arriving at this result takes a lot of specifications and 'ifs,' but this last equation is generally true for quantum mechanics as well as classical mechanics. However, it is not always true that measurement B completely specifies the state at that instant, so that there may actually be an entire set of measurements that may be made and which gives different values of b. In this case, the classical result would be

$$P_{ac} = \sum_b P_{ab}P_{bc} = \sum_b P_{abc}. \tag{1.41}$$

Result (1.41) is always true in classical mechanics. But in quantum mechanics, this is not the case, and it is sometimes false. The fact is that the probability is defined somewhat differently in quantum mechanics, and one has to use (1.39). The individual probabilities are then defined as

$$P_{ab} = |\psi_{ab}|^2, \quad P_{bc} = |\psi_{bc}|^2, \quad P_{ac,q} = |\psi_{ac}|^2, \tag{1.42}$$

where the subscript q has been added to the last term to indicate the quantum probability, as it may differ from the classical value. Indeed, the classical law following from the second term in (1.41) must be replaced by

$$\psi_{ac} = \sum_{b} \psi_{ab}\psi_{bc}. \tag{1.43}$$

In quantum mechanics, if result (1.43) is correct, then it is quite likely that the last term in (1.41) is incorrect. The logical error is the assumption that as the quantum system moved from a to c, it had to pass through b, so that any measurement B had to give a value of b.

If an attempt is made to measure B between experiments A and C, it may be concluded that the last term of (1.41) is correct. Even if the apparatus is set up to measure B but is not used, so that only the A-to-C correlation is recorded, this is still true in the Feynman view [62]. But this can still generate controversies in the orthodox interpretation, as Schrödinger has pointed out [69], as in this orthodox theory the measurement is *not* fully specified when and where it happens so that one can argue that the measurement is not done made until it is communicated to the world. Indeed, this is an essential part of the cat paradox that will be futher elaborated in section 5.2.1 of chapter 5.

Hence, there is a difference between the quantum result (1.42) and the classical result (1.41), depending upon whether or not an attempt is made to measure B, at least in the preferred view of quantum mechanics. This is because no matter how weakly one tries to make the measurement, it still disturbs the system. In the preferred interpretations, though, it does not collapse the entire wave function into the classical world, as discussed in section 1.3. Thus, (1.43) is a typical expression that arises within the wave interpretation of matter. The probabilities of moving from a to c through many possible states b without trying to measure these states are then represented by the square of the sum of several complex quantities, one for each route that is possible. The important point by which (1.42) and (1.41) differ is the fact that the magnitude squared of the left-hand side of (1.43) is *not* equal to the sum of the magnitude squared of the terms within the summation. That is, the magnitude operation sits outside the summation term on the right-hand side of (1.43). Phase interferences between different terms within the summation can lead to variations in the result. The result follows from the well-known Schwarz inequality that even holds in classical probability:

$$|\psi_{ab}|^2 = |\sum_{b} \psi_{ab}\psi_{bc}|^2 \leqslant \sum_{b} |\psi_{ab}|^2 \sum_{b} |\psi_{bc}|^2. \tag{1.44}$$

The fact that phase interference phenomena can determine the probability of (1.39) has some rather 'unexpected' consequences for the individual behavior of systems. For example, in figure 1.1, each trajectory can be considered as an individual quantum system when only one particle is injected at a time. Since the set of all trajectories has to reproduce the interference phenomena present in the ensemble probability, each trajectory must 'unexpectedly' bend, even without encountering any apparent physical obstacles in its path, in order to concentrate on positions where the probability is higher.

1.4.2 Particles in quantum field theory

Quantum field theory was developed continuously over a long period (which is still going on) and began with Heisenberg's early work. In principle, the idea was to bring some semblance of quantum mechanics into Einstein's relativistic theory and to apply it to electromagnetic fields and Maxwell's equations. An important point is that it is easy to measure different fields at different spatial points independently, provided that the space–time points for each measurement lie outside each light cone. That is, the measurements are space-like within relativity theory. Within this view, the measurements lie on a space-like surface in the three-dimensional space [70]. At each point on this surface, the field quantities are measured and determine that the quantum state of the fields is $\psi[t(\mathbf{r})]$. However, it is additionally assumed that all the field quantities at these positions are *operators* which commute with the Hamiltonian. This commutation is a necessary fact for the operators to be capable of being measured together. A new issue has now arisen. In normal Schrödinger wave theory, the wave function is a normal number, called a *c*-number, in that it is not an operator. Here, the wave function is now an operator as well, and it is termed a *field operator*. This is sometimes referred to as second quantization, and the fields are now thought of as operators. This has carried over to many-body theory, in which the representation is thought of as an occupation number formalism, since each fermion state can hold a single electron of a definite spin [71]. These field operators satisfy the commutators

$$[\psi(\mathbf{r}), \psi^{\dagger}(\mathbf{r}')] = \psi(\mathbf{r})\psi^{\dagger}(\mathbf{r}') - \psi^{\dagger}(\mathbf{r}')\psi(\mathbf{r}) = \delta(\mathbf{r} - \mathbf{r}'), \tag{1.45}$$

for bosons, and

$$\{\psi(\mathbf{r}), \psi^{\dagger}(\mathbf{r}')\} = \psi(\mathbf{r})\psi^{\dagger}(\mathbf{r}') + \psi^{\dagger}(\mathbf{r}')\psi(\mathbf{r}) = \delta(\mathbf{r} - \mathbf{r}'), \tag{1.46}$$

for fermions.

It is convenient at this point to revert to nonrelativistic terminology, as nearly all semiconductor devices of interest live in this world. We still have the fields, and the quantity $A_{\mu}(\mathbf{r})$ is the μth component of the vector potential, for example, but it could be any of the electromagnetic fields and is usually the electric field for our purposes. Similarly, $\psi_{\alpha}(\mathbf{r})$ is the αth component of the electron or hole spinor field operator, although the spin itself is usually unimportant for the current considerations (although it becomes important for spin field-effect transistors (SpinFETs) or for devices involving the spin Hall effect, both of which have attracted interest in recent years [72]). In addition to the various operators, the quantum system is said to have a state Φ. The nature of this state is decided by the number of particles present within the state [70], that is by the presence or absence of interparticle interactions. Note here that particles have been decided to be existing quantities, whether or not they accompany the quantum wave or are localized wave packets. The state itself evolves according to the Schrödinger picture.

Julian Schwinger introduced Green's functions to the world of quantum field theory [73]. These Green's functions describe the propagation of the wave function from one temporal point on our space-like surface, upon which the wave function lives, to a subsequent point on this surface. In this regard, Schwinger's Green's functions describe methods for calculating the perturbations that arise from an effective external potential. This also includes the many-body interactions. On the other hand, Feynman uses his integral summation of the various paths followed by particles (in which each path is weighted by an integral of the Lagrangian) during this propagation. Nevertheless, both Schwinger and Feynman seek to preserve the basic symmetry between past and future. That is, they invoke a principle of time-reversal symmetry. This is an important point, as real devices are usually nonlinear and are certainly not time-reversal symmetric, a point we return to in the next section.

Improvements on the Green's functions have been put forward by many in the field, and the leading proposals are the so-called nonequilibrium Green's functions. However, far too many of these proposals still retain the approach used by Schwinger, with its assumption of equality between past and future. Nevertheless, the use of field operators that began with quantum field theory continues to be applied in many, if not most, approaches to quantum transport in semiconductor devices. It is important to note that many approaches to this topic still rely upon particles as the basic units for which transport is evaluated. These various approaches will be discussed in the next section.

1.5 Various approaches for semiconductor devices

Generally, one normally thinks of the operation and performance of a semiconductor device in terms of the motion of electrons or holes, as well as the resulting space charge and self-consistent potentials. This motion of carriers within the device is simulated using classical or semiclassical methods. It is fair to ask, then: just what makes quantum transport approaches different from these classical efforts? Certainly, the mathematics is somewhat different, as it arises from the equations used to simulate the transport. In certain cases, this mathematics is much more complicated. In reality, however, it is the physics and the physical effects that occur within the device and that are *due to quantum mechanics* that must therefore be handled by quantum mechanics. Certainly, this includes the spatial quantization that can occur in a small device or in spatially compact regions of a larger device (such as the inversion layer in a metal–oxide–semiconductor field-effect transistor). But quantum effects also give rise to phase interference, as discussed above for the two-slit experiment.

Interference is a property of the electrons or holes when they are treated as waves. It turns out that the scattering or diffraction of an electron around a charged impurity can give quite similar interference effects. Quantum interference and entanglement are some of the most important aspects of *pure* quantum transport [87]. They cannot be obtained using a semiclassical treatment of the particles. Self-interference and the existence of vortices has already been discussed in connection with (1.18) and (1.19). While vortices are well known in classical hydrodynamics, quantum waves create vortices as well, especially in many-body interactions. Such self-interference and spatial quantization can arise not only from transverse confinement but also from longitudinal resonances in the current flow directions caused by momentum mismatches at the interfaces between the channel and the contact regions.

Quantum transport must differ from semiclassical transport in the semiconductor device in the treatment of these truly quantum effects which lead to observable behavior. No perturbatively obtained extra potential can give rise to this interference behavior. Moreover, no perturbative treatment of impurity scattering, even in quantum mechanics, can exhibit the interferences that are expected (especially in coulomb scattering). Much of this behavior can occur over regions the size of the thermal de Broglie wavelength [74]. In a Si device at room temperature, this length is a few nanometers (nm). In the modern world, some feature sizes are much less than this, and quantum effects are expected to be prevalent in the transport through these modern devices. As a result, any quantum mechanical representation must meet all the requirements of a self-contained and consistent mathematical theory. It also has to correctly reflect the laws of nature, and, in this respect, it is not that different from classical simulations. But the underlying quantum physics can be much deeper.

There are basically four common approaches to the treatment of quantum transport. These are: (1) direct solution of the Schrödinger equation, (2) the density matrix, (3) Wigner functions, and (4) nonequilibrium Green's functions (NEGFs). It is not our purpose to treat these in depth here (see for example a recent review on computational methods [75]). Rather, we will give a brief introduction to each and leave fuller discussions to the appropriate chapters that occur in the remainder of this book.

1.5.1 Basic beginnings

In discussions of semiconductor devices, whether these discussions relate to the experimental or theoretical aspects, some general concepts are used to discuss the quantum effects. This is because the device has many different regions, some of which may be attributed to the internal parts and others that are best described as part of the environment. Consider a simplified version of the device illustrated in figure 1.4. The upper section may be considered to represent the physical domain of the device, but even this is somewhat simplified, as the importance of the circuit and the interactions between the device and the circuit are not shown, even though it is known that the circuit can dominate the device, especially in microwave circuits. Nevertheless, this device illustration shows the major parts that constitute a field-effect transistor. These parts are then assigned to the appropriate parts of the simulation domain illustrated in the lower panel.

The longitudinal direction, lying horizontally from left to right along the current flow direction in both parts of the figure, is the most important dimension here and is taken to be the x-axis. The basic parts shown and to be discussed are mostly the same, regardless of whether this is a planar FET, a FinFET, or even a dual-gate FET. Both the physical device (upper panel of the figure) and the simulation domain

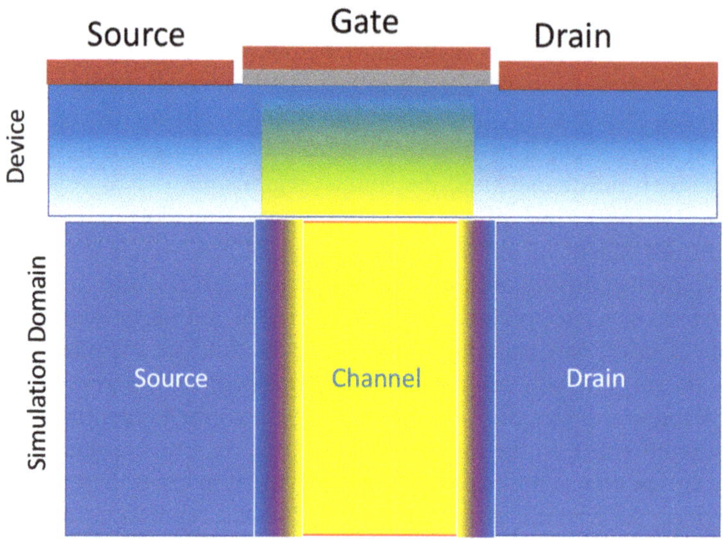

Figure 1.4. Various domains to be considered in the quantum transport simulation of a field-effect transistor. (a) The physical domain that is normally considered in simulation and modeling. (b) The real simulation domain, including interfaces, that must be considered.

(lower panel) can be split into five distinct sections. The obvious ones are the source, channel, and drain. The channel, which is not indicated in the upper panel, is the region beneath the gate and between the source and drain, as indicated in the lower panel. This is the region of the inversion layer in the FET. It is also important to recognize that there are two transition regions, one between the source and the channel and the second between the channel and the drain. These transition regions are extremely important, and they may be much larger than indicated in this simple drawing. For example, it has been shown that the high kinetic energy that carriers bring into the drain region may take as much as 20–30 nm to dissipate as the carriers thermalize [76]. This means that failure to carry any detailed simulation far enough into the drain may well miss important device physics. The drain and source regions are *not* simply 'contacts.' Important quantization occurs in these regions, especially in nanowire devices, and the transition regions have to relate this quantization to that occurring in the channel. In fact, in some basic approaches to quantum transport, as we will see below, one begins and ends with the quantization in these source and drain regions.

The above is just the device. Of great importance are the local internal potentials that control the charge density and the current density through the device. That is, proper simulation has two parts: (a) determining the transport (movement of charge) through the device under the applied voltages (see figure 1.5), and (b) a self-consistent determination of the local potentials and fields that arise from the inhomogeneous charge within the device. This latter involves solving Poisson's equation along with the transport equations. These potentials are shown in figure 1.5. In the upper panel, the physical device illustration is repeated from figure 1.4, in

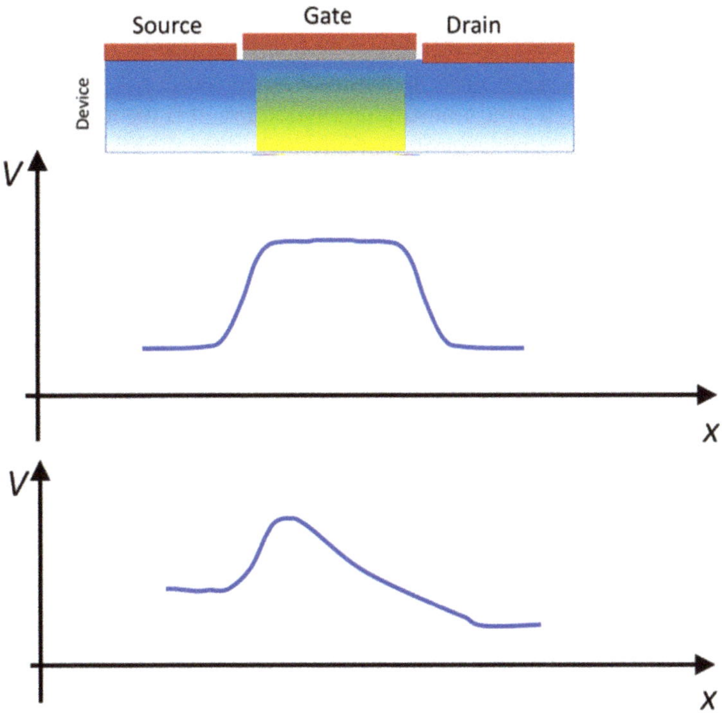

Figure 1.5. (a) The physical device schematic from figure 1.4 is shown for comparison with the potentials. (b) The potential variation along the device with no applied biases. This potential arises from the *p–n* junctions at the source and drain ends of the channel. (c) The potential variation along the channel when positive drain and gate voltages are applied.

order to position the potential relative to the device parts. In the central panel of figure 1.5, the variation in the potential magnitude from source to drain is indicated in the absence of any bias. While this may seem unusual, one has to consider that, in an *n*-channel metal oxide semiconductor FET (MOSFET), the source and drain are *n*-type materials, while the channel region is typically *p*-type material. This means that there are *p–n* junctions between the channel and the source and/or drain. Thus, there is a band-bending potential such as that of the central panel that arises from the depletion regions of these *p–n* junctions. Such a variation is also present in the absence of dopants in the channel region (the so-called junction-free FET) but the potential amplitude is less than in a doped channel region. Finally, the lower panel indicates the potential changes when positive gate and drain voltages are applied, and one should note that this now implies the presence of a significant longitudinal electric field in the channel region. The peak in the potential in the lower panel is less than that in the central panel due to the gate bias, as the gate bias creates the inversion layer. The local peak near the source end serves to meter the current into the channel. It must always be remembered that potential variations require charge inhomogeneities (a potential jump requires a dipole of charge). Consequently, even in the absence of bias, the device is *not* in equilibrium but in a sort of steady state with the *p–n* junction depletion. All of these space–charge regions have to be incorporated (and described accurately) in any simulation. While this remains true for classical simulations, even more is required for quantum simulations [77].

While we have indicated a natural simulation domain, it is incomplete. While it may work well enough for particle-based simulations in classical situations, one has to probe a little deeper in quantum simulations. This is because full consideration of the wave properties of the carriers is needed in order to see effects such as interference and entanglement. And in the consideration of wave propagation and confinement, different quantization boundaries are likely to occur in the different regions of the device that are pictured in figures 1.4 and 1.5. Thus, it is now conceivable that matching conditions must be determined and utilized between different regions of the device. This means that the Hamiltonian has a great many different terms.

The quantum problem in the simulation domain is usually handled by partitioning the domain into various parts, just as in the physical device. This partitioning was discussed in quantum theory at least as early as the work of Löwdin [78, 79]. In order to explain this adequately, we consider a quasi-two-dimensional device which is homogeneous in the direction normal to the surface. As is usual in solving the Poisson equation, this two-dimensional region is divided into slices in the transverse direction (normal to the current flow but in the surface plane). Hence, at each slice, the transverse quantization can be determined by solving the Schrödinger equation subject to the transverse boundary conditions. This determines the wave function at each point in the transverse slice, and this can rapidly be transformed into the amplitude of each transverse mode in the momentum representation. It is no problem to expand this to a full three-dimensional device, and this approach has been applied to MOSFETs at room temperature with the inclusion of all scattering processes [80]. These slices are illustrated in figure 1.6. The number of such slices is not limited to those shown, as these should be taken to be illustrative of the actual number of slices. Each slice must be coupled to its neighbor by a transfer matrix (while the transfer matrix method of solution is well-known in quantum mechanics,

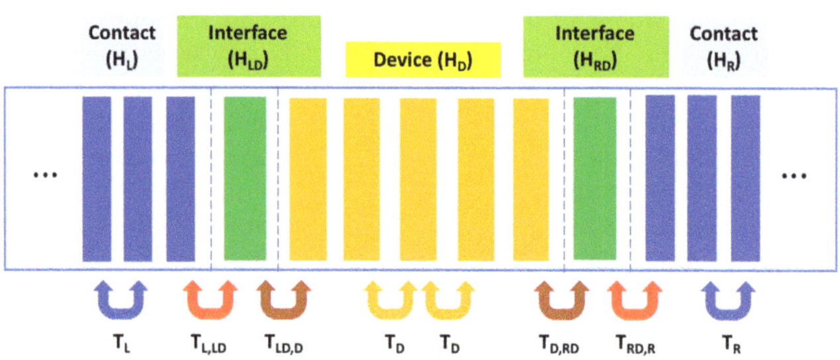

Figure 1.6. The device structure is broken into slices, each of which is subject to transverse quantization. The number of slices shown here is illustrative and does not represent the totality of the slices necessary for any single device. The sections of the device are indicated along the top, along with the components of the diagonal of the total Hamiltonian. The off-diagonal parts of the Hamiltonian are shown along the bottom of the device, and these reside in the first off-diagonal area, so that the total Hamiltonian is tri-diagonal.

it is usually unstable for large numbers of slices, and a preferred method is based upon the scattering matrix approach [81, 82], which is often found in microwave studies). The five basic regions discussed in connection with figure 1.4 may be seen in figure 1.6. The central device region is shown in orange, while the 'contacts' are in blue. The left-hand contact corresponds to the source in an FET or the emitter in a bipolar transistor. This left-hand contact must have a sufficient number of slices that the leftmost slice is in 'equilibrium.' This means that there is a homogeneous density in this slice and its neighbors. A similar requirement exists for the rightmost contact, which may be the drain in an FET or the collector in a bipolar transistor. This rightmost contact has to be sufficiently thick and must have the appropriate number of slices to ensure that that the carriers have relaxed to an equilibrium distribution at the far right.

The device itself, i.e. the central region, must have enough slices to capture the detailed physics of the transport and self-consistent charge density and potential within this region. The transition regions, shown in green, cannot usually be simulated using a single slice. The grids used in numerical solutions are governed by certain mathematical requirements for the grid size, and the transitions have to incorporate sufficient slices to meet these conditions. Nevertheless, these transition regions are represented by a single slice in figure 1.6, but it must be remembered that this is merely illustrative. These various transfer matrices couple to the diagonal parts of the total Hamiltonian. The diagonal terms are indicated along the top of the figure, while the transfer terms are aligned along the bottom of the figure. These lead to the total Hamiltonian, which may be written as

$$\begin{bmatrix} H_L & T_L & 0 & 0 & 0 & 0 & 0 \\ T_L^\dagger & H_L & T_{L,LD} & 0 & 0 & 0 & 0 \\ 0 & T_{L,LD}^\dagger & H_{LD} & T_{LD,D} & 0 & 0 & 0 \\ 0 & 0 & T_{LD,D}^\dagger & H_D & T_{D,RD} & 0 & 0 \\ 0 & 0 & 0 & T_{D,RD}^\dagger & H_{RD} & T_{RD,R} & 0 \\ 0 & 0 & 0 & 0 & T_{RD,R}^\dagger & H_R & T_R \\ 0 & 0 & 0 & 0 & 0 & T_R^\dagger & H_R \end{bmatrix}. \tag{1.47}$$

The central H_D is also a tri-diagonal matrix with the term H_D along the diagonal, the term T_D along the first off-diagonal above, and the term $T_D^\dagger$ along the first off-diagonal below, so that the off-diagonal terms keep the total Hamiltonian Hermitian. While the total Hamiltonian may be directly inverted for the entire device, this is usually a difficult proposition. For example, with a three-dimensional device and a grid size of $100 \times 100 \times 100$, the order of this matrix is 10^6. This is solvable, but usually only by large-scale computing clusters. A more common method is to iterate the solution from the two ends, and this is the rationale for dividing the entire structure into slices. For example, this begins with solving just the Hamiltonian of the leftmost slice of figure 1.6. In three dimensions, this would be just 100×100, or a matrix of order 10^4, which is considerably more manageable these days, even with laptop computers.

1.5.2 Landauer and beyond

When one develops the mode structure of the transverse slices as discussed in the previous section, it becomes possible to write a relatively simple formula for the conductance through the structure. This formula has a rich history, as it originally arose from considerations of tunneling structures. In momentum space, it may be written as [83]

$$I = \frac{e}{4\pi^2\hbar}\iint dk_y dk_z \int_0^\infty dE_x T(E_x)\left[f\left(E_{xL}, k_y, k_z\right) - f\left(E_{xR}, k_y, k_z\right)\right]. \tag{1.48}$$

The integration is over the transverse momentum k_y, k_z and the longitudinal (current direction) energy E_x, with the corresponding one-dimensional density of states for this energy scale. The function $T(E_x)$ is the tunneling coefficient between the two sides, and this is a bilateral function that is independent of the particle tunneling direction. The functions $f\left(E_{xL}, k_y, k_z\right)$ and $f\left(E_{xR}, k_y, k_z\right)$ are the distribution functions on the left-hand side and the right-hand side, respectively, of the tunneling region. These two functions depend differently upon the longitudinal energy due to the presence of a bias voltage between the two ends of the device. The density of states for the longitudinal energy does not appear, nor does the longitudinal velocity of the particles, as these two quantities cancel in quasi-one-dimensional tunneling (or transport of any kind) [84].

Whether in the two-dimensional situation discussed in the previous section or in three dimensions, the transverse integrations are replaced in quantum transport by a summation over the transverse quantum modes in the quasi-one-dimensional case of current along a single axis. This leads us to the Landauer formula for degenerate contact regions [85]:

$$I = \frac{e}{\pi\hbar}\int_{-\infty}^{\infty} dE \sum_{l,r} | T_{lr} |^2 \left[f\left(E - \mu_l\right) - f\left(E - \mu_r\right)\right]. \tag{1.49}$$

Here, the energy scales have been rationalized to each other and it is the Fermi levels, μ_l and μ_r that are offset by the applied bias voltage. The T_{lr} is now the transmission from a mode l in the left contact region to a mode r in the right contact region. It is exceedingly important that the mode selection is sufficiently far to the left to be in a region that is essentially in equilibrium, and similarly, the mode selected on the right must be sufficiently far to the right to be in a region that is also in equilibrium. Otherwise, much important information is lost or ignored. This point was especially emphasized by Landauer [85] in order to properly account for the role played by the charge dipoles in the interface region as well as for particle relaxation in the output contact region, both of which were discussed above.

Some researchers have expressed the view that this formula is limited to low-temperature ballistic systems, but this is not the case. If properly applied, (1.49) is quite universal, and the determination of the transmission between modes can even handle the presence of significant intermode scattering and room-temperature behavior [80]. Thus, it is important to keep this in mind, as many, if not most, of the methodologies for quantum transport eventually use the Landauer formula. However, in semiconductors, there is an additional factor that must be incorporated due to the fact that these are not metals. We have to account for the fact that different modes have different propagation velocities. Thus, the transmission terms are modified by these velocities as follows:

$$|T_{lr}|^2 \rightarrow \frac{v_n}{v_m} |t_{nm}|^2, \tag{1.50}$$

which involves the nth mode and its velocity on the left and the mth mode and its velocity on the right. The integration in (1.49) involves only the energies between the two Fermi levels (left and right) at low temperatures but the entire energy range at higher temperatures to account for the broadening of the Fermi–Dirac distributions.

Using the mode structure means that we are relying upon the wave functions that exist in the device. The density matrix may be introduced as

$$\rho(\boldsymbol{x}, \boldsymbol{x}', t) = \sum_{i,j} a_{ij}\psi^{\dagger}(\boldsymbol{x}, t)\psi(\boldsymbol{x}', t). \tag{1.51}$$

This function depends upon two positions and two wave functions, as indicated. If we expand each wave function into a series of basis functions which represent, for example, the transverse modes, as in (1.28), the density matrix may be rewritten as

$$\rho(\boldsymbol{x}, \boldsymbol{x}', t) = \sum_{i,j} a_i^{\dagger} a_j \varphi_i^{\dagger}(\boldsymbol{x}, t)\varphi_j(\boldsymbol{x}', t)e^{i(E_i - E_j)t/\hbar}. \tag{1.52}$$

It is clear from this form that the diagonal terms of the density matrix merely represent the probabilities of occupancy at each energy level (the integration over position produces a delta function that requires the two positions to be equal when computing the total probability). On the other hand, the off-diagonal terms oscillate rapidly due to the difference in energies, and this rapid oscillation usually leads one to ignore these terms. However, in the presence of interactions, the energies can acquire imaginary parts, and these parts may not vanish at times but contribute to entanglements between different modes. This is an important part of quantum transport.

The Landauer equation and the density matrix are brought to our attention here because the use of wave functions is important and these functions actually provide the basis for all other approaches to quantum transport, which are discussed in the remainder of this section. Direct treatments of the transport via mode-matching methods, such as the scattering matrix mentioned above, have not involved the use of particles, at least to date. Certainly, the diagonal terms of the density matrix may be thought of as an accumulation of particles, although their occupation numbers are restricted by the uncertainty principle. If this idea is extended to using particles as numerical representations of the wave function, one could conceive of a particle approach to solving the density matrix for transport. In this case, the off-diagonal terms would represent phase differences between the diagonal particles, with interactions leading to imaginary parts of the phase. The fact that no one has proceeded along this line (to our knowledge) is mainly due to the fact that the density matrix is a function only of position. Classical simulations usually occur in phase space, so that quantum transport methods that rely also on phase space have been preferred and developed. This is the basis of the Wigner functions and to a somewhat lesser extent the Green's functions, which are discussed in the next two sections.

1.5.3 Wigner functions

Normally, one uses the Boltzmann equation in classical transport physics. This yields a distribution function that is defined in position, momentum, and time. That is, the distribution is defined in a 6+1 dimensional coordinate system (three for momentum, three for position, and one for time). From the beginnings of quantum mechanics, there has always been a recognition that a phase space formulation would allow an easier comparison of quantum and classical physics. Such a formulation was developed quite early by Wigner [86]. Generally, the Wigner function is obtained from the density matrix through the introduction of center-of-mass and difference coordinates defined as follows:

$$\begin{aligned} \boldsymbol{R} &= \frac{1}{2}(\boldsymbol{x} + \boldsymbol{x}'). \\ \boldsymbol{s} &= \boldsymbol{x} - \boldsymbol{x}' \end{aligned} \tag{1.53}$$

The density matrix may then be Fourier transformed on the difference coordinate $\boldsymbol{s}$ as follows:

$$f_W(\boldsymbol{R}, \boldsymbol{p}, t) = \frac{1}{h^3}\int d\boldsymbol{s}\rho\left(\boldsymbol{R} + \frac{\boldsymbol{s}}{2}, \boldsymbol{R} - \frac{\boldsymbol{s}}{2}, t\right)e^{i\boldsymbol{p}\cdot\boldsymbol{s}/\hbar}. \tag{1.54}$$

If the operator O in (1.22) is taken to be the density matrix, the equation becomes known as the Liouville equation. We will see in detail in chapter 11 that this leads to an equation of motion for the Wigner function that may be expressed as (in the absence of any interactions or scattering and for a homogeneous material)

$$\frac{\partial f_W}{\partial t} + \frac{\boldsymbol{p}}{m^*}\cdot\frac{\partial f_W}{\partial \boldsymbol{R}} - \frac{1}{h^3}\int d\boldsymbol{p}'\, W(\boldsymbol{R}, \boldsymbol{p}')f_W(\boldsymbol{p} + \boldsymbol{p}') = 0, \tag{1.55}$$

where

$$W(\boldsymbol{R}, \boldsymbol{p}') = \frac{1}{i}\int d\boldsymbol{s}\exp\left(\frac{\boldsymbol{s}\cdot\boldsymbol{p}'}{\hbar}\right)\left[V\left(\boldsymbol{R} + \frac{\boldsymbol{s}}{2}\right) - V\left(\boldsymbol{R} - \frac{\boldsymbol{s}}{2}\right)\right], \tag{1.56}$$

and V is the electrostatic potential that appears in the Hamiltonian. The term $W(\boldsymbol{R}, \boldsymbol{p}')$ is termed the *Wigner potential* in most applications. This potential is nonlocal, as can be seen from the term in brackets in (1.56). Scattering may easily be included by adding a gain–loss term to the right-hand side of (1.55), but some caveats are necessary to keep this from being a simple semiclassical process, as will be discussed in chapter 11.

In equilibrium, the Wigner function is positive definite. That is, the ground state is always positive definite. Yet the Wigner function is not a positive definite function under far-from-equilibrium conditions [87], i.e. the conditions for transport. This latter are usually considered to be a result of the uncertainty principle or the the result of the appearance of entanglement in the case of a multiparticle Wigner function. When the Wigner function is integrated either over position or momentum, a positive definite function results that is the square magnitude of the wave functions in momentum or position, respectively. This gives the probabilities, which are positive definite and thus satisfy normal quantum mechanics.

As mentioned, the onset of negative excursions of the Wigner function away from equilibrium is viewed as the appearance of correlation and/or entanglement [88]. This correlation or entanglement is directly connected to the off-diagonal elements of the density matrix, as discussed in the previous section. These terms lead to oscillatory behavior in the Wigner function that produces the negative excursions. As a result, it is possible to identify where quantum corrections appear by comparison with the classical results for a problem.

1.5.4 Green's functions

Equation (1.28) gives us an expansion of the wave function in a set of basis functions. These basis functions could, for example, be the set of transverse functions in a nanowire or the two-dimensional device discussed in section 1.1. This expansion also allows us to find the temporal development of an evolving wave function, such as we need in order to discuss transport or the transient response of a semiconductor device. Let us begin with a situation in which there is an initial condition for the wave function itself, which is described by $\psi(\boldsymbol{x}, 0)$. This can now be used with (1.28) to find the coefficients, as follows:

$$a_i = \int d\boldsymbol{x}\varphi_i^\dagger(\boldsymbol{x})\psi(\boldsymbol{x}, 0). \tag{1.57}$$

This expression for the coefficient is now reinserted into (1.28) to give

$$\psi(\boldsymbol{x}, t) = \sum_i \int d\boldsymbol{x}'\varphi_i^\dagger(\boldsymbol{x}')\psi(\boldsymbol{x}', 0)\varphi_i(\boldsymbol{x})e^{-E_i t/\hbar}. \tag{1.58}$$

The last equation can be rewritten in terms of a *propagator kernel* as

$$\psi(\boldsymbol{x}, t) = \int d\boldsymbol{x}'K(\boldsymbol{x}, \boldsymbol{x}'; t, 0)\psi(\boldsymbol{x}', 0). \tag{1.59}$$

Such a propagator kernel describes how the initial condition evolves into the current state of the wave function. By comparing the last two equations, it may be seen that the propagator kernel satisfies the definition

$$K(\boldsymbol{x}, \boldsymbol{x}'; t, 0) = \sum_i \varphi_i^\dagger(\boldsymbol{x}')\varphi_i(\boldsymbol{x})e^{-E_i t/\hbar}. \tag{1.60}$$

It is important to note that the initial time does not have to be $t = 0$. It can be any arbitrary time, even a time *after* the time on the left-hand side of (1.60). Such a later time would be acausal but is not forbidden mathematically and will be useful in this section. The reader should be reminded that it was Feynman who suggested that the positron (our hole in semiconductors) was an electron traveling backwards in time [89].

It was Schwinger who formally connected the propagator kernel to the idea of Green's functions, already well known in the mathematics of differential equations [73]. Initially, these were independent of time, as he was concerned with the equilibrium state, but they certainly extended to time variation, although he assumed time reversibility in his treatment and hence stuck to the equilibrium system. This equilibrium was maintained, even in the presence of perturbations, by the equality of $t \to \pm\infty$. The perturbations were formally handled by the existence of a unitary operator that involved the perturbing Hamiltonian exponentially [90], although this operator was developed much earlier [63].

When the system goes out of equilibrium, there is no real knowledge of the distribution function. Classically, one has to solve the Boltzmann equation, which itself is an approximation. Similarly, the quantum distribution is no longer the Fermi–Dirac distribution and one must solve NEGFs. The NEGFs differ from the simple propagator of (1.60), as new functions must be found to describe the nonequilibrium distribution [91, 92]. Whether with the propagator above or with NEGFs, the wave functions ψ and $\psi^\dagger$ go beyond the simple forms, as these are usually discussed as field operators, as introduced in section 1.2. To begin, the propagator (1.60) is used to describe two basic equilibrium Green's functions

$$\begin{aligned} G_r(\boldsymbol{x}, \boldsymbol{x}'; t, t') &= -i\Theta(t - t')\langle\{\hat{\psi}(\boldsymbol{x}, t), \hat{\psi}^\dagger(\boldsymbol{x}', t')\}\rangle \\ G_a(\boldsymbol{x}, \boldsymbol{x}'; t, t') &= i\Theta(t_0 - t)\langle\{\hat{\psi}^\dagger(\boldsymbol{x}', t'), \hat{\psi}(\boldsymbol{x}, t)\}\rangle \end{aligned}. \tag{1.61}$$

These two functions are termed the *retarded* and *advanced* Green's functions, respectively (the difference is denoted by the subscripts). Notice that the retarded function runs forward in time, while the advanced function runs backward in time. In a homogeneous system (which is no longer dependent upon $\boldsymbol{R}$ or the equivalent average time $T = t + t'$), the Fourier transforms ($\boldsymbol{x} - \boldsymbol{x}' \to k$, $t - t' \to \omega$) give rise to the spectral density

$$A(k, \omega) = -\mathrm{Im}\{G_r(k, \omega) - G_a(k, \omega)\}. \tag{1.62}$$

This latter quantity relates the energy to the momentum of the quantum state. Classically, this relation is a delta function asserting that, for example, $E = p^2/2m^*$, in which the momentum $p = \hbar k$ and m^* is the effective mass in the semiconductor. In quantum transport, this simple classical relationship is no longer correct as a result of an important quantum effect—state broadening. This is clearly a result of the uncertainty principle between position and momentum. Normally, equilibrium systems are interpreted as being at a temperature of zero. Thus, all states below the Fermi level are full and all states above the Fermi level are empty (the filled states are often referred to as the filled Fermi sea). The Green's functions then determine the transport of an excitation over a certain distance (given by the two positions and times in the arguments) to the location where it is annihilated (or destroyed). Hence, one may think of the retarded Green's function as discussing the excitation of an electron to a state above the Fermi level at $\boldsymbol{x}'$, t', followed by its propagation to $\boldsymbol{x}$, t where it is annihilated (recombines with a hole). Similarly, the advanced function describes the motion of the hole left behind.

If we now return to the problem of the distribution, two new Green's functions are introduced in nonequilibrium systems. These are

$$\begin{aligned} G^{<}(\boldsymbol{x}, \boldsymbol{x}'; t, t') &= i\langle \hat{\psi}^{\dagger}(\boldsymbol{x}', t')\hat{\psi}(\boldsymbol{x}, t)\rangle \\ G^{>}(\boldsymbol{x}, \boldsymbol{x}'; t, t') &= -i\langle \hat{\psi}(\boldsymbol{x}, t)\hat{\psi}^{\dagger}(\boldsymbol{x}', t')\rangle \end{aligned}. \tag{1.63}$$

These two Green's functions are known as the 'less than' and 'greater than' Green's functions, respectively, as indicated by the superscripts. Note that the terms on the right-hand side do not incorporate the anti-commutators appearing in (1.61). With only the direct products of the two field operators, these functions are *c*-numbers and are usually referred to as correlation functions. While these two functions are thought to be conjugates of one another, it is not accurately known whether this is true in the far-from-equilibrium situation. The 'less-than' function can be thought of as first annihilating an electron (from an occupied state, of course) and then recreating it elsewhere. Hence, one may conclude that the 'less-than' function is proportional to the density of electrons. It has been hypothesized that it retains the same form as that found near equilibrium [92]. This hypothesis, or ansatz, known as the Kadanoff–Bahm ansatz (or KBA) relates the 'less-than' function to a distribution function as follows:

$$G^{<}(k, \omega) = f(\omega)A(k, \omega). \tag{1.64}$$

However, it is important to note that this is not thought to preserve causality, so that a modified version has been suggested [93] (in the untransformed domain, which remains more physical and contains all the position and time variables so that it also applies to inhomogeneous systems):

$$G^{<}(\boldsymbol{x}, \boldsymbol{x}'; t, t') = -G_r(\boldsymbol{x}, \boldsymbol{x}'; t, t')\rho(\boldsymbol{x}, \boldsymbol{x}', t') + \rho(\boldsymbol{x}, \boldsymbol{x}', t)G_a(\boldsymbol{x}, \boldsymbol{x}'; t, t'), \tag{1.65}$$

which is termed the generalized KBA (or GKBA).

In the application of NEGFs to semiconductor devices, it is often found that (1.64) is used. Generally, an integro-differential equation is found for the 'less-than' function, and then (1.64) is used to give an equivalent integro-differential equation for the density matrix [94]. While this may be useful near equilibrium, it does not satisfy the need for causality mentioned above. But it does give something closer to the Boltzmann equation. On the other hand, if (1.65) is used, then the resulting equation for the density matrix is certainly non-Markovian (it contains memory of the initial state) and must still be carefully evaluated to give something similar to a distribution function. This is a major complication for the use of Green's functions. Thus, it is hard to see how a particle simulation could be used to evaluate the resulting integro-differential equation. However, there are even more problems than this with the use of Green's functions in simulations of semiconductor devices. These will be dealt with in chapter 12.

1.6 An outline of this book

It may be seen from the table of contents that this book has been divided into three sections as a method to aid the reader and those who would like to implement the particles methods that are discussed here. Certainly, a discussion of the foundations of quantum mechanics is far from the scope of this book. However, as a book that provides models used to make predictions about quantum transport, in principle, we need to select a quantum theory to elaborate such predictions. For a realist researcher, this theory must somehow be linked to a particular view of what constitutes a particle. This ties us to various interpretations of what the quantum theory entails. Particles can easily be used without this understanding.

This book on quantum transport adopts neither a purely realistic nor a purely empirical view. It is a balance of the two options, but these options assume that particles have a reality without regard to any subsequent measurements. In any case, the models presented in this book are inspired more by the views of de Broglie and colleagues and by Schrödinger, both discussed above. In this sense, the word 'particle' refers to either of these interpretations. The particle is assumed to exist in the ordinary physical space and to be possessed of its own reality. We strongly believe that complementing the development of transport models with a simple and concrete view of what the particle represents helps the reader in better understanding quantum device performance, and it improves our capabilities for predicting unexpected phenomena. This is, we believe, the great novelty of this book. Along the way, we will avoid referring to the orthodox quantum theory as the 'quantum theory' because, as we have said, there are many other quantum theories as good as the orthodox one. However, we must clarify that the present book is not linked to any single quantum interpretation because, as we have mentioned, computational models are useful for making predictions independently of the interpretation.

Part I is devoted to the basic fundamentals of quantum transport and ultrasmall semiconductor devices. This first chapter has discussed the problems with the interpretation of quantum mechanics itself, the problems with the basics of quantum transport methods, and the problem of deciding just what a particle really is. Chapter 2 will deal with problems that arise from considerations of the fundamentals of quantum transport. These include the role of quantization in ultrasmall semiconductor devices, just what the concept of a measurement means, and just what current and noise mean in these small quantum structures. Chapter 3 then deals with the general problem of a small device coupled to its environment. Here, we deal with the generality of open quantum systems and with wave function behavior in these devices. We also discuss the extra problems that arise in high-frequency devices.

Part II of the book deals with the general problem of modeling and simulating semiconductor devices and the strains this puts on quantum mechanics. Chapter 4 summarizes the types of semiconductor devices that have to be dealt with in the real world and how these present particular problems in quantum transport. Chapter 5 then deals with how the particular problems seen in devices are incorporated into the quantum world and their inclusion in quantum transport. Chapter 6 deals with electronic band structure and with the modifications required for ultrasmall structures. This becomes increasingly important as this band structure is often dealt with directly in these small structures in order to further improve the quantum transport. Chapter 7 then brings us to solving Poisson's equation in small structures which are quite likely strongly nonlinear. Particular problems arise when the grid size chosen for numerical solutions of Poisson's equation is comparable to things like the de Broglie thermal wavelength and the coherence lengths of the particles themselves. Finally, chapter 8 raises the questions of going beyond Poisson's equation, which is an electrostatic solution, to the specific problems of high-frequency devices; this involves complete and complementary solutions of Maxwell's equations.

It is in part III of this book that we finally arrive at particle-based approaches to the simulations of quantum devices. Chapter 9 reviews the ensemble Monte Carlo approach to transport, which has become the standard approach to classical simulations. It also has already been used in quantum transport simulations, especially with the Wigner function. Chapter 10 addresses the use of quantum potentials, from whatever source they may originate, and particle approaches based upon Bohmian mechanics, as hinted in section 1.3. Chapter 11 then deals with Wigner functions and phase-space quantum transport. This chapter is the heart of current particle-based quantum transport. Finally, chapter 12 deals with the problems of NEGFs in modern devices, and why they generally fail to adequately treat the physics of semiconductor devices.

Appendix A On quantization and second quantization

Quantum mechanics is generally viewed as a discussion of systems in which the space or the motion undergoes quantization. Normally, this means spatial constriction of e.g. the wave function and the development of a basis set that is composed of the possible solutions to the Schrödinger equation. In general, the time-independent Schrödinger equation is just one form of a Sturm–Liouville equation in space [95]. Quantum mechanics goes beyond the classical solutions of these types of equations in that the momentum and the position are considered to be operators that satisfy an uncertainty relationship and a commutator such as (1.21):

$$xp - px = i\hbar, \tag{A.1}$$

in which the momentum is given as a differential operator in the position representation as

$$p = -i\hbar\frac{\partial}{\partial x}. \tag{A.2}$$

The solutions to the equations, the basis functions, are combined to give the total wave function as follows:

$$\psi(x) = \sum_i a_i\phi_i(x). \tag{A.3}$$

In this last expression, $\phi_i(x)$ is one of the basis functions, while a_i is related to the fraction of that basis function present in the total wave function. In the next section, this approach is explained for the harmonic oscillator, showing how the exact basis functions depend upon the exact nature of the confining potential which really determines the type of Sturm–Liouville equation that is of interest. The harmonic oscillator is chosen for this example for two reasons. First, it is one of the few problems that can be solved exactly. Second, it is heavily used as the solution for quantum field theory, in which the concept of photons results from expanding each momentum component of the fields as a harmonic oscillator. Similarly, the harmonic oscillator is heavily used in condensed-matter physics, for example for lattice vibrations. In this latter problem, the motion of the atoms around their equilibrium positions is developed via Fourier transforming the Hamiltonian into the collective modes in momentum space and then treating each momentum state as its own harmonic oscillator [16]. Thus, a quantum unit of energy for a particular mode and momentum is a phonon.

In sections 1.4 and 1.5, a totally different notation was introduced. In these sections, it was pointed out that the basis functions $\phi_i(x)$ could actually be treated as operators with their own commutator relationships. This can be approached in two ways. The first is as stated, i.e. the basis functions are operators. The second is to treat the coefficients a_i as operators. Both of these approaches find extensive usage and are referred to as *second quantization.* The second approach is used somewhat more often and is also known as the *occupation number representation.* Again, this will be discussed in a subsequent section for the harmonic oscillator, where it provides a very quick method for solving the Schrödinger equation with a minimum of detailed mathematics. Field operators themselves will be treated in the last section of this appendix. Since these approaches are used extensively in quantum transport, it is important to learn them early.

A.1 The harmonic oscillator

Perhaps the simplest problem in classical mechanics is a mass on a spring, in which the vibrations of the mass are an example of sinusoidal behavior The total energy for such a simple system is

$$E = \frac{p^2}{2m} + \frac{m\omega^2 x^2}{2}, \tag{A.4}$$

in which p is the momentum of the mass m in its motion, x is the position of the mass about the center (stable) point, and ω is radian frequency of the oscillation. In quantum mechanics, this energy goes into the Hamiltonian, which is expressed as

$$H = -\frac{\hbar^2}{2m}\frac{d^2}{dx^2} + \frac{m\omega^2 x^2}{2}, \tag{A.5}$$

with p being replaced by (A.2). This now leads to the time-independent Schrödinger equation

$$-\frac{\hbar^2}{2m}\frac{d^2\psi}{dx^2} + \frac{m\omega^2 x^2}{2}\psi = E\psi. \tag{A.6}$$

The mass on a spring has the same motion as a pendulum, and both are referred to as 'harmonic oscillators.' The form of this equation is well known in mathematical physics, and the solutions are Hermite polynomials. But obtaining these solutions is exceedingly messy mathematically and normalizing them also becomes messy [96]. The usual approach for the latter is usually through the generating functions. Here, the details are ignored, and we concentrate on the important results. The easy part is to introduce reduced position coordinates as follows:

$$\xi = \sqrt{\frac{m\omega}{\hbar}}x, \tag{A.7}$$

for which (A.6) becomes

$$\frac{d^2\psi(\xi)}{d\xi^2} + \left[\frac{2E}{\hbar\omega} - \xi^2\right]\psi(\xi) = 0. \tag{A.8}$$

At this point, the messiness begins, as one can observe that, if the second term in the square brackets were not present, the solution would be a simple exponential in the reduced position. But, finding the full basis sets is difficult. The solutions are Hermite polynomials, the first few of which are

$$\begin{gathered} H_0(\xi) = 1, \qquad H_1(\xi) = 2\xi, \\ H_2(\xi) = 4\xi^2 - 2, \qquad H_3(\xi) = 8\xi^3 - 12\xi, \\ H_2(\xi) = 16\xi^4 - 48\xi^2 + 12, \; H_5(\xi) = 32\xi^5 - 160\xi^3 + 120\xi. \end{gathered} \tag{A.9}$$

These are plotted in figure A.1. As already suggested by the first term in the square brackets of (A.8), the energy is quantized as

$$E = \left(n + \frac{1}{2}\right)\hbar\omega. \tag{A.10}$$

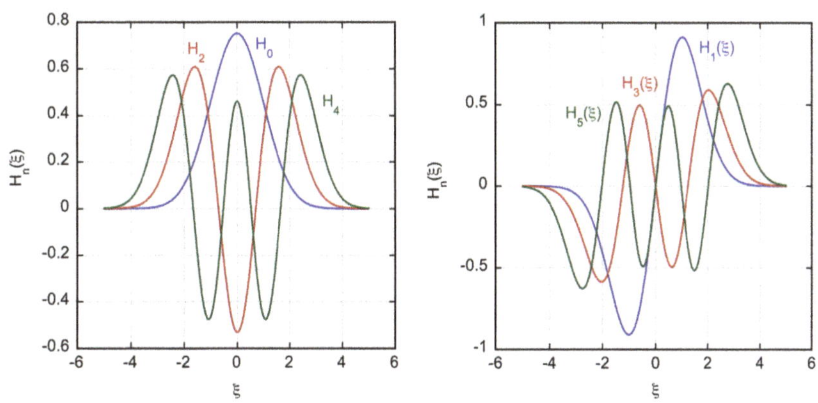

Figure A.1. The lowest three normalized harmonic oscillator wave functions, in reduced coordinates, are shown for the even functions (left panel) and the odd (functions).

One may note that the Hermite polynomials with even order are all even polynomials of the position, and the polynomials of odd order are odd polynomials of the position. As is normal for a second-order differential equation such as the Sturm–Liouville equations, there are two independent solutions. In this case, one of these is a summation over the even-order polynomials, while the second is a summation over the odd-order polynomials. A typical normalized basis function is then given by

$$\varphi_n(x) = \frac{1}{\sqrt{2^n n!}}\left(\frac{m\omega}{\pi\hbar}\right)^{1/4} e^{-m\omega x^2/2\hbar} H_n\left(\sqrt{\frac{m\omega}{\hbar}}x\right). \tag{A.11}$$

Obtaining the normalization factors in front is a task that presents its own difficulty and degree of messiness. For these reasons, the details have been omitted here, but are well known in the literature [96].

A.2 The operator approach to the harmonic oscillator

As may be expected from the discussion of second quantization, this may lead to an easier method of evaluating the wave functions and solving (A.8). For this purpose, (A.8) is rewritten as

$$\frac{1}{2}\left[\xi^2 + p_\xi^2\right]\psi(\xi) = \frac{E}{\hbar\omega}\psi(\xi), \tag{A.12}$$

where

$$p_\xi = -i\frac{\partial}{\partial\xi} \tag{A.13}$$

is the reduced momentum. Knowledge of complex variables leads us to consider defining a pair of operators given by

$$\begin{aligned} a &= \frac{1}{\sqrt{2}}\left(\xi + ip_\xi\right), \\ a^\dagger &= \frac{1}{\sqrt{2}}\left(\xi - ip_\xi\right). \end{aligned} \tag{A.14}$$

Since the reduced position and reduced momentum are not expected to commute, it is now necessary to consider how these two new operators commute and then to begin to understand just what the operators do. First, consider the products of the two operators:

$$\begin{aligned} a^\dagger a &= \frac{1}{2}\left(\xi - ip_\xi\right)\left(\xi + ip_\xi\right) = \frac{1}{2}\left(\xi^2 + p_\xi^2 - 1\right), \\ aa^\dagger &= \frac{1}{2}\left(\xi + ip_\xi\right)\left(\xi - ip_\xi\right) = \frac{1}{2}\left(\xi^2 + p_\xi^2 + 1\right). \end{aligned} \tag{A.15}$$

Thus, we arrive at the commutator relation

$$[a,a^\dagger] = aa^\dagger - a^\dagger a = 1 \tag{A.16}$$

and the Hamiltonian becomes

$$H = \left(a^\dagger a + \frac{1}{2}\right)\hbar\omega. \tag{A.17}$$

Since the spacing of the energy levels is $\hbar\omega$, according to (A.10), it is recognized that the product $a^\dagger a = n$, where n is the order of the harmonic oscillator basis function and thus labels the energy level of interest. Because of this relationship, this second quantization is often called the *occupation number representation.*

In order to find out just what these operators do, it is necessary to play with them a bit. The product $a^\dagger a$ is a c-number; that is, it is just a conventional number and not an operator. Thus, if a basis function is subject to this product, it produces just a c-number prefactor, as follows:

$$a^\dagger a\varphi_n(\xi) = \lambda_n\varphi_n(\xi). \tag{A.18}$$

The opposite ordering also produces a c-number, but a different one, which can be determined from the commutator (A.16). To begin, consider using this operator product on a product of one of the operators and the basis function, as follows:

$$\begin{aligned}(a^\dagger a)a^\dagger\varphi_n(\xi) &= a^\dagger(a^\dagger a + 1)\varphi_n(\xi)\\ &= a^\dagger(\lambda_n + 1)\varphi_n(\xi) = (\lambda_n + 1)a^\dagger\varphi_n(\xi).\end{aligned} \tag{A.19}$$

Thus, the operator $a^\dagger$ *raises* the energy of the state by moving up the ladder of harmonic oscillator states. Thus, it raises the occupation and creates an excitation in the harmonic oscillator. It is consequently called a creation operator. Performing the same magic with the other operator gives

$$\begin{aligned}(a^\dagger a)a(\xi) &= (aa^\dagger - 1)a\varphi_n(\xi)\\ &= a(\lambda_n - 1)\varphi_n(\xi) = (\lambda_n - 1)a\varphi_n(\xi).\end{aligned} \tag{A.20}$$

The operator a *lowers* the energy of the state by moving down the ladder of harmonic oscillator states. Thus, it decreases the occupation number and annihilates an excitation in the harmonic oscillator. It is consequently called an annihilation operator.

To proceed further, the expectation value of the operator product is considered for a given basis state, as follows:

$$\begin{aligned}\langle a^\dagger a\rangle &= (\varphi_n(\xi), a^\dagger a\varphi_n(\xi))\\ &= (a\varphi_n(\xi), a\varphi_n(\xi)) \geqslant 0.\end{aligned} \tag{A.21}$$

Here, the parentheses indicate an integration over the reduced coordinates. In this integration, the terms before the comma indicate that an adjoint or complex conjugate is used, so that this integral is that of the squared magnitude of the function $a\varphi_n(\xi)$; therefore, the result is necessarily positive semidefinite. As a result of this, there must be a lowest value of n, which is taken to give the ground state $\varphi_0(\xi)$ corresponding to the zero-order Hermite polynomial. Operating on this ground state with the annihilation operator cannot produce another state, so must yield zero

$$a\varphi_0(\xi) = 0. \tag{A.22}$$

The first line of (A.14), along with (A.13), now leads to the first-order differential equation

$$a\varphi_0(\xi) = \frac{1}{\sqrt{2}}\left(\xi - \frac{\partial}{\partial \xi}\right)\varphi_0(\xi) = 0. \tag{A.23}$$

This trivially leads to

$$\varphi_0(\xi) = e^{-\xi^2/2}, \tag{A.24}$$

which is the unnormalized zero-order basis function.

To determine the normalization that must occur, there are two steps. First, a set of constants must be considered. To determine these properly, the unreduced coordinates are inserted into the wave function, so that

$$\varphi_0(x) = C_0 e^{-m\omega x^2/2\hbar}, \tag{A.25}$$

which is a standard Gaussian that has the following normalization:

$$C_0 = \left(\frac{m\omega}{\pi\hbar}\right)^{1/4}. \tag{A.26}$$

There is then a multiplier from one basis function to the next. Consider, for example, the case in which each basis function is written as $C_n\varphi_n(\xi)$, so that

$$\begin{aligned}(a\varphi_n(\xi), a\varphi_n(\xi)) &= (\varphi_n(\xi), a^\dagger a\varphi_n(\xi)) = nC_n^2 \\ &= (\varphi_{n-1}(\xi), \varphi_{n-1}(\xi)) = C_{n-1}^2.\end{aligned} \tag{A.27}$$

This means that

$$C_n = \frac{1}{\sqrt{n}}C_{n-1}, \tag{A.28}$$

and the complete normalized function may be written as

$$\varphi_n(x) = \left(\frac{m\omega}{\pi\hbar}\right)^{1/4}\frac{1}{\sqrt{(n+1)!}}(a^\dagger)^n\varphi_0(x). \tag{A.29}$$

In this last line, the real-space operators are written as

$$\begin{aligned}a &= \sqrt{\frac{m\omega}{2\hbar}}\left(x + i\frac{p}{m\omega}\right), \\ a^\dagger &= \sqrt{\frac{m\omega}{2\hbar}}\left(x - i\frac{p}{m\omega}\right).\end{aligned} \tag{A.30}$$

The use of these raising and lowering operators provides a much easier method of evaluating the basis functions and the Hamiltonian. It turns out that the harmonic oscillator potential (the last term in (A.4)) is not the only potential for which this works. The quadratic potential of (A.4) is only one of a class of potentials that allow the use of second quantization to factor the Hamiltonian in the manner done above [97, 98]. This is now referred to as supersymmetric quantum mechanics but should not be confused with the supersymmetry in quantum field theory. Rather, this field derives from a simple supersymmetric system consisting of a single spin-1/2 particle moving in a one-dimensional potential [99], such as the potentials mentioned here. A general discussion of supersymmetric quantum mechanics in terms of the second quantization operators may be found in [100]. This approach has been useful for solving for the density distribution in double-gate MOSFETs [101]. Here, the operators have been developed for bosons. The discussion of spin and fermion operators will be deferred to appendix A (chapter 5).

A.3 Field operators

As mentioned above, the field operators were introduced in section 1.2. In this section, it is desired to expand the definitions given in that section in order to gain a little more understanding. The development of field operators flows naturally from the appearance of second quantization. In general, a wave function can be written as an expansion of the basis functions, as in (A.3). In a future discussion of spin states, the basis functions will also include a spin function, but that is not necessary here. The introduction of second quantization means that (A.3) can be rewritten as

$$\psi^\dagger(x) = \sum_i a_i^\dagger \varphi_i^*(x), \tag{A.31}$$

so that the creation operator appears in the expansion and converts the entire wave function into a field operator. The coefficient on the right-hand side is a creation operator, while the basis function complex conjugate is used. This produces a *creation* field operator. In essence, this field operator changes e.g. the ground state into an occupied state (or states). These are now operators and not simply wave functions.

The concept of 'field' used here is precisely that of electromagnetic fields in mathematics. Here, a field is a mathematical quantity that has a defined value at every point in space, the position variable space. One can go further, since the basis functions work as unit vectors in the Hilbert. Thus, the field operator can be considered to be a vector field in that a vector direction is determined by the coefficients for the field operator on the left-hand side of (A.31). These field operators also satisfy a commutator relationship, which may be defined as

$$\begin{aligned}[]
[\psi^\dagger(x), \psi(x')] &= \psi^\dagger(x)\psi(x') - \psi(x')\psi^\dagger(x) \\
&= \sum_{i,j} a_i^\dagger \varphi_i^*(x) a_j \phi_j(x') - \sum_{i,j} a_j \phi_j(x') a_i^\dagger \varphi_i^*(x) \\
&= \sum_{ij} \varphi_i^*(x)\phi_j(x')\delta_{ij} = \delta(x - x')
\end{aligned} \tag{A.32}$$

This last line is just the closure of a complete set. The commutation of the field operators ensures that the basis functions are properly normalized and orthogonal to one another, as required by solutions to any of the Sturm–Liouville equations.

The relationships for the field operators as creation and annihilation operators follow from the discussion of (A.19) and (A.20), so that the field operator $\psi^{\dagger}(x)$ creates a particle at position x. Whereas the operators $a_i^{\dagger}$ and a_i create and annihilate the occupation of the appropriate basis function, the total field operator creates the excitation (or annihilation) of a particle at a specific position. This has the same relationship as that of Fourier transformation, where the summation over the various Fourier components produces the value at a position in real space. More about the field operators will appear in appendix A (chapter 5), where the discussion turns to the second quantization of fermions.

References

[1] Newton I S 1686 *Philosophiae Naturalis Principia Mathematica* (London: Pepys)
[2] Boltzmann L 1868 *Wiener Ber.* **58** 517
[3] Boltzmann L 1872 *Wiener Ber.* **66** 275
[4] Planck M 1901 *Ann. Phys.* **309** 553
[5] Newton I A 1704 *Opticks: Or a Treatise on the Reflexions, Refractions, Inflexions, and Colours of Light* (London: Smith and Walford)
[6] Young T 1807 *A Course of Lectures of the Natural Philosophy and Mechanical Arts* 1 Lecture (London: J Johnson)
[7] Maxwell J C 1865 *Phil. Trans. Royal Soc. London* **155** 459
[8] Taylor G I 1909 *Proc. Cambridge Phil. Soc.* **15** 114
[9] Bohr N 1913 *Phil. Mag., Ser. 6* **26** 1
[10] Lagrange J-L 1788 *Mécanique Analytique* (Paris: Desaint)
[11] Hamilton W R 1835 *Phil. Trans. Royal Soc. London* **125** 95
[12] Bogoliubov N N 1946 *J. Phys. Sowjet.* U **10** 256
[13] Born M and Green H S 1946 *Proc. Roy. Soc. London* A **188** 10
[14] Kirkwood J G 1946 *J. Chem. Phys.* **14** 180
[15] Yvon J 1937 *Act. Sci. Ind.* 542, 543 (Paris: Herman)
[16] Ferry D K 1991 *Semiconductors* (New York: Macmillan)
[17] Selberherr S 1984 *Analysis and Simulation of Semiconductor Devices* (Wien: Springer)
[18] Ferry D K 2020 *Semiconductors: Bonds and Bands* 2nd edn (Bristol: IOP Publishing)
[19] Leclerc G-L 1733 *Hist. L'Acad. Roy. Sci.* (Paris: Royal Imprint) p 43
[20] Metropolis N and Ulam S 1949 *J. Am. Stat. Assoc.* **44** 335
[21] Metropolis N 1987 *Los Alamos Science* special issue p 125
[22] Kurosawa T 1966 *Proc. Int. Conf. Phys. Semicon., Kyoto J. Phys. Soc. Jpn.* **21** 424
[23] Hockney R W and Eastwood J W 1981 *Computer Simulation Using Particles* (New York: McGraw-Hill)
[24] Rutherford E 1911 *Phil. Mag., Ser. 6* **21** 669
[25] Nagaoka H 1904 *Phil. Mag., Ser. 6* **7** 445
[26] Sommerfeld A 1916 *Ann. Phys.* **51** 1
[27] Einstein A 1917 *Verh. Deutsch. Phys. Gesell* **19** 82
[28] Brillouin L 1926 *J. Phys. Radium* **7** 353
[29] Keller J B 1958 *Ann. Phys. (New York)* **4** 180

[30] Berry M V and Tabor M 1976 *Proc. Roy. Soc.* A **349** 101
[31] Gutzwiller M C 1990 *Chaos in Classical and Quantum Mechanics* (New York: Springer)
[32] Heisenberg W 1925 *Z. Phys.* **33** 879
[33] Born M and Jordan P 1925 *Z. Phys.* **34** 858
[34] Dirac P A M 1925 *Proc. Roy. Soc. London* A **109** 642
[35] Dirac P A M 1926 *Proc. Roy. Soc. London* A **112** 661
[36] Heisenberg W 1927 *Z. Phys.* **43** 172
[37] de Broglie L 1923 *Nature* **112** 540
[38] Cushing J T 1994 *Quantum Mechanics* (Chicago: University of Chicago Press)
[39] de Broglie L 1924 *Phil. Mag., Ser. 6* **47** 446
[40] Einstein A 1917 *Phys. Z.* **18** 121
[41] Jones S 2008 *The Quantum Ten* (Toronto: Thomas Allen)
[42] Schrödinger E 1926 *Ann. Phys.* **79** 361
[43] Schrödinger E 1926 *Ann. Phys.* **79** 489
[44] Schrödinger E 1926 *Ann. Phys.* **79** 734
[45] Zettl A 2005 *Sturm–Liouville Theory* (Providence, RI: American Mathematical Society)
[46] Madelung E 1927 *Z. Phys.* **40** 322
[47] Bohm D 1952 *Phys. Rev.* **85** 166
[48] Philippidis C, Dewdney C and Hiley B J 1979 *Il Nuovo Cim.* **52B** 15
[49] Marian D, Zanghì N and Oriols X 2016 *Phys. Rev. Lett.* **116** 110404
[50] Kennard E H 1928 *Phys. Rev.* **31** 876
[51] Barker J R 2008 *AIP Conf. Proc.* 995 104
[52] Barker J R and Martinez A 2004 *J. Comput. Electron.* **3** 401
[53] Griffiths R B 2001 *Consistent Quantum Theory* (Cambridge: Cambridge University Press)
[54] Hartle J B 1998 *Phys. Scripta* **T76** 67
[55] Omnès R 1994 *The Interpretation of Quantum Mechanics* (Princeton, NJ: Princeton University Press)
[56] Griffiths R B 2017 *Phys. Rev.* A **96** 032110
[57] Holland P R 1993 *The Quantum Theory of Motion* (Cambridge: Cambridge University Press)
[58] Brush S G 1967 *Graduate* J **7** 477
[59] Gell-Mann M 1979 *The Nature of the Physical Universe: 1976 Nobel Conf.* ed D Huff and O Prewett *(New York)* (Wiley Interscience) p 29
[60] Selleri F 2003 *American Institute of Physics: Oral History Interviews* https://www.aip.org/history-programs/niels-bohr-library/oral-histories/28003-1
[61] Feynman R P 1965 *The Character of Physical Law* (Cambridge, MA: MIT Press) p 129
[62] Feynman R P 1948 *Rev. Mod. Phys.* **20** 367
[63] Dyson F J 1949 *Phys. Rev.* **75** 1736
[64] Oriols X and Ferry D K 2021 *Proc. IEEE* **109** 955
[65] Ferry D K 2021 *Hot Carriers in Semiconductors* (Bristol: IOP Publishing)
[66] Shaw M P, Grubin H L and Solomon P R 1979 *The Gunn–Hilsum Effect* (New York: Academic)
[67] Gunn J B 1963 *Solid State Commun* **1** 88
[68] Ridley B K 1963 *Proc. Phys. Soc.* **82** 954
[69] Schrödinger E 1935 *Naturwissenschaften* **23** 807 823 844
[70] Dyson F J 1949 *Phys. Rev.* **75** 486

[71] Dirac P A M 1927 *Proc. Roy. Soc.* A **114** 243
[72] Ferry D K *et al* 2022 *Semicond. Sci. Technol.* **37** 043001
[73] Schwinger J 1951 *Proc. Natl Acad. Sci.* **37** 452
[74] Fetter A L and Walecka J D 1971 *Quantum Theory of Many-Particle Systems* (New York: McGraw-Hill)
[75] Weinbub J and Kosik R 2022 *J. Phys.: Condens. Matter* **34** 163001
[76] Ramey S M and Ferry D K 2004 *Semicond. Sci. Technol.* **19** S238
[77] Sverdlov V, Undersboeck E, Kosina H and Selberherr S 2008 *Mat. Sci. Eng.* R **58** 228
[78] Löwdin P-O 1951 *J. Chem. Phys.* **19** 1396
[79] Löwdin P-O 1963 *J. Mol. Spectr* **10** 12
[80] Gilbert M J, Akis R and Ferry D K 2005 *J. Appl. Phys.* **98** 094303
[81] Ando T 1991 *Phys. Rev.* B **44** 8017
[82] Usuki U *et al* 1995 *Phys. Rev.* B **52** 8244
[83] Duke C B 1969 *Tunneling in Solids* (New York: Academic)
[84] Ferry D K, Goodnick S M and Bird J P 2009 *Transport in Nanostructures* 2nd edn (Cambridge: Cambridge University Press)
[85] Landauer R 1957 *IBM J. Res. Develop.* **1** 223
[86] Wigner E 1932 *Phys. Rev.* **40** 749
[87] Weinbub J and Ferry D K 2018 *Appl. Phys. Rev.* **5** 041104
[88] Ferry D K and Nedjalkov M 2018 *The Wigner Function in Science and Technology* (Bristol: IOP Publishing)
[89] Feynman R P 1949 *Phys. Rev.* **76** 749
[90] Schwinger J 1960 *Proc. Natl Acad. Sci.* **46** 883
[91] Baym G and Kadanoff L P 1961 *Phys. Rev.* **124** 287
[92] Kadanoff L P and Baym G 1962 *Quantum Statistical Mechanics* (New York: W. A. Benjamin)
[93] Lipavský P, Ŝpiĉka V and Velický B 1986 *Phys. Rev.* B **34** 6933
[94] Mahan G D 1984 *J. Phys.* F **14** 941
[95] Morse P M and Feshbach H 1953 *Methods of Theoretical Physics* (New York: McGraw-Hill) ch 5
[96] Ferry D K 2001 *Quantum Mechanics* 2nd edn (Boca Raton, FL: CRC Press)
[97] Sukumar C V 1985 *J. Phys. A: Math. Gen.* **18** 2917
[98] Cooper F, Khareb A and Sukhatme 1995 *Phys. Rep.* **251** 267
[99] Witten E 1981 *Nucl. Phys.* B **188** 513
[100] Ferry D K 2001 *Quantum Mechanics* 3rd edn (Boca Raton, FL: CRC Press)
[101] Shishir R S and Ferry D K 2008 *J. Comput. Electron* **7** 14

Quantum Transport in Semiconductor Devices
Simulation using particles
David K Ferry, Xavier Oriols and Josef Weinbub

Chapter 2

The microscopic world and microscopic properties

The study of semiconductors can reasonably be attributed to the ancient Greeks, who rubbed amber with cloth to study electrical effects. The name, though, seems to come from Michael Faraday's work with silver sulfide in 1833 [1]. The first actual device seems to be the rectifier made from a semiconductor–metal structure that was observed by Braun in 1874 [2]. While such diode detectors were certainly a major part of microwave systems in World War II, the development of the devices we would usually call semiconductor devices began after the war with the invention of the bipolar transistor [3, 4]. Rapid progress began with the integrated circuit [5], and evolution then seemed to follow Moore's law [6]. Today, there are a great many types of semiconductor devices. Although silicon (Si) remains the principal material, a great many types of materials are currently used in a wide variety of devices. Discussion of the main devices will be left to chapter 4, but one cannot ignore the fact that the dominant device today is the metal–oxide–semiconductor field-effect transistor (MOSFET). It is used by the tens of billions in each integrated circuit, which itself is manufactured daily in the tens of millions. This current state is the result of rigorous attention to Moore's law.

The growth in technology known as Moore's law is basically the idea that the dimensions of semiconductor devices decrease from one technological generation to the next, with the result that the number of transistors on a single chip increases exponentially [6], although this law is basically an economic law rather than a physical law [7, 8]. These integrated circuits are mainly digital, which means that the transistors used are switching transistors. These transistors need well-defined stable states, which define the logic levels. These voltages are determined by the electrical ground (zero voltage and a reference) and the bias voltages. The transition between these two states is not of great concern as long as it happens quickly and reliably, although the actual switching time is an important metric. Switching transistors are not the only requirement, however. Amplifiers are also needed, which

doi:10.1088/978-0-7503-5237-6ch2

should exhibit linear behavior at both small and large signal levels. Amplifiers require analog signals for this purpose. In microwave applications, these are rarely Si-based devices, but are often III–V devices.

The structure of the MOSFET has undergone a great deal of change since its first use in an integrated circuit. The downsizing of the device that has arisen from following Moore's law has led to new structures, new materials, and new configurations. For example, as the size of the MOSFET decreased, a growing problem appeared, namely that the mobility continued to drop. This drop was due to the effect of surface-roughness scattering at the oxide–semiconductor interface [9] as well as an increase in impurity scattering due to increased doping of the substrates. The first change made to overcome this decrease in mobility was to introduce strain into the channel [10]. Eventually, one had to address the problem of the oxide becoming too thin, which led to the use of a new gate material, namely HfO_2 [11]. Finally, continued evolution forced one to turn the FET on its edge, leading to the FinFET [12]. The future may well lie in gate-all-around quantum wire FETs [13, 14]. What this has all led to is that nanowire structures defined by dimensions in the nanometer range have become the heart of modern MOSFETs. This immediately means that the wire structures are quantized in the transverse dimensions and there are quantum mismatches between the quasi-one-dimensional wires and the three-dimensional source and drain contacts in these devices. We have arrived in the quantum transport world, whether or not we are ready for it.

But the problem is more complex. Through most of the evolution of the integrated circuit, transistors have been downsized through the use of a strict scaling relationship [15] in which the electrostatics are maintained. All dimensions are reduced by the same factor, and the voltages and doping densities are adjusted to maintain the electrostatics. Such an approach ignores the parasitics which have become more important as the device gets smaller and the other devices get closer. While the electric fields can be scaled, capacitances begin to be dominated by edge effects and then do not continue to scale properly. The device can no longer really be considered in isolation, as it sits in an environment consisting of other devices. The device cannot be isolated from this environment [16]. There is also the question of charge granularity that arises from small numbers of dopant atoms. If one considers a small device in a semiconductor region 20 nm long, 20 nm high, and 5 nm thick, doped at 10^{18} cm^{-3}, there are, on average, only two dopant atoms in the volume. The exact location of those dopants then becomes critical, as they affect the device current more when they are close to the source.

Thus, we see that quantum effects are going to arise in a great many ways merely because the distances involved in the details of the device are all on the size of the de Broglie wavelength. Hence, quantization will appear in many guises. In this chapter, we will try to establish the ground rules by which these quantum effects can be simulated. That is, we want to talk about the environment and subtle problems associated with the relationships between charge, current, and potential. There is also the major problem of what constitutes a measurement, and what is meant by current and current noise. These will be dealt with here.

2.1 The measurement problem in quantum mechanics

As was discussed in chapter 1, the orthodox theory of quantum mechanics holds that reality occurs only when a measurement is made. In such a view, quantum objects (such as electrons) are all part of some nebulous many-particle wave function, yet have no properties at all. All of these properties magically appear when the measurement is made. To make matters worse, this orthodox theory does not define exactly what constitutes a measurement and just which interactions lead to this measurement. To understand why, in such a measurement, a well-defined position of the electron can arise (for example, in a tunneling problem, the electron is located to the left or the right of the barrier), one needs to introduce an extra equation of motion that describes the evolution of the final state [17]. This new equation of motion can only be invoked during the measurement process and it has the effect of selecting only a single eigenstate from the classical set. Another view of this process is that the measurement *projects* the wave function onto the classical basis state of the measurement. Determining the extra equation is equivalent to defining the projection operator necessary for this process. Only then does the property of the electron become well defined, and only then can it be given by the eigenvalue of the eigenstate. If we stop measuring the system, the wave function has already collapsed, and the new (post-collapse) state may evolve into a new superposition. The well-defined position of the electron is then lost; it has become undefined once again. At this point, readers might wonder what has happened to the electrons and their properties that are used to explain the behavior of electronic devices. Are they still present or not?

Given the confusion generated by the above description, it is clear why Feynman commented that [18] '...no one understands quantum mechanics...' when discussing the orthodox viewpoints. The understanding of causality in the real world leads to the alternative interpretations of quantum mechanics that were also discussed in chapter 1. These theories provide the alternate world in which reality exists, especially when we need to follow a set of particles moving through a semiconductor device. With this reality, it is possible to believe that good physical theories explain, or at least provide clues about, this very reality of our comprehensible world. Access to the reality underlying the mathematics of the theory allows a simple, rigorous, and intuitive understanding of the phenomena occurring inside an electron device. Certainly, there is no problem at all in assuming that the measurement modifies the system (in an electronic circuit, adding an ammeter modifies the value of the current). That is, it is fairly well known that measurements do disturb the system, and our *natural* intuition accepts such practical modifications. What is unacceptable is that the measurement itself dictates when the system becomes real.

Nevertheless, even in the causal reality, the measurement of any property of the particle is accompanied by a perturbation of the wave function and the particle property that is being measured. For example, this may be through the velocity field $\boldsymbol{v}(\boldsymbol{x}, t)$, or through the simultaneous measurement, at time t, of the position $\boldsymbol{x}(t)$. However, the simultaneous measurement of these two properties still retains an uncertainty, as they are adjoint variables and subject to Heisenberg's principle. This should not be a surprise, as the same thing is true classically. Only now, this uncertainty is considered to be fundamental and not a simple measurement problem. It is important to emphasize that the fact that if something cannot be directly measured, this does not mean that it is not real. Yet the quantum reality behind these views is that the set of well-defined trajectories (one trajectory for each particle) moving in ordinary three-dimensional space is causal in the sense that their evolution is determined by a velocity and the existence of initial conditions, among other possible properties. Such a velocity can be extracted from the wave function solution of the Schrödinger equation, without stochastic collapse, and with a quantum potential that modifies movement from the classical case, as discussed by Kennard [19]. In particular, if a particle is sent toward a two-slit apparatus, its initial position determines which slit it passes through, as seen in figure 1.1 in chapter 1. Obviously, if only one of the slits is open, the motion of the particle is different, and no interference effects are observed. It is also true that, if some measuring apparatus interacts with the particle, either before or after it passes through slit, its trajectory is modified. There is no magic here, and this latter fact shows that there is a reverse action of the apparatus on the particle. Despite the fact that the measurement can modify the particle trajectory, it has no role in defining the reality of the particle. This topic will be revisited in more of a quantum setting in following sections, especially in sections 2.4 and 2.5.

2.2 Measurements and the environment

A semiconductor device and its surrounding environment form a complex system that is composed of multiple parts. Such a system has parts that are typically both nonlinear and inhomogeneous. In chapter 1, we pointed out that this locates our quantum semiconductor device within a surrounding environment. It was also discussed there that measurements of the device must, by necessity, be done in the environment and not internally in the device. This does not arise on fundamental grounds but because the device is often incredibly small (for example, on the nanometer scale) and that the important area of the device is difficult to reach with any reasonable experimental probe. This led to the ideas, also presented in chapter 1, that the environment must interact with the quantum states within the actual device if the properties of these states are to be measurable. In addition, there is the likelihood that the environment may affect the nature of these states as well as generating additional quantum states that would not exist in the isolated device. Here, we illustrate each of these two interaction directions using effects that have occurred in integrated circuits and in individual devices.

2.2.1 Drain-induced barrier lowering

The first of these effects is a well-known process in MOSFETs that is usually referred to as drain-induced barrier lowering (DIBL) [20]. In this effect, drain bias is applied in the environment, as the actual position at which the bias is imposed is separated from the heart of the device by a considerable part of the drain structure (see, e.g. figure 1.5). As the drain area is usually fairly highly doped, the series resistance of this area usually does not reduce the potential by much. However, the injection of carriers into the MOS channel is governed by the potential difference between the source and gate potentials at the source end of the channel. The barrier height over which carriers are injected is controlled by this gate–source potential difference. Normally, the channel region is relatively highly doped, so that there is enough scattering in the channel of the injected carriers by these ionized dopants as well as by phonons that the drain voltage is 'screened' from this barrier. That is, the scattering localizes carriers within the channel, and this charge strongly affects the potential variation from the source to the drain such that the drain potential does not affect the gate–source potential difference; see the lower panel in figure 1.6. The peak in the potential shown in this figure is at the source end of the channel. With the source normally grounded, the positive gate voltage lowers this peak and allows more electrons to enter the channel. Normally, there is enough charge in the channel that the drain potential does not affect this peak. But in modern devices where the channel region (the region under the gate, shown in yellow in figure 1.4) is lightly doped, or even undoped, the ionized impurity scattering is extremely weak. If the channel is short and ballistic transport begins to occur, the screening effect of the scattering is significantly reduced. The source–gate potential barrier then begins to be affected by the drain potential, which, in turn, can act to lower this barrier, letting more current into the channel [21]. In a Si nanowire MOSFET, the transition from scattering-dominated to ballistic transport can occur in as little as 8–10 nm [22], so that this effect can become quite important. Since the drain voltage is applied some distance from the actual active part of the device, we consider this to be an environmentally induced effect.

The contacts themselves are also a complex and poorly understood region of the device. For example, if the drain is sufficiently heavily doped, the potential does not vary much through this region. On the other, it remains possible for the derivative of the electric field to change as one moves from the fully metallic contact to the semiconductor drain region. Such a change in the derivative of the field requires charge accumulation to occur, and this has to be considered in defining anything related to contact resistance. We will return to this in section 2.3.

2.2.2 Light emission

The observation of light emission from direct-gap III–V devices would not surprise anyone. But the observation of light coming from Si devices, particularly light that has sub-bandgap photon energies, is a surprise. This is because Si has an indirect bandgap and is not expected to be an emitter of sub-bandgap radiation. Light emission has been seen in Si *p*–*n* junctions under conditions of diode breakdown [23, 24], but this emission has photon energies above the bandgap. So, it was surprising when light was observed from MOSFETs operating under hot electron conditions [25]. The spectrum of the light was shown to be Maxwellian and to have a temperature in the range of 1000–1500 K for a device with a 1.5 μm channel length; the lower temperatures occurred at higher values of the gate voltage [26]. Spatial studies showed that this light emission was coming from the drain end of the gate.

Surprisingly, this emission was observed over the range of 0.55–1.55 eV, which meant that some light was sub-bandgap in origin [27]. The latter authors also showed that this background light emission had a different spectrum from electron–hole recombination radiation.

To understand how this sub-bandgap radiation can occur, one has to understand the role that hot carriers play in the MOSFET. If one looks at the details of the Si band structure (which will be discussed in chapter 6), the minima of the conduction band lie along the Δ lines near the X point—the zone edge at (100). At this zone edge, the lowest conduction band is degenerate with the second conduction band. Since the momentum is only defined within a reciprocal lattice vector, electrons reaching X can be transported to negative values of X or appear to progress across the zone boundary into the second zone *and the second conduction band.* As carriers are heated in an electric field, they move up the band to the zone boundary and then continue into this higher-lying band. Here, they can begin to undergo a direct optical transition back to the lower conduction band. This involves the emission of a photon that lies within the observed spectrum. There is also no reason for this emission to be cut off at the bandgap. Eventually, it was determined that the likely source of the light emission was transitions within the conduction band itself [28–30]. The observation of below-bandgap radiation was a supporting sign of this intra-conduction-band optical transition. That is, these light-emitting transitions involved only a single type of carrier—the electrons in the *n*-channel devices.

When the MOSFET is embedded in an integrated circuit, the light emission described above can affect other devices within the circuit. This becomes an interaction between the light-emitting device and its environment. This is because the light does not remain in the device, but can propagate throughout the integrated circuit. For example, these photons have been observed to excite carriers in the substrate, which then affect an adjacent device by modifying the substrate bias [31]. In addition, the light can propagate around an isolating oxide and affect the neighboring devices [32]. Thus, light emission from one device can modify the environment of the entire integrated circuit and can be a problem that must be addressed.

In addition, if the fields are high enough near the drain end of the MOSFET, impact ionization can occur, as mentioned above for *p–n* junctions. This produces light emission and additional carriers. While the additional electrons head to the drain, the holes produced are more likely to move into the substrate, where the substrate current now creates an additional voltage drop in the substrate. This then changes the substrate bias seen for other devices on the circuit. Hence, there are many conditions which lead one device to affect other devices in its neighborhood.

2.3 Landauer and contacts

In the early days of semiconductor devices, Rolf Landauer presented an approach to transport that was dramatically different from the microscopic kinetic theory based on the Boltzmann equation that is normally used [33, 34]. In this approach, one computes the conductance by simply determining the transmission of a mode from an input reservoir to a similar mode in an output reservoir. These modes in the source and drain were discussed in section 1.6. As the transverse dimension is limited, the Schrödinger equation leads to a set of transverse modes that arise from the size quantization in these regions as well as in the channel. The number of these modes that are occupied in the source (or the drain) depends upon the carrier population and the Fermi energy. The calculation of the transmission probability from one end of the device to the other is very similar to the computation of a tunneling probability, except that there is no requirement that the process be one of tunneling. It could equally well be scattering that moves the carriers from one transverse mode to another during their flight from source to drain. The only real constraint is lateral confinement and the number of these transverse modes. That is, the carriers in the reservoirs are described by their intrinsic Fermi–Dirac distributions, and any applied potentials appear as a shift of the relative energies (which would shift one Fermi level relative to the other). Of course, there could also be temperature differences between the two ends. Landauer originally considered that the transport was ballistic, but this is not required. If scattering is present, it must be described specifically as a transfer of the electron from one internal mode to another within the channel. This goes well beyond Landauer's initial interpretation and provides the ability to compute transport directly using solutions of the Schrödinger equation, whether these be via the wave function, the Wigner function, or even Green's functions.

2.3.1 The Landauer formula

If it is assumed that ballistic transport occurs in the semiconductor structure, then a simple result can be obtained. In actual practice, this result carries over to non-ballistic situations once the transmission coefficient has been defined. To compute the current passing through the structure, only the charge that occupies each mode need be determined, namely the charge for the energy range between E_F and $E_F +$ eV, where E_F is the Fermi level at the source and V is the applied potential (at the drain). Only carriers in this range of energies can move from source to drain, although a nonzero temperature broadens the distributions and spreads the energies somewhat. For the moment, though, this energy range is sufficient to get the desired result. When the applied bias potential is small, then the excess charge can be estimated for a one-dimensional transverse mode as follows:

$$\delta Q \approx \frac{e^2 V}{2} \rho_{1d}(E_F) = e^2 V \sqrt{\frac{m^*}{2\pi^2 \hbar^2 E_F}}. \tag{2.1}$$

Here, $\rho_{1d}(E_F)$ is the one-dimensional density of states at the Fermi level in the source. Notice that only one half of this density is used, since only those electrons that are traveling from source to drain (those defined as having positive momentum) are of interest. In order to compute the current, (2.1) is multiplied by the carrier velocity at the Fermi energy

$$v_g = v_F = \frac{1}{\hbar} \frac{dE}{dk} \bigg|_{E=E_F} = \frac{\hbar k_F}{m^*} = \sqrt{\frac{2E_F}{m^*}}, \tag{2.2}$$

and this leads to the current

$$I_c = \delta Q v_g = e^2 V \sqrt{\frac{m^*}{2\pi^2\hbar^2 E_F}} \sqrt{\frac{2E_F}{m^*}} = \frac{2e^2}{h} V. \tag{2.3}$$

Notice that the energy has dropped out of the equation. This is a result of the velocity canceling the energy dependence of the density of states, which always occurs in one-dimensional situations. The current is now independent of the energy, but this becomes more complicated when multiple modes are considered, as discussed below. The total current is found from the number of modes that are occupied, which is

$$I_{\text{total}} = NI_c = N\frac{2e^2}{h}V. \tag{2.4}$$

From this last expression, the conductance is obtained as follows:

$$G = \frac{I_{\text{total}}}{V} = N\frac{2e^2}{h}. \tag{2.5}$$

It is this last expression that usually called the Landauer formula. The conductance per mode is quantized in units of $2e^2/h$ (~77.28 μS) with a resulting magnitude that depends only upon the number of occupied channels. As is well known for devices, in saturation, the transport depends only upon the injection of carriers from the source to the channel. Hence, only the number of modes in the source region is important for saturation.

In the above, no consideration was given to the fact that an electron in one mode in the source might actually wind up in a different mode in the drain. This is not reflected in the formula (2.5). Nor does (2.5) offer the possibility of partial transmission of a mode. The transitions from one mode to another are even more important when scattering within the channel region is considered. Consequently, it is more intuitive to adapt (2.5) to the real situation in which partial transmission of a mode is possible. To do this, (2.4) is rewritten as

$$I_{\text{total}} = \frac{2e^2}{h}VT(n), \tag{2.6}$$

where $T(n)$ is the transmission of the n channels in the source, given by

$$T(n) = \sum_{m=1}^{\infty} T_{mn}, \tag{2.7}$$

and T_{mn} is the probability that electrons entering the channel in source mode n are scattered to drain mode m prior to exiting the device.

2.3.2 Charge and contacts

Some readers might jump to the conclusion that purely ballistic transport should yield infinite conductance rather than the finite value of (2.6). What actually occurs is that the transition between the reservoir and the channel itself has a nonzero resistance, which is the contact resistance between the channel and the reservoirs. Normally, the source and drain reservoirs are very wide regions that are considered to be in equilibrium, at least in the parts of the source and drain far from the contact. The contact resistance can arise from a mode mismatch between these latter regions and the channel, but charge accumulation at the contact also contributes resistance. The contact resistance goes directly to the question of where the voltage drop across the device actually occurs. The voltage, or applied potential between the source and drain, exists between these two regions, but actually varies between them. In a normal resistor, the voltage should drop uniformly along the length of the constriction, but this still requires unusual behavior in the contacts, which leads to contact resistance even in this simple case.

Let us consider the ballistic case as an example, so that there is no scattering within the channel. The voltage drop across the channel is considered to create a constant electric field, such as that seen in figure 2.1. In reality, the voltage is not dropped smoothly across the device, and this leads to charge accumulation at the contacts. This, in turn, leads to variations in the voltage along the length. As a result, the field in the center of the constriction is quite small, and the voltage drops are forced to occur in the transition regions. Suppose we take the case of a linear voltage drop across the constriction, as shown in figure 2.1. There is now a constant electric field in the channel region. The potential drop will look like that shown in the

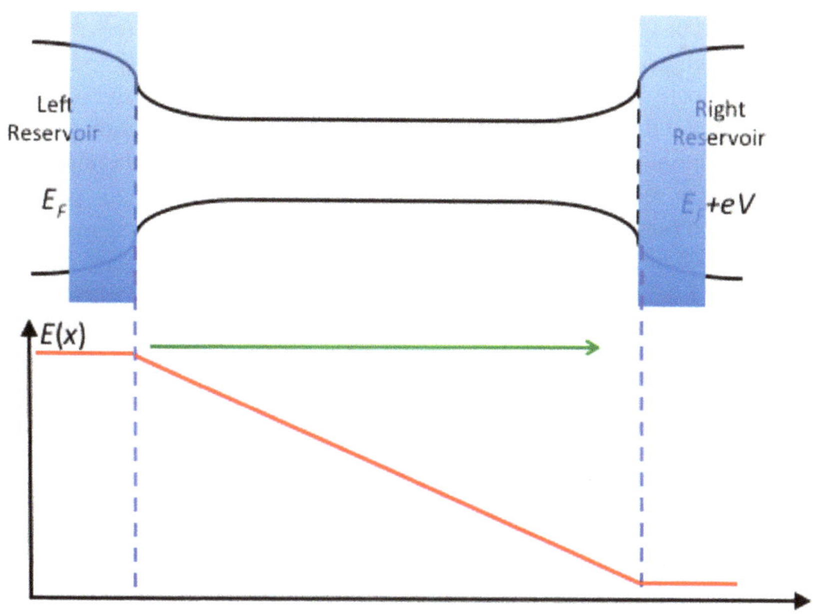

Figure 2.1. The energy $E(x)$ (negative of the potential) is plotted against the length of a channel structure. An electron entering from the source at the left and traveling ballistically gains energy as it moves through the channel, exiting at the drain with an additional energy defined by the applied voltage, as indicated by the green arrow. Reproduced from [35]. © IOP Publishing Ltd. All rights reserved.

bottom panel of figure 2.1. But notice that, at the two points where the dashed line has been extended from the transition region, the derivative of the electric field is discontinuous. Such a discontinuity in the electric field derivative requires that a sheet of excess charge be present at this discontinuity. This follows from Gauss' law and is expressed in Maxwell's equations. In fact, the charge at the left must be positive, while the charge at the right must be negative. Hence, there must be a accumulation of electrons at the left and an depletion of electrons at the right. If electrons are the major charge carriers, then the left-hand reservoir is the source through which the electrons enter the structure. The accumulation of charge at the transition point contributes to the contact resistance.

There is still a problem for a real device, such as a MOSFET, with the potential as drawn in figure 2.1. The linear potential drop leads to an electric field which can be quite large. For example, 1 V across 25 nm results in an average field of 400 kV cm^{-1}. This is large enough to create impact ionization in Si [36]. This field accelerates the carriers that enter from the source. This can be seen in the lower panel of figure 2.1, where the electron energy is plotted. If a ballistic carrier is injected from the left contact, it gains energy from the electric field. That is, the ballistic electron travels at constant energy (the green arrow in the figure) and the potential energy that exists at the source, relative to the drain, is converted to kinetic energy as the carrier travels. By the time it reaches the drain at the right-hand side, it will have gained kinetic energy given by eV. A more important aspect is introduced by Kirchhoff's current law. As the electrons are accelerated by the electric field, their increase in velocity requires that the carrier density must decrease to preserve constant current. Kirchhoff's current law requires the current to be uniform throughout the structure. Since the cross-sectional area does not change, this requires the product of density and velocity to be a constant value. As the velocity rises, the density must decrease. In a semiconductor, the density is usually set by the doping, but in a gated device, this carrier depletion is a common occurrence, as the injected carriers in an FET spread out below the channel as the potential energy is reduced along the channel (we will see this again below and in figure 2.3). As the number of carriers drops, this creates an additional space charge in the channel, which in turn requires that the potential be nonlinear, i.e. *the potential drop cannot be a linear one*. The true form must be obtained from Poisson's equation. The result of this argument is that the linear potential drop shown in figure 2.1 cannot be correct if the carriers are moving via ballistic transport and are constrained to a quasi-one-dimensional channel. A linear potential drop only can occur if there is sufficient scattering to ensure that the carriers move with a near-equilibrium energy that follows the decrease of energy shown in the lower panel of figure 2.1. Our conclusion, then, can only be that the electric field must essentially be near zero in a channel of constant cross-section, if the carriers are to move via ballistic transport. This is shown in figure 2.2.

As a result of the above argument, the real potential drop must be divided between the source and drain transition regions, as shown in figure 8.2. While the two voltage drops are drawn as being nearly the same, there is no real requirement that this be the case. In fact, it is usually assumed that most of the discontinuity is assigned to the source end of the channel. Ballistic transport can now occur through

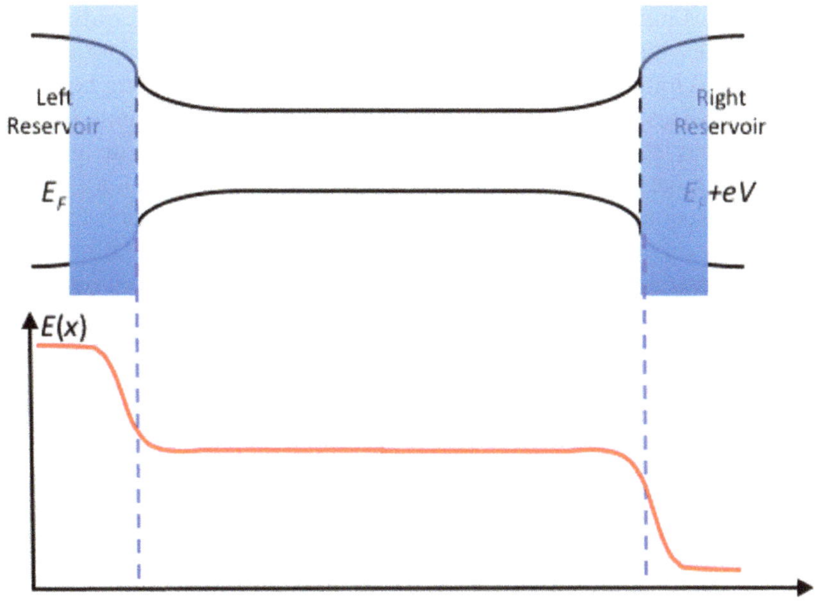

Figure 2.2. In the case of ballistic transport, Kirchhoff's current law requires that the potential be dropped at the two contact regions as shown (and discussed in the text). This leads to dipole charge concentration at these two potential transitions. Reproduced from [35].

the channel without the carriers gaining excess energy from the applied bias. At the same time, the potential drops in the transition regions now require that *dipole* charge densities exist at each transition. The potential drops can only occur through the existence of such dipole charge densities. The end result of all of this discussion is that for the potential in a device to vary smoothly throughout the device, the density must be inhomogeneous, and this requires a solution to Poisson's equation. Thus, it is necessary to couple Poisson's equation to the Schrödinger equation. This allows Landauer's equation to be used to determine the conductance, even under far-from-equilibrium conditions.

Recall that the point was expressed that, even with ballistic transport, the conductance would not be infinite. This can be illustrated by rewriting (2.6) as follows:

$$\frac{1}{G} = \frac{h}{2e^2}\frac{1}{NT} = \frac{h}{2e^2N} + \frac{h}{2e^2}\frac{1-T}{NT} \equiv \frac{1}{G_\mathrm{C}} + \frac{1}{G_\mathrm{D}}. \tag{2.8}$$

The resistance has been divided into a pair of resistances. One is the contact resistance $1/G_\mathrm{C}$, which describes the transitions at the contacts and can never be zero, especially if the normal resistance of the device is nonlinear [37]. The second is a device resistance $1/G_\mathrm{D}$, which can actually vanish in the absence of scattering. The contact resistance is connected with the charge distributions that determine the actual voltage drops of figure 2.2.

The Landauer formula was developed by treating the transport in a manner in which the energy relaxation actually occurred in the reservoirs (true ballistic transport). This is not really required, as mentioned above. The Landauer formula provides a quite rigorous framework in which to analyze the current through nanostructures as long as current is conserved and the reservoirs can be analyzed in terms of modes. This allows us to determine transmission coefficients for transport between these modes. Determining this transmission, especially in non-tunneling cases, is very suitable for simulation by particles. This can be done for a device using the Schrödinger equation [38], or via a classical approach in which the Landauer formula is easily shown to be the Duke tunneling formula in three dimensions [39]. As long as the scattering can be described by moving a particle from one channel to another, then any level of scattering can be incorporated within the Landauer formula [38]. This does not imply that energy is not transferred to the crystal lattice in the scattering process. Rather, only particle and current conservation are required for the validity of the Landauer formula. Further equations are required if we desire to examine the energy flow within the nanostructure.

2.3.3 Casting the MOSFET in Landauer form

We now want to examine the MOSFET as an example of how the common equation for the MOSFET can be connected to the Landauer formula. This will clearly demonstrate how the behavior of the common MOSFET is not that different from that of the quasi-one-dimensional channel, or an equivalent quasi-two-dimensional channel, as discussed above. The normal textbook gradual channel approximation formula for the MOSFET is given by an expression for the drain current in terms of the various potentials applied to the device, as follows:

$$\begin{aligned} I_D &= \frac{W\mu C_{ox}}{L}\left(V_G - V_T - \frac{V_D}{2}\right)V_D \\ &= \frac{W\mu C_{ox}}{2L}[(V_G - V_T)^2 - (V_G - V_T - V_D)^2] \end{aligned} \tag{2.9}$$

Here, W and L are the gate width and length (along the channel), respectively, C_{ox} is the gate capacitance per unit area, μ is the electron mobility, V_T is the threshold voltage where charge begins to accumulate in the inversion layer, and V_G and V_D are the gate and drain bias voltages (the source is usually taken to be grounded), respectively. In the second line of (2.9), the voltage terms have been rearranged so that the first term in the square brackets is the source–gate potential difference, while the second term in the square brackets is the drain–gate potential difference. With this latter form, it may be seen that the device current is the difference between the current injected at the source and a reverse current injected at the drain. Saturation occurs when the current injected at the drain is zero, and this is normally referred as the 'pinch-off point,' where the second term in brackets zero.

In writing the equation in this way, its connection to the Landauer formula becomes obvious. The first term in the square brackets represents particles entering the device from the source (left) electrode, while the second term represents particles entering the device from the drain (right) electrode. Thus, the source and drain are our contact regions, and the particles flow exactly as in the Landauer formula. Here, the potential reference has been taken to be that of the source itself, and this is equivalent to setting $E_F = 0$ in figure 2.1.

With no potential applied to the gate and drain electrodes, the two terms in the last line of (2.9) cancel each other and there is no current flow through the device. Such a cancellation it still the case even with a gate voltage applied, as long as no drain voltage is applied. The gate voltage changes the properties of the channel but does not upset the detailed balance of particle flow through the structure. However, when a small drain voltage is applied, the second term in the square brackets is reduced relative to the first term, and a small current begins to flow. That is, the two injection currents from the two ends of the device are no longer equal. As the drain potential is raised, the current increases. Once the second term is reduced to zero (it is not allowed to go negative, as this violates the conditions under which it was derived), the current saturates through the device at the normal value:

$$I_{D,\,sat} = \frac{W\mu C_{ox}}{2L}(V_G - V_T)^2. \tag{2.10}$$

When a drain potential is applied, the drain region is lowered in energy relative to the source region, just as shown in figures 2.1 and 2.2. Saturation occurs when this energy difference is sufficiently large that electrons in the drain can no longer surmount the barrier to reach the source. No matter whether the transport is ballistic or diffusive (when it includes a lot of scattering), the energy regions in the source and drain are much the same. What differs between these two cases is how the energy drops between the source and drain in response to solving Poisson's equation.

The flow of particles in an FET illustrates this. In figure 2.3 [40], the energy profile (negative of the voltage) is shown for a high-electron-mobility transistor, but the behavior is the same in an MOSFET (the details of various devices will be discussed more fully in chapter 4). The device in the figure uses an InGaN (relatively narrow bandgap) channel grown on a GaN substrate and an AlGaN layer (wide bandgap) that plays the role of the oxide in an MOSFET. The channel carriers are represented by the black line of dots near the source end of the device. As these carriers reach the pinch-off point, note that they spread out into the depth of the device away from the AlGaN layer. They eventually reach the drain and are once more ordered in the channel region. The spread of the carriers under the gate is required by Kirchhoff's current law as the velocity rises in the channel. This behavior is common to all FETs.

In order to examine the resistance of the MOSFET, the second line of equation (2.9) is rewritten in terms of an 'equilibrium' current and is actually the saturation current, as follows:

$$I_{eq}^{+} = I_{D,\,sat} = \frac{W\mu C_{ox}}{2L}(V_G - V_T)^2. \tag{2.11}$$

Equation (2.9) now becomes

$$I_D = I_{eq}^{+}\left[1 - \left(1 - \frac{V_D}{V_G - V_T}\right)^2\right]. \tag{2.12}$$

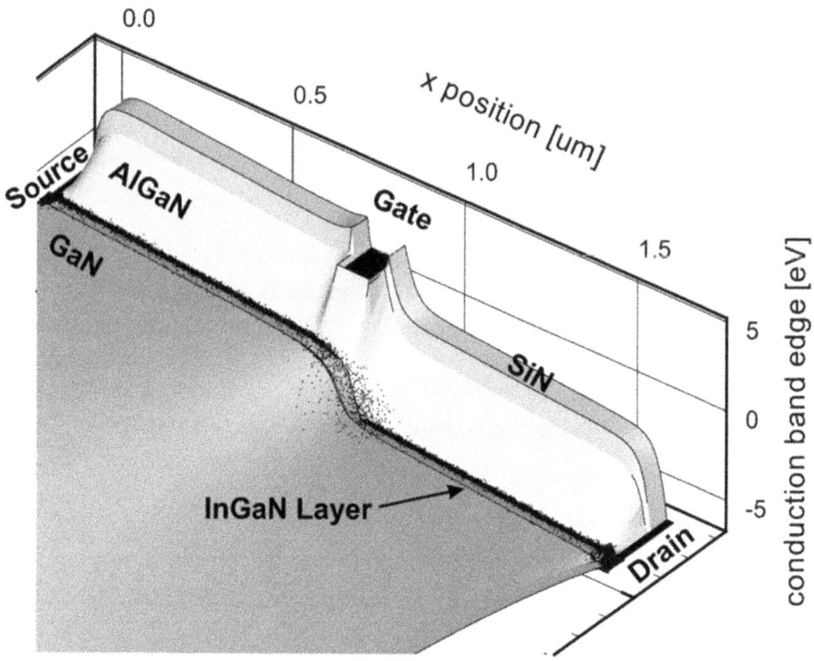

Figure 2.3. The conduction band profile and charge distribution (black dots) corresponding to a drain voltage of 6 V. This is a high-electron mobility transistor in which the InGaN is the channel layer and the AlGaN layer plays the role of the oxide in a MOSFET. © 2010 IEEE. Reprinted, with permission, from [40].

When the drain potential is small, linear response gives the following resistance (seen as a contact resistance)

$$R_C = \frac{V_D}{I_D} \sim \frac{V_G - V_T}{2I_{eq}^+}. \tag{2.13}$$

Even for a very short device with ballistic transmission, this minimum resistance appears exactly as $1/G_C$. Such a resistance characterizes the transition region between the pure source and the channel (just as in the Landauer approach) and appears regardless of whether the transport is described by the mobility or is ballistic. In fact, this resistance describes a barrier between the source and the channel [37]. The current flux of equation (2.11) arises from the carriers entering from the source with sufficient kinetic energy to overcome this barrier between the source and the channel. It is important to note that this flux is actually metered by the gate potential, so that this is a gate-controlled barrier, as depicted in the lowest panel of figure 1.4. Once the carriers overcome this barrier, they flow downhill to the drain, regardless of the nature of the transport. Whether the transport is ballistic or diffusive only determines the shape of the potential landscape between the source and the drain through Poisson's equation. The presence of scattering actually affects this potential landscape and serves to isolate the drain from the source. In a ballistic device, this isolation does not occur, and the source–channel barrier can be affected by the drain voltage, an effect known as drain-induced barrier lowering, as discussed above.

2.4 Microscopic properties and the measurement of classical systems

In almost all books on classical (or semiclassical) simulation of electron devices, there is no discussion about measurements or how they might affect the device. In fact, this is also true for general books on classical mechanics. On the contrary, most books on quantum mechanics spend at least a chapter in carefully explaining the nature of a measurement and the consequences of measuring a quantum system [18, 41–47]. The reason that the topic of measurement is not explained in books on classical mechanics is because the macroscopic and microscopic properties of a classical system have well-defined values at any time, whether or not they are measured. For example, the positions and velocities of electrons inside a device (or any other macroscopic property deduced from them, such as electrical current) are well defined (even when a measuring apparatus is not included). In fact, the values of such positions and velocities can be perfectly predicted from Hamilton's equations (1.8) mentioned in section 1.1 of chapter 1.

How do we define a *good* measuring apparatus for a classical system? Well, it can be roughly defined as any procedure in the laboratory that gives a value of the property of the system that coincides with the value predicted by the above-mentioned classical Hamilton equations (even though the measuring apparatus is not included in these equations). Thus, for those interested in the simulation of electron devices using a classical or semiclassical framework, the details of the measuring procedure used in the laboratory are totally irrelevant. In such classical or semiclassical backgrounds, one always assumes that a *good* measuring apparatus exists. Of course, the task of experimentalists in the laboratory who are trying to obtain the values predicted by the theory is not always easy. The way in which a measuring apparatus gets information about the system is by interacting with it. Without such interaction, the measuring apparatus and the system become independent systems, and the measuring apparatus cannot provide any reliable information about the physical properties of the system. Consider the measurement of the currents and voltages in the circuits depicted in figure 2.4 from a classical perspective. Without the voltmeter, the current $I(t)$ in the resistor in panel (a) connected to the battery is simply given by Ohm's law. However, when a voltmeter is connected, the new circuit in panel (b) is, in fact, different from the original. In the new situation, the current flowing through the resistor is $I_1(t)$, which is different from $I(t)$ because of the current $I_2(t)$ flowing through the voltmeter. The task of experimentalists is to ensure that the voltmeter of figure 2.4 is good enough (i.e. that is has a high enough input impedance) that the current $I_2(t)$ is so small that it becomes almost negligible, giving $I(t) \approx I'(t) \approx I_1(t)$. The value of the current obtained experimentally from this circuit, shown in panel (b), will then coincide (to within a negligible error) with the value predicted by Ohm's law in panel (a).

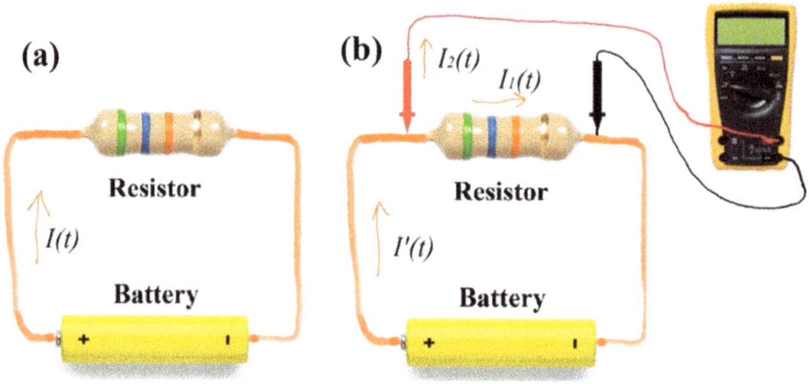

Figure 2.4. (a) The current $I(t)$ flowing through the resistor of a circuit without a measuring apparatus. (b) The same circuit with a voltmeter added to experimentally determine the predicted values of the current and voltage in circuit (a).

There are other measurements, for example, those looking for the properties of electronic devices at hundreds of gigahertz or a few terahertz, for which directly measuring the properties of the system is not as easy as for the case discussed above. At frequencies above a few gigahertz, the measurement apparatus drawn in figure 2.4(b) (just a voltmeter) is no longer a *good* measuring apparatus for many reasons. One reason for this, for example, is that the cables are no longer considered to be ideal elements (with a constant voltage drop along them) as was assumed in figure 2.4(b). More complicated experimental setups (see the one shown in figure 2.5) are needed to measure the properties of electronic devices at such high frequencies. In figure 2.5, the electronic device is known as the device under test (DUT). A vector network analyzer (VNA) is connected through coaxial cables (used in microwave measurements) to a printed circuit board that contains the DUT. At high frequencies, if the properties of the DUT alone are to be measured, it must be de-embedded from the values measured by the VNA. One must remove (calibrate out) the non-negligible effects of the cables, the printed circuit board, and the VNA itself. The so-called de-embedding techniques are among the most successful techniques used in this type of measurement [48]. They basically require the assumption that the coaxial cables, the printed circuit board, and the VNA are parts of a larger system. From a knowledge of the equations for currents and voltages in the cables and printed circuit board, it is possible to know which parts of the values measured at the VNA belong to the DUT and which parts are just spurious/parasitic effects that originate from the measurement setup. All these de-embedding techniques are of almost no interest to those merely involved in the simulation of electron devices. It is always assumed that the DUT has well-defined values of current and voltage (linked to the microscopic properties of the electrons) even if experimentalists are unable to directly measure such voltages and currents in the DUT when working with devices at terahertz frequencies.

Another relevant example of the empirical difficulties suffered by experimentalists in observing the properties predicted by classical theory appears in the so-called

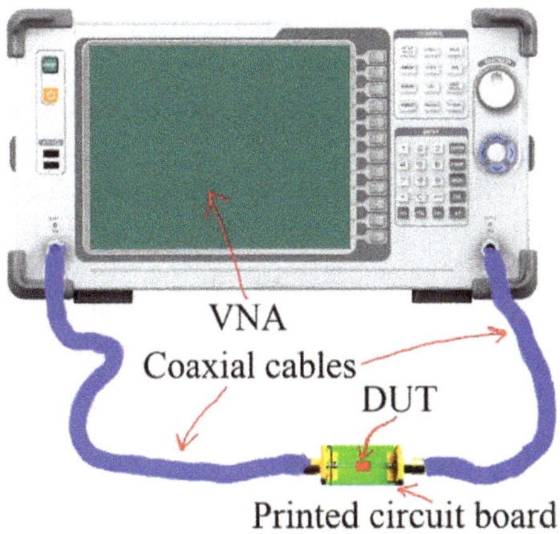

Figure 2.5. An experimental setup for the measurement of electronic devices at frequencies higher than a few gigahertz. The system to be measured is the electronic device labeled DUT. The printed circuit board (green), the coaxial cables (blue), and the VNA are part of what is considered as the measuring apparatus.

Monte Carlo solution of the Boltzmann equation, which will be studied in chapter 9 [49]. From a theoretical point of view, the classical Monte Carlo technique provides the position and velocities of all electrons inside the active region. The possibility of developing a *good* apparatus that provides such information, or the use of the de-embedding technique mentioned above, is not a solution here because one needs a huge number (on the order of the Avogadro number) of data points corresponding to the velocity and the position of each electron inside the active region. Again, the fact that the microscopic information about the positions and velocities of all electrons in the active region is not accessible to experimentalists is not a problem when dealing with classical simulations. From a classical point of view, all electrons in the device's active region have well-defined properties even if the experimentalists have enormous technological difficulties in measuring them. It will be seen in the next section that the quantum explanation of a measurement, in the orthodox view, is dramatically different.

2.5 Microscopic properties and the measurement of quantum systems

The measurement of a quantum system is, by far, a topic much more controversial than the measurement of a classical system. Obviously, the measurement of a quantum system in the laboratory has similar problems to those mentioned in figures 2.4 and 2.5 for classical systems. However, what is radically new in the quantum system is that there is still a major controversy among those involved in the foundations of quantum mechanics about what really constitutes a measurement in a quantum system. All these discussions belong to the so-called *quantum measurement problem* [50–53]. As mentioned earlier, since this is a book on quantum transport, the issue of quantum measurement has already been present in various sections of chapter 1 and briefly introduced in section 2.1 of this chapter. Here, we aim to classify the different explanations of quantum measurements available in the literature. There are different quantum theories available in the literature that provide different explanations/solutions of the quantum measurement problem. Depending on the explanation/solution adopted, quantum theories can be divided into two groups.

In one group, there are quantum theories postulating that a quantum system has no microscopic properties by itself when it is not explicitly measured. These theories postulate that it is the fact of obtaining a measurement (also known as the observation) that *creates* the microscopic properties of the quantum system. These quantum theories are named *quantum theories with observers* because the microscopic properties of the world are *not real* without the action of *observers*. Certainly, these quantum theories with observers (as creators of the quantum reality) are dramatically different from the classical theory used to explain the circuits of figures 2.4 and 2.5. The orthodox theory [41, 42, 54, 55] explained in chapter 1 (also known as the Copenhagen theory) is, obviously, the most famous example of such a type of quantum theory. Notice that there is no question here that classical dynamics are different from quantum dynamics, but what interests us is that which is *real and exists* in the quantum world. Jordan, who was one of the physicists working with Niels Bohr to define the new orthodox view of the quantum world in the 1920s, once claimed that 'observations not only disturb what has to be measured, they produce it.' He claimed that 'we ourselves produce the results of measurements.' Does one really need to be so extreme about what *is real or exists* in order to understand the quantum world?

In the second group, there are quantum theories which defend the principle that a quantum system has microscopic properties independently of the fact of whether or not they are measured; this is the case for classical theory. Among others, the already mentioned hydrodynamics or Bohmian theory [45–47, 56–58], the consistent histories [59–62], and spontaneous collapse theories [63, 64] are examples of this type of quantum theory. These quantum theories are named *quantum theories without observers*. This is not because they have no observer, but because the observer does not determine what *is real or exists* in the quantum system. Notice that this viewpoint about the role of the observer is very similar to the role of the ammeter in the classical circuit of figures 2.4 and 2.5: the ammeter perturbs the value of the electrical current, but the electrical current *is real or exists* in the circuit with or without the presence of the ammeter.

In the next two subsections, the measurement of the electrical current in a quantum system is presented from the viewpoints of these two groups of theories. It will be seen that while the mathematics of both approaches are quite similar, the views of quantum reality they offer differ dramatically [65, 66].

2.5.1 Measurements according to quantum theories with observers

To simplify the mathematical burden in the discussion of the orthodox measurement, a quantum system representing the active region of an electron device is represented by the wave function $\psi(x)$, which involves only one degree of freedom, x. The orthodox theory postulates that such a quantum system has a well-defined value of the electrical current, given by the (eigen)value I, only when the wave function is described by an eigenstate, $\phi_I(x)$. In the literature, this link between quantum reality and measurements is known as the *eigenstate–eigenvalue link* [67]. In fact, it is well known that an arbitrary wave function $\psi(x)$ can be written as a superposition of several eigenfunctions $\phi_{I_i}(x)$ of the electrical current (each one with its own eigenvalue I_i), as follows:

$$\psi(x) = \sum_i c_{I_i}\phi_{I_i}(x), \tag{2.14}$$

where $c_{I_i} = \int \phi_{I_i}(x)\psi(x)dx$ is a complex number that quantifies the importance of each eigenfunction $\phi_{I_i}(x)$ in $\psi(x)$. Due the linearity of the Schrödinger equation, a time evolution of $\psi(x)$ in (2.14) following the Schrödinger equation in (1.24) will never destroy the above superposition of different currents I_i. Thus, the Schrödinger equation in (1.24) is not enough to explain why a quantum system prepared using the superposition (2.14) can yield a well-defined value of the current I in the laboratory when the electrical current is measured.

The orthodox theory needs another corollary, different from the Schrödinger equation in (1.24), to determine the electrical current. This new law is named *collapse law* or *reduction state law* [41, 42, 67]. The fact that the measured value of the current is I also implies that the initial wave function $\psi(x)$ in (2.14) has suffered from the following nonlinear evolution

$$\psi(x) \to c_I\phi_I(x). \tag{2.15}$$

Notice that the rest of the wave function $\sum_{i \neq I} c_{I_i}\phi_{I_i}(x)$ that appears in (2.14) simply *disappears* from the right-hand side of (2.15) during the measurement. Appendix A.1 gives more mathematical details about this collapse by including a quantum description of the measuring apparatus.

As a result of the above collapse, the final value of the current I obtained in the laboratory is random but follows the probability distribution $P_I = \int | c_I\phi_I(x) |^2 dx = |c_I|^2$, according to the so-called *Born law* [42]. This remains a requirement now that $\int | \phi_I(x) |^2 dx = 1$. If needed, the collapsed state in (2.15) can easily be normalized by dividing it by the square root of $P_I = |c_I|^2$. In the bra-ket notation, the collapsed state (2.15) can be written as a result of the following *projection* operator $| \phi_I\rangle\langle\phi_I |$ that acts on the state (2.4) rewritten in this notation as $| \psi\rangle = \sum_i c_{I_i} | \phi_{I_i}\rangle$. The projected form becomes

$$\sum_i c_{I_i} |\phi_{I_i}\rangle \to |\phi_I\rangle\langle\phi_I | \left(\sum_i c_{I_i} | \phi_{I_i}\rangle \right) = c_I | \phi_I\rangle. \tag{2.16}$$

Here, the orthogonality of the eigenstates $\langle \phi_I |, \phi_{I_i}\rangle = \delta_{I,\,I_i}$ has been utilized in the reduction.

In summary, the orthodox theory postulates that only after the process of measurement, when the collapse in (2.15) takes place, does the system get a well-defined value I of the current. In other words, according to the orthodox theory, the observer not only *observes* the system, but he/she/it also *creates* the microscopic value of the current I while doing so. Before the measurement takes place, i.e. when the state is as described by (2.14), there is no well-defined value of the current for the system. It is important to specify that the orthodox theory is not saying that there is a well-defined value of the current I in (2.14) before the measurement that is ignored by humans. Rather, the orthodox theory is saying that in (2.14), there is no value of the current at all. The difficulty of accepting this view lies in whether or not one really needs to be so extreme in order to understand the quantum world?

This type of quantum measurement is named a *strong* or *direct* measurement. The mathematical details of a *weak* or *indirect* measurement [68], which is more appropriate for discussing the measurement of the electrical current, will be explained in appendix A.2. The conclusions about the role of the observer as a creator of the reality in the orthodox theory remain unchanged, whatever the nature of the measurement. Among those working in the foundations of quantum mechanics, the orthodox explanation of the measurement process is not very satisfactory. This dissatisfaction is the origin of the abovementioned *measurement problem* [50–53]. Among others, one criticism of the orthodox measurement is that it does not explain what an observer (or a measuring apparatus) is [53]. In the example of the DUT measured by a VNA in figure 2.5, are the coaxial cables and the printer circuit board part of the measuring apparatus, or are they part of the quantum system? What are the differences between the electrons in the DUT (system) and those in the cables (measuring apparatus)? If there are no clear answers to these questions, how can one know whether to apply unitary evolution (2.14) or non-unitary evolution (2.15) in a quantum system? When does the measurement take place? Is communication of the measured outcome by a human being needed to ensure that it qualifies as a measurement?

This discussion about two different laws for describing a systems (with and without measuring apparatus) is dramatically different from what happens in classical mechanics, in which the fundamental equations that govern the dynamics of the degrees of freedom of the system are never changed by the presence of a measuring apparatus. In fact, it is clear from classical mechanics that the division between system and apparatus is an *artificial* one that can be useful for numerical computations, but it cannot have a deep fundamental role. These reflections, which seem so obvious in classical systems, are not true in quantum systems if one follows the postulates of the orthodox theory. The orthodox theory really needs, at the most *fundamental* level, a spatial and temporal division between a quantum system and a measuring apparatus. Such a division becomes of fundamental importance in order to know whether the unitary law (2.14) or the non-unitary law (2.15) is to be used to describe an interaction between subsystems. Surprisingly, the orthodox theory does not specify where and when such divisions appear.

2.5.2 Measurements according to quantum theories without observers

As already mentioned in chapter 1, many of the great fathers of quantum theory, including Einstein, Schrödinger, de Broglie, Bohm, and Bell had many difficulties in accepting the orthodox explanation of the measurement process [45, 69]. As explained above, there are other quantum theories that avoid the conceptual difficulties related to the orthodox explanation of measurement, while still achieving full empirical agreement with the laboratory results. Within these latter quantum theories without observers, the measurement process does not require any ad hoc rule. The measurement is treated as any other interaction, as happens when measurement is described in classical mechanics. The collapse becomes just an effective way of talking and does not have fundamental implications.

To illustrate the quantum theories without observers, consider the experiment depicted in figure 2.6. Here, a single electron impinges upon a tunneling barrier. In panel (a), the (unitary) evolution of the wave function solution of the Schrödinger equation is shown. After interacting with the barrier, the wave function splits into a transmitted component and a reflected component. But we know that, in such experiments, the electron is detected either on the left or the right of the barrier and not as a fraction on each side! The orthodox theory solves this deep disagreement between the Schrödinger equation and the experimental results by saying that the measurement process cannot be described by the (unitary) Schrödinger equation itself but by the new (non-unitary) evolution that is termed the *collapse law*. If position detectors are placed on the left and the right of the barrier, then the eigenstates linked to such a measuring apparatus, $\phi_l(x)$, are either wave functions fully located on the left or on the right. Thus, the collapse law occurring at time t_1 implies that the transition $\psi(x) \rightarrow c_l\phi_l(x)$ mentioned in (2.15) is such that the wave function is fully transmitted, as depicted in panel (b), or fully reflected, as in panel (c). With the help of this figure, the measurement process of these quantum theories can easily be understood without observers. For example, within the Bohmian theory [46, 47, 56], apart from the wave function, there is a particle, depicted as a solid circle in the figure, that has a well-defined position that defines the real location

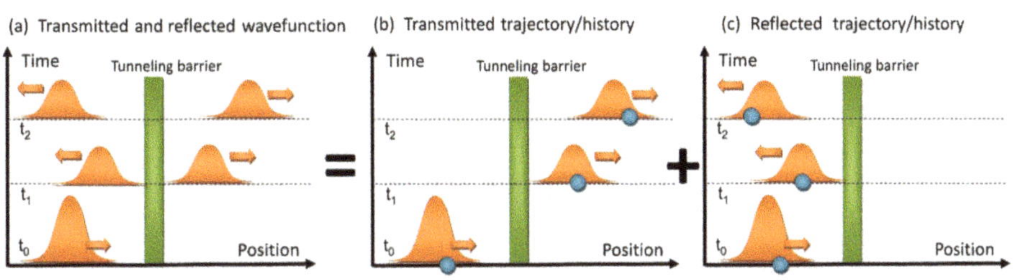

Figure 2.6. A schematic representation of an electron impinging on a tunneling barrier. The electron is prepared at time t_0. Times t_1 and t_2 can be interpreted as the times when the electron is measured. (a) The wave function solution of the Schrödinger equation alone. (b) The wave function (and trajectory) of a transmitted particle electron. (c) The wave function (and trajectory) of a reflected particle electron.

of the electron at all times. The role of the wave function is merely to guide this particle [70]. Thus, since the position of the particle is always well defined, the particle (i.e. the electron) is either transmitted or reflected, regardless of the fact that the wave function solution of the Schrödinger equation splits into a reflected component and a transmitted component. By construction, the particle guided by the wave function cannot be located both on the left and on the right of the barrier!

At the practical level, after time t_1, the particle (i.e. the electron) is guided only by the transmitted component in figure 2.6(b) or the reflected component in figure 2.6(c). Thus, it is necessary to use only one such component to know how the wave function guides the electron at time t_2, regardless of the *fundamental* wave function depicted in figure 2.6(a) as the solution of the Schrödinger equation. For all practical purposes, it may be assumed that the wave function suffers an *effective* collapse at time t_1. Notice, however, that there is a dramatic difference between this *effective* collapse and the *fundamental* collapse implicit in the orthodox theory. In the former, the reality of the position of the electron does not depend on the measurement. On the contrary, in the orthodox theory, it is the collapse itself that gives reality to the position of the electron.

Quantum theories without observers do not imply that the measurement process (the observer) does not influence the system. They only say that the fact of obtaining a measurement does not create the microscopic properties of the system. Of course, as shown in figures 2.4 and 2.5, the microscopic properties in the active region of the device are often affected by the measuring setup. Thus, the measurement itself affects the system properties. If it is desired to know all the positions and velocities of the electrons, it is technologically impossible to obtain such a huge amount of microscopic information. The fact that such a large amount of information about the microscopic properties of electrons is not available in a measurement does not imply that such microscopic properties cannot be real and cannot be predicted within these quantum theories without observers (as readily happens with the Monte Carlo solution of the Boltzmann equation [49]).

An additional point merits discussion here to fully appreciate the Bohmian explanation of quantum measurement. All measuring apparatus ultimately presents the measured output as the position of a pointer, dial, pixel on a screen, or similar indicators. Thus, strictly speaking, even when measuring observables other than position, the only ‘value’ that the experimentalist has genuinely ‘observed’ is the location of the pointer or similar indicator. For instance, when one claims to have measured the spin of an electron, one is essentially saying that the pointer of the apparatus designed for measuring spin has changed its position. This is the fundamental reason why a theory capable of accurately addressing the evolution of all particle positions (in particular those of the of pointers) can effectively explain any type of measurements in the laboratory, as all measurements inevitably culminate in a pointer’s or indicator’s position. The Bohmian theory’s great merit lies in elucidating this privileged role of positions in understanding how measurements are conducted in a laboratory [47].

In summary, quantum theories with or without observers naturally give the same empirical results in the laboratory. The difference between the two groups of quantum theories does not lie in the measured properties, but in those properties of the quantum system that are not measured [68, 71, 72]. In the orthodox theory, the unmeasured properties simply do not exist, while in the quantum theories without observers, such unmeasured properties are perfectly well-defined.

2.5.3 What physical reality says about measurements

The previous two subsections described the importance of observers in one group of quantum theories and their role in the second group of quantum theories. Which should one believe? To assess the answer, it is necessary to ask the question about what measurement can really tell us about the quantum system. It turns out that experiments can tell us a great deal about the electronic structures of, for example, the semiconductors used in semiconductor devices.

First, however, it is necessary to realize that whether the system is classical or quantum mechanical, the system or DUT is no longer the closed reversible system treated in quantum mechanics textbooks. Rather, it has to be embedded in an environment, as all measurements are normally made in this environment, and not in the interior of the system. This is already apparent from figure 2.4, where the ammeter and voltmeter (with its extra wires) constitute the environment. Once the ammeter is in the circuit, the voltage can no longer be determined exactly. Once the voltmeter is in the circuit, the current passing through the DUT can no longer be determined exactly. Measurement thus requires the ammeter to have very low impedance and the voltmeter to have very high impedance. This problem carries over to the quantum world. This environmental problem has already been briefly discussed in section 1.3.1 and section 2.2. It will be discussed in more detail in chapter 3. Nevertheless, the presence of the environment does not reduce the problem of ascertaining what the measurement does to the quantum system.

Perhaps one of the most important tools for studying semiconductor (and other) materials is photoemission spectroscopy [73]. Using a synchrotron source, high-energy photons (in the ultraviolet or x-ray energy range) are sent into a sample of the semiconductor in order to excite electrons to energies at which they are capable of being emitted from the material. The energy of these electrons is then analyzed to determine the energy structure within the material from which they were excited. The energy analysis can tell one about the core levels and their energies and it can provide considerable information about the valence and conduction bands. Today, much more information is obtained by analyzing both the energy and the momentum of the emitted electrons using angle-resolved photoemission spectroscopy (ARPES) [74]. This latter approach can almost completely determine the energy structure of the valence band. However, its detailed structure is not the main point of the present discussion; it is merely an example.

The point is that the semiconductor sample can be measured on day one. The experimentalist can return on a later day and again make measurements, obtaining essentially the same electronic properties. He/she can put the sample in a shipping container and send it halfway around the world to be measured by colleagues who will obtain essentially the same results. This leads to the conclusion that the quantum mechanical structure is not destroyed by the measurements. Of course, a strict orthodox believer can still argue that the mentioned experimental facts align with the notion that the entire band structure of materials (involving the interactions of atoms with electrons) doesn't exist at all. According to this viewpoint, the structure is formed only when an experimentalist measures it. However, finding any appealing advantage in this orthodox perspective over the idea that the band structure is always there becomes very difficult, especially when considering the goal of describing how electron devices work.

Consider an *s*-state of one of the core levels in the atoms that makes up the semiconductor sample. This state contains two electrons that are correlated due to their spin properties: each must have the opposite spin to the other. In fact, the actual spin of each of the two electrons is unknown. But the Schrödinger equation gives an idea of the probability of each spin for one of the electrons. The question that arises from the previous two subsections is whether these two electrons actually do *not* have precise spin values (section 2.5.1), or is it rather the case that they have precise spin values but these values are not known *a priori* by the observer (section 2.5.2). By exciting one of the electrons to the photoemission level, the spin of that electron can be determined by using polarized photons (the polarization, or spin, of the photon, determines which spin becomes excited). Just as in other such experiments, the spin of the remaining electron is known once the photoexcited electron has been measured. Did it just *jump* into this spin state once the measurement was made, or did it already have this spin state before the measurement? The answer to this question lies within the ontological viewpoint that is accepted in the philosophy of quantum mechanics. In other words, it relies upon whether one accepts section 2.5.1 or section 2.5.2. In the Bohmian theory, for example, since the reality of the quantum system is also defined by the position, there is no fundamental jump into the final spin state because the Bohmian position determines the spin state at all times, even before the measurement.

2.6 Single-time measurements

The experiment described in figure 2.6 may also be used to explain the measurement of the electrical current $I(t)$ in electronic devices. Consider a flux of independent electrons, not merely a single one, that impinge upon the barrier from left to right. All electrons are described by the same wave function $\psi(x)$, except that they are differentiated by the preparation time t_0, which is different for each electron. Thus, there is a flux of electrons, all initiated from the same state, impinging upon the barrier one after another. In the laboratory, the steady-state current, I_{DC}, also known as direct current (DC) because it flows in only one direction in the device, is computed by time averaging the measured value of the total current $I(t)$ in a *unique device* during a large (ideally infinite) period of time τ so that *many electrons* are tunneling through the barrier. This average is given by

$$I_{\text{DC}} = \lim_{\tau \to \infty} \frac{1}{\tau} \int_0^{\tau} I(t)dt. \tag{2.17}$$

In the prediction of the DC current (2.17) is typically calculated while assuming that the system is ergodic, so that a time average is equivalent to an ensemble average. It is important to know that this ergodic theory is not valid in transient situations, nor it is usually valid in nonequilibrium situations [75]. Thus, the measurement must be done after a steady state is reached and the value I_{DC} in (2.17) is computed from an average of all possible values of the current I_i that are measured at only one particular time t_1 over an *ensemble of N (identical) tunneling barriers, each one with one electron injected.* Thus, (2.17) becomes

$$I_{DC} = \sum_{i=1}^{M} I_i \cdot P(I_i), \tag{2.18}$$

where the index $i = 1, \ldots, M$ indicates the individual values of the electrical current that can occur at time t_1 when the electron is measured in the different samples. The probability $P(I_i)$ of measuring the current I_i is defined as the ratio of the number N_i of samples that gives I_i divided by the total number N of samples, i.e. $P(I_i) = N_i/N$. As in the case of τ in (2.17), we assume here that N is very large (ideally infinite). When the ergodic theorem is invalidated by transient behavior, (2.18) is always the correct way to determine the current estimate.

From a mathematical point of view, i.e. one that is independent of any interpretation of the quantum phenomena, the (eigen)value I_i can be linked to the eigenfunctions $\phi_i(x)$. Both satisfy the equation $\hat{I}\phi_i(x) = I_i\phi_i(x)$, where $\hat{I}$ is a current operator that will be discussed later. The entire set of eigenfunctions is an orthonormal base of the Hilbert space, so that the wave function in figure 2.6(a) can always be decomposed as shown in (2.14) with $c_i = \int \phi_i(x)\psi(x)dx$. Then, (2.18) can be rewritten as

$$I_{DC} = \sum_{i=1}^{M} I_i \cdot P(I_i) = \int \psi^*(x)\ \hat{I}\ \psi(x)\ dx, \tag{2.19}$$

where the orthonormal property of the eigenstates $\int \phi_j^*(x)\phi_i(x)dx = \delta_{ij}$ has been used and the definition of the (Born) probability is $P(I_i) = N_i/N = |\, c_i \,|^2$. From a practical point of view, the right-hand side of (2.19) is extraordinarily useful because it allows the prediction of I_{DC} without knowledge of the eigenstates $\phi_i(x)$. Only the operator $\hat{I}$ and the wave function solution $\psi(x)$ of the unitary Schrödinger equation (before the measurement) are required. Even an assumption of the collapse law mentioned in (2.15) is not required.

Unfortunately, the simplicity of the right-hand side of (2.19) has had the undesired effect of silencing the few who have criticized the orthodox collapse law in electron devices [66]. At first sight, it seems that the right-hand side of (2.19) really provides a pre-measurement value of the electrical current belonging to $\psi(x)$. This is only true for quantum theories without observers, but it is not at all true within the orthodox view. According to section 2.5.1, the system described by $\psi(x)$ has no current at all. It is only when measuring the system that the observer collapses the system into $\phi_I(x)$, randomly creating the value of the electrical current I in the quantum system. The simplicity of

(2.19) arises because the measurement of the system is made only once, so that the system evolution after this first measurement is unimportant. When dealing with multi-time measurements in the next subsection, it will become apparent that the unavoidable role of the collapse law in the measurement cannot be avoided. Nevertheless, the importance of this measurement time is readily apparent in measurements.

As an example, it is possible to see intervalley transfer in a semiconductor subjected to an electric field using ARPES, described in section 2.3. ARPES can be performed with femtosecond resolution [76]. In this process, the momentum appears in the angle at which the electrons are emitted. In a recent study, GaAs was probed with ultrafast pulses with photon energies between 1.6 and 2.4 eV, and then probed with time-resolved ARPES [77]. The electrons injected into the conduction band equilibrate between themselves in momentum, but not in energy, in a few tens of femtoseconds. Relaxation of the carriers takes place on a much longer timescale and is governed by the interaction between the electrons and a variety of phonons, especially for photoelectron densities less than 5×10^{17} cm^{-3}. These electrons do not have sufficient energy to be photoemitted. So, the photoemission photons are used to probe this distribution by raising these hot electrons to the photoemission threshold. Then, using the energy resolution, only photoelectrons from this distribution are studied. The results of experiments on GaAs are shown in figure 2.7. In panel (a), one can see the initial excitation of the electrons near the Γ point in the conduction band, but some of these electrons are already relaxing into the proper band structure in the Γ valley. The solid and dashed white lines are the energy bands in the directions toward L and X. In panel (b), a larger slice of the Brillouin zone is shown with the L valleys. These latter valleys are already being populated at 20 fs after the pump pulse. Populations are shown as a function of energy in (c) and (d) at both 20 fs and 70 fs after the pump. Panels (e)–(i) show the temporal distributions for these two valleys at several particular energies (all measured from the conduction band minimum at Γ). This provides a clear illustration of the scattering to the satellite valleys and the dynamic information that can be obtained using modern measurement techniques. Yet it is clear that these dynamic measurements are obtained without information about any collapse, just as discussed in the previous paragraph. This femtosecond distribution and the ARPES study, as well as other ultrafast techniques, are discussed in [78].

There remains an important point to discuss in expression (2.19): what is the operator $\hat{I}$ that has been used to predict the steady-state current? How is it determined? In fact, the operator $\hat{I}$ is a method of representing the measuring apparatus. However, the theory of section 2.5.1 does not give us a clear algorithm with which to find such an operator. This lack of a proper definition is addressed in the literature by making an educated guess about which type of operator will best yield predictions that are in agreement with empirical results obtained in the laboratory. In the case of steady-state current, the way to anticipate such an operator is to know that it is desired to quantify the number of electrons per unit time crossing a surface S located at position x_S, as depicted in figure 2.8 (this surface could be the electron detector in the experiments of figure 2.7). The required

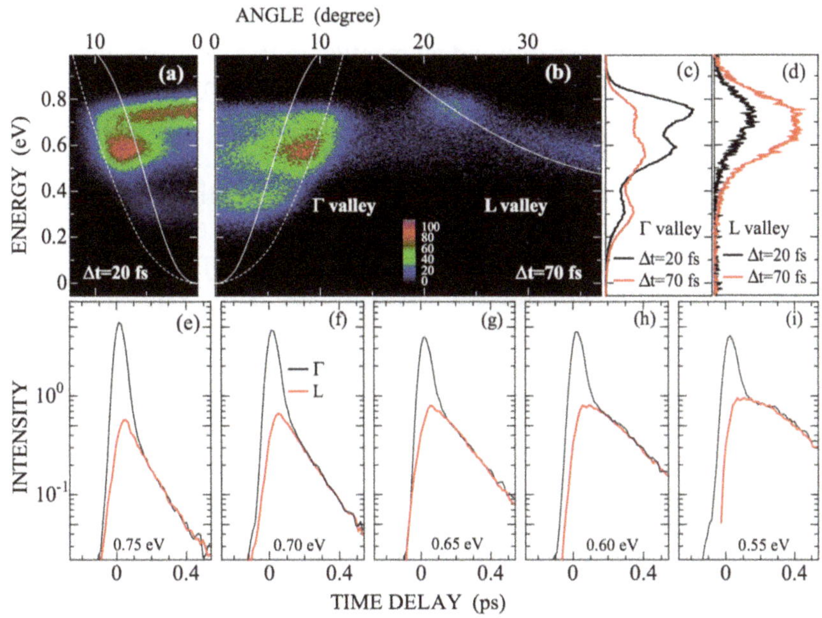

Figure 2.7. (a) A photoemission image of hot electron distribution in the Γ valley mapped in energy and momentum space, 20 fs after excitation by *s*-polarized 2.3 eV photons. (b) An image of the hot electron distribution 70 fs after excitation, showing carriers in the L valley as well ($\theta > 13°$). In (a) and (b), the color scale indicates the photoemission intensity and all energies are referenced to the conduction band minimum. The solid and dashed curves represent the energy dispersions in the Γ–L and Γ–X directions, respectively. The angle-integrated, energy-resolved distributions 20 and 70 fs after excitation are shown in the (c) Γ valley and (d) L valleys. Even at 20 fs, the population of the L valley is non-negligible because of finite pulse widths and ultrafast scattering rates. (e)–(i) The time-resolved populations of the Γ and L valleys at specific energies in the bands. The populations of the L valleys have been multiplied by a scale constant to bring them to the same order of magnitude as those of the Γ valleys. Reprinted with permission from [77], copyright 2016 by the American Physical Society.

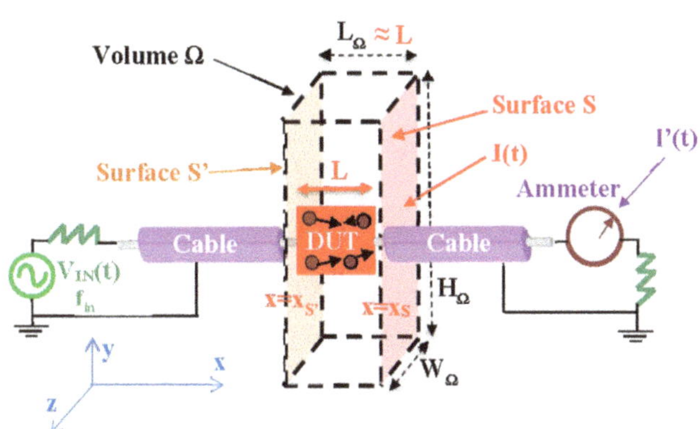

Figure 2.8. A schematic plot of the measurement setup of figure 2.5 indicating the volume Ω and the surface S used for the computation of the total (particle plus displacement) current.

operator can then be related to some function of the velocity of the electron crossing x_S (such as determining the energy and momentum of this electron). It is known that such an operator can be written in Dirac's bra-ket notation as

$$\hat{I} = | x_S\rangle\langle x_S | \hat{p}_x + \hat{p}_x | x_S\rangle\langle x_S |, \tag{2.20}$$

where $| x_S\rangle\langle x_S |$ is a projection operator that gives the position operator x_S for any state and $\hat{p}_x$ is the momentum (velocity) operator. A sum is added to (2.20) in order to build a proper Hermitian operator with real (i.e. noncomplex) eigenvalues. In the bra-ket notation, such an operator provides the expectation value $I_{DC} = \langle\psi | \hat{I} | \psi\rangle$ in (2.19), which is then rewritten as

$$\begin{aligned} I_{DC} &= \langle\psi | \int dx \, | x\rangle\langle x | \hat{I} \int dx' |x'\rangle\langle x'|\psi\rangle \\ &= -e \int \psi^*(x, t)\langle x | \hat{I} | x\rangle\psi(x, t) dx = -e \; J_x(x_S, t) \end{aligned} \tag{2.21}$$

Here, the relationship $\int dx|x\rangle\langle x | = \hat{1}$ is used as a resolvent of the delta function in Hilbert space, and

$$J_x(x, t) = \frac{\hbar}{m} Im\left(\psi^*(x, t)\frac{\partial\psi(x, t)}{\partial x}\right). \tag{2.22}$$

This result is easily shown to be another form of the normal current operator defined in most quantum mechanics textbooks.

Obviously, an identical result for the I_{DC} in (2.21) can be obtained from a quantum theory without an observer (see section 2.5.2). For such theories, the operator $\hat{I}$ discussed in (2.20) is not needed at all because, in these quantum theories, the electrical current exists independently of whether or not it is actually measured. For example, in the Bohmian theory [46, 47], the current of one electron can be defined as the product of its charge $-e$ and its velocity. The contribution to the current at x_S made by an electron at position $x(t)$ is simply

$$J_x(x_S, t) = v_x(x(t), t)\delta(x_S - x(t)). \tag{2.23}$$

Here, the Bohmian velocity is given by

$$v_x(x, t) = \frac{J_x(x, t)}{| \psi(x, t)|^2}. \tag{2.24}$$

This expression coincides with expression (1.38) when $J_x(x, t)$ is given by (2.22). In this case, finding the current requires the use of an ensemble over all possible electrons crossing the surface S at x_S at time t. Since it is assumed here that all electrons are prepared with the same wave function, the probability distribution of electrons at time t is given by $| \psi(x, t)|^2$. Thus, the steady-state current is given by

$$I_T = -e \int v_x(x, t) \; \delta(x_S - x) \, | \psi(x, t)|^2 \, dx = -eJ_x(x_S, t), \tag{2.25}$$

which exactly reproduces (2.21). It is important to note that result (2.25) has been obtained without any mention of the measurement process (or measuring operator) because these quantum theories without observers do not need a measurement to make predictions about the microscopic properties of the electrons. In other words, according to the Bohmian theory, the value in (2.25) is the ensemble value of the current of an electron with or without a measuring apparatus.

Quantum theories without observers allow one to recover the original idea discussed in the classical measurements, namely that a measuring apparatus is any procedure in the laboratory that gives a value of the property of the system that coincides with the value predicted by the theory *when no measuring apparatus is present.* Of course, in the quantum regime (just as in the classical regime) one frequently cannot avoid the possibility that the measuring apparatus provokes a perturbation of the system (see the discussion of figure 2.4 in section 2.3). But the quantum measurements have no further implications in these quantum theories without observers. Sometimes, the perturbation of the system by the measurement can be minimized through the use of an appropriate measuring protocol (such as assuring the impedances of the ammeter and voltmeter in figure 2.4). The weak value, discussed below at the end of appendix A.2, is a modern laboratory-based attempt to compute the *microscopic property of a quantum system when no measuring apparatus is present.* Ironically, this simple understanding of what constitutes a quantum measurement is incompatible with the orthodox theory because the latter requires, as a fundamental postulate, that there is no microscopic property *when no measuring apparatus is present.*

2.7 Multi-time measurements

Up until this point, it has seemed that the wave function $\psi(x)$ solution of the Schrödinger equation (1.24) is sufficient for the computation of the current $I(t)$. This is not generally true. If a second measurement of the same quantum device is performed, after the first measurement, one would realize that the strategy of using *only* the wave function solution of the Schrödinger equation (without including the collapse law) to compute the current $I(t)$ can give predictions that disagree with experimental results. For example, the collapse law (in its fundamental or effective form) has a prominent role in the predictions of quantum noise, i.e. the fluctuations of a current around its steady value. In general, such fluctuations computed as

$$\Delta I = \sum_{i=1}^{M} \sqrt{(I_i - I_{\mathrm{DC}})^2 \cdot P(I_i)} \tag{2.26}$$

does not vanish. The ergodic interpretation of (2.18) would suggest that no multi-time measurement is needed for evaluating (2.26). However, expression (2.26) cannot be evaluated in the laboratory because it implies that the measuring apparatus is able to capture *all* the dynamics of the electrons (no matter how fast they are). However, all electronic devices only work properly up to a certain cutoff frequency [79]. Therefore, electronic measuring systems (understood as a low-pass filter at some point) are not able to measure the all-frequency noise appearing in

(2.26), but only the noise whose components have a frequency lower than some cutoff frequency. Therefore, what is measured in the laboratory is the noise up to the cutoff frequency of the electronic setup. Mathematically, this noise is computed from the power spectral density, which can be related through a Fourier transform to the time correlation function [79].

The time correlation function is the ensemble-averaged value of an event defined by measuring the current $I_i(t_1)$ at time t_1 and the current $I_j(t_2)$ at time $t_2$and relating them through

$$\langle I(t_2)I(t_1)\rangle = \sum_j\sum_i I_j(t_2)\cdot I_i(t_1)\cdot P\big(I_j(t_2),\, \mathrm{I}_i(t_1)\big). \tag{2.27}$$

The probability $P\big(I_j(t_2),\, \mathrm{I}_i(t_1)\big)$ of this two-time measurement process has to be determined with care because it involves obtaining empirical information about the system at two different times. In order to simplify the discussion, it may be assumed that a strong measurement can be made, and this is expressed using a bra-ket notation. If the first measurement at time t_1 produces the eigenvalue $I_i(t_1) \equiv I_1$, the non-unitary evolution of the initial state $|\, \psi(t_1)\rangle$ during the time of the measurement due to the collapse law in (2.16), can be written using the projector operator as

$$|\, \psi_{I1}(t_1)\rangle = \big|\, \phi_{I1}\big\rangle\big\langle\phi_{I1}\, \big|\psi(t_1)\big\rangle. \tag{2.28}$$

Equation (2.28) does not preserve the proper norm, since the parts of the initial wave function that do not correspond to the eigenstate $|\, \phi_{I1}\rangle$ are eliminated. At a later time, the measured state $|\, \psi_{I1}(t_1)\rangle$ in (2.28) evolves with the unitary evolution operator $\hat{U}(t_1,\, t_2) = \exp\,[iH(t_2\, -\, t_1)/\hbar]$ as follows:

$$|\, \psi_{I1}(t_2)\rangle = \hat{U}(t_1,\, t_2)\, |\, \psi_{I1}(t_1)\rangle. \tag{2.29}$$

Now, $|\, \psi_{I1}(t_2)\rangle$ is no longer an eigenstate of the eigenvalue I_1 because of the linear evolution described by the unitary operator. The evolution in (2.29) is just the expression of the Schrödinger equation in the bra-ket notation. The subscript I_1 on the wave functions just reminds the reader that such states have already suffered a nonlinear collapse at the earlier time t_1. Finally, the system is measured again to get the (eigen-)value of the current $I_j(t_2) \equiv I_2$ with the equivalent of (2.28), as follows:

$$\big|\, \psi_{I1,\, I2}(t_2)\big\rangle = \big|\, \phi_{I2}\big\rangle\big\langle\phi_{I2}\, \big|\psi_{I1}(t_2)\big\rangle. \tag{2.30}$$

The Born rule says that the probability of this three-step process $P(I1,\, I2)$ is just the norm of the final state

$$P(I1,\, I2) = \big\langle\psi(t_1)\big|\, \phi_{I1}\big\rangle\big\langle\phi_{I1}\, |\, \hat{U}^{\dagger}(t_1,\, t_2)\, \big|\, \phi_{I2}\big\rangle$$

$$\times\, \big\langle\phi_{I2}\, |\, \hat{U}(t_1,\, t_2)\, \big|\, \phi_{I1}\big\rangle\big\langle\phi_{I1}\, \big|\psi(t_1)\big\rangle. \tag{2.31}$$

This can be written as a function of the initial state at time t_1 by substituting (2.28) into (2.29) and then substituting (2.29) into (2.30). The mandatory perturbation of the state due to the measurement is the reason why simulating the measurement process plays a fundamental role in determining the noise, even with low- (or zero-) frequency noise. The development from (2.28) to (2.30) has been performed using the so-called strong measurement approach. As has been mentioned, there are other types of measurements (weak measurements) that imply softer collapses (see appendix A.2) below.

Of course, the multi-time measurement described above is consistent with the orthodox theory, where the collapse law is of fundamental importance. This can alternatively be explained by quantum theories without observers. Referring back to figure 2.6(b), an electron prepared at time t_0 on the left of the barrier is guided by a wave function $\psi(x, t)$ that is a solution of the Schrödinger equation, just as depicted in figure 2.6(a). Such an electron that is transmitted at time t_1 remains transmitted at time t_2. The velocity of the transmitted electron computed using (2.24) at time t_2 is only determined by the transmitted wave function. Equivalently, in figure 2.6(c), an electron prepared at time t_0 on the left of the barrier is guided by a wave function $\psi(x, t)$ that is also a solution of the Schrödinger equation, just as in the previous case. The initial position of the electron is now different from the that in figure 2.6(b) so that the electron is reflected at time t_1. At time t_2, it remains reflected. The velocity of this reflected electron computed using (2.24) at time t_2 is only determined by the reflected wave function. It can thus be seen that the Bohmian theory can explain multi-time measurements without invoking any operator or any fundamental collapse law.

2.8 Displacement current

At this point, one could argue that no practical contribution to our understanding of electron quantum devices arises from the quantum theories without observers. Only different interpretations of the results have been discussed (with explanations, more or less digestible, about the quantum reality: the so-called quantum ontology) It is worth discussing a further example of the practical utility of such theories. Quantum theories without observers deal with different mathematical frameworks, and they can provide better tools for some practical applications. In principle, the computation of alternating current (AC) and its frequency components requires the knowledge of $I(t)$ at all times (or at time steps as small as the inverse of the highest frequency of interest). Today's state-of-the-art electronic devices have dimensions on the nanometer scale and operate at frequencies of up to hundreds of gigahertz or even terahertz [80, 81]. At such time and length scales, a full quantum treatment of the electrical current is mandatory.

But what is measured when the VNA of figure 2.5 is used to determine the current $I(t)$ at 100 GHz? Certainly, it is not the number of particles crossing a surface S that belongs to the active region in figure 2.8, because the ammeter (the measuring apparatus) is very far from such a surface. The symbol $I(t)$ in figure 2.8 refers to the current crossing the surface S of the active region, while the symbol $I'(t)$ refers to the current in the ammeter. Are both currents equal? At high frequencies, the total current includes not only the particle, or conduction current (flux of particles) but also the displacement current (the time derivative of the electric field). This is necessary so that the value $I(t)$ predicted at the surface S coincides with the value $I'(t)$ measured far from the surface S [82–85]. The total current is given by

$$I(t) = -e\iint J_x(x_S, x, y, t)ds + \iint \varepsilon\frac{\partial E(x_S, x, y, t)}{\partial t}ds, \tag{2.32}$$

where ε is the (homogeneous for simplicity) dielectric permittivity, and $ds = dy\,dz$. The first term is the particle (or conduction) current, while the second term is the displacement current. Note that, according to figure 2.8, the (ensemble) particle current density $J_x(x_S, x, y, t)$ is in three-dimensional space, even though it is the current density crossing a surface S.

The orthodox operator for the particle current of section 2.5.1 can be computed according to (2.20) after generalization to a three-dimensional scenario. The electric field $E(x_S, x, y, t)$ is the solution of the Poisson equation with proper boundary conditions. But what is the orthodox operator and what are the eigenstates of the displacement current? Should such eigenstates be linked to the time-dependent electric field inside the active region? This electric field depends on Maxwell's equations, which involve the charges of all electrons in a time-dependent framework. This orthodox theory postulates that such an operator exists, but it does not specify just what defines this operator in such a specific case. Of course, this does not mean that there is no orthodox operator with which to compute the displacement current, only that it is very difficult to find its general expression.

At this point, one must ask whether the quantum theories without observers (section 2.5.2) can provide a definition of either the total or the displacement current. For such a goal, it is very useful to refer to the well-known classical Ramo–Shockley theorem [86, 87]. This theorem links the dynamics of electrons with the total current present on a surface of the active region for a two-terminal device under the assumption that electrons have well-defined trajectories and uses a semiclassical treatment of the electromagnetic fields. Within the Bohm theory, these electrons also have trajectories describing their motion [57]. In this case, it is acceptable to use the classical Ramo–Shockley approach to reach a simple definition of the total quantum (particle plus displacement) current [84]. It can be shown that the total current on a surface S, belonging to the volume Ω, can be well approximated by

$$I(t) = -e\sum_{i=1}^{N(t)}\frac{v_{x,i}(\mathbf{r}, t)}{L}, \tag{2.33}$$

where $N(t)$ is the number of electrons inside the active region at time t and $v_{x,i}(\mathbf{r}, t)$ is the Bohm velocity of the ith electron in the transport direction x. Notice that electrons contribute to the total current across the surface S while moving inside the active region (even when they are not crossing this surface). Outside the active region, in the (metallic) contacts, the electric field of the electrons is screened and its contribution to the total current neglected.

Expression (2.33) also provides a trivial understanding of why the cutoff frequency in electron devices is related to the transit time inside the active region. When the voltage conditions are changed in a device, a transient appears in the current of the device until all electrons have had enough time to adapt to the new scenario. This transient time is related to the time needed for *old* electrons (entering the active region under the old voltage conditions) to abandon the active region and be replaced by *new* electrons (entering under the new conditions). Such a substitution is achieved when the last *old* electron (the one that entered the active region just before the voltage change) has just left the active region. Thus, the time needed for this last old electron to traverse the active region is the transit time. The inverse of this transit time is the cutoff frequency. This will be discussed in somewhat more detail in chapter 8.

Everything discussed here in connection with the cutoff frequency is quite trivial for a quantum theory without observers. However, the language used is unacceptable from the viewpoint of section 2.5.1, because it has been assumed that electrons have positions even though they are not being measured. How, then, does the orthodox theory predict the transit time of an electron in the active region without assuming a position for each electron? It is difficult for the orthodox theory to answer such a request, and this explains why the *tunneling time problem* has been (and is still today) a hot topic of debate during the last century [88–90].

Appendix A Weak and strong measurements

This chapter has discussed the so-called *strong (or direct)* measurement as well as the so-called *weak* measurement. The strong measurement is usually discussed when the measuring apparatus is directly coupled to the measured system. However, in many cases, the measuring apparatus is not directly coupled to the measured system but to another subsystem that connects the measuring apparatus and the measured system. In fact, this second type of measurement is what is plotted in figure 2.8. In this figure, the ammeter is not directly measuring the current in the DUT but the electrical current flowing through the cable (not through the system). Of course, there is a reasonably close correspondence between the current in the cable and that in the DUT. This second type of measurement is called *weak (or indirect)* measurement. In this appendix, the weak and strong measurements and a quantum description of the measuring apparatus are of interest.

A system coupled to an environment and to a measuring apparatus (such as the ammeter in figure 2.8) is schematically illustrated in figure A.1. When there is no measuring apparatus, as in panel (a), the orthodox theory indicates that there is no value of the current in the device active region, I, or in the environment, I'. Only when an ammeter is connected directly to the device active region using the strong measurement approach depicted in panel (b) is the value of I defined according to this theory. On the contrary, in the weak measurement approach (as discussed later) the measurement of the current is performed in the environment instead of in the active region, so that only the value I' in the environment is real/exists, as depicted in panel (c).

A.1 Von Neumann's strong measurements

From the very beginning, those who defended the orthodox explanation of the measurement have tried to provide a more physics-based explanation of just what

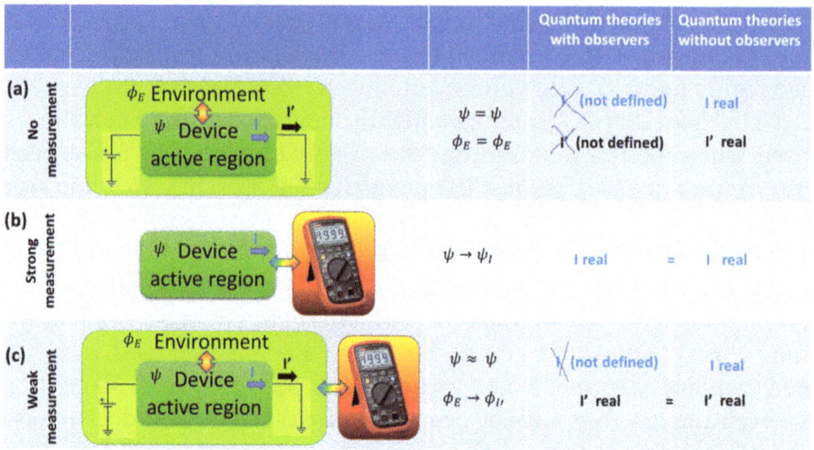

Figure A.1. Illustrations of measured and unmeasured properties according to quantum theories with and without observers in different scenarios: (a) without a measuring apparatus, (b) using a strong measurement, and (c) using a weak measurement.

the collapse law in (2.15) means. As early as 1932, von Neumann [41] proposed an explanation for the orthodox measurement that included the wave function of the ammeter as $\phi_M(x_M)$, where x_M indicates the degree of freedom of the ammeter (e.g. the pointer of the ammeter). At first sight, it seems that the orthodox theory allows one to describe the measuring system using the same language as that used to describe the unitary evolution of a quantum system. But at some point, the non-unitary collapse law must still be invoked.

The total system–apparatus wave function can initially be described as $\Psi(x, x_M) = \psi(x)\phi_M(x_M)$, although this form indicates that there are no initial interactions between the quantum system and the measuring system. However, as indicated in (2.14), this initial system–apparatus wave function can be rewritten as

$$\Psi(x, x_M) = \left(\sum_i c_{I_i}\phi_{I_i}(x)\right)\phi_M(x_M). \tag{A.1}$$

Following a unitary evolution with a Hamiltonian that contains an interaction term (thus mixing the system and the measurement apparatus) of the form $H_I = -i\hbar\beta I_i \partial/\partial x_M$, von Neumann showed that this state (A.1) evolves into

$$\begin{aligned}\Psi(x, x_M) &= \sum_i c_{I_i}\phi_{I_i}(x)\phi_M(x_M - I_i) \\ &= \sum_i c_{I_i}|\phi_{I_i}\rangle \int \phi_M(x_M) \mid x_M\rangle dx_M\end{aligned}. \tag{A.2}$$

Here, the second line expresses the wave function in the bra-ket notation. The time evolution arising from the interaction term H_I during a time interval τ can be written as the unitary operator $\hat{U}(\tau) = \exp(-i\hat{H}_I\tau/\hbar)$. For simplicity, one may assumed that the other terms of the Hamiltonian (kinetic energies) are not relevant during this interaction. A Taylor expansion of $\hat{U}(\tau)$ using $H_I = -i\hbar\beta I_i \partial/\partial x_{\mathrm{M}}$ in the position representation gives

$$U(\tau) = 1 - \tau\beta I_i\ \partial/\partial x_{\mathrm{M}} - \tau^2\beta I_i^2\ \partial^2/\partial x_{\mathrm{M}}^2 + \ldots. \tag{A.3}$$

This interaction now operates almost entirely upon the measurement system wave function $\phi_{\mathrm{M}}(x_{\mathrm{M}} - \tau\beta I_i)$. The parameter $\beta = 1/\tau$ may be adjusted to fit the final result in (A.2). This unitary evolution from (A.1) to (A.2) is known as a 'pre-measurement' or 'channelization' process in the literature [90]. If it is assumed that the wave function of the measuring apparatus is very close to a delta function $|\ \phi_{\mathrm{M}}(x_{\mathrm{M}} - I_i)\ |^2 \approx \delta(x_{\mathrm{M}} - I_i)$, then there is a perfect correlation in (A.2) between each well-defined value of the pointer I_i and each eigenvalue $\phi_i(x)$.

Up to this point, the explanation of the pre-measurement process is the same for any quantum theory. Only a unitary evolution has been assumed. The conceptual difficulties (revolving around the *quantum measurement problem* mentioned above) are exactly why the expression in (A.2) can maintain a superposition of measurement pointers (the currents I_i), when only one pointer (one current I) becomes real/exists in a laboratory measurement. The solution that is invoked for quantum theories with observers is different from the solution for quantum theories without observers.

So the theories now diverge; here, the orthodox theory is followed. Following the orthodox postulates, during the measurement process, a collapse of (A.2) occurs. The (unnormalized) wave function that appears after the random collapse is given by

$$\begin{gathered} \Psi(x, x_{\mathrm{M}}) \to \Psi_I(x, x_{\mathrm{M}}), \\ \sum_i c_{I_i}\phi_{I_i}(x)\phi_{\mathrm{M}}(x_{\mathrm{M}} - I_i) \to c_I\phi_I(x)\phi_{\mathrm{M}}(x_{\mathrm{M}} - I). \end{gathered} \tag{A.4}$$

On the right-hand side of (A.4), a well-defined value for I has appeared, while I_i on the left-hand side was a variable. The probability (Born law) of such a random process occurring is

$$P(I) = \iint |\ \Psi_I(x, x_{\mathrm{M}})|^2\ \ dx\ \ dx_{\mathrm{M}} = |\ c_I\ |^2, \tag{A.5}$$

where $\int |\ \phi_I(x)\ |^2\ dx = 1$ and $\int |\ \phi_M(x_{\mathrm{M}} - I)\ |^2\ dx_{\mathrm{M}} = 1$ have been used. If the measurements are repeated for an ensemble of identical devices, all with the same initial wave function $\Psi(x, x_{\mathrm{M}}) = \psi(x)\phi_{\mathrm{M}}(x_{\mathrm{M}})$ mentioned above, different currents I_i with probability $P(I_i)$ are obtained. The expectation value is again $\sum_{i=1}^{M} I_i \cdot P(I_i) = I_{\mathrm{DC}}$, which recovers the previous results (2.18) and (2.19).

The type of measurement explained above corresponds to the strong (or direct) measurement process mentioned in figure A.1(b). However, as indicated in the introduction to this appendix, this strong (or direct) measurement does not properly represent the measurement of the electrical current that happens in electronic devices, where the ammeter is not placed inside the active region but quite far from it, as indicated in figure A.1(c).

The orthodox collapse law invoked in (A.4) is eliminated as a real physical process when quantum theories without observers are used. For example, if we assume that the quantum system is defined by the additional Bohmian variables for the system, $x(t)$, and for the measuring apparatus, $x_M(t)$, then the position $x_M(t)$ of the pointer indicates the measured value. The position $x_M(t)$ can only correspond to one 'channel,' $|\phi_M(x_M - I_i)|^2 \approx \delta(x_M - I_i)$, indicating the well-defined value of the measured current I. By construction, in a strong measurement, there is no spatial overlap between the different measurement channels. In other words, the different summands in (A.4) do not overlap in the configuration space because the terms $|\phi_M(x_M - I_i)|^2$ do not overlap. In the simplified example of figure 2.6, the transmitted and reflected wave packets do not spatially overlap; therefore, the Bohmian trajectory can only be either reflected or transmitted. It cannot be both.

A.2 Von Neumann's weak or indirect measurements

The same wave function of the system $\psi(x) = \sum_i c_{I_i}\phi_{I_i}(x)$ is utilized again. As indicated in figure A.1(c), the system has no interaction with the measurement apparatus, even though it still interacts with an environment (represented by the cables, the batteries, etc). The wave function of the environment is decomposed into a superposition of M' eiegnstates $\phi_E(x_E) = \sum_j b_{I_j}\phi_{I_j}(x_E)$, where $\phi_{I_j}(x_E)$ is a basis function of the environment's Hilbert space with an eigenvalue I_j. The coefficient may be shown to be $b_{I_j} = \int \phi_{I_j}(x_E)\phi_E(x_E)dx_E$, but it will be discussed in more detail below once the environment function $\phi_E(x_E)$ has been defined. The initial system–environment wave function can then be written as

$$\Psi(x, x_E) = \sum_{i,j} c_{I_i}\phi_{I_i}(x)b_{I_j}\phi_{I_j}(x_E). \tag{A.6}$$

Just as in the previous part of this appendix in (A.2), a system–environment interaction $H_I = -i\hbar\beta I_i \partial/\partial I_j$, which represents a correlation (entanglement) between the current of the system and the current of the environment, leads to a wave function

$$\Psi(x, x_E) = \sum_{i,j} c_{I_i}\phi_{I_i}(x)b_{(I_j-I_i)}\phi_{I_j}(x_E). \tag{A.7}$$

Notice that this differs from (A.4) in the sense that there can be overlap between the different summands of (A.7). A measuring apparatus (ammeter) that measures the current I_j of the environment (not of the system) is now introduced. Just as in (A.1), the wave function of the ammeter is defined as $\phi_M(x_M)$, where x_M is the pointer of the ammeter. The entire wave function is channelized using the same procedure that led to (A.2). A *weak measurement of the system is just a strong measurement of the environment*. The entire wave function after the 'channelization' is now

$$\begin{gathered}\Psi(x, x_E, x_M) \rightarrow \Psi_{I'}(x, x_E, x_M) \\ \sum_{i,j} c_{I_i}\phi_{I_i}(x) b_{(I_j - I_i)}\phi_{I_j}(x_E)\ \phi_M\left(x_M - I_j'\right) \rightarrow \sum_i c_{I_i}\phi_{I_i}(x) b_{(I' - I_i)}\phi_{I'}(x_E)\phi_M(x_M - I').\end{gathered} \tag{A.8}$$

In this approach, the 'collapsed' wave function on the right-hand side of (A.8) still involves a superposition of system eigenstates $\phi_i(x)$. Thus, the orthodox interpretation of the weak measurement in (A.8) is that the 'current' in the environment has a well-defined value I', but the current in the system I_i remains undefined, as depicted in figure A.1(c).

Just as in appendix A.1, one can alternatively use quantum theories without observers to reach the same effective result as that of (A.8) without invoking the collapse. In all cases, the probability of this measurement process is

$$\begin{aligned}P(I') &= \iiint |\, \Psi_{I'}(x, x_E, x_M)\,|^2 \ \ dx\ dx_E\ dx_M \\ &= \sum_i |\, c_{I_i}|^2 |\, b_{(I' - I_i)}\,|^2 .\end{aligned} \tag{A.9}$$

If $|\, b_{(I' - I_i)}\,|^2 = \delta_{I',\, I_i}$, then (A.9) is simply converted into $P(I') = |\, c_I\,|^2$, as was done in (2.15), when dealing with a strong measurement. The interesting case of the weak measurement discussed here occurs when

$$|\, b_{(I' - I_i)}\,|^2 = N_b \cdot \exp\left(-\frac{(I' - I_i)^2}{2\sigma_E^2}\right). \tag{A.10}$$

Here, N_b is a normalization constant and σ_E defines a (large) dispersion in possible values. The determination of the expectation value $\left\langle I' \right\rangle = \int_{-\infty}^{\infty} I' P(I') dI'$ using (A.9) and (A.10) then gives

$$\begin{aligned}\langle I' \rangle &= \int_{-\infty}^{\infty} I' \sum_{i=1}^{M} |\, c_{I_i}|^2\ N_b \cdot \exp\left(-\frac{(I' - I_i)^2}{2\sigma_E^2}\right) dI' \\ &= \sum_{i=1}^{M} I_i \cdot P(I_i) = I_{DC}.\end{aligned} \tag{A.11}$$

The evaluation $\int_{-\infty}^{\infty} I' N_b\ \exp[-(I' - I_i)^2 / 2\sigma_E^2] dI' = I_i$ has been used, so the weak measurement of the system gives a perfect value of the direct current in the system. When $\sigma_E \rightarrow \infty$, then the approximation $|\, b_{(I' - I_i)}\,|^2 \approx N_b$ can be used in (A.10) so that the wave function of the system on the right-hand side of (A.8) is roughly given by $\psi_{I'}(x) \approx \sum_i c_{I_i}\phi_{I_i}(x) N_b$. In conclusion, this leads to the information about the (expectation) value of the system while causing almost no perturbation of the wave function of the system. No collapse of the system wave function has taken place because the system wave function, after the weak measurement of the current, is still a superposition of all eigenstates $\phi_i(x)$. This type of weak measurement is what was represented in figure A.1(c). Notice that the orthodox interpretation of this weak value states that the ammeter gives a well-defined value of the current, despite the fact that the system has no well-defined current!

Finally, it should be pointed out that the so-called *weak values* are based on a combination of this weak measurement with a strong one; they provide empirical values of some *exotic* magnitudes, such as the velocity of electrons [68, 71, 72]. A more detailed discussion of weak values is provided in chapter 10 and in appendix A of that chapter.

References

[1] Faraday M 1833 *Experimental Researches in Electricity* **Ser. IV** paragraphs 433–9 republished as *Phil. Trans. Roy. Soc. (London)* **123**
[2] Braun F 1874 *Ann. Phys. Pogg* **153** 556
[3] Bardeen J B and Brattain W H 1950 *US Patent* 2,524,035EJ
[4] Bardeen J B and Brattain W H 1948 *Phys. Rev.* **74** 230
[5] Kilby J 1976 *IEEE Trans. Electron Devices* **23** 648
[6] Moore G E 1965 *Electronics* **38** 114
[7] Ferry D K 2008 *Science* **319** 579
[8] Rupp K and Selberherr S 2011 *IEEE Trans. Semicon. Manuf.* **24** 1
[9] Prange R E and Nee T-W 1968 *Phys. Rev.* **168** 779
[10] Thompson S E, Armstrong M and Auth C *et al* 2004 *IEEE Electron Device Lett.* **25** 191
[11] Mistry K, Allen C and Auth C *et al* 2007 *Proc. Int. Electron Devices Meeting* (New York: IEEE Press) pp 247
[12] Hisamoto D, Hu C and Bokor J *et al* 2000 *IEEE Trans. Electron Devices* **47** 2320
[13] Barraud S, Lapras V and Previtali B *et al* 2017 *Proc. Int. Electron Devices Meeting* (New York: IEEE Press) pp 29.2
[14] Song T, Rim W and Kim H *et al* 2021 *IEEE Solid State Circuits Conf. Technical Digest (New York)* (IEEE) pp 24.3
[15] Dennard R *et al* 1974 *IEEE J. Solid -State Circuits* **9** 256
[16] Barker J R and Ferry D K 1980 *Solid-State Electron* **23** 531
[17] Faye J 2014 *Stanford Encyclopedia of Philosophy* ed E N Zalta (Redwood City, CA: Stanford University Press)
[18] Feynman R 1965 *The Character of Physical Law* (Cambridge: MIT Press) pp 129
[19] Kennard E H 1928 *Phys. Rev.* **31** 876
[20] Tvisidis Y *Operation and Modeling of the MOS Transistor* 2nd (New York: Oxford University Press)
[21] Ferry D K, Akis R and Gilbert M J 2007 *J. Comput. Theor. Nanosci.* **4** 1149
[22] Akis R, Gilbert M and Ferry D K *J. Phys.: Conf. Ser.* **38** 87
[23] Chynoweth A G and McKay K G 1956 *Phys. Rev.* **102** 369
[24] Goetzberger A and Stephens C 1961 *J. Appl. Phys.* **32** 2646
[25] Tam S, Hsu F C and Ko P K *et al* 1983 *IEEE Electron Device Lett.* **4** 386
[26] Toriumi A, Yoshimi M and Iwase M *et al* 1987 *IEEE Trans. Electron Devices* **34** 1501
[27] Herzog M and Koch F 1985 *Appl. Phys. Lett.* **53** 2620
[28] Bude J and Yoshii A 1992 *Phys. Rev.* B **45** 5848

[29] Hublitz K and Lyon S A 1992 *Semicond. Sci. Technol.* **7** B567
[30] Lacaita A L, Zappa F, Bigliardi S and Manfredi M 1993 *IEEE Trans. Electron Devices* **40** 577
[31] Tam S, Hsu F and Ko P *et al* 1982 *IEEE Electron Device Lett.* **3** 376
[32] Akers L A and Walker M 1985 *Physica* **134B** 116
[33] Landauer R 1957 *IBM J. Res. Develop.* **1** 223
[34] Landauer R 1970 *Phil. Mag.* **21** 863
[35] Ferry D K 2020 *Transport in Semiconductor Mesoscopic Devices* (Bristol: IOP Publishing Ltd) 2nd edn
[36] Ferry D K 2021 *Hot Carriers in Semiconductors* (Bristol: IOP Publishing)
[37] Kroemer H 1968 *IEEE Trans. Electron Devices* **15** 819
[38] Gilbert M J, Akis R and Ferry D K 2005 *J. Appl. Phys.* **98** 094303
[39] Duke C B 1969 *Tunneling in Solids* (New York: Academic)
[40] Marino F A, Faralli N and Palicios T *et al* 2010 *IEEE Trans. Electron Devices* **57** 353
[41] von Neumann J 1932 *Mathematische Grundlagen der Quantenmechanik* (Berlin: Springer)
[42] Landau L D and Lifschitz E M 1981 *Quantum Mechanics, Non-Relativistic Theory* (Boston, MA: Butterworth-Heinemann)
[43] Ballentine L E 1990 *Quantum Mechanics* (Englewood Cliffs, NJ: Prentice-Hall)
[44] Howard M W and Milburn G J 2009 *Quantum Measurement and Control* (New York: Cambridge University Press)
[45] Ferry D K 2019 *The Copenhagen Conspiracy* (Singapore: Pan Stanford)
[46] Dürr D and Teufel S 2008 *Bohmian Mechanics: The Physics and Mathematics of Quantum Theory* (Berlin: Springer Science & Business Media)
[47] Oriols X and Mompart J 2019 *Applied Bohmian Mechanics: From Nanoscale Systems to Cosmology* (Singapore: Pan Stanford)
[48] Mangan A M *et al* 2006 *IEEE Trans. Electron Devices* **53** 235–41
[49] Jacoboni C and Lugli P 1989 *The Monte Carlo Method for Semiconductor Device Simulation* (Wein: Springer)
[50] Herbert N 1984 *Quantum Reality* (New York: Anchor Books)
[51] Zeh H D 1970 *Found. Phys.* **1** 69
[52] Maudlin T 1995 *Topoi* **14** 7
[53] Bell J S 1990 *Phys. World* **3** 33
[54] Heisenberg W 1925 *Z. Phys.* **31** 879
[55] Born M and Jordan P 1925 *Z. Phys.* **34** 858
[56] Bohm D 1952 *Phys. Rev.* **85** 166
[57] Madelung E 1927 *Z. Phys.* **40** 322
[58] Holland P R 1993 *The Quantum Theory of Motion* (Cambridge: Cambridge University Press)
[59] Griffiths R B 2003 *Consistent Quantum Theory* (Cambridge: Cambridge University Press)
[60] Hartle J B 1998 *Phys. Scr.* **T76** 67
[61] Omnès R 1994 *The Interpretation of Quantum Mechanics* (Princeton, NJ: Princeton University Press)
[62] Griffiths R B 2017 *Phys. Rev.* A **96** 032110
[63] Ghirardi G C, Rimini A and Weber T 1986 *Phys. Rev.* D **34** 470
[64] Bassi A *et al* 2013 *Rev. Mod. Phys.* **85** 471
[65] Svensson B E 2013 *Quanta* **2** 18
[66] Jacobs K and Steck D A 2006 *Contemp. Phys.* **47** 279

[67] Gilton M J R 2016 *Stud. Hist. Phil. Mod. Phys.* **55** 92
[68] Aharonov Y, Albert D Z and Vaidman L 1988 *Phys. Rev. Lett.* **60** 1351
[69] Oriols X and Ferry D K 2021 *Proc. IEEE* **109** 955
[70] de Broglie L 1923 *Nature* **112** 540
[71] Traversa F L *et al* 2013 *Phys. Rev.* A **87** 052124
[72] Kocsis S *et al* 2011 *Science* **332** 1170
[73] Suga S *et al* 2021 *Photoelectron Spectroscopy: Bulk and Surface Electronic Structures* (Zurich: Springer Nature)
[74] Smallwood C M, Kaindl R A and Lanzara A 2016 *Europhys. Lett.* **115** 27001
[75] Price P J 1979 *Semicond. Semimet.* ed R K Willardson and A C Beer 14 (New York: Academic) pp 249–308
[76] Sjakste J *et al* 2018 *J. Phys.: Condens. Matter* **30** 353001
[77] Tanimura H *et al* 2016 *Phys. Rev.* B **93** 161203
[78] Ferry D K 2021 *Appl. Phys. Rev.* **8** 021324
[79] Oriols X and Ferry D K 2013 *J. Comput. Electron* **12** 317
[80] Ashley T *et al* 2003 *Int. Semiconductor Device Research Symp.* (New York: IEEE) pp 196–7
[81] Deal W R *et al* 2010 *Proc. IEEE Compound Semiconductor IC Symp.* (New York: IEEE) pp 1–10
[82] Eisenberg B, Oriols X and Ferry D 2017 *Mol. Based Math. Biol.* **5** 78
[83] Eisenberg B 2017 *Hung. J. Ind. Chem.* **44** 1
[84] Marian D, Zanghì N and Oriols X 2016 *Phys. Rev. Lett.* **116** 110404
[85] Eisenberg B, Oriols X and Ferry D 2022 (under review)
[86] Ramo S 1939 *Proc. IRE* **27** 584
[87] Shockley W 1938 *J. Appl. Phys.* **9** 635
[88] Hauge E H and Støvneng J A 1989 *Rev. Mod. Phys.* **61** 917
[89] Jonson M 1992 *Quantum Transport in Semiconductors* ed D K Ferry and C Jacoboni (New York: Springer) pp 193–238
[90] Pandey D *et al* 2021 *Phys. Rev.* A **103** 052219

Chapter 3

Many-body open systems outside thermodynamic equilibrium

The general approach to quantum mechanics is that it is usually considered to be a closed system which is, in any respect, in equilibrium with its background, although no connection with this background is ever discussed. Even more narrowly, there is usually no discussion at all of the temperature—as if it were not an important parameter. But transport does not involve closed systems! If current is to flow, as it must in semiconductor devices in the 'on' state, a phase transition to irreversible behavior is required to occur [1]. The quantum mechanics of open systems is more difficult, not least because of the need to consider initial and boundary conditions. While the early work on semiconductors and semiconductor devices was discussed in chapter 2 in terms of the efforts of Faraday [2] and Braun [3], serious work on markedly nonequilibrium devices began with the work of Eric von Hippel in 1933 [4] and Herbert Fröhlich in 1937 [5] on the electrical breakdown of gasses and solids. It was extended to the electrical breakdown of silicon shortly after World War II [6].

The studied breakdown in semiconductors is not merely a nonequilibrium phenomenon, but is really considered to take place *far from equilibrium.* In both bulk semiconductors and in semiconductor devices, this far-from-equilibrium behavior has come to be called hot-carrier behavior [7]. By the term 'hot carriers,' we mean that the carrier system has acquired a quasi temperature that is larger than the lattice temperature—there is far more kinetic energy in the carrier system than is the case in equilibrium. The driving force for this is the electric field to which the material or device is subjected. In this sense, the carrier system is no longer in thermal equilibrium—in particular, it is no longer in equilibrium with the lattice—and one has already made an assumption that a temperature can describe the electron system. Nevertheless, one has to deal with the manner in which this excess energy flows from the carriers to the environment, which is often through the phonon system.

doi:10.1088/978-0-7503-5237-6ch3

In general, the transport of these hot carriers is quite nonlinear and the conductance of the material is now a function of the electric field. This nonlinear conductance does not arise from higher-order terms in the field itself, even though that approach is often taken, but is usually derived from the fact that the distribution function is itself a function of the electric field. The situation is more complicated, as the hot carriers can also drive the phonon system out of equilibrium, since the energy transferred to the phonons through the carrier–phonon interaction is larger than can be dissipated by the subsequent transfer to the cold lattice environment.

It is now easy to understand that the treatments of semiconductor devices, whether of classical or quantum origin, are not simple undertakings. Any real device is a complicated open system that is nowhere near thermal equilibrium. Moreover, because of the carriers, the phonons, and the driving forces, this is also a many-body problem that can be quite difficult to analyze in great detail. In addition, there are difficult spatial and temporal inhomogeneities that complicate the task. In this chapter, we will try to address some of the general problems that arise from considering many-body systems, especially those that are open and require initial and boundary conditions.

3.1 The many-body problem in quantum mechanics

Treating transport in any system is often quite difficult, and this can especially be the case with semiconductor devices. Consider the simple fact that the semiconductor material has more than 10^{22} atoms/cm^3 and more than 10^{22} electrons/cm^3. It also has interactions between each atom and every other atom, between each electron and every other electron, and between each atom and every electron. This is basically an unsolvable problem. Certainly, there are approximations that simplify the problem greatly, otherwise we would know nothing about the material other than what could be discovered from measurements. The most important approximation is to assume that the material has a crystal structure. Once this structure is known, the periodicities that arise place considerable limitations on the possible solutions that could be put into a theory. We certainly know that one can write down the total energy for the closed system, as given quantum mechanically by the Hamiltonian

$$H = H_{\mathrm{L}} + H_{\mathrm{e}} + H_{\mathrm{e-L}}, \tag{3.1}$$

where the electronic portion is

$$H_{\mathrm{e}} = \sum_i \frac{p_i^2}{2m_0} + \sum_{i,j\neq i} \frac{e_i e_j}{4\pi\epsilon_0 x_{ij}}, \tag{3.2}$$

and the atomic portion is

$$H_{\mathrm{L}} = \sum_j \frac{P_j^2}{2M_j} + \sum_{r,s\neq r} \frac{Q_r Q_s}{4\pi\epsilon_0 X_{rs}}. \tag{3.3}$$

In each of these two equations, the first term on the right is the kinetic energy of the atomic motion, with the uppercase letters corresponding to the atoms. The charge of the electron is $-e$. The second terms on the right of (3.2) and (3.3) are the interaction term between the electrons and the atoms, respectively. The double subscripted term in the denominator is the distance between the two electrons or atoms that are interacting. Care must also be taken not to double count the interactions; sometimes, this is avoided by letting both indices run over all values and adding a factor of 1/2 in front of these two terms. The interaction between the electrons is usually the coulomb force that exists between two charged bodies. While this is still true for the atoms, their interaction is often described by mechanical forces. The third term in (3.1) for the interaction of electrons with atoms will be described later.

3.1.1 The Born–Oppenheimer approximation

The most important approximation now comes from treating these three terms separately. To understand this separation, it is important to understand that it is just not possible to solve the entire problem exactly. Instead, the approximation is the adiabatic approximation, which arises from the recognition that the electrons and the atoms move on different timescales. This is often called the Born–Oppenheimer approximation [8], in which the atomic motion and the electronic motion are treated separately. When we investigate the electronic motion, the atoms are assumed to be moving too slowly to consider, so they are treated as if they are frozen rigidly to the lattice sites defined by the crystal structure. This leads to the energy bands that will be discussed in chapter 6. In this way, we ignore the atomic motion and treat the presence of the atoms as only a rigid shift in the energy. Hence, the energy bands appear for a rigid, periodic potential provided by the atoms stuck in their places. On the other hand, when we examine the interactions of the slowly moving atoms, we consider that the electrons are so fast that they instantaneously follow the atomic motion. Thus, the electrons appropriate to an atom are frozen to it—they adiabatically adjust to the atomic motion.

Similarly, the electronic motion is ignored when the lattice dynamics—the motion of the atoms—is studied. The electrons are assumed to move much faster than the atoms, as they have much smaller masses, and therefore follow the atoms rapidly. While this assumption is universally made, it is not quite a foregone conclusion that it really applies. For example, the vibrational energy of the atoms in the optical modes can be in the 10–100 meV range for various materials (e.g. in GaAs, the upper longitudinal (LO) mode at the zone center is 36 meV). This is certainly approaching, or above, the thermal energy of a conduction band electron at room temperature (e.g. $3k_BT/2$ is 52 meV at room temperature). The upper LO mode in materials such as GaAs is a polar mode because of the inequality of the two atomic charges; it certainly interacts with the electronic vibrations. This latter interaction is called plasmon–phonon coupling and is important in many semiconductors. Nevertheless, the approximation of separating the two motions is almost universal in the field, more because it is necessary to reach any usable solutions.

Finally, what is still important to us is the small vibration of the atoms about their average positions that the electrons can actually see. This is sometimes a small effect and is therefore treated by perturbation theory; this is the electron–phonon interaction that gives us the scattering properties (this is the subject of chapter 5). The final term on the right-hand side of (3.1) is just this interaction between the electrons and the atoms. One of these is the plasmon–phonon interaction discussed in the previous paragraph and is not small. But the latter is usually treated separately from the above electronic and atomic solutions [9]. The remaining interactions are usually considered to be small and are treated by perturbation theory. The interaction term between the electrons and the atoms can be expressed as

$$H_{\mathrm{e-L}} = -\sum_{i,s} \frac{e_i Q_s}{4\pi\varepsilon_0 \, |\boldsymbol{x}_i - \boldsymbol{X}_s|}. \tag{3.4}$$

This term can be rewritten as

$$H_{\mathrm{e-L}} = -\sum_{i} e_i V(\boldsymbol{x}_i) \tag{3.5}$$

where

$$V(x_i) = \sum_{s} \frac{Q_s}{4\pi\varepsilon_0 \, |\boldsymbol{x}_i - \boldsymbol{X}_s|} \tag{3.6}$$

is the potential of the atom with which the electron is interacting. This can be separated into two terms: (i) the first is the part due to the exact position of the atoms residing precisely on the lattice positions (their average positions, in a sense) and (ii) the motion of the atoms about this position which can perturb the electronic properties of the electrons. The first term is already treated in the determination of the electronic band structure found using (3.2). It is the second term, the motion of the atoms around their lattice sites, that is important for the scattering consideration. This second term is the electron–phonon interaction that primarily leads to scattering processes that are treated by perturbation theory. In quantum transport, even this perturbation theory can lead to a real part of the energy associated with the interaction that can cause energy shifts of the electron states as a whole (beyond the loss of a phonon's energy). This is a true many-body shift of the energy.

3.1.2 Quantum statistics

Most of us have grown up with semiconductors and are used to treating them in a relatively simple way in which energy is related to the carrier momentum through the relation

$$E(k) = \frac{p^2}{2m^*} = \frac{\hbar^2 k^2}{2m^*}, \tag{3.7}$$

where the wave momentum is denoted by $\hbar k$ and m^* is the effective mass of the carrier. Certainly, many are familiar with the extension to hyperbolic bands described by $\boldsymbol{k} \cdot \boldsymbol{p}$ theory and the presence of non-parabolic bands. While there is some belief that this represents the introduction of quantum mechanics into semiconductors, this really remains semiclassical behavior. One has to understand that the electronic bands computed using (3.2) and discussed in chapter 6 actually mostly ignore the second term on the right-hand side of (3.2). That is, the electronic structure is for a mostly empty band that contains at most a single electron. This approach really ignores the many-body nature of these energy bands. The result is that the density of states is a simple function of energy (or momentum) under an assumption of the correctness of (3.7). In three dimensions, this density of states is well known to be simply

$$\rho(E) = \frac{1}{2\pi^2}\left(\frac{2m^*}{\hbar^2}\right)^{1/2}\sqrt{E}. \tag{3.8}$$

But this simple relationship is dependent upon the validity of (3.7). This function is plotted in figure 3.1(a) and is a simple curve relating k to E.

In fact, the many-body interactions that arise in quantum mechanics destroy this simple view of the energy structure and the density of states. The idea of the spectral density was introduced in chapter 1 via (1.77) as a relationship between momentum and energy given by $A(k, E)$ (with the energy defined from the frequency via Planck's formula). Classically, this relation is a delta function in which

$$A(k, E) \sim \delta\left(E - \frac{\hbar^2 k^2}{2m^*}\right) \tag{3.9}$$

for a parabolic band. When interactions are present, whether they are carrier–carrier interactions or carrier–phonon scattering, the spectral density deviates from this simple delta function. The energy is no longer a simple function of the momentum, but the two (energy and momentum) are separate quantum variables. In quantum mechanics, the interactions arising in (3.2) and (3.3) take the system from a one-electron picture to the many-body picture. These interactions lead to changes in the single-particle energy through a single term, which is called the *self-energy*. When the interactions are present, the energy is shifted from (3.7) to a form given as

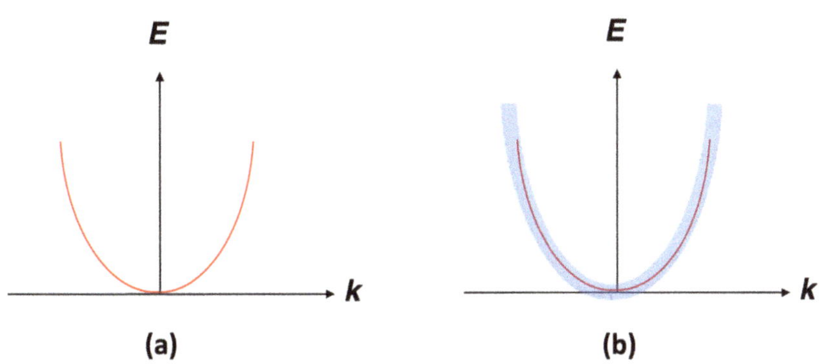

Figure 3.1. (a) The density of states (red line) for the classical picture in which each value of the wave number k has a unique energy. (b) In the quantum case, the broadening of the state density described through the spectral density gives rise to a range of values of energy for a given k and vice versa. This is shown pictorially by the blue width overlaid on the classical value. No energy shift due to the real part of the self-energy is shown here.

$$E(k) = \frac{\hbar^2 k^2}{2m^*} - \Sigma(k), \tag{3.10}$$

where the last term is the self-energy. The self-energy causes the delta function to broaden into a Lorentzian line shape, as follows:

$$A(k, E) = \frac{2\mathrm{Im}\{\Sigma(k)\}}{\left[E(k) - \frac{\hbar^2 k^2}{2m^*} - \mathrm{Re}\{\Sigma(k)\}\right]^2 + [\mathrm{Im}\{\Sigma(k)\}]^2}. \tag{3.11}$$

In full quantum field theory, it is often found that the real part of the self-energy actually diverges. Such infinities have long been a problem in this area, but it has been recognized that they can usually be removed by renormalizing the system with an effective mass correction [10, 11]. Much like the wave function, the spectral density has a normalization that requires [12]

$$\frac{1}{2\pi\hbar} \int dEA(k, E) = 1. \tag{3.12}$$

From the spectral density, the density of states can now be expressed as a single function of the energy by integrating out the momentum. Thus, one obtains

$$\rho(E) = \frac{1}{\pi\hbar} \int \frac{d^3\mathbf{k}}{8\pi^3} A(k, E). \tag{3.13}$$

This is no longer the sharp line shown in figure 3.1(a). Rather, the state population at a given value of k actually has a spread of energies (and vice versa). This is indicated pictorially in figure 3.1(b). This density of states is now not for a single electron in an empty band, but for that electron in the great density of other electrons and atoms. If we say that (3.9) defines a thin shell of energy in momentum space, then (3.13) describes a broad range of possible values that lie both on and off the thin shell. Thus, the effects of spectral broadening are often referred to as 'off-shell effects.'

As satisfying as the above descriptions are, there is a serious problem with them. We recall from chapter 1 that in the study of transport, we are interested in the movement of a particle, or, more particularly, a wave function from $\psi(\boldsymbol{x}, t)$ to $\psi(\boldsymbol{x}', t')$. From the discussion around the center-of-mass variables (1.68), k is the Fourier transform on $\boldsymbol{x} - \boldsymbol{x}'$, that is, on the variable $\boldsymbol{s}$. Similarly, it is understood that ω, or the energy $E = \hbar\omega$, is the Fourier transform on $t - t'$. What is lost is the dependence of everything on $\boldsymbol{R} = (\boldsymbol{x} + \boldsymbol{x}')/2$ and $T = (t + t')/2$. The fact is that the development above is for a system that is homogeneous in both space and time. Semiconductors are homogeneous for neither of these quantities. They have strong spatial variations and are operated at reasonably fast timescales. The above abandonment of the space and time variations is most usually found in the use of Green's functions, as mentioned in chapter 1. To see the importance of these quantities, we will consider the quantum requirements imposed on treatments of the interaction terms.

3.1.3 Quantum effects in scattering

In classical transport, a collision between a carrier and a phonon is assumed to be entirely local. That means that the collision occurs at a single point in space and in time. The details of the collision, however, are usually worked out via perturbation theory using the quantum details of atomic motion [7]. Quite often, this view is carried over to quantum transport without further consideration by assuming that the quasi-classical approach is still valid. But one has to remember that the many-body situation in quantum devices involves two positions and times, as discussed in the previous paragraph. This means that even in quantum transport, the collision is much more difficult, as it is necessary to modify the normal semiclassical scattering functions to account for off-shell effects and collisional broadening. In quantum transport, scattering from a phonon, for example, does not have to be local; the event may spread over a small spatial distance. This arises because the collision is not instantaneous but takes a few femtoseconds to complete [13, 14]. Even though this is short, the particle can be accelerated by the intracollisional field effect. In addition, the particle also moves a bit due to the particle's own velocity, even without any additional acceleration by the field. For example, if the particle is moving at 10^7 cm s^{-1}, then during a 5 fs collision duration, it travels some 0.5 nm. While this is small, consider the situation for a modern nanosheet metal–oxide–semiconductor field-effect transistor (MOSFET) [15, 16] in which the nanowire channels have dimensions of 2×8 nm and a length of some 20 nm. Even a 0.5 nm shift in the particle's position during the scattering is significant in this case.

In this nonlocal scattering, it is important to recognize the evolution of the quantum system. Even in perturbation theory, the scattering matrix elements assume complex values as a result of quantum interference (discussed further below) and the fact that the system is subjected to the uncertainty principle. The negative parts of the resulting self-energy are direct indications of the regions where uncertainty becomes important. The nonlocality in both space and time arises because the collision is a two-step process. This two-step process is required whether one is dealing with scattering by a photon or a phonon. This means that optical photon absorption and carrier scattering by a phonon both involve this two-step process. While the photon is envisioned as a particle, it is the electromagnetic wave aspect that is important in creating the electron–hole pair. Similarly, with the phonon, it is the lattice wave that creates the interaction. In each case, the wave has an amplitude that plays the important role. In optical absorption, it is the vector potential which fulfills this role. With the phonon, it is the lattice displacement that fills the role, and this displacement creates a modification of the band structure which generates a potential.

Perhaps the earliest study of the time necessary for the interaction was that of Geltman, who wished to find out just how long it took for a photon to ionize an atom [17]. In ionizing an atom (or absorbing a photon in a semiconductor) from an incident electromagnetic field, this field first creates a correlation between the initial (ground or valence band) and final (excited or conduction band) states; this, in turn, may result in Rabi oscillations between these two states [18]. This means that the unoccupied quantum state in the conduction band becomes entangled with the occupied state in the valence band. This entanglement or correlation is referred to as the polarization. The first step of the electromagnetic wave of the photon is to create this polarization. The conjugate field then destroys this correlation, resulting in the completed process. The electron might remain in the valence band as the photon goes on its way. However, in semiconductors, it usually results in the electron transitioning from the valence band to the conduction band, ultimately creating the electron and hole (or the condensed exciton). Although the electromagnetic field (photon) creates the interaction, in an energy-conserving process, it is the energy of the absorbed and annihilated photon that is thought to be transferred in creating the electron–hole pair in the absence of any many-body effects. The entire process takes only a few femtoseconds and is governed by the semiconductor Bloch equations [19–21].

With the phonon, the process is entirely the same except that it is the lattice motion that generates the field which interacts with the carriers. This field also creates the polarization between the initial and final states of the carrier before and after the scattering event [14]. But for the phonons as well as for the photons, standard Fermi golden-rule treatment in perturbation theory involves the magnitude-squared matrix element and a long time limit. While each matrix element is one step in the overall two-step process, the long time limit ignores the details of this process and leads to an energy-conserving delta function. This cannot be done in the quest for quantum transport in modern small semiconductor devices, and many researchers have sought a better way to handle the nonlocality of the process in both space and time. In particular, methods for handling this nonlocality have been suggested for particle simulations of quantum transport [22–24]. To properly include the role of the finite collision duration, we need to account for two effects: (i) the collision duration itself with its resulting energy shift and (ii) the position change during the collision.

One useful approach is based upon the theoretical formulations of Levinson for the Wigner function [25] but is not limited to this approach. In this process, one half of the phonon momentum is passed to the path at each of two time steps. For a virtual process in which the scattering is not completed, this half of the momentum is then withdrawn at the second vertex so that no net momentum is passed to the path. If the scattering process is completed, however, the net momentum change is $\hbar q$, just as in the classical case. This provides a method by which a scattering event can be inserted into the path of a particle. A phase factor is also passed to the particle, which is proportional to

$$\cos\left[q(r - r') + \omega_q(t - t')\right], \tag{3.14}$$

where ω_q is the frequency of the phonon (or photon in the case of optical absorption). For a virtual process (in which no scattering actually occurs), this half of the momentum is withdrawn at the second vertex, so no net momentum is passed to the path. If the scattering process is completed, however, the net momentum change is $\hbar q$, just as in the classical case.

There is still the problem of identifying the two positions and the two times in (3.14); e.g. identifying the change in position and time during the collision. As discussed above, the scattering is nonlocal and involves a time duration. In the center-of-mass coordinate and time shifts of discussed in equation (1.53) of chapter 1. In these transformations and the resulting Fourier transforms, the momentum is connected to the direction normal to the axis defined by the two spatial points. Hence, the time duration is connected to that required to transit between the two positions at the corresponding momentum. These different variables are thus tied together in a type of self-consistent manner. This has suggested another approach to the actual evaluation of the nonlocality in space and time. As remarked previously, the collision duration itself is usually only a few femtoseconds [14]. In this time, the change in position that occurs as a result of the collision duration may be approximately written as [26]

$$\delta x = \left[v(t) - \frac{e\boldsymbol{F} \cdot v(t)}{m^* v(t)} \frac{\tau_c}{2} \right] \tau_c. \tag{3.15}$$

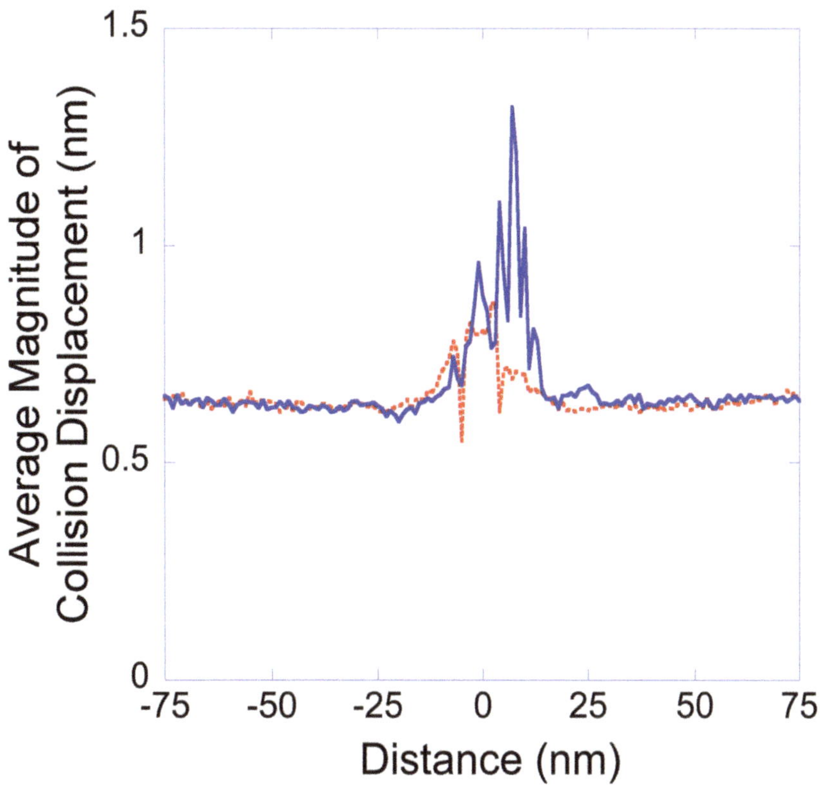

Figure 3.2. The root-mean-square displacement of a particle during a phonon scattering event as a function of position in an RTD. The tunneling structure is centered around $x = 0$. The solid curve (blue) represents a bias of 0.5 V, while the dotted curve (red) represents the absence of any bias.

In this expression, $\boldsymbol{F}$ is the electric field and τ_c is the collision duration. The first term is just simple displacement due to the particle's motion. The second term represents the intracollisional field effect. Notice the scalar product in the numerator of (3.15). For the present, the particle is represented by a wave packet as discussed in section 1.5. This packet is centered on the expected position of the particle throughout its evolution. During the scattering interaction, one may assume that there is an overlap between the wave packets at the two positions separated by δx. The two wave functions may be taken to be Gaussian in shape, so that the correlation between the initial and final positions leads to an overlap that varies as

$$I(\delta x) = \frac{1}{\sqrt{4\pi\sigma^2}} e^{-(\delta x)^2/4\sigma^2}, \tag{3.16}$$

where the broadening σ is taken to be the thermal de Broglie wavelength

$$\sigma \sim \sqrt{\frac{\hbar^2}{12m^* k_B T}} \sim 2 \text{ nm}. \tag{3.17}$$

The probability for a given shift is then

$$P(\delta x) = | I(\delta x) |^2 = \frac{1}{4\pi\sigma^2} e^{-(\delta x)^2/2\sigma^2}. \tag{3.18}$$

The direction of the shift in position is determined by the relation between the velocity and the electric field. This provides a method of incorporating the displacement in time and space for the scattering process. There is still the problem of the collision duration. When the mean time is known, one can estimate the position shift from (3.15) and then evaluate the probability of such a shift using (3.18). Generally, the collision duration may be considered to be exponentially distributed and may then be estimated by

$$\tau'_c = -\tau_c \ln(r), \tag{3.19}$$

where τ_c is the collision duration as determined by direct calculations [14] and r is a random number between zero and one.

As an example, this has been studied for particle simulations (connected to Wigner functions but applicable to any particle simulation of quantum transport) for a resonant tunneling diode (RTD) [26]. The RTD structure contains 3 nm AlGaAs barriers 0.3 eV high and a GaAs 5 nm well. The device is symmetrically located in a 30 nm spacer region (also made of GaAs) with a doping level of 10^{16} cm^{-3}. The emitter and collector contact regions are each 60 nm and are doped to 10^{18} cm^{-3}. In this structure, the barriers will test the energy relation (3.15) during scattering events in their vicinity. Statistics are determined for the energy changes and displacements during many thousands of scattering events and are collected for 2 ps during each bias point. In figure 3.2, the average value of the displacement during the collision is plotted as a function of position within the RTD, with $x = 0$ taken to be the center of the quantum well itself. Away from the tunneling region, the root-mean-square (rms) displacement is only about 0.6 nm, which agrees well with the simple calculation at the beginning of this section. It should be pointed out that the average displacement is actually zero, as both positive and negative displacements can occur in both terms of (3.15), since the average value of the current is zero without any bias. But in the tunneling region, the rms displacement can be much larger. This is because the field is considerably larger around the tunneling structure, where most of the potential drop occurs. And in this region, the displacement is peaked in the direction of the electric field.

The increase in displacement near the tunneling region can lead to some interesting and unexpected results. These are related to the possibility that a particle could actually pass over the barrier *during the scattering time*. This particle would also absorb or emit a phonon. Without the presence of nonlocal effects during the collision, it would be reflected. But if the field is sufficiently high, this particle might gain enough energy to overcome the barrier and enter a state which goes directly over the barrier. This arises from the positional displacement. During such a collision, the particle may emit or absorb a phonon. If it emits a phonon, it then ends up with a lower kinetic energy in its final state. The carrier may also absorb a phonon and still end up with a lower kinetic energy in its final state. The difference between these two processes is a difference in the energy that must be absorbed from the electric field during the collision. The complementary process also has to occur, so if the positional displacement takes the particle into the barrier region, yet the absorbed energy exceeds the barrier height, then the particle has to lose the potential energy represented by the barrier.

3.2 Open systems—interaction with the environment

The complications that arise from the device being embedded within an environment are apparent when we consider the measuring process. The entire system becomes quite a complex system. This means that there are multiple parts to the overall system, and these parts may all interact with one another. More importantly, the system is far from equilibrium when the device is operating. The device system is extremely irreversible, since excitation follows a different path through the phase space of variables than the relaxation of the system. There may often be hysteresis in the operation. While we have discussed much of the quantum nature and the properties desired for the quantum theory of interest, particularly in chapter 1, there is much more to be said, especially about the connections between the device and its environment.

3.2.1 Interaction with the environment

It is clear that the desired system (or device in our case) interacts with the environment in which it is located. As discussed above, this interaction goes both ways. One method of examining this is to realize that, in an open system, the environment imposes what may be called super-selection rules upon the device, and these may cause some of the quantum states within it to be decohered by strong interaction with the environment, although a great many quantum states remain [27].

These interactions are part of the information that resides in the *classical* correlations that exist between the device and the environment, so that this is an environment-induced process that maintains a significant number of quantum states within the device, a process termed einselection [28] (short for environment-induced super-selection). These remaining quantum states include a particular set of states known as *pointer* states that have only weak interactions with the environment.

It must be recognized that all measurements of the device are actually performed in the environment. Thus, measuring quantum properties means that some of the remaining quantum states, the pointer states, must have some interaction with those states strongly coupled to the environment if they are to have effects that are measurable. States which have absolutely no coupling to the environmental states are unmeasurable, but this does not mean that they do not exist. The pointer states are characterized by their robustness, despite the existing environment, and also by an ability to create 'offspring' states (e.g. higher harmonics) through which they advertise information about themselves throughout the spectrum. This makes it possible for different observers (measurements) to see the same information about these states. This latter process is termed *quantum Darwinism* [29]. The upshot of all of this is that in order to measure a quantum system objectively, it is necessary to design the system in such a manner that the quantum-to-classical transition is observable in the measurements.

We want to illustrate how the environment and device/system interaction can be handled. First, we recognize that there will be a set of at least quasi-classical states of the environment and a set of quantum states of the device/system. These states are all contained in what is called a density matrix. Both the classical states, which can be described via the solution of a boundary-value problem for the closed environment, and the quantum states, determined from the Schrödinger equation for the closed device/system, can be written together in terms of a 'wave function' extended to the actual dimensionality of the environment and device. The density matrix can then be written as

$$\rho(\boldsymbol{x}, \boldsymbol{x}', t) = \sum_{i,j} a_{ij}\psi^{\dagger}(\boldsymbol{x}, t)\psi(\boldsymbol{x}', t). \tag{3.20}$$

Since (3.20) has a matrix form, the device density matrix can be separated from the total density matrix, in the absence of interactions, by simply tracing out the unwanted variables. Thus, we can write the device density matrix as

$$\rho_{\mathrm{D}}(\boldsymbol{x}, \boldsymbol{x}', t) = \mathrm{Tr}_{\mathrm{E}}\{\rho(\boldsymbol{x}, \boldsymbol{x}', t)\}, \tag{3.21}$$

where the subscripts D and E refer to the device and the environment, respectively. In a similar manner, one could instead trace out the device variables to obtain just the environment density matrix $\rho_{\mathrm{E}}(\boldsymbol{x}, \boldsymbol{x}', t)$. Thus, it can be seen that, in the absence of interactions, the total density matrix is just the tensor product of these two partial density matrices.

In the presence of interactions between the environment and the device, these simple trace operations do not always continue to work. To see this, a beginning is made by defining the total Hamiltonian as

$$H = H_{\mathrm{D}} + H_{\mathrm{E}} + H_{\mathrm{int}}, \tag{3.22}$$

where the three terms define the Hamiltonians for the device, the environment, and the interactions, respectively. Two crucial steps are now required in order to obtain the equation for the evolution of the set of desired states in the device, which can be considered here to be the pointer states. The first is to project these desired states from the entire density matrix via a projection superoperator [30–32]. These commutator-generating superoperators are denoted by P, $Q = 1 - P$, $P + Q = 1$, where the second term is the orthogonal projection onto the undesired states. The second step is the trace over the environmental variables, as discussed above. The details of these operations have been illustrated many times (see e.g. [33]) but these details are not essential to the current argument. When these two steps are complete, the result for the evolution equation of the pointer states is

$$i\hbar\frac{\partial \rho_{\mathrm{PD}}}{\partial t} = (H_{\mathrm{PD}} + H'_{\mathrm{int}})\rho_{\mathrm{PD}} + \Sigma_{\mathrm{int}}\rho_{\mathrm{PD}}, \tag{3.23}$$

where the additional subscript 'P' refers to the projected part of the device variables, and the modified interactions are

$$H'_{\mathrm{int}}\rho_{\mathrm{PD}} = \mathrm{Tr}_{\mathrm{E}}\{PH_{\mathrm{int}}\rho\}. \tag{3.24}$$

The last term on the right-hand side of (3.23) involves the interaction-induced self-energy of the device, and this includes any scattering processes arising from the environment, which of course includes the phonons. This last term is, in longer form,

$$\Sigma_{\mathrm{int}}\rho_{\mathrm{PD}} = \mathrm{Tr}_{\mathrm{E}}\{C\rho\} = \frac{i}{\hbar}\mathrm{Tr}_{\mathrm{E}}\left\{\int_0^t dt' PHQe^{-iQHQt'/\hbar}QHP\rho(t - t')\right\}. \tag{3.25}$$

The various parts of the term can be recognized. PHQ and QHP pick out the matrix elements that represent interactions between the pointer state and the rest of the system, the exponential term leads to the energy-conserving delta function in the long time limit (after the integration), and the entire expression is temporally retarded, as expected when one does not impose Boltzmann's *Stosszahlansatz.* Notice here that the projected device density matrix cannot be separated from the full expression in either (3.24) or (3.25). This means that not only do environmental effects appear within the projected device but that device variables appear in the environment, a result that is of course necessary for measurements. However, there is also the possibility that new states which couple the environment to the device will appear in the open interacting system. A very simple system will be considered to illustrate the pointer states and some environment-induced states.

3.2.2 An example of pointer states

The example to be considered is an open quantum dot, as shown in panel (a) of figure 3.3 and also schematically in the inset of panel (b). The dot is fabricated on a GaAs/AlGaAs heterostructure and defined by a set of metal gates (gray in the figure) [34, 35].

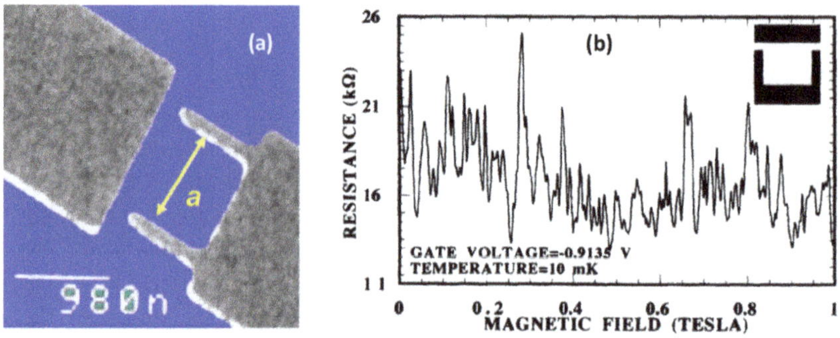

Figure 3.3. (a) A colorized TEM image of an open quantum dot, with dimension *a*. The gray areas are the metal gates and the dimension bar shows that the dot has a submicron size. (b) Resistance oscillations for transport through the dot as a function of the applied magnetic field (normal to the plane of the dot). (a) Reprinted from [35] with permission from IOP Publishing. All rights reserved. (b) Reprinted with permission from [34], Copyright (1994) by the American Physical Society.

It has a characteristic dimension a. The term 'open' means that the two openings shown in the figure are not tunneling structures but they allow one or more quasi-one-dimensional modes to pass, as these openings are open quantum point contacts (QPCs) [36, 37]. In panel (b), the magnetoresistance of the transport through the dot is shown as a function of the magnetic field. From the value of the peak resistance, it is apparent that slightly more than perhaps two modes are propagating through each of the two QPCs. The magnetoresistance shows oscillations in amplitude that are quasi periodic in the magnetic field. Here, the magnetic field is rather small, and under higher magnetic fields, there is a transition into the edge state regime appropriate for the quantum Hall effect. The electrical size of the dot is smaller than the physical size indicated in the figure due to depletion around the edges of the metal gates. In fact, the electrical size can be estimated by using a very high magnetic field in which there are only one or a very few edge states and observing the Aharonov–Bohm effect arising from trapped edge states within the dot [38].

Initially, it was supposed that the oscillations observed in the open quantum dots were the result of what are known as universal conductance fluctuations that arise from the random doping nature of impurities in these materials [39]. But these oscillations are too regular, and the dopant concentration is too small, especially with modulation doping, for such an interpretation. Further studies have clearly shown that these oscillations are related to the intrinsic energy structure of the quantum dot, even after it has been opened to the environment [40]. Indeed, it is found that there is a dominant, recurring wave function that periodically appears in the dot in the form of both a magnetic field and energy [41]. The energy structure can be seen in figure 3.4, where a smaller 0.3 μm dot is considered. Here, the vertical scale is the energy over a small range, and this energy corresponds to the Fermi energy in the open dot. The horizontal scale is the applied magnetic field, which is oriented normal to the dot. In panel (a), the eigenenergy spectrum found for the *closed* dot is shown as function of energy and magnetic field. In panel (b), the simulated quantum magnetoconductance (see [40], or section 3.3, for the simulation

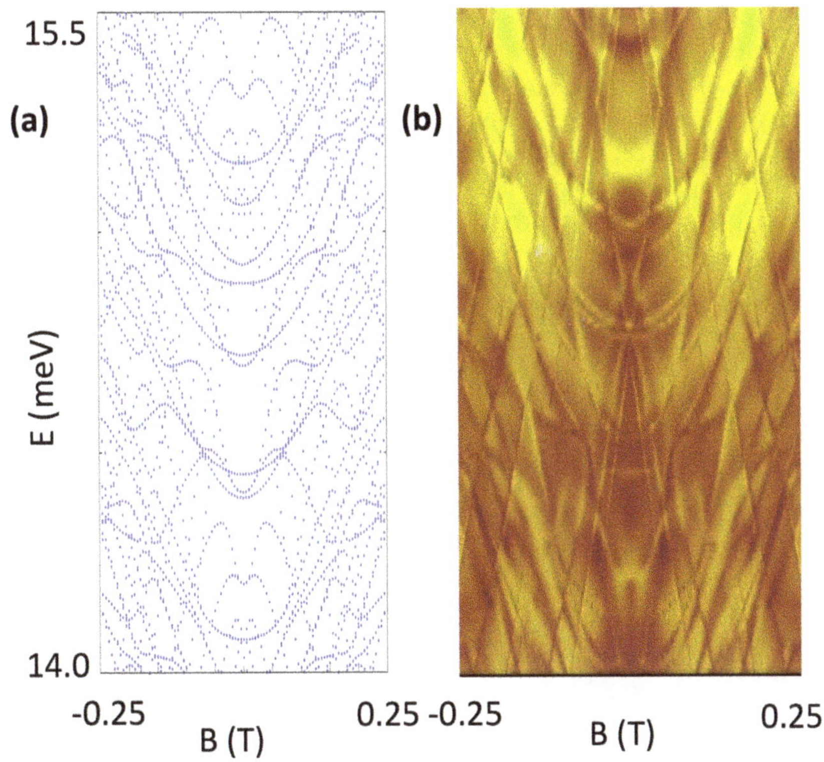

Figure 3.4. (a) The quantum bound states for a 0.3 μm closed quantum dot are plotted as a function of the energy of the state and the magnetic field present. (b) The simulated magnetoresistance of the dot for the same conditions as panel (a) but with the dot open to transport. Darker colors correspond to higher conductances. Reproduced from [41].

method) for transport through the open dot is shown for the QPCs that are open to non-tunneling transport. The similarities between the two panels are obvious, although the conductance also shows considerable transport for decohered states induced by the environment. The darkness of the lines in panel (b) corresponds to the height of the conductance level [41]. Nevertheless, in both panels, one can ascertain both linear lines moving mainly vertically and parabolic lines whose energy obviously varies as the square of the magnetic field. These lines are the residual resonances arising from the bound states of the dots and represent where the pointer states lie for a given energy and magnetic field.

A dominant wave function reappears along the lines of the quasi-periodic resonances shown in figure 3.4(b). This wave function is shown in figure 3.5 for the same 0.3 μm dot size as that used for figure 3.4. Sometimes, this recurrent wave function is referred to as a 'scar' that originates from a classical trapped orbit in the dot. While such scars are often considered to arise from unstable orbits [42], these wave functions are very stable, and this is thought to arise from the enhancement of the wave function around a periodic orbit characteristic of the classical system [43]. Stability arises from the fact that this wave function appears throughout the energy spectrum, so that each appearance reinforces it, and this is thought to be a tenet of quantum Darwinism [29]. This connection to the classical world is important in the understanding of the pointer states.

One may notice in figure 3.5 that there are underlying oscillations in the wave function. At first glance, these may be thought to be connected to the lateral modes of the square dot. For example, it is easily possible, either classically or quantum mechanically, to expand the properties of the dot in a two-dimensional Fourier series such as

$$\psi(x, y) = \sum_{i,j} b_{ij} \sin\left(\frac{n\pi x}{a}\right) \sin\left(\frac{m\pi y}{a}\right). \tag{3.26}$$

One then recognizes

$$\begin{aligned} k_{xn} &= \frac{n\pi}{a}, \\ k_{ym} &= \frac{m\pi}{a}, \end{aligned} \tag{3.27}$$

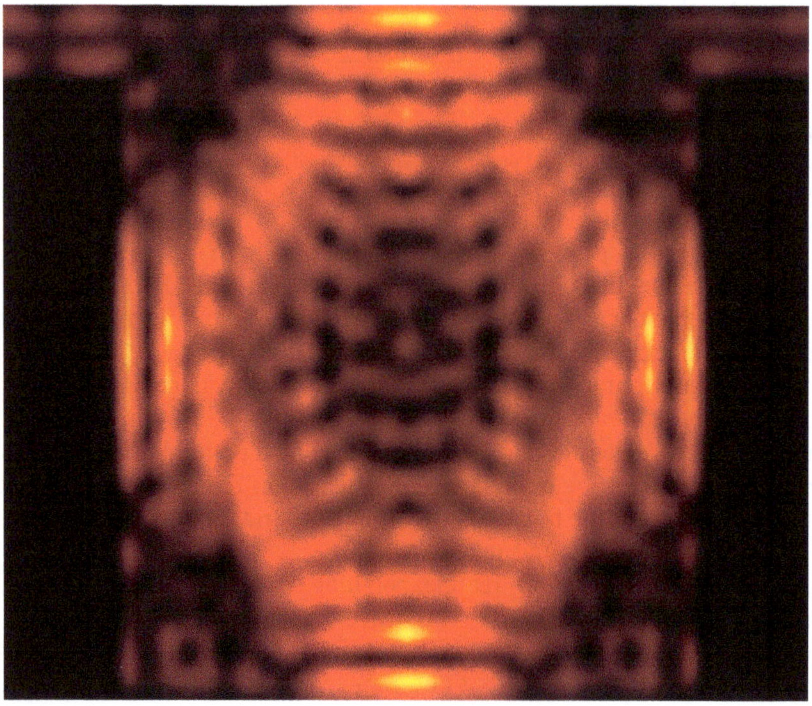

Figure 3.5. A diamond scar for a 0.3 μm quantum dot for which the QPCs are located at the top edge (these may be faintly seen in the image). This scar reappears throughout the spectrum of figure 1.4(b) along the quasi-resonance lines and is quite stable when the conditions correspond to one of the resonances. Reproduced from [35].

as the lattice vectors in momentum space for a periodic structure. In fact, however, the assumption of the Fourier series (3.26) already introduces periodicity (the Fourier integral should be used for aperiodic structures). In fact, the periodicity observed in figure 3.5 arises almost completely from the grid that is adopted for the simulation. Once a periodic grid is assumed, it is almost impossible to ignore the presence of this periodicity. This actually speaks to the advantage of using particles in real space for simulations, even for quantum systems, which is that such lattice-induced periodicities do not appear. On the other hand, condensed matter studies almost always begin with the periodic lattice of a crystal. It is also important that one recognizes that two-dimensional systems, such as the quantum dot discussed here, have some peculiarities that arise from this periodicity. Using the normal periodicity of the lattice (or of the grid), the two-dimensional surface can be folded into a two-torus (a doughnut shape). An important result of this is that the magnetic field forces any trajectory into certain dependences upon this periodicity. A stable wave function only arises if the trajectory closes upon itself (is non-chaotic), and this requires the number of turns around each axis to be rationally related [44, 45]. Hence, there are very few specific values of magnetic field and energy for which wave functions can actually be found. While this is a relatively dense set, the background chaos represents an even larger set. We can illustrate this by comparing classical and quantum simulations for the open dot.

An array of open quantum dots is considered in which the lithographically defined dimension of each dot is around 0.4 μm [46]. After edge depletion is accounted for, this leads to an electrically defined size of ~0.3 μm, as used above. The experiments were carried out at a carrier density of 2.4×10^{11} cm^{-2} ($E_F = 8.5$ meV) and an electron mobility of 1.24×10^6 cm^2 V^{-1} s^{-1}. It is fairly well known that the self-consistent potential due to the gates is approximately quadratic, even in the quantum situation, so that a confining potential for the dots may be assumed to be of the form [47]

$$V_{dot} = \frac{m^*}{2}\left[\omega_x^2(x - x_{0,i})^2 + \omega_y^2 y^2\right], \tag{3.28}$$

where the two frequencies are adjusted to give the proper dot size at the experimental Fermi energy, and $x_{0,i}$ is the center of the ith dot along the x-axis, which is the central line of the dot array. The QPCs are defined by a different potential which has the shape [47]

$$V_{QPC} = \frac{m^*}{2}\left[-\omega_{x,\,PQC}^2(x - x_{Q,j})^2 + \omega_y^2 y^2\right] + V_{0,\,QPC}, \tag{3.29}$$

which describes a saddle potential and for which the potential height at the center is the potential of the last term on the right. The x-frequency is adjusted to give the proper length of the QPC as well as to smoothly couple to the dot potential. The height of $V_{0,\,QPC}$ is adjusted to set the proper number of modes propagating through the QPC to the experimental values. The classical equations of motion for these potentials lead to a pair of coupled modes in a magnetic field, given by the frequencies [38] in a QPC

$$\omega_{\pm}^2 = -\beta' \pm \sqrt{\beta'^2 + \omega_{x,\,\mathrm{PQC}}^2 \omega_y^2}, \tag{3.30}$$

with

$$\beta' = \frac{1}{2}\left(\omega_{x,\,\mathrm{PQC}}^2 - \omega_y^2 - \omega_c^2\right). \tag{3.31}$$

In the dots themselves, these hybrid frequencies are given by

$$\omega_{\pm}^2 = \beta \pm \sqrt{\beta^2 - \omega_x^2 \omega_y^2}, \tag{3.32}$$

with

$$\beta = \frac{1}{2}\left(\omega_x^2 + \omega_y^2 + \omega_c^2\right). \tag{3.33}$$

Here, $\omega_c = eB/m^*$ is the cyclotron frequency. Detailing the frequencies is important, as there is an almost universal nature to the transport that results. Different effects occur for different ratios of the characteristic potential frequencies and the cyclotron frequency.

The classical Poincaré phase space plot is generated along the central line of the dot array ($y = 0$). In this plot, the x-position is plotted versus the x-velocity for the specific time at which the trajectory passes through this $y = 0$ plane with a positive velocity in the y-direction. A magnification of part of the first dot Poincaré phase space plot is shown in black and white in the left portion of figure 1.6 and in blue and white in the right portion of the figure. The random set of dots corresponds to the sea of chaos, where no periodic trajectories exist [47]. On the other hand, a set of closed lines exists along the central line near the right-hand end of the plot. These sets of concentric circles correspond to a set of energy levels for periodic orbits which are attracted to the 'center' at the center point of the circles, forming a Kolmogorov–Arnold–Moser (KAM) island [48]. This attractor corresponds to the set of pointer states given in figure 3.5. To show this, the wave function of figure 3.5 is converted to a Husimi plot which provides a Gaussian smoothing of the wave function [49]. The smoothing function is

$$\xi(x, y) = \left(\frac{1}{\pi\sigma^2}\right)^{1/4} \exp\left(-\frac{x^2 + y^2}{2\sigma^2} + i\boldsymbol{k} \cdot \boldsymbol{r}\right) \tag{3.34}$$

and the Husimi function becomes

$$H(\boldsymbol{r}, k) = \left| \int d\boldsymbol{r}' \psi(\boldsymbol{r}') \xi(\boldsymbol{r} - \boldsymbol{r}', \boldsymbol{k}) \right|^2. \tag{3.35}$$

This Husimi function is simply a positive-definite smoothing of the Wigner phase space function. The Husimi function is superimposed upon the Poincaré phase space plot in the right portion of figure 3.6 and appears as the colored parcels (red denotes the maximum amplitude). It is clear that the Husimi function is exactly centered upon the KAM island that arose from the classical simulations. As a quantum wave

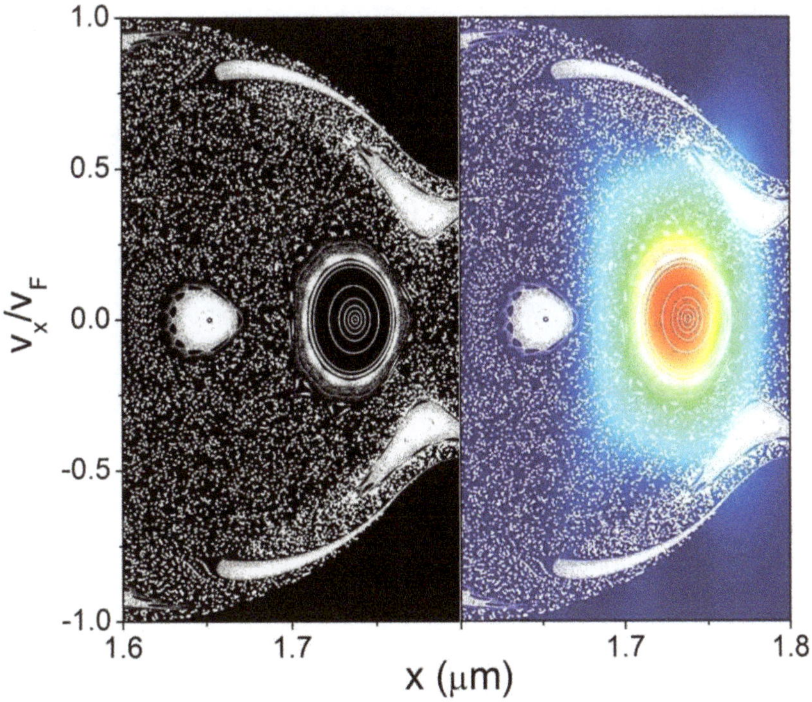

Figure 3.6. The Poincaré phase space plot (described in the text) for a portion of a single dot in an array. The black and white plot on the left and the blue and white plot on the right are identical. The color overlay represents the Husimi function (3.35) for the pointer state of figure 1.5, which lies exactly on the KAM island of stable closed trajectories. The comparison here clearly shows the connection between the classical state and the quantum state. Reproduced from [38].

function is much broader than a simple classical trajectory, so the Husimi function is more spread out than the simple points of the KAM island. Nevertheless, it is clear that the pointer state in figure 3.5 has connections to the classical trajectories in the KAM island.

But if the pointer states are isolated from the environment, how are they incorporated into the conductance? This is accomplished by the clearly nonclassical effect of phase space tunneling from an environmentally decohered state to/from the pointer state [50]. By varying the size of the dot and the applied gate voltage, agreement between the theoretical oscillatory frequency and the experimentally observed frequency was obtained. This agreement again confirms the connection between the pointer states and the measurement.

3.2.3 Environment-induced states

In addition to the natural set of pointer states in the dot array, there are back-scattering states that arise in the magnetoresistance, in which a particle follows a curved orbit and strikes the sides of the dot in a manner that leads to it being back-scattered into the entrance QPC [47, 51]. It is also possible for an internal

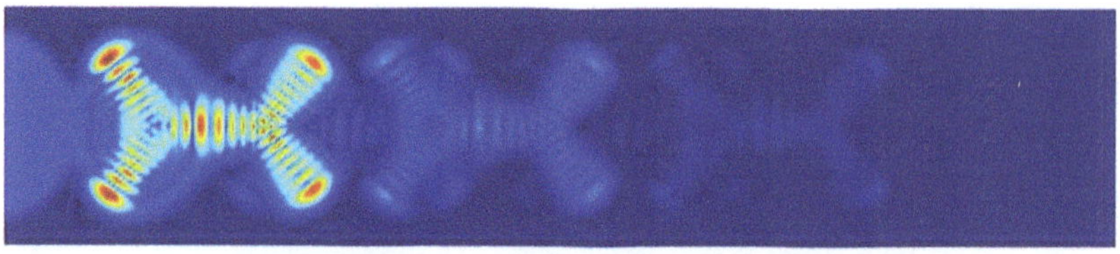

(a)

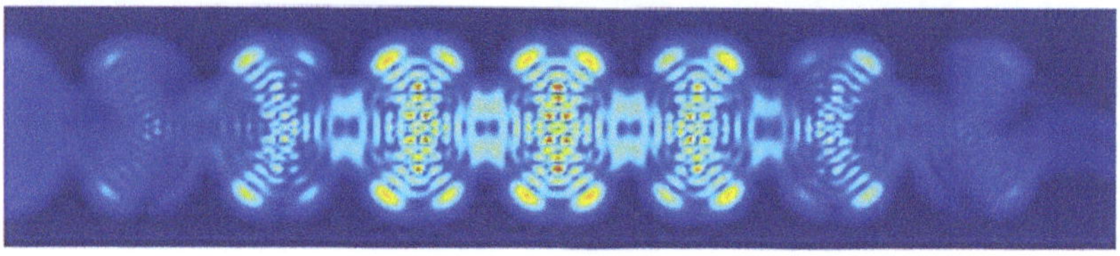

(b)

Figure 3.7. Environmentally induced quantum states in a seven-dot array [46]. (a) A bipartite state that couples the first and second dot, although remnants of an extension into the third dot may be seen as well. (b) A more complex state that couples dots 2–6 is shown. These dots are thought to be stabilized by quantum Darwinism, due to their existence throughout the group of modes propagating through the QPCs [52].

back-scattering state to occur, which results from a orbit trapped between two dots within the array. The instigation of such a trapped orbit arises from the presence of the environment [52]. In figure 3.7, such environmentally induced states are shown for two instances. Panel (a) shows a bipartite state that couples two adjacent dots in the dot array. A more complex state that actually couples five of the dots in the array is shown in panel (b). In particular, the bipartite state of panel (a) is quite stable and appears in each mode that propagates through the QPC, suggesting that it is in fact stabilized by quantum Darwinism [52]. Classical simulations show that a bow-tie-shaped closed orbit is centered on the QPC between the two dots of this panel. The frequencies for the classical and quantum orbits correspond and again show a strong correlation between the classical particle trajectories and the quantum wave functions.

3.3 Wave functions for open systems

Generally, one thinks of the wave function as being limited to a closed quantum system. An active region of an electronic device interacting with the environment is an open system. The quantum theories for open systems [53] are often based on the idea that an open system cannot, in general, be described by a wave function. The proper quantum tool to describe such a system is the so-called reduced density matrix. On the contrary, it is known that many electron transport simulation techniques (e.g. the scattering states in the Landauer model discussed in chapter 1) actually do describe electrons in open systems with wave functions. Fortunately, in many scenarios (but not always), the wave function of the open system can be interpreted as the physical quantum state of the open system but is 'conditioned' by some property of the environment. For example, conditional wave functions, from the Bohmian perspective, was discussed in section 3.1. The same concept appears in the so-called relative states, which are defined as states of the open system that assume a knowledge of the states of the environment (because of measurement or decoherence in the framework of the pointer states discussed in section 3.2). These ideas will be developed in section 3.4. For the moment, it is sufficient to conclude that there are reasonable arguments in most quantum scenarios to assume that a physical description of an open system is possible with wave functions.

To proceed, one has to think about the manner in which classical microwave systems are solved for the propagation of signals that are actually waves. In practice, one computes the propagating signal in the active region, and this is conditioned by a 'knowledge' of the properties of the signal in the input and output contacts. This uses the scattering matrix for microwave systems, which has been adapted to quantum waveguides. When the system of interest is large, it is often computationally more useful to use a recursive approach. In this situation, the solution starts with the boundary conditions that will be imposed on the system. The properties of each following slice are determined by those of the preceding slices, as the wave propagates from one contact region to the other. Here, the interface matching between adjacent slices is governed by the scattering matrix [54, 55], an adaptation borrowed from microwave theory, as mentioned. The transmission through the entire structure is computed at the contacts so that the conductance may be evaluated using the Landauer formula from section 2.3. Since this approach may be applied to systems whose dimensions vary along the channel, or in which scattering processes are considered, the transmissions are handled via (1.50) as well as the approach of section 2.3. This approach was used in all the quantum simulations of section 3.2.

At this point, a relevant question is to ask just how one 'knows' the properties of the wave function in the contacts for the Landauer approach? For a classical system, this question seems totally irrelevant because the propagating signal has some well-defined properties in the contacts. But in a quantum system, from an orthodox point of view, the extra knowledge of some properties of the wave function in the contacts (or the environment or to whatever the 'device' is connected) needs to be mediated by a type of measurement, as explained in the appendix of chapter 2. Somehow, as will be discussed in section 3.4, the Markovian assumption tries to solve these difficulties. At the end of the day, what can be discussed here overlooks the lack of microscopic properties that are provided by the orthodox theory.

3.3.1 The scattering matrix approach

The structure is considered to be composed of slices, according to the description of figure 1.6. The number of such slices is taken to be $N+2$, with each slice given an index s, which runs from 0 to $N+1$. The first and last slices are the boundary slices. Each slice has a transverse grid with $M+1$ grid points, although grid points 0 and $M+1$ are the lateral boundary points in which the wave function is assumed to vanish. Thus, the slice matrix is only of dimension $M \times 1$. The stability of this approach arises from the Lippmann–Schwinger equation [56], as the modal solutions to the Lippmann–

Schwinger equation maintain their orthogonality through the scattering process [56]. The slice wave function is written as

$$\psi_s = \begin{bmatrix} \psi(M, s) \\ \vdots \\ \psi(2, s) \\ \psi(1, s) \end{bmatrix}. \tag{3.36}$$

Note also that the single subscripted wave function is the *slice* wave function, as opposed to the *site* (value located on a grid point) wave function which has two variables. The Schrödinger equation can now be rewritten as a matrix equation that relates the slice vectors

$$H_{0,s}\psi_s - t\psi_{s+1} - t\psi_{s-1} = EI\psi_s, \tag{3.37}$$

where t is the diagonal matrix whose elements couple adjacent grid points and/or slices and are given by the so-called hopping energy

$$E_t = \frac{\hbar^2}{2m^*a^2} \tag{3.38}$$

(and these elements should not be confused with the transmission matrix elements, which have subscripts). In (3.37), the Hamiltonian for the isolated slice is written as

$$H_{0,s} = \begin{bmatrix} (V_{M,s} - 4E_t) & E_t & 0 & & 0 & \\ E_t & (V_{M-1,s} - 4E_t) & \cdot & & & \\ \cdot & \cdot & \cdot & & & \\ & & & \cdot & \cdot & 0 \\ & 0 & & \cdot & (V_{2,s} - 4E_t) & E_t \\ & & & \cdot & E_t & (V_{1,s} - 4E_t) \end{bmatrix}, \tag{3.39}$$

where the upper right block and lower left block are identically zero, and I is an identity matrix. The individual elements of the slice wave function are the corresponding site wave functions. These transverse wave functions for the width confinement are found from the corresponding slice Hamiltonian (3.39) with $s = -1$. This process produces the site wave functions and the slice wave function, which is doubly degenerate with the ability to propagate to the right and to the left. These wave functions are found by solving the equation

$$\det[T_0 - \lambda I] = 0, \tag{3.40}$$

where T_0 is a matrix whose rank is twice that of the Hamiltonian (3.39) and is given by

$$T_0 = \begin{bmatrix} U_+ & U_- \\ \lambda_+ U_+ & \lambda_- U_- \end{bmatrix}. \tag{3.41}$$

Here, $U_{\pm}$ are matrices whose rank is equal to that of the Hamiltonian and whose columns give the beginning slice wave functions in the site representation; they are found by solving the Schrödinger equation for the single slice. These matrices give the site-to-mode transformations of the various wave functions and also the modes of the transverse confinement. The positive and negative signs correspond to positive (right-directed) propagation and negative (left-directed) propagation. Similarly, the $\lambda_{\pm}$ are the corresponding eigenvalues. There is a corresponding Fermi energy (the simulation is at zero temperature at this point), so that modes whose eigenvalues are below the Fermi energy propagate, while those whose eigenvalues lie above the Fermi energy are evanescent. This is then propagated forward to slice zero. Once we obtain slice zero in the mode representation, we can proceed with the recursion.

To carry out the recursion, everything is put into the scattering matrix form. The site-to-mode transformation creates slice wave functions (3.36) in which each element corresponds to a mode in the system. These modes can be either propagating or evanescent. To proceed, two new matrices are created, and these are the 'scattering matrices' C_1^s and C_2^s at the general site s. The recursion relation for scattering matrices is expressed as

$$\begin{bmatrix} C_1^{s+1} & C_2^{s+1} \\ 0 & 1 \end{bmatrix} = \begin{bmatrix} 0 & I \\ -I & t^{-1}(H_{0,s} - EI) \end{bmatrix} \begin{bmatrix} C_1^s & C_2^s \\ 0 & 1 \end{bmatrix} \begin{bmatrix} I & 0 \\ P_{1s} & P_{2s} \end{bmatrix}, \tag{3.42}$$

where the first matrix on the right-hand side is denoted by T_s, and H_{0s} is the slice Hamiltonian (3.39). For the zero-temperature formulation, C_1^s is set to a unit matrix, but for a finite-temperature simulation, the individual elements are given by the value of the Fermi–Dirac distribution for the energy of the corresponding eigenvalue. Similar values are given to C_2^s *at the right-hand end of the system.* Equation (3.42) must now be satisfied, and this thus allows us to determine the propagation matrices P_{1s} and P_{2s} through the recursions

$$\begin{aligned} C_1^{s+1} &= P_{1s} = -P_{2s}(-I)C_1^s \\ C_2^{s+1} &= P_{2s} = \left(-IC_2^s + t^{-1}(H_{0,s} - EI)\right)^{-1}. \end{aligned} \tag{3.43}$$

The unit matrix (the $-I$ term) is actually the 2,1 element of the $t^{-1}t$ matrix, which is different if a magnetic field is applied and time reversal symmetry is broken.

At the end of the sample, the waves are transformed back into the mode representation, and the transmission is used to determine the conductance. One determines the wave function at the output in terms of the mode number and its values at the various sites. Thus, we determine that at site i and mode m, the contribution to the wave function is $\psi_N(i, m)$. This can now be back propagated to find the equivalent wave functions at each slice via

$$\psi_j(i, m) = P_1^j + P_2^j \psi_{j+1}(i, m), \tag{3.44}$$

so obviously this recursion now proceeds from right to left. The velocities for each mode are found from the imaginary part of the eigenvalue for that mode. At the same time, the density can be determined at each site from the square magnitude of the wave function at that site (initially, the normalization is to unity, but this can be understood to be in terms of the density in the initial contact) [40].

When we use a vanishingly small bias between the two reservoirs (between the source and the drain), then we need only compute the transmission and then use the Landauer formula. However, when a real, nonzero level of bias is applied, such as in a real semiconductor device, the situation becomes more difficult on several levels. First, we have to compute both a source-derived flow and a drain-derived backflow, just as for the MOSFET in equation (2.9). Both flows must be computed to determine the total wave function and the density at each grid point, since the nonzero bias drives the system out of equilibrium. This implies that we then need to use this density to self-consistently solve for the potential within the device, as the value of the potential at each grid point goes into the Hamiltonian equation (3.39). This creates an iterated loop that solves the quantum transport simulation and then solves for the self-consistent potential. This is repeated for each step until convergence is obtained. The scattering matrix approach is well suited to this process, as the density is a natural output of the simulation through equation (3.44). In a real device, we have to account for the temperature at each end of the device, as we can no longer assume that each mode has unity amplitude in the reservoir. As mentioned above, the amplitude is determined from a Fermi–Dirac distribution at the temperature of the source. The drain reservoir can be at a higher temperature, as this is ultimately where the last of the excess carrier energy is deposited.

3.3.2 An introduction to scattering

It is quite simple to modify the recursion method to include scattering by an on-site self-energy that is added to the diagonal elements of (3.39), just as in section 3.1. The self-energy represents the interactions that arise from various scattering processes. It has both real and imaginary parts and it is the latter that are of interest for dissipative scattering. In semiconductors, the scattering is generally weak and is traditionally treated by first-order, time-dependent perturbation theory. This process yields the common Fermi golden rule for scattering rates, which may generally be found in any quantum mechanics textbook. The real part of the self-energy can generally be ignored for the phonon interactions, and the real part that arises from the carrier–carrier interactions is incorporated into the solutions of Poisson's equation. In the many-body formulations of the self-energy, the latter is usually expressed as a two-site function [57]

$$\Sigma(\boldsymbol{r}_1, \boldsymbol{r}_2), \tag{3.45}$$

where the two positions are usually three-dimensional vectors. Time has not been shown here, but it can easily be added to account for the scattering time discussed in section 3.1.3. Since we are using transverse modes in the recursion, this expression is rewritten as

$$\Sigma\left(i, s; i', s'; \boldsymbol{r}_1, \boldsymbol{r}_2\right). \tag{3.46}$$

As usual, the i and s (and their primed versions) are the transverse site and mode, respectively, and correspondingly the x's represent positions along the nanowire. We can then introduce a center-of-mass transformation, as in (1.53), in position (here we use ξ for the difference variable, as s has been taken already to denote the slice)

$$\boldsymbol{R} = \frac{\boldsymbol{r}_1 + \boldsymbol{r}_2}{2}, \quad \xi = r_{1x} - r_{2x}, \tag{3.47}$$

and then Fourier transform on the difference variable to give

$$\Sigma(i, s; i', s'; \boldsymbol{R}, k_x) = \frac{1}{2\pi} \int d\xi e^{ik_x\xi} \Sigma(i, s; i', s'; \boldsymbol{R}, \xi). \tag{3.48}$$

The center-of-mass position $\boldsymbol{R}$ remains in the problem because the mode structure may change as one moves along the channel. The left-hand side of this last equation is the self-energy computed by the normal scattering processes as computed in quantum wells and quantum wires [12, 58]. It is sufficient to compute these using Fermi's golden rule, which is an evaluation of the bare self-energy in (3.48), rather than incorporating more elaborate many-body effects. Since the recursion is in the site representation, we have to reverse the Fourier transform in (3.48) to get the x-axis (along the wire) variation. The mode-to-site unitary transformation is then used to get the self-energy in the form necessary for the recursion. The imaginary part that arises from Fermi's golden rule gives the expression (for each scattering process)

$$\mathrm{Im}\left\{\Sigma(i, s; i', s'; \boldsymbol{R}, k_x)\right\} \equiv \hbar \left(\frac{1}{\tau}\right)_{i, s}^{i', s'}. \tag{3.49}$$

It is the latter calculated scattering rate that is a function of the x-directed momentum (which is related, in turn, to the energy of the carrier) which can be thought of locally as a quantum wire. This scattering rate must be converted to the site representation via a unitary transformation. At each site, the correction due to scattering is added to the local potential by the term

$$\Gamma(s, l, \eta) = \mathrm{Im}\{\Sigma\} = U_s^{\dagger} \left(\frac{\hbar}{\tau}\right)_{i, s}^{i', s'} U_s, \tag{3.50}$$

where U_s is the mode-to-site transformation matrix discussed in (3.41). The actual scattering rates are computed in the normal manner [59, 60]. This methodology, which is based upon the scattering matrix recursion technique, has been applied to a number of both ballistic and scattering approaches to gated silicon quantum wire structures with excellent results [61]. In these approaches, it was possible to directly determine the diffusive-to-ballistic crossover length.

3.3.3 The simulation of a small MOSFET

There are many effects that arise in short-channel MOSFETs. When the transport is fully quantum, then the transitions from the low-dimensional channel to the larger three-dimensional source and drain play a role. When one of these latter interfaces is crossed, there is a discontinuity in the carrier momentum, and since this occurs at both source and drain, quasi-bound states (often called resonances) can occur in the longitudinal wave function for carriers in the channel. At certain biases, the drain current can have large peaks that are almost like resonant tunneling peaks. These may be seen in figure 3.8, which is a quantum simulation of ballistic transport in a Si MOSFET with a 9.8 nm channel length [62] at room temperature. The channel cross-section is 18.5 nm wide and 6.5 nm deep. The channel area is doped as a *p*-type semiconductor at an impurity concentration of 5×10^{18} cm^{-3}. A quick calculation shows that there are only about six impurities, on average, in the channel. This causes the current to exhibit sizable fluctuations as potentials are varied [12]. The source and drain regions are doped as n-type semiconductors at a concentration of 6×10^{19} cm^{-3} and are 18.5 nm wide, 10.32 nm in length, and again 6.51 nm in depth. While the source, drain, and channel regions have an equal width and depth, there are still potential discontinuities at the transitions, so that there are still momentum mismatches, as discussed above.

In panel (a), the characteristics are shown for six different device doping configurations as the gate voltage is varied for a fixed drain voltage (10 mV); the dopants are randomly placed according to the doping concentration and the grid size in the simulation. The random peaks are the conductance fluctuations for atomistic doping when only a few dopants are present. Since there are so few dopants in the channel, their exact positions introduce considerable variability. Each

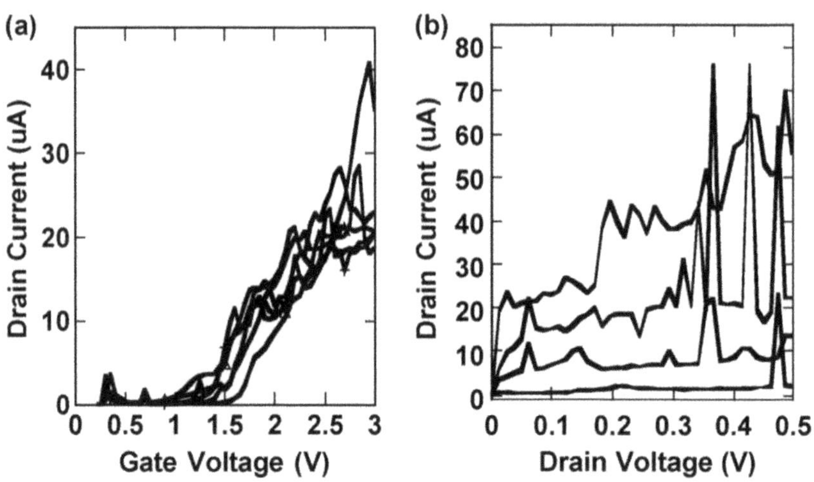

Figure 3.8. (a) Gate characteristics for six devices. The fluctuations are caused by interference due to the random positions of the dopants, discussed further in the text. (b) Drain characteristics for one device. The smaller fluctuations are due to the dopants, but the large peaks at the two lowest gate voltages are longitudinal resonances. © 2005 IEEE. Reprinted, with permission, from [62].

initialization of the doping concentration gives a different set of positions for these six dopants and therefore provides different fluctuation behavior. This is evident in the figure.

In panel (b), the drain characteristics are shown for a single device (that is, for one doping configuration) as the drain voltage is varied for a fixed gate voltage. Four different gate voltages (1.5, 2.0, 2.5, and 3.0 V) are shown. While the small fluctuations are caused by the dopant, the large peaks are the result of longitudinal resonant levels arising from the mismatches at the source and drain transitions. In a sense, these can be considered as source–drain resonant tunneling. These are effects which have to considered in device design, and efforts are needed to reduce such effects.

3.3.4 Simulation with particles

To our knowledge, very little work has used particles to simulate the quantum wave function in open systems. But there is no reason not to use particle approaches, as they may be even faster than the recursive approaches mentioned above. Of course, one particle method follows the hydrodynamic approach of section 1.2.3. In this latter approach, particles follow their natural trajectories under the constraints of both the classical potentials and the quantum potential (1.34). This means that the wave function must be found in order to determine the quantum potential, so this implies a particle method which is related to the wave function. However, it is a little more complicated than simply using particles to simulate the wave function. And it differs in that the hydrodynamic particles are real particles that are accompanied by the wave function, whereas the particles that simulate solutions to the Schrödinger equation are thought to be wave packets. Nevertheless, we can compare the two for a simple transverse barrier situation.

In figure 3.9, a quantum wave guide with a single transverse barrier is shown; the longitudinal (current direction) is labeled as x and the transverse direction indicated by y. The device has a width of 100 nm and a length of 300 nm [63]. The wave functions are found for this device by solving via the recursive approach above, and the quantum potential is found from these wave functions. Obviously, the barrier creates mismatches in the various modes, so it is important for that reason. In panel (a), particle trajectories are shown for their motion under both the confining and quantum potentials, while the amplitude of the quantum potential is indicated by the color background. The inhomogeneous nature of the quantum potential follows from the differing propagation of the several modes that can move in the areas away from the barrier. The Fermi energy in the simulation is 18.84 meV, which gives rise to five propagating modes. Only two modes can propagate through the constriction. One can see that none of the trajectories cross, which is a physical constraint even for ballistic motion in phase space. The trajectories also bend around the barrier, which one expects for the introduction of wave diffraction. Bunching of the particles is also obvious at the exit end of the structure. This bunching follows for the same reason as that shown in figure 1.1.

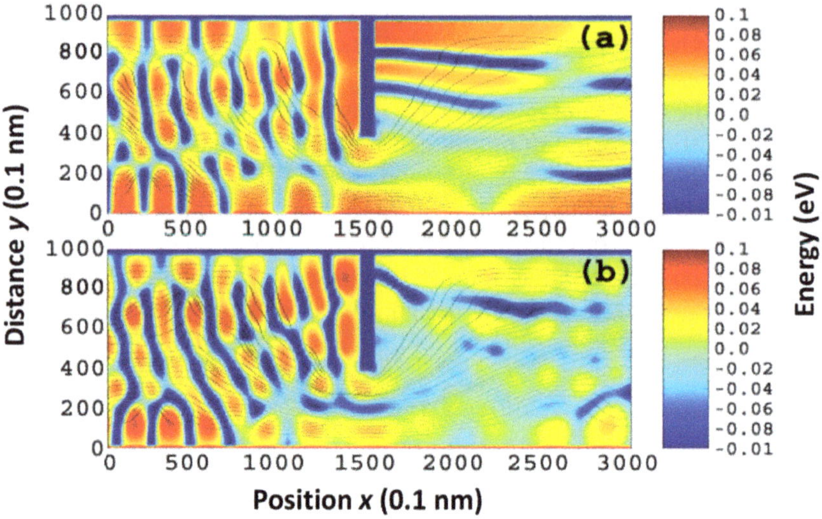

Figure 3.9. Particle trajectories computed from their individual initial velocities and the local and quantum potentials. (a) Trajectories for only the quantum and barrier potentials. The color-coded background represents the amplitude of the quantum potential, while the black lines are the trajectories of typical particles. (b) Trajectories that also are subject to the self-consistent potential arising from the local density variations. Reprinted from [63], Copyright (2000), with permission from Elsevier.

Obviously, the inhomogeneities in the quantum potential give rise to inhomogeneities in the density in the structure. In panel (b), the self-potential is determined from the local carrier density, and the simulation is repeated. The differences are small, but the bunching at the output is more pronounced. Moreover, the local variations in the trajectories are smoothed to some extent.

It is possible to actually simulate just the wave functions using particles. In this case, each of the particles in the ensemble is initialized at the entrance with a position, velocity, and mode number. Obviously, this ensemble is distributed through the various occupied modes (at energies near or below the Fermi energy) according to the Fermi–Dirac distribution. It is then allowed to propagate through the structure according to the confinement along the structure. Where there are discontinuities, such as at the barrier in figure 3.9, one has to move the carrier from a mode that no longer propagates. This means it has to transfer to either one of the propagating modes or to an evanescent mode, according to the probabilities set by the local Fermi–Dirac distribution. But for particles that transfer to an evanescent mode, a fraction must be transferred to another propagating mode as the distance from the discontinuity increases. Thus, far from the discontinuity, all the particles have returned to propagating modes. There is also a probability of back-scattering at the discontinuity, and the same rules apply for those particles propagating back toward the entrance (otherwise the transmission could not be computed). Nevertheless, the various transfers are governed by conservation of total energy and velocity across the interface. Notice that it is not momentum that is conserved across the interface, but velocity. This is to satisfy current continuity (the product of charge density and velocity). This velocity condition leads to the boundary condition between the regions to the left and right of the discontinuity [64]:

$$\frac{1}{m_{\rm L}^*}\frac{\partial \psi_{\rm L}}{\partial x}=\frac{1}{m_{\rm R}^*}\frac{\partial \psi_{\rm R}}{\partial x}, \tag{3.51}$$

where x is the direction normal to the interface and the m^* are the effective masses on the two sides of the discontinuity. This transfer boundary condition is required whether it is wave functions or particles that are traversing the interface. Obviously, in a particle simulation, the density of particles in the transverse position can simulate the sum of the various mode wave functions. Determining derivatives, though, is a difficult numerical task that will be treated in more detail in chapter 7.

3.4 Microscopic equations of motion for particles

As explained in section 2.5, the orthodox postulates place strong restrictions on the information available about the microscopic properties of electrons. In this regard, different strategies have been developed to overcome these restrictions. For example, the Landauer approach assumes that the wave function in the output reservoir is composed of transmitted plane waves whose behavior is proportional to e^{ikx}. An intransigent orthodox follower could ask how Landauer knows that the microscopic property of the momentum of the electron is $p = \hbar k$, if no momentum measurement has been conducted? The answer is that one understands e^{ikx} to be a wave function that properly defines the electron inside the device and is 'conditioned' to the fact that a measurement of the momentum of such particle gives the value $p = \hbar k$. But what happens if the measurement of the momentum is not effectively done?

As we will discuss below in section 3.4.4, the property of Markovian behavior for a quantum system relaxes the orthodox postulate of having no microscopic properties without measurement [53]. In addition, as pointed out in chapter 2, there are other quantum theories that accept the existence of the microscopic properties of electrons in the absence of measurement. This is a great advantage of these quantum theories without observers that is not always well appreciated. Before opening a discussion of Markovian behavior, the concept of a 'conditional' wave function that is offered by the Bohmian theory is addressed [65–67].

3.4.1 Bohmian conditional wave functions

In truth, the exact solution for the many-body wave function is numerically intractable, and alternative approximate solutions are needed. For example, the Born–Oppenheimer approximation, mentioned in section 3.1, separates the entire wave function into two parts, one wave function for electrons and one for atoms. When dealing with the wave function of the electrons, the positions of the atoms are no longer degrees of freedom but parameters given by their equilibrium (frozen) position. In this regard, the electronic wave function can be considered to be a wave function of electrons that is 'conditioned' to a particular set of atomic positions. But, as explained in section 3.1, the band structure (computed from the frozen position of the atoms) is an approximation that needs to be improved by some perturbative technique to recover the fact that atoms move around their equilibrium positions (this is the origin of phonons). Is there any fundamental rigorous and general way of performing such a separation of the degrees of freedom? This is exactly the question that will be answered in this section.

The previous Born–Oppenheimer approximation was inexact because atoms are not frozen but vibrate around their equilibrium positions. So, it seems that a better solution lies in the use of time-dependent trajectories for the atoms instead of frozen positions. However, such trajectories are not accepted as rigorous parameters within the orthodox framework. At the end of the day, the problem again is that the orthodox theory denies the possibility of having microscopic positions for the atoms (unless they are explicitly measured). Fortunately, we have seen in chapter 2 that there are other quantum theories in which particles have well-defined microscopic properties independently of the measurement. One of these quantum theories (without observers) is Bohmian mechanics [65, 67], in which the concept of a 'conditional' wave function appears 'naturally.'

The elements of Bohmian theory [66–68] that define a quantum system with N particles are two. The first element is the same many-body wave function solution of the Schrödinger equation that has been discussed in section 3.1, which is given as

$$\Psi(x_1, \dots, x_N, t) = \Psi(x_1, x_E, t). \tag{3.52}$$

To simplify the notation, the definition $x_E \equiv \{x_2, \dots, x_N\}$ is used when needed. The second element is a set of trajectories, one for each degree of freedom, denoted by

$$\left\{x_1^j(t), x_2^j(t), \dots, x_N^j(t)\right\} \equiv \left\{x_1^j(t), x_E^j(t)\right\}. \tag{3.53}$$

This wave function (3.52) can typically be prepared in the laboratory, while the trajectories in (3.53) cannot be fixed in the laboratory (unless the positions are measured, but then Ψ transforms into a product of delta functions). In some experiment j, the trajectory $x_1^j(t)$ is transmitted, while in another experiment j', the trajectory $x_1^{j'}(t)$ is reflected. In other words, the Bohmian theory is ontologically deterministic but empirically unpredictable [66, 67]. The different trajectories representing different experiments are indicated by the superindex j and selected at the initial time t_0 according to the probability distribution [66]

$$|\Psi(x_1, \dots, x_N, t_0)|^2 = \frac{1}{M}\sum_{j=1}^{M}\delta(x_1 - x_1^j(t_0))\cdots\delta(x_N - x_N^j(t_0)). \tag{3.54}$$

If the set of trajectories $\left\{x_1^j(t_0), x_E^j(t_0)\right\}$ for a large ensemble $j = 1, \dots, M \to \infty$ is selected from the probability distribution in (3.54) at time t_0, then the distribution $|\Psi(x_1, \dots, x_N, t)|^2$ can be satisfied at any other time t by counting the positions of the particles. The reason that the Bohm and Copenhagen theories are empirically equivalent is due to the equivariance condition [67, 68] implicit in (3.54).

The relevant point in the discussion here is that the Bohmian theory provides the wave function (3.52) plus the positions of the particles in (3.53). The latter are not present in the orthodox machinery. From this additional knowledge of the positions, one can directly construct the so-called Bohmian conditional wave function [65–67], as follows:

$$\varphi_1^j(x_1, t) = \Psi(x_1, x_E^j(t), t) = \Psi(x_1, x_2^j(t), \ldots, x_N^j(t), t). \tag{3.55}$$

The wave function $\varphi_1^j(x_1, t)$ is a slice of the many-body wave function $\Psi(x_1, x_E, \mathrm{t})$ obtained by evaluating all degrees of freedom along the (Bohmian) trajectory while leaving x_1 unevaluated. This conditional wave function $\varphi_1^j(x_1, t)$ has the great merit of giving exactly the same velocity for the particle x_1 that one gets from the total wave function $\Psi(x_1, x_E, \mathrm{t})$ [65–67]. In polar form, $\varphi_1^j(x_1, t) = R_1^j(x_1, t)e^{iS_1^j(x_1, t)/\hbar}$ with $S_1^j(x_1, t) = \mathrm{S}(x_1, x_E^j(t), t)$ is the phase of the total wave function $\Psi(x_1, x_E, \mathrm{t})$, and one has

$$v_1^j(x_1^j(t), t) = \frac{1}{m}\frac{\partial S_1^j(x_1, t)}{\partial x_1}\Bigg|_{x_1 = x_1^j(t)} = \frac{1}{m}\frac{\partial \mathrm{S}(x_1, x_E, t)}{\partial x_1}\Bigg|_{x_1 = x_1^j(t); x_E = x_E^j(t)}. \tag{3.56}$$

The velocity depends only on the spatial derivative in the x_1-direction, which can be equivalently computed from $\varphi_1^j(x_1, t)$ or $\Psi(x_1, x_E, \mathrm{t})$. If we consider the degree of freedom x_1 to be a system and the rest as the environment, then $\varphi_1^j(x_1, t)$ can be considered to be a wave function of an open system in the most formal way. The fundamental reason that it is rigorous to call $\varphi_1^j(x_1, t)$ a wave function of an open system is because the trajectories in (3.53) are ontologically well-defined for any circumstance within Bohmian mechanics [65–67].

The exact equation of motion of the conditional wave function $\varphi_1^j(x_1, t)$ in (3.55) has been shown to be given by [69]

$$i\hbar\frac{\partial \varphi_1^j(x_1, t)}{\partial t} = \left[-\frac{\hbar^2}{2m}\frac{\partial^2}{\partial x_1^2} + U_1(x_1, x_E^j(t)) + G_1(x_1, x_E^j(t), t) + iJ_1(x_1, x_E^j(t), t)\right]\varphi_1^j(x_1, t), \tag{3.57}$$

where $U_1(x_1, x_E^j(t))$ is the potential appearing in the Hamiltonian that depends on the set of all particles. The other potentials, $G_1(x_1, x_E^j(t), t)$ and $J_1(x_1, x_E^j(t), t)$, also depend on all the particles and they are quantum correlation potentials defined in [69]. The evaluation of both $G_1(x_1, x_E^j(t), t)$ and $J_1(x_1, x_E^j(t), t)$ needs the derivative of the full wave function Ψ along the environmental coordinates, so that educated guesses are needed for the practical use of (3.57) [69]. In section 10.5 of chapter 10, some practical examples of the application of the conditional wave functions to time-resolved quantum transport will be presented.

3.4.2 The reduced density matrix and the statistical density matrix

The idea of a conditional wave function is so useful that it reappears in different ways in other approaches. Here, a formal strategy to recover this idea without invoking

Bohmian mechanics will be explained using the concept of a *relative* state [70]. To develop the explanation, the many-body wave function in (3.52) is rewritten here as

$$\Psi(x, x_E, t) = \sum_{i,j} c_{i,j}(t)\varphi_{S,i}(x, t)\phi_{E,j}(x_E, t). \tag{3.58}$$

The label $x \equiv x_1$ is used unless identification of the particle is needed. The wave functions $\varphi_{S,i}(x, t)$ refer to the system and $\phi_{E,j}(x_E, t)$ to the remainder of the degrees of freedom that form the environment. The variables x and x_E can be interpreted either as two degrees of freedom or as two vectors encoding many degrees of freedom. Since an orthodox quantum state is exclusively defined by a wave function (with no additional *hidden* variables), all experiments are modeled using the same wave function and there is no need for a superindex (such as those used for the Bohmian conditional wave function).

When (3.58) is used, the expectation value of an arbitrary operator $A(x)$ acting on the system, $\langle A\rangle = \int dx\, dx_E \Psi^*(x, x_E, t)A(x)\Psi(x, x_E, t)$, can be written as

$$\begin{aligned}\langle A\rangle = \sum_{i,j,i',j'} c^*_{i',j'}(t)c_{i,j}(t) \int dx_1\; \varphi^*_{S,i'}(x, t)A(x)\varphi_{S,i}(x, t)\\ \times \int dx_E \phi^*_{E,j'}(x_E, t)\phi_{E,j}(x_E, t)\end{aligned}. \tag{3.59}$$

There is entanglement between the system and the environment in (3.58) in the sense that whole wave function cannot be written as a product of one system wave function multiplied by one environmental wave function [53]. The system and its environment are inseparable. In these circumstances, the formal orthodox description of the quantum system alone (without the degrees of freedom of the environment) requires the use of the so-called reduced density matrix

$$\begin{aligned}\rho_{red}(x', x, t) &= \int dx_E \Psi^*(x', x_E, t)\Psi(x, x_E, t)\\ &= \sum_{i,i'} \varphi^*_{S,i'}(x, t)\varphi_{S,i}(x, t)a_{i',i}(t)\end{aligned}, \tag{3.60}$$

where $a_{i',i}(t) = \sum_{j,j'} c^*_{i',j'}(t)c_{i,j}(t) \int dx_E \phi^*_{E,j'}(x_E, t)\phi_{E,j}(x_E, t)$. Within the reduced density matrix, it is a straightforward procedure to show that

$$\langle A\rangle = \int dx\; \left[A(x)\rho_{red}(x', x, t)\right]_{x'=x}. \tag{3.61}$$

The different summations present in $\rho_{red}(x', x, t)$ in (3.60) are identified by the subindexes i and i'. One can now write a matrix using i as an index in the rows and i' in the columns. In this matrix, the diagonal elements $i = i'$ are termed the *populations* (or occupation probabilities) and are given by $|\varphi_{S,i}(x, t)|^2\, a_{i,i}(t)$. The off-diagonal terms for $i \neq i'$ are called the *coherences* (or *entanglements*) and are given by $\varphi^*_{S,i'}(x, t)\phi_{S,i}(x, t)a_{i',i}(t)$. Since the populations have a much simpler physical interpretation than the coherences, it is reasonable to wonder whether it possible to describe an open system using a density matrix (3.60) with populations only? As explained below, the concept of relative state (borrowed from Zurek's work [27, 28] and Everett's relative state interpretation [71, 72]) can help in this approach.

Some extra information about the system can be obtained from the state $\Psi(x, x_E, t)$ in (3.58) if one knows that the environment is described only by the state $\phi_{E,j0}(x_E, t)$. The many-body wave function (3.58), which is conditioned to this extra information about the environment, can then be written as

$$\begin{aligned}\Psi_{j0}(x, x_E, t) &= \left(\sum_i c_{i,j0}(t)\varphi_{S,i}(x, t)\right)\phi_{E,j0}(x_E, t) \\ &= \phi_{S,j0}(x, t)\phi_{E,j0}(x_E, t)\end{aligned} \quad (3.62)$$

where the state of the system *conditioned* to this new information about the environment is the *relative state* [53, 70–72] given by $\phi_{S,j0}(x, t) = \sum_i c_{i,j0}(t)\varphi_{S,i}(x, t)$. In fact, using (3.62), one can rewrite the original state $\Psi(x_1, x_E, t)$ in (3.58) as

$$\Psi(x, x_E, t) = \sum_j \Psi_j(x, x_E, t) = \sum_j \phi_{S,j}(x, t)\phi_{E,j}(x_E, t). \quad (3.63)$$

The many-body wave function $\Psi(x, x_E, t)$ in (3.63) is just a mathematical (Schmidt) decomposition into different 'channels' or 'branches' and has no new physical interpretation.

However, one may easily realize that expression (3.63) has the same (premeasurement) structure as the many-body wave function discussed in the 'channelization' of the wave function during a measurement (see expression (A.2) in an appendix of chapter 2). If $\phi_{E,j0}(x_E, t)$ in (3.62) corresponds to the state linked to the pointer state of the measuring apparatus, indicating the value $j0$, then the relative state $\phi_{S,j0}(x, t) = \sum_i c_{i,j0}(t)\varphi_{S,i}(x, t)$ corresponds to the physical state describing the system after that measurement. Since a measurement assumes that there is no further interference between the different 'channels,' there is no interference between the different relative states $\phi_{S,j}(x_1, t)$ present in (3.63). This means that the reduced density matrix in (3.63), written with the index j in the rows and j' in the columns, has no coherences, only populations

$$\rho_{\text{stat}}(x', x, t) = \sum_j \phi^*_{S,j}(x', t)\phi_{S,j}(x, t). \quad (3.64)$$

where $a_{j',j}(t) = \int dx_E \phi^*_{E,j'}(x_E, t)\phi_{E,j}(x_E, t) = \delta_{j,j'}$, because there is no overlap between different 'channels.' This type of reduced density matrix (3.64) is, in fact, named a statistical density matrix [53]. It physically corresponds to a system that is described by $\phi_{S,1}(x, t)$ in some experiments and by $\phi_{S,2}(x, t)$,in others. Generally, for an open quantum system that is not measured, $\rho_{\text{red}}(x', x, t) \neq \rho_{\text{stat}}(x', x, t)$, and one needs to use (3.60) instead of (3.64) unless additional arguments can be used to reach (3.60), as will be discussed below.

3.4.3 Thermalization, decoherence, and pointer states

A careful reader will have noticed that in the transition $\rho_{\text{red}}(x', x, t) \rightarrow \rho_{\text{stat}}(x', x, t)$, from (3.60) to (3.64), the fundamental ingredient is, in fact, not the process of measurement itself but the assumption that the states of the environment, $\phi_{\text{E},j}(x_{\text{E}}, t)$ and $\phi_{\text{E},j'}(x_{\text{E}}, t)$, do not interfere. In other words, the prerequisite for reaching a physical description of an open system as a statistical matrix is that the different j-'channels' must not interfere with each other. Thus, one may wonder whether the transition $\rho_{\text{red}}(x', x, t) \rightarrow \rho_{\text{stat}}(x', x, t)$ can be formulated without invoking the measurement process but instead by using the modern theories of quantum thermalization or decoherence [53, 70].

In fact, this point of view on the importance of decoherence in determining the evolution of systems entangled with environments has already been explained in section 3.1, which mentioned the works of Zurek on that topic [27–29]. A particular example is discussed in section 3.2. Here, a much simpler example related to thermalization (decoherence) will be presented. In this simple example, the environment alone is a closed or isolated system that does not exchange particles or energy with the outside world. Thus, the many-body wave function for the environment alone (ignoring the system degree of freedom) can be written here as

$$\Psi(x_{\text{E}}, t) = \sum_j e^{-iE_j t/\hbar} c_j \varphi_j(x_{\text{E}}), \tag{3.65}$$

where $\left\{e^{iE_j t/\hbar}\varphi_j(x_{\text{E}})\right\}$ is the set of energy eigenstates of the environment that forms a basis for the states. The density matrix of (3.65) can be written in this energy representation using $c_j c_{j'}^* = e^{i\theta_{j,j'}} \left| c_j c_{j'}^* \right|$, as

$$\rho(x_{\text{E}}', x_{\text{E}}) = \sum_{j,j'} e^{i(E_{j'}-E_j)t/\hbar} e^{i\theta_{j,j'}} \left| c_j c_{j'}^* \right| \varphi_{j'}^*(x_{\text{E}}')\varphi_j(x_{\text{E}}), \tag{3.66}$$

where the indexes j, j' are now used to indicate the rows and columns of the matrix. The diagonal terms ($j = j'$) are time-independent populations, while the off-diagonal terms ($j \neq j'$) are time-dependent coherences. At time $t = 0$, one gets $e^{i(E_{j'}-E_j)t/\hbar} = 1$ in (3.66) so that the different $e^{i\theta_{j,j'}}$ determine the initial shape of the wave function in (3.65). These phases $\theta_{j,j'}$ are a 'memory' of the initial state encapsulated in the ($j \neq j'$) coherences. If the number of energy eigenstates in the reservoir is large enough and the separation between eigenenergies $E_{j'} - E_j$ is not periodic (but random), after some time (termed the equilibration time) t_{eq}, one can consider that the different terms in the coherences are basically random numbers around zero $e^{i(E_{j'}-E_j)t/\hbar+\theta_{j,j'}} = \cos\left((E_{j'} - E_j)t_{\text{eq}}/\hbar + \theta_{j,j'}\right) + i\sin((E_{j'} - E_j)t_{\text{eq}}/\hbar + \theta_{j,j'})$, where the information $\theta_{j,j'}$ is just *hidden* in these random numbers. It is important to emphasize that the different coherences in (3.66) do not become zero for $t < t_{eq}$, they just become random with no perceptible correlation among them.

The expectation value of an arbitrary operator $A(x_E)$ given by $\langle A \rangle = \int dx_E \Psi^*(x_E, t) A(x_E) \Psi(x_E, t)$ can be rewritten as

$$\langle A \rangle = \sum_j A_j \mid c_j \mid^2 + \sum_{j, j' = j} e^{i(E_{j'} - E_j)t/\hbar} e^{i\theta_{j,j'}} \mid c_j c_{j'}^* \mid A_{j,j'}, \tag{3.67}$$

with $A_j = \int dx_E \varphi_j^*(x_E) A(x_E) \varphi_j(x_E)$ and $A_{j,j'} = \int dx_E \varphi_{j'}^*(x_E) A(x_E) \varphi_j(x_E)$. One can now apply the fact that the coherences of the density matrix become basically random numbers for $t > t_{eq}$ to conclude that (3.67) is a sum of random (positive and negative) numbers; therefore, $\sum_{j,j'=j} e^{i(E_{j'} - E_j)t/\hbar} e^{i\theta_{j,j'}} \mid c_j c_{j'}^* \mid A_{j,j'} \approx 0$. Thus, the expectation value for $t > t_{eq}$ becomes $\langle A \rangle \approx \sum_j A_j$. The final expectation value is time independent, and it is said that the system has thermalized after $t > t_{eq}$ [53, 73]. One can then easily find that the thermalized value of the same expectation values $\langle A \rangle \approx \sum_j A_j$ can be obtained using the expression

$$\langle A \rangle \approx \sum_j A_j = \int dx_E \left[A(x_E) \rho_{stat}(x_E', x_E) \right]_{x'=x}, \tag{3.68}$$

with the definition of the statistical density matrix [53] given by

$$\rho_{stat}(x_E', x_E) = \sum_j \mid c_j \mid^2 \varphi_j^*(x_E') \varphi_j(x_E). \tag{3.69}$$

This statistical density matrix in (3.69) has only the diagonal elements ($j = j'$) of the original density matrix in (3.66). Notice that the information $\theta_{j,j'}$ about the initial shape of the quantum system has disappeared in (3.69). In this sense, one can say that the environment eliminates the 'memory' of the initial state.

At first sight, the reader may find an apparent paradox in the fact that the Schrödinger equation governing the evolution of the wave function (3.65) is a microscopic time-reversible equation, while the evolution from (3.65) to (3.69) is time irreversible. This apparent paradox disappears because the physical density matrix of the closed system in (3.66) is still computed from the time-reversible wave function (3.65) with nonzero coherences. What has been demonstrated here is that, *for all practical purposes*, when dealing with the evaluation of expectation values after thermalization, which gives $\langle A \rangle \approx \sum_j A_j$, one can use $\rho_{stat}(x_E', x_E)$ as the density matrix describing the environment [73].

As mentioned in section 3.1, Zurek [27–29] and others [53, 70] developed more sophisticated scenarios that included the system, the measuring apparatus, and the environment in order to show how particular 'channels' $\phi_{S,j}(x, t) \phi_{E,j}(x_E, t)$ in (3.63) evolve and become disentangled from other 'channels' using arguments about decoherence similar to the ones mentioned above. In simple words, the reduced density matrix can be converted into a statistical density matrix not because of any collapse law, but because of the decoherence. When the operator of the measuring apparatus commutes with the Hamiltonian of the apparatus/environment, the states $\phi_{E,j}(x_E, t)$ are named pointer states [53, 70] because they remain for all time as an indication of the measuring apparatus (see also the discussion in section 3.1).

Finally, it is important to mention that decoherence alone does not explain the entire measurement process that was explicitly described in chapter 2. The statistical density matrices in (3.64) or (3.69) include all possible relative states of the system, but in a measurement, only one state remains. In simple words, the statistical density matrix is sufficient to evaluate the expectation value over an ensemble of measured experiments but not to discern which state of the system results in an individual experiment [74]. As explained in chapter 2, in the orthodox language, the collapse law (randomly) specifies which final state is obtained in a particular experiment. In the Bohmian approach, it is the trajectory arising from one 'channel' and not from other non-overlapping 'channels' that specifies the relevant state after a measurement [66, 67].

3.4.4 Markovian and non-Markovian open systems

One has now all the elements necessary to discuss the concept of quantum Markovian behavior [53, 75]. The Markovian description of an open system assumes that the evolution of a reduced density matrix can be described as a quantum state suffering, at every time interval Δt, a process of decoherence as described in section 3.4.3 (which can also be understood to be a hypothetical measurement process). Thus, the system can be described by a reduced density matrix that evolves without interference among the 'channels' so that $\rho_{red}(x', x, t) \approx \rho_{stat}(x', x, t)$ from time t to time $t + \Delta t$. The tendency of the system to mix different 'channels' is avoided by another decoherence phenomenon that occurs at time $t + \Delta t$ such that $\rho_{red}(x', x, t) \approx \rho_{stat}(x', x, t)$ also during the time from $t + \Delta t$ to $t + 2\Delta t$. Of course, the time interval Δt has to be larger than the time t_{eq} needed to thermalize so that it is reasonable to assume that $\rho_{red}(x', x, t) \approx \rho_{stat}(x', x, t)$ most of the time.

In fact, such continuous monitoring of pointer states coupled to different relative states can be used to derive the so-called Lindblad master equations [75], which are dynamical equations for the reduced density matrix of a system in a Markovian environment. A Markovian system also allows us to develop the stochastic Schrödinger equations [76]. From time t to $t + \Delta t$, the system can be described by a relative state $\phi_{S,j}(x, t)$ that changes to another state $\phi_{S,j'}(x, t)$ at time $t + \Delta t$ due to the decoherence phenomena (a hypothetical measurement). An ensemble over the different $\phi_{S,j}(x, t)$ present in (3.64) recovers the evolution of the reduced density matrix. Note that the subindex j changes over time, providing additional microscopic information. Such a pure state trajectory for a Markovian environment can always be physically interpreted in the orthodox explanation through a pure unraveling of the reduced density matrix. In the Bohmian view, these states $\phi_{S,j}(x, t)$ are, in fact, the conditional wave functions mentioned in (3.55) [76–78].

The essential element in the description of a Markovian system is the fact that the different 'channels' do not interfere. Thus, the evolution of $\phi_{S,j}(x, t)$ is not suddenly affected by another state $\phi_{S,j'}(x, t)$ from another 'channel.' In this sense, the evolution of the system has no 'memory' of the other 'channels.' However, what if we had an environment that did not meet the condition of no interference between different 'channels' implicit in (3.69) or (3.64)? The process of decoherence $\rho_{red}(x', x, t) \rightarrow \rho_{stat}(x', x, t)$ now needs some time t_{eq} to be effective, as seen in (3.67). Thus, the time-resolved description of an open system with a very small time step (high-frequency resolution) cannot be considered to be a Markovian one [73, 79]. It is important to mention that the concept of the conditional wave functions $\varphi_1^j(x_1, t)$ discussed in (3.55) is always valid, since the extra information about the trajectories in (3.53) used to define it does not depend at all on the fact of the environment being Markovian or not or whether the time resolution of the system in the modeling is very small. This a great advantage of the Bohmian conditional wave function which seems to remain mainly unnoticed in the literature [76–78].

3.5 The macroscopic world and thermodynamics

In section 3.4, an important effort has been made to show how we can acquire additional information about the microscopic properties of electrons. However, too much microscopic information can also be inconvenient in the modeling of some devices. For example, in section 3.3, it is shown that the statistical density matrix in (3.69), without all the microscopic information of the non-diagonal elements present in the reduced density matrix in (3.66), is enough to reproduce all the expectation values after some thermalization time. The idea of disregarding microscopic details while only keeping information that can be used to provide good predictions of expectation values is at the origin of statistical mechanics and thermodynamics.

Thermodynamics is a scientific discipline that explains complex systems through macroscopic properties, avoiding the need to discuss microscopic details. Historically, the thermodynamic laws were developed only for systems in so-called thermodynamic equilibrium. In recent years, however, thermodynamics as a scientific theory has evolved to systems outside thermodynamic equilibrium [80]. The so-called classical irreversible thermodynamics, under the hypothesis of local equilibrium, borrows most of the concepts and laws of equilibrium thermodynamics and applies them to nonequilibrium systems. Thus, given the flexibility of thermodynamics as a scientific discipline, it is always possible to construct a branch of thermodynamics that has the ability to predict almost everything. In other words, thermodynamics is becoming a *science of everything* [81]. In any case, it has the great merit of being able to explain complex phenomena occurring in systems with densities of 10^{22} particles/cm^3 using very few macroscopic parameters.

The typical strategy used in most classical or quantum transport models lies in dividing the entire device into a reservoir and an active region, as depicted in figure 3.10. Figures 2.1 and 2.2 in chapter 2 as just particulars examples within the

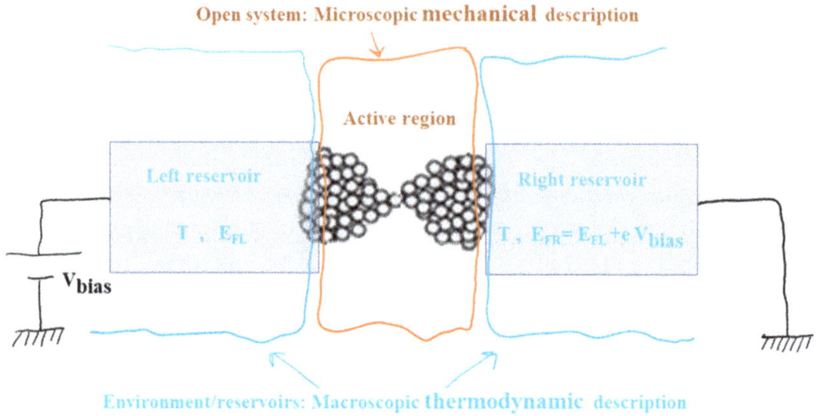

Figure 3.10. The common strategy for modeling electronic devices is to approach the dynamics of electrons in the active region using fundamental classical or quantum mechanical laws. The reservoirs, typically modeled through thermodynamic laws, determine the flux of particles and energy between the active region and the exterior.

Landauer model. For some devices, more reservoirs can be connected to the active region without changing the arguments presented in this section. One can expect that the dynamics of electrons in the active region is more convoluted than the behavior of electrons in the reservoirs, so that careful modeling of electrons is needed in the active region. Typically, one avoids the use of thermodynamic arguments in the active region and prefers the use of fundamental classical or quantum laws. Basically, this book focused on how to model the dynamics of electrons in the active region using fundamental quantum laws. The reservoirs, however, provide the boundary conditions for the flux of energy and particles from the active region to the exterior. The modeling of reservoirs is also necessary, and it is usually done through thermodynamics parameters.

The reservoirs (also called contacts in the literature as well as in this book) are typically much bigger than the active region; they therefore involve such huge numbers of electrons that their direct modeling is impossible. The battery depicted in figure 3.10 implies that there is a net current circulating through the reservoirs. Thus, strictly speaking, the reservoirs are not systems in thermal equilibrium. However, as mentioned above, the thermodynamics arguments are flexible enough that they can also be applied to scenarios outside thermodynamic equilibrium. One can reasonably assume that the electrons in the reservoirs and the battery have a behavior quite similar to that of the electrons in a thermal reservoir described in section 3.4. It is said that reservoirs are in quasi-thermal equilibrium. In fact, because of the different potentials at the borders of the battery, given an applied bias V_{bias}, the potential energy of an electron in the left reservoir is higher than in the right reservoir.

Strictly speaking, the thermalized density matrix described in (3.69) is an equilibrated density matrix, not a thermalized density matrix [73]. The discussion in section 3.3 addresses a closed (or isolated) system, i.e. one that does not interchange energy or particles with its surroundings. Thus, the expected probability of the different eigenstates in (3.69) after $t > t_{eq}$, corresponding to a thermal density matrix, is a microcanonical distribution. Strictly speaking, the simple and didactic development given in section 3.3 only shows that $\langle A \rangle \approx \sum_j A_j$ after $t > t_{eq}$ but not that the distribution of the different energy eigenstates follows a microcanonical distribution, that is $|c_j|^2 = cte$. In the microcanonical regime, the coherence becomes irrelevant and the thermalized density matrix has no 'memory' at all of the initial state $\Psi(x, x_E, t)$ in (3.65). When the system becomes thermalized after an exchange of energy with its surroundings, the probabilities of the different energy eigenstates of the statistical density matrix are expected to follow a canonical distribution. Finally, a grand canonical ensemble (also known as the macrocanonical ensemble) represents the density matrix of a thermalized system that interchanges particles and energy with its surroundings. Since the reservoirs in figure 3.10 interchange energy and particles with the active region, this last grand canonical ensemble is the one that better characterizes the process of thermalization/decoherence of the reservoirs in electronic devices.

When we use the description of reservoirs given by a grand canonical (or macrocanonical) ensemble, we assume that the distribution of the different energy states of the density matrix of the reservoir are described by a Fermi–Dirac distribution, which is characterized by a quasi-Fermi level and a temperature depicted in figure 3.10. In fact, because of the battery with a potential of V_{BIAS} mentioned previously, the left-hand quasi-Fermi level E_{FL} is not equal to the right-hand quasi-Fermi level $E_{FR} = E_{FL} + eV_{BIAS}$, as depicted in figure 3.10.

Some final remarks seem necessary here. The 'thermalized' density matrix in (3.69) is just an approximation to the real reduced density matrix in (3.66). In other words, the non-diagonal elements of the density matrix in (3.66) do not disappear; they just become random, and finally irrelevant, for the computation of expectation values. However, the non-diagonal elements are still there as part of the description of the full wave function in (3.52). In other words, the microscopic description given by (3.66), or by the many-body wave function $\Psi(x, x_E, t)$, is the correct description of the quantum system at the most fundamental level. This remark is pertinent because sometimes in the literature there are arguments that seem to indicate that all reservoirs (or contacts) depicted in figure 3.10 have to be 'thermal' reservoirs, by construction, under all circumstances. This is not always true [79]. For example, in sections 3.3 and 3.4, it has been shown that thermalization is a process that requires some time to complete. Thus, a reservoir biased with a very high-frequency AC voltage cannot be a thermalized system because it has no time to complete thermalization during a significant change in the external conditions. Even at a low frequency, if we want to model time-dependent electrons inside the active region, the thermalized density matrix can be misleading. It seems from the expression $\rho_{stat}(x_E', x_E) = \sum_j \varphi_j^*(x_E')\varphi_j(x_E)$ that each electron in the reservoir is perfectly described by an energy eigenstate. Thus, it also seems that an electron inside the active region should have a time-independent energy eigenstate. This modeling can be reasonable when looking for DC predictions, but it becomes unphysical when we model quantum transport using a time resolution comparable to the electron transit time inside the active region. In such cases, one needs to remember that the correct physical description of the electron is given by $\Psi(x, x_E, t)$ in (3.52), where coherences are still relevant. A time-dependent conditional wave function is necessary to describe the high-frequency behavior of electrons inside the active region. Another example for which thermalized reservoirs are not pertinent is in the modeling of devices used for quantum computing, for which the most important issue is avoiding decoherence, which means preventing the reservoirs from behaving like thermal reservoirs. In any case, the design of electronic devices without the assumption of thermal reservoirs seems to be a quite interesting and unexplored avenue.

References

[1] Prigogine I 1955 *Introduction to the Thermodynamics of Irreversible Processes* (Chicago, IL: Thomas C C Publisher)
[2] Faraday M 1833 *Experimental Researches in Electricity* **Ser. IV** 433–9 paragraphs repub lished as *Phil. Trans. Roy. Soc. (London)* **123** 519
[3] Braun F 1874 *Ann. Phys. Pogg.* **153** 556
[4] von Hippel E 1933 *Zeitschr. Elekrochem.* **39** 506
[5] Fröhlich H 1937 *Proc. Roy. Soc.* A **160** 230–41
[6] Fröhlich H and Seitz F 1950 *Phys. Rev.* **79** 526
[7] Ferry D K 2021 *Hot Carriers in Semiconductors* (Bristol: IOP Publishing)
[8] Born M and Oppenheimer J R 1927 *Ann. Phys.* **389** 457
[9] Ferry D K 2020 *Semiconductors: Bonds and Bands* (Bristol: IOP Publishing) 2nd edn
[10] Schwinger J 1948 *Phys. Rev.* **73** 416
[11] Schwinger J 1948 *Phys. Rev.* **74** 1439
[12] Ferry D K, Goodnick S M and Bird J P 2009 *Transport in Nanostructures* (Cambridge: Cambridge University Press) 2nd 525 pp
[13] Lipavsky P *et al* 1991 *Phys. Rev.* B **43** 6650
[14] Bordone P, Vasileska D and Ferry D K 1996 *Phys. Rev.* B **53** 3846
[15] Seon Y, Chang J, Yoo C and Jeon J 2021 *Electron* **10** 180
[16] Mochizuki S *et al* 2021 *Proc. IEEE Electron Dev. Mtg.* (New York: IEEE) paper 2.3
[17] Geltman S 1977 *J. Phys.* B **10** 831
[18] Haan S L and Geltman S 1982 *J. Phys.* B **15** 1229
[19] Lindberg M and Koch S W 1988 *Phys. Rev.* B **38** 3342
[20] Haug H 1988 *Optical Nonlinearities and Instabilities in Semiconductors* ed H Haug (San Diego, CA: Academic)
[21] Kuhn T and Rossi F 1992 *Phys. Rev.* B **46** 7496
[22] Reggiani L, Lugli P and Jauho A P 1987 *Phys. Rev.* B **36** 6602
[23] Pascoli M, Bordone P, Brunetti R and Jacoboni C 1998 *Phys. Rev.* B **58** 3503
[24] Bertoni A, Bordone P, Brunetti R and Jacoboni C 1999 *J. Phys.: Condens. Matter* **11** 5999
[25] Levinson I B 1970 *Sov. Phys. JETP* **30** 362 http://jetp.ras.ru/cgi-bin/e/index/e/30/2/p362?a=list
[26] Shifren L and Ferry D K 2003 *Phys. Lett.* A **306** 332
[27] Zurek W H 2003 *Rev. Mod. Phys.* **75** 715
[28] Zurek W H 1982 *Phys. Rev.* D **26** 1862

[29] Blume-Kohout R and Zurek W H 2008 *Phys. Rev. Lett.* **101** 240405
[30] Ferry D K 1991 *Semiconductors* (New York: Macmillan)
[31] Nakajima S 1958 *Prog. Theor. Phys.* **20** 948
[32] Zwanzig R 1960 *J. Chem. Phys.* **33** 1338
[33] Barker J R and Ferry D K 1980 *Solid-State Electron.* **23** 531
[34] Bird J P *et al* 1994 *Phys. Rev.* B **50** 18678
[35] Ferry D K *et al* 2011 *Semicond. Sci. Technol.* **26** 043001
[36] van Wees B J *et al* 1988 *Phys. Rev. Lett.* **60** 848
[37] Wharam D A *et al* 1988 *J. Phys. C: Solid State Phys.* **21** L209
[38] Ferry D K 2020 *Transport in Semiconductor Mesoscopic Devices* (Bristol: IOP Publishing) 2nd
[39] Marcus C M *et al* 1994 *Semicond. Sci. Technol.* **9** 1897
[40] Akis R, Ferry D K and Bird J P 1996 *Phys. Rev.* B **54** 17705
[41] Akis R and Ferry D K 1998 *Semicond. Sci. Technol.* **13** A18
[42] Heller E J 1984 *Phys. Rev. Lett.* **53** 1515
[43] Fishman S, Georgeot B and Prange R E 1996 *J. Phys. A: Math. Gen.* **29** 919
[44] Harper P G 1955 *Proc. Phys. Soc.* A **68** 874
[45] Hofstadter D 1976 *Phys. Rev.* B **14** 2239
[46] Elhassan M *et al* 2004 *Phys. Rev.* B **70** 205341
[47] Brunner R *et al* 2007 *Phys. Rev. Lett.* **98** 204101
[48] Gutzwiller M C 1990 *Chaos in Classical and Quantum Mechanics* (New York: Springer)
[49] Husimi K 1940 *Proc. Phys.-Math. Soc. Jpn.* **22** 246
[50] de Moura A P S *et al* 2002 *Phys. Rev. Lett.* **88** 236804
[51] Ochiai Y *et al* 1998 *Phys. Low-Dimens. Semicond. Struct.* **1998** 109
[52] Brunner R *et al* 2008 *Phys. Rev. Lett.* **101** 024102
[53] Breuer H-P and Petruccione F 2002 *The Theory of Open Quantum Systems* (Oxford: Oxford University Press)
[54] Ando T 1991 *Phys. Rev.* B **44** 8017
[55] Usuki U, Saito M, Takatsu M, Kiehl R A and Yokoyama N 1995 *Phys. Rev.* B **52** 771
[56] Merzbacher E 1970 *Quantum Mechanics* 2nd (New York: Wiley)
[57] Fetter A L and Walecka J D 1971 *Quantum Theory of Many-Particle Systems* (New York: McGraw-Hill)
[58] Kotylar R *et al* 2004 *Appl. Phys. Lett.* **84** 5270
[59] Ferry D K *et al* 2016 *Nanoscale Silicon Devices* ed S Oda and D K Ferry (Boca Raton, FL: CRC Press) 1–36 pp
[60] Gilbert M J, Akis R and Ferry D K 2005 *J. Appl. Phys.* **98** 094303
[61] Akis R, Ferry D K and Gilbert M J 2009 *J. Comput. Electron.* **8** 78
[62] Gilbert M J and Ferry D K 2005 *IEEE Trans. Nanotechnol.* **4** 355
[63] Shifren L, Akis R and Ferry D K 2000 *Phys. Lett.* A **274** 75
[64] Bastard G 1981 *Phys. Rev.* B **24** 5693
[65] Dürr D, Goldstein S and Zanghì N 1992 *J. Stat. Phys.* **67** 843
[66] Oriols X and Mompart J 2019 *Applied Bohmian Mechanics: From Nanoscale Systems to Cosmology* (Singapore: Pan Stanford)
[67] Dürr D and Teufel S 2009 *Bohmian Mechanics: The Physics and Mathematics of Quantum Theory* (Berlin: Spinger)
[68] Holland P R 1993 *The Quantum Theory of Motion* (Cambridge: Cambridge University Press)

[69] Oriols X 2007 *Phys. Rev. Lett.* **98** 066803
[70] Brasil C A and Castro L A 2015 *Eur. J. Phys.* **35** 065024
[71] Wheeler J A 1957 *Rev. Mod. Phys.* **29** 463
[72] Everett H 1957 *Rev. Mod. Phys.* **29** 454
[73] Destefani C F and Oriols X 2023 *Phys. Rev.* A **107** 012213
[74] Schlosshauer M 2007 *Decoherence and the Quantum-to-Classical Transistion* (New York: Springer)
[75] Wiseman H M and Milburn G J 2009 *Quantum Measurement and Control* (New York: Cambridge University Press)
[76] Pandey D *et al* 2019 *Entropy* **21** 1148
[77] Wiseman H M and Gambetta J M 2008 *Phys. Rev. Lett.* **101** 140401
[78] Gambetta J M and Wiseman H M 2003 *Phys. Rev.* A **68** 062104
[79] Oriols X and Nikolic H 2023 *Eur. Phys. J. Plus* **138** 250
[80] Jou B, Casas-Vazquez J and Lebon G 1999 *Rep. Prog. Phys.* **62** 1035
[81] Maugin G A 1999 *The Thermodynamics of Nonlinear Irreversible Behaviour* (Singapore: World Scientific)

Part II

General modeling considerations

IOP Publishing

Quantum Transport in Semiconductor Devices
Simulation using particles
David K Ferry, Xavier Oriols and Josef Weinbub

Chapter 4

An overview of semiconductor devices

4.1 Introduction

Some degree of quantum transport has been present in semiconductor devices for quite some time. However, quantum mechanics may be involved at several different levels. The only true, and new, quantum properties in devices (or, indeed, in semiconductors themselves) are the appearance of the wave nature of the electron (instead of its normal particle properties) and discretization of the wave momentum due to the periodicity of the crystal lattice. The basic band structure of the semiconductor material arises from this latter periodicity of the lattice, and from this band structure arises the concept of the effective masses of the quasi particles (the electrons and holes). The effective masses, in turns, allow one to treat the carriers as if they were classical objects. This approach is successful provided that the quasi particles remain within a single band.

The extension of classical concepts to the quantum world was reinforced by Kennard [1], who demonstrated that the Schrödinger equation could be manipulated in a manner such that the potential to which the carriers responded was the normal classical potential plus an additional quantum potential. This quantum potential is already present in the band structure that yields the effective mass. However, in size quantization situations, new contributions to the quantum potential arise, and these will be discussed in chapter 10. However, when one goes beyond simple drift–diffusion models of transport and begins to consider the physics underlying the scattering of carrier lattice vibrations, the real scattering probabilities are found from a quantum mechanical treatment. On the other hand, confinement in small structures is not really a quantum effect, although this confinement leads to transport that deviates from bulk transport. Rather, such confinement is merely the imposition of new boundary properties, and confinement effects have been studied for more than two centuries. Nevertheless, it seems to be assumed that confinement leads to quantum transport, perhaps because confinement is more acute on the size scale of quantum effects. And the manner in which this confinement can

doi:10.1088/978-0-7503-5237-6ch4

modify the band structure does lead to observable quantum effects that change the transport. However, one may then recognize that semiconductor devices have had no need to be described by quantum transport over most of the decades for which they have existed. But the change of the band structure that results from adding strain to the device also serves to modify the effective mass and electronic structure [2].

In the following sections, a brief history of the evolution of the field-effect transistor (FET) is presented. A discussion of various quantum effects and how they may be observed in the transport then follows. In these discussions, the reader should notice how quantum transport differs from classical transport. The final sections then describe the various devices in more detail and explain how quantum transport is important in each of them.

4.1.1 An example—the evolution of the FET

The Lilienfeld patent for the metal–oxide–semiconductor field-effect transistor (MOSFET) [3] is considered to be the first conceptualization of an FET. However, it is not all that clear in the patent description, and a later patent by Heil [4] describes the device more clearly. History suggests that early work at Bell Telephone Laboratories attempted to make an MOSFET using Ge, but this material lacks a stable oxide. Perhaps this led the researchers to the point contact transistor (two metallic points attached to a Ge layer) [5] developed by Bardeen and Brattain [6, 7]. Shockley would subsequently develop the junction transistor (diffused or grown layers doped differently) [8]. Achieving the first working FET was to take somewhat longer, and required much work on surface passivation and control. Atalla seems to have been the first to stabilize the Si surface via the growth of a thermal oxide [9]. With this oxide, the door was open for he and Kahng to create a MOSFET (around 1960). The advantages of the surface-oriented MOSFET were recognized by many others; this eventually led to its use in integrated circuits and then to patents being filed in 1959 [10, 11].

The coupling of both *n*-channel and *p*-channel MOSFETS together as a single logic gate appeared in 1962 [12, 13]. Known as complementary MOS (or CMOS), this provided low-power gates and eventually dominated the development of integrated circuits. A major advantage that CMOS possessed lay in its planar layout (the plane of the current flow is the surface plane, whereas the current in a junction bipolar transistor is normal to the surface plane). This planar layout is very important for lithographic processing and was one of the causes of the rapid growth in density.

While the early integrated circuits were, and continue to be, digital, other applications exist for FETs, such as analog amplifiers. In these devices, one desires a linear (amplifier) behavior in small- or large-signal circuits. Such analog applications notably appear in microwave applications. In many cases, these are III–V devices, particularly the high-electron-mobility transistor (HEMT). The highest-frequency transistors produced to date have been InP MOSFETs [14] and InP-based, InAs channel HEMTs [15]. Today, GaN-based devices have led to excellent power HEMTs [16]. A general discussion of the detailed structures of these devices will be left to later sections, such as the discussion of the MOSFET in section 4.3.

4.1.2 How does quantum transport arise?

Quantum effects and quantum transport have been discussed and debated for years. Their use in FETs is just a new aspect of this discussion. One part of this is the actual physics of the particle (or quasi particle), be it electron or hole. The size of the classical electron has been investigated for decades without reaching a firm value. In fact, in classical mechanics, it is usually described as a 'point' particle. This will not work for a quantum particle because of the uncertainty principle. That is, the size of the wave packet representation in momentum space is coupled to the physical size of the quasi particle in quantum mechanics, which can be easily estimated from the de Broglie wave length [17]. This problem will be considered in depth in chapter 10, but the arguments made there suggest that the size is of the order of 3–7 nm. This becomes a problem if one thinks of a 5 nm critical dimension in a device. This is already true for a simple MOSFET, for how can a ~3–7 nm electron be stuffed into the classical size of the inversion layer, which is only about a ~1 nm potential well? In fact, this just cannot be done, and a self-consistent solution of Poisson's and Schrödinger's equations must be pursued [18].

Pursuing the self-consistent potential mentioned in the last paragraph gives rise to other effects in an *n*-channel Si MOSFET. One is the splitting of the sixfold degenerate valleys of the Si conduction band. These valleys are ellipsoids of revolution, so that each has a heavy mass direction and a small mass direction. The two valleys whose heavy mass directions are normal to the oxide–semiconductor interface form one set of quantum levels. These levels are lower in energy than those of the other four valleys that have the light mass normal to the interface. Nearly all MOSFETs exhibit this quantization. The quantization is not observed in normal operation for several reasons. First, the transport direction is parallel to the interface, not normal to it, so the quantization is a second-order effect. Second, the use of tensile strain to increase the separation between the twofold valleys and the fourfold valleys is now common; this is done to raise the effective mobility in the MOSFET (by keeping the electrons in the sub-bands with the smallest effective mass).

Quantization can result whenever confinement effects become comparable to the extent of the quasi-particle wave function. Quantum effects are thus important in today's ultrasmall devices. The quantum effects may even change the nature of the FET from a surface-oriented device to a bulk-oriented device. This would have the added benefit of reducing scattering from the rough interfaces. In a similar manner, quantum transport becomes important when these size scales begin to have the same order of magnitude as the mean free path of the particle. The importance of this size scale is that the optical phonon scattering (in Si) responsible for transfer among the equivalent valleys of the conduction band breaks any coherence that may be attached to the wave function. When this coherence is broken, classical transport takes over. The mean free path is a measure of the coherence length, especially when the scattering is dominated by these intervalley phonons. On the shorter spatial scale, coherence affects the transport, and this is where quantum transport, which describes this coherence, is important. This was discussed for small structures in chapter 3 and is illustrated for the coherence present around impurity scattering in the next chapter in figure 5.1.

Another concept that has been around for some time is tunneling, which is thought to be a true quantum effect. Tunneling in FETs became a concern when particles began passing through the gate oxide. In the presence of a high electric field across the gate oxide, the barrier between the gate metal and the FET channel becomes trapezoidal or triangular in nature, leading to Fowler–Norheim tunneling [19]. This has been a known problem in thin dielectrics for almost as long as MOSFETs have existed [20]. The trend toward smaller devices has consequently led to an increase in direct tunneling through this oxide. The adoption of dielectrics with higher values of the dielectric constant (the so-called high-κ dielectrics) was a direct response to the problem of gate tunneling in very thin oxides. These dielectrics allow a thicker dielectric that has the same capacitance as the thin oxide. More quantum effects are appearing, and in ever greater quantities, as devices continue to evolve. It is not at all evident that each effect can be treated in isolation, one at a time, since quantum mechanics is a notoriously nonlocal theory. Many of the quantum effects are quite likely to interact with one another and may well change our understanding of how FETs operate.

4.1.3 How do quantum transport and classical transport differ?

It is important to understand that many forms, or interpretations, of quantum mechanics exist, as was discussed in chapter 3. For example, Bohr was interested in doing away with causality and determinism [21]. There are alternative interpretations of quantum mechanics, and most engineers and device scientists tend to follow these alternatives [22]. It must be remarked that most devices and our understanding of them rely upon causality and determinism. Quantum mechanics that allows this has real particles that follow real trajectories. This is how one understands that semiconductor devices work. It is then that the '...additional quantum potential...' described by Kennard [1] provides the difference between quantum transport and classical transport. This quantum potential is defined by the wave function itself, and it introduces both interference and coherence effects that appear in the transport. The interference also creates entanglement, which is described as that magical property that is unique to quantum mechanics and is essential in quantum computing [23]. An example of this will be seen in chapter 5, where a particle can interfere with itself as it passes a single impurity (see figure 5.1) [24]. This can arise because of the wave nature of the particle, as it is the wave that exhibits the interference. Classically, the particles do not behave differently from a normal coulomb interaction between two charges. However, quantum mechanically, this interference can affect the resulting self-consistent potential. The presence of such interference requires quantum transport approaches.

Thus, one can summarize the need for quantum transport as a need to properly handle interference and coherence effects that can affect particles during their transport within the device. As remarked previously, there are many approaches to quantum transport, and they are discussed in the following chapters. But one important aspect remains in all approaches—the approach must respect and describe the interferences and coherences that can occur in these devices. Some of the most obvious interferences and coherences that can actually occur in experimental measurements are described in the following sections of this chapter. Others are left to the following chapters. However, one conclusion is that if the interferences and coherences are not adequately described by a given approach to quantum transport, then that approach is not a meaningful path to follow!

We emphasize that our point in discussing the necessity of incorporating interferences and coherences into the description of electron devices is that quantum laws differ from classical ones. However, a physical theory encompasses not only laws but also elements to which these laws are applied. These elements are referred to as the theory's ontology. In this book, we contend that quantum laws diverge from classical laws, yet this doesn't necessarily mean that quantum ontology must differ from the familiar classical ontology.

4.2 Diodes and bipolar junction transistors

Quantum effects in *p–n* junction diodes and bipolar transistors tend to occur only when the physical size is small. In such small structures, many effects can occur that require quantum mechanics for their treatment. One such case occurs when the various regions of either device are very heavily doped. Here, the valence band in the *p*-region and the conduction band in the *n*-region can overlap in energy. This occurs when the two materials are degenerately doped so that the Fermi energy resides inside the bands and not in the gap. This situation may be found in both tunnel diodes and laser diodes. Indeed, in the latter case, quantum coherence is critical to the laser operation.

Another situation that arises from heavy doping is a physically small depletion region in the *p–n* junction itself. In this case, the very high electric fields affect any scattering process that occurs in the junction region, whether it is a diode or a bipolar transistor. When the field is this high, acceleration of the particles may occur *during the scattering* event. This is known as the intracollisional field effect [25, 26] and dramatically changes the scattering probabilities. This effect has been discussed and treated via quantum transport for decades now.

Finally, the third major situation in which quantum mechanics becomes important is when the base region is physically small within a bipolar junction transistor. If the transistor is an *n–p–n* transistor, then the base is an electron quantum well for which the collector and emitter potentials form the confining potential. This is important, as the operation of the device depends upon electron injection into the base region. Similarly, in a *p–n–p* device, in which the operation depends upon the injection of holes into the base, the base itself is a hole quantum well. Thus, quantization effects in the base affect the operation because they affect the statistics of the injected particles. This, in turn, leads to a need for quantum transport, due both to the level quantization and to the change in the diffusivity of the carriers. The latter arises from the former—the change in the level statistics leads to changes in the scattering, which is the basis for diffusivity (or the so-called diffusion coefficient).

As discussed in the introduction, these devices depend upon whether the transport is diffusive (limited by scattering) or ballistic (in which there is no scattering).

Whether the transport is one or the other depends upon the physical length scale of the depletion regions in the device. In degenerate material, the diffusivity is defined mainly by the Fermi velocity and is given by

$$D = \frac{v_F^2 \tau}{d}, \tag{4.1}$$

where τ is the reciprocal scattering rate, v_F is the particle's Fermi velocity, and d is the appropriate dimensionality of the system considered. The dimensionality arises because the root-mean-square (rms) value of the velocity at the Fermi energy is determined by this Fermi energy in the normal manner. However, the directed velocity is usually a fraction of the total of v_F^2, and this fraction is given by the dimensionality. In nondegenerate material, the only change in (4.1) is to replace the Fermi velocity with the thermal velocity. Hence, when the quantization leads to discrete energy levels within the base quantum well, then both the scattering rate and the dimensionality are affected. This in turn leads to variations in the diffusivity and an underlying need for quantum transport.

4.3 The MOSFET

A conceptual MOSFET is shown in figure 4.1. The actual structure of such a device has undergone considerable change since the development of the early MOS integrated circuits mentioned in the introduction to this chapter. The introduction mentioned that the advantage possessed by the CMOS structure lies in the planar layout of the devices. As mentioned, this is a major advantage in lithography for large-scale integration, and this has led to exponential growth in this integrated circuit technology. Normally, the growth is measured by the number of transistors on the chip and is gauged by Moore's law [27]. In this progression, the individual dimensions of a device decrease from one generation to the next, while the dimensions of the chip remain constant or increase. This has the result that the

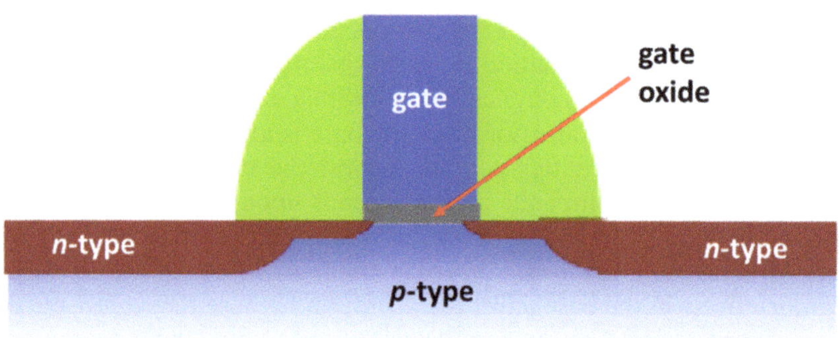

Figure 4.1. A schematic diagram of a planar MOSFET. The source is assumed to be on the left and the drain is assumed to be on the right-hand side of the device. The gate is used to invert the density in the p-type region at the interface with the oxide, making an *n*-type connection between the source and the drain. Reprinted from [30].

number of transistors on a single chip increases exponentially. Moore's law is basically an economic law rather than a physical law [28, 29], as the cost is determined by the fabrication of the chips. If the cost remains relatively constant for a given chip size, then the individual devices become exponentially cheaper.

Integrated circuits are, by and large, digital, so that the transistors used are switching transistors that require well-defined and stable states. These states define the logic levels and are usually specified as the ground potential and the bias voltage. Transitions between these two states are not of much concern as long as they happen quickly and reliably. Throughout the period of chip evolution, the downsizing of each device has led to the introduction of new structures, new materials, and new configurations, all of which have gradually changed how figure 4.1 looks. This evolution also occurred in analog devices, although this was more concerned with maintaining the linearity of device operation.

As the size of switching devices has reduced over the years, several growing problems appeared. These largely affected the mobility, which continued be degraded as the devices were downscaled. Consider the basic structure of a planar MOSFET, as shown in figure 4.1 [30]. For an *n*-channel device, the source and drain regions are donor-doped to be *n*-type (as shown in the figure), while the base region is acceptor-doped to be *p*-type (or in the modern case, it may be left undoped). Typically, the source is on the left of the figure and the drain is on the right. The technology used to fabricate such a transistor is reviewed in [31]. The gate is biased with a positive voltage to draw electrons to the interface between the oxide and the *p*-type substrate, thus inverting the population to make an *n*-type channel at this interface. This channel connects the source and drain, thus establishing a current path with good mobility. The density is modulated by the gate voltage, which thereby controls the current. The mobility of the carriers is important, and the degradation of mobility can be attributed to surface-roughness scattering at the oxide–semiconductor interface [32] as well as to an increase in impurity scattering due to the increased impurity doping in the substrates. To overcome this mobility decrease, the first change was to introduce strain into the channel [33]. Introducing *tensile* strain into the *n*-channel splits the sixfold degenerate conduction band valleys (as discussed above). That is, the additional quantization in the inversion channel, which lies in the (001) plane of Si, splits the six valleys into a fourfold set and a twofold set. In the twofold set, the heavy mass of the ellipsoidal valleys lies normal to the interface, and thus the twofold set has a lower quantization energy than the fourfold set of valleys. While this naturally occurs in the inversion channel, strain raises the separation between these two valleys and provides a much lower transverse mass for the channel mobility. In *p*-channel devices, *compressive* strain is used in order to warp the valence band in a manner that effectively lowers the effective mass and increases the mobility [34]. Introducing this differing strain may be accomplished by growing a SiN layer over the gate stack (the green and dark blue regions above the channel in figure 4.1) in *n*-channel devices. A SiGe alloy is often used for the source and drain regions (red in the figure) in *p*-channel devices. Later, device designers had to go further and address the problem of the oxide becoming too thin, an effect which led to tunneling through the oxide. This latter problem was addressed by introducing a new gate structure featuring HfO_2 [35]. Finally, continued evolution forced designers to turn the normally horizontal FET on its edge, leading to the FinFET around 2011 [36]. Future evolution is open to discussion, but gate-all-around quantum wire FETs are thought to be one way forward [37, 38].

It was already mentioned that switching devices are different from analog devices. Switching devices need to make the on-to-off transition and the off-to-on transition rapidly. The linearity of the transistor between these two end points is not particularly relevant. However, leakage current in the off state can create a significant problem with heat dissipation in integrated circuits [39]. Even an off current of 1 nA is too large for a circuit with 10^{10} transistors! The transition from the planar FET to the FinFET was due in part to the need to control this off current. The ability to pinch the leakage current using two opposing voltages from either side of the fin [36] better controls this leakage current. A tri-gate version (with a gate on the top of the fin) is illustrated in figure 4.2 [40]. The source and drain of this device are ion implanted. After the fin is formed by etching, it is oxidized and the gate deposited on both sides and the top. A typical shape of the fin itself is shown in figure 4.3 [41]. In this case, a layer of SiGe has been grown on top for use in a *p*-channel FinFET. More recently, there has been a move toward gate-all-around quantum wires, or nanosheet 'wires,' for a similar reason. An important point in all these changes is that the base silicon area is the economic driver for Moore's law [28, 29]. Thus, the effective width of the transistor must increase in order for the transistor to offer an improvement over the basic planar transistor. In figure 4.4, the nanowire and nanosheet transistors are compared with the FinFET device [42]. The effective gate width is W_{eff} and its definition is shown at the bottom of each of the three types of transistors in the figure. The brown areas represent silicon, the blue

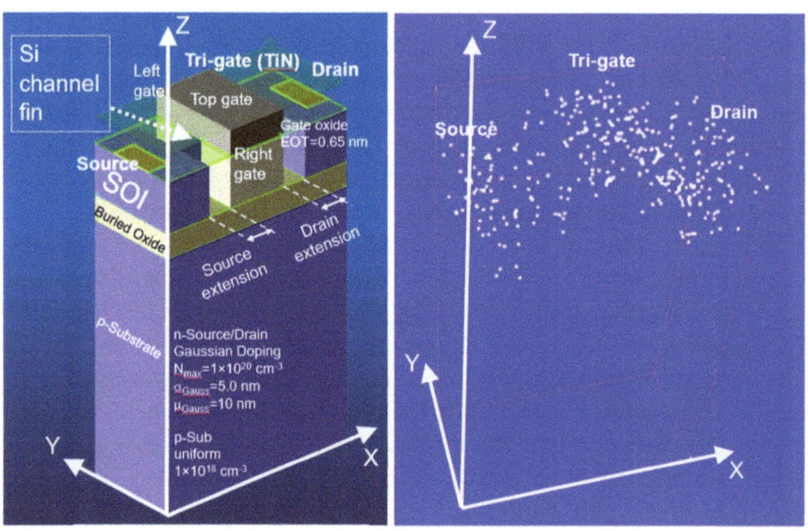

Figure 4.2. A three-dimensional view of (a) the structure of a 25 nm silicon-on-insulator tri-gate FinFet and (b) the spatial distribution of the ion implanted impurities throughout the device. Reproduced from [40].

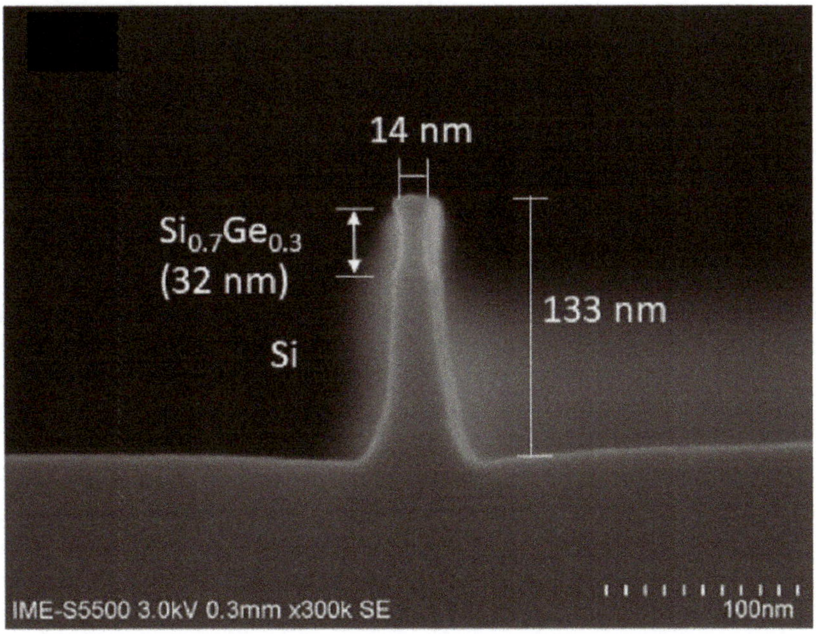

Figure 4.3. A cross-sectional view of a Si/$Si_{0.7}Ge_{0.3}$ fin obtained by plasma etching. The top layer of SiGe is used for the actual transistor. Reproduced from [41].

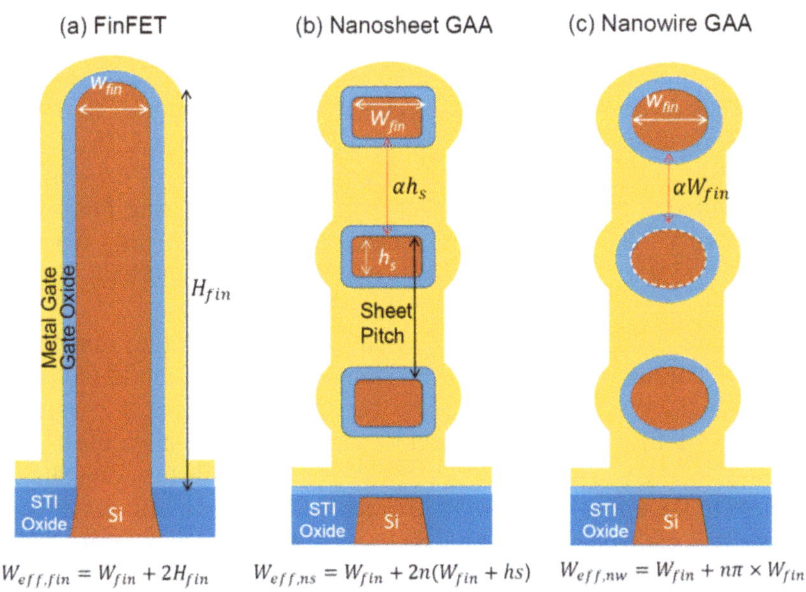

Figure 4.4. A comparison of (a) a finFET, (b) a vertically stacked nanosheet FET, and (c) a vertically stacked nanowire FET. The latter two are gate-all-around structures. Each has the same physical width W_{fin} for comparison, while the effective gate length W_{eff} is shown at the bottom of each panel. Reproduced from [42].

areas represent oxide, and the gold areas represent metal gates. Stacked nanowires generally do not offer an improvement over the finFET, but the nanosheet device does so because the lateral width of the nanosheets is increased.

In an analog device, the linearity of the gain over the voltage range between the on and off states is of primary importance. Saturation at the end points is not nearly as important. The device is nearly always on, and it is biased to a point that provides as large a voltage swing as possible while yielding as linear a gain as possible. These are quite different requirements than those of the switching transistor. This is further complicated in large-signal devices by a need to deliver significant power. Usually, if the power is increased too far, the gain drops, an effect known as *gain compression.* The gain compression that occurs at high power is also a strong consideration in device design. This means that the design of a microwave power device is quite different from that for a low-power linear amplifier design [14] and still different again in a mixed-mode device [43].

In the early years, most of the evolutionary advances in downsizing MOSFETs did not rely upon simulation and modeling, but rather the intuition of the engineers and scientists. More recently, this situation has changed. Simulation and modeling have become important to modern devices, especially as the size has continued to be reduced. This has become truer as the increased role of parasitics in the design has become apparent. That is, through most of the evolution of Moore's law, downsizing has followed a strict scaling relationship [44] in which the electrostatics are maintained. All dimensions are reduced by a common factor, and the voltages and doping densities are adjusted to maintain the electrostatics. This approach ignores the role of parasitics, which become more important as the device gets smaller and other devices appear much closer. While electric fields can be scaled, capacitances begin to have significant edge effects. Hence, they do not continue to scale properly. Further, a single device can no longer be considered in isolation. A device sits in an environment consisting of other devices and circuit elements. This produces an environment in which a single device cannot be isolated [45]. Scattering becomes more complicated, whether it is interface roughness scattering [46, 47], remote phonon scattering [48], or even device–device interactions arising from coupling between devices [49]. It is not apparent how many of these effects have been included within widely used simulation and modeling packages.

There is a concern over granularity, such as that inherent in randomly placed dopants. If one considers a small device in a semiconductor region 10 nm long, 40 nm wide, and 5 nm thick, doped to 10^{18} cm^{-3}, there are, on average, two dopants in the volume. Within an FET channel, the exact locations of these dopants are critical. They are far more effective when close the source, as they affect the turn-on voltage. When they are close to the drain, they have little effect. At higher impurity levels, a problem exists in which an electron or hole can interact with several impurities at one time [50]. In this situation, one cannot compute an accurate screening function. One cannot even establish that there is any time between collisions. The only clear fact is that multiple scattering can dominate [51]. Other factors, such as rough-interface scattering can also be affected by this granularity, through the inhomogeneity that spreads throughout the device. This can be especially problematic when it couples with quantum effects.

4.3.1 Inversion layers

Probably the first example of quantization in a MOSFET arises in the inversion layer upon which device operation depends. If one uses a classical treatment of the inversion layer for a MOSFET, the carrier density is found to decay exponentially with the surface potential away from the surface; that is, the peak of the density is at the surface. On the other hand, in a quantum treatment, the wave function, and therefore the density, must almost vanish at the surface (or oxide interface). The density peak is then well away from the surface. These quantum effects at surfaces and interfaces have been studied since the late 1960s and early 1970s. The inversion layers at the Si/SiO_2 interface are a classic example. In fact, this quantization and the resulting quasi two-dimensional behavior were used for the first observations of the quantum Hall effect [52, 53]. With quantum effects, the carriers are no longer simple localized objects but are represented by wave functions. A localized carrier has a quantum wave packet, and this packet becomes deformed when confined within small structures. This is definitely not classical behavior. While the positive gate voltage (an n-channel device is considered here) attracts the charge to the surface, the nature of the quantum effect is such that the charge has to be nearly zero at the oxide interface, as mentioned above. The (the time-averaged over many electrons) local charge is represented by $\rho(z) = -e \mid \psi(z) \mid^2$, where $\psi(z)$ is the wave function. Since the wave function must nearly vanish (due to the large offset potential of the oxide), the charge must similarly vanish. Moreover, the potential formed between the oxide potential and the band bending within the semiconductor leads to a quantum well that limits the motion normal to the surface. One or more sub-bands are formed in this quantum well. The energy–momentum relationships for motion parallel to the interface now are quasi two-dimensional. The actual shape of this potential along with the corresponding wave functions are self-consistently determined by solving Schrödinger's equation and Poisson's equation together [18].

Quantization effects in the inversion layer are illustrated in figure 4.5, where the potential and two sub-band energies are shown [54]. The self-consistent potential is labeled as $V_{eff}(z)$. The potentials are obtained by including the normal Hartree potential found using the Poisson equation, an exchange-correlation correction, and an image force [18]. The thick curves are calculated with the quantum exchange-correlation self-energy correction, while the thin curves do not include this correction. Here, it was assumed that the p-type substrate was doped to 2.8×10^{15} cm^{-3} and the inversion density was 4×10^{12} cm^{-2}. The exchange-correlation potential is a many-body correction that accounts for the carrier–carrier interaction, and the form used was taken from Hedin and Lundqvist [55]. It may be observed that the wave function peaks approximately 1.5–2.0 nm from the surface and then decays. The two major observables that arise from the quantization are that the peak in the wave function that is set back from the interface and the nonzero energy minimum of the lowest sub-band, which appears to be about 45 meV from the minimum of the conduction band. These two changes affect the gate capacitance of the MOSFET [56].

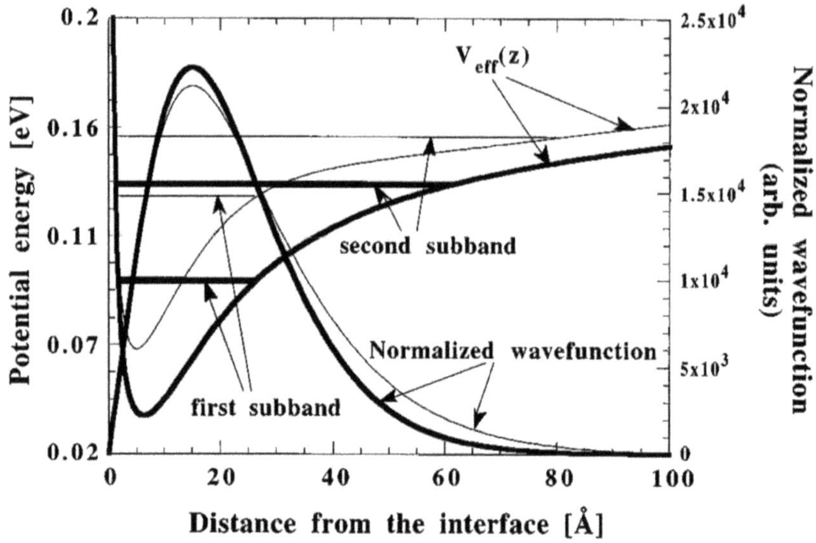

Figure 4.5. The calculated energy band profile and the wave function for (100) Si. The thick lines denote the inclusion of the exchange-correlation potential, while the thin lines ignore this many-body correction. The potential is modified to account for an effective form as described in chapter 10. Reprinted from [54], with the permission of AIP Publishing. © American Vacuum Society.

With this quantization in the quantum well, the energy levels depend inversely on the effective masses of the carriers normal to the interface in each valley. Hence, in a (001) surface of the Si, carriers in the pair of (001) valleys exhibit the heavy ~$0.98m_0$ longitudinal mass, while those in the other four valleys exhibit the light ~$0.19m_0$ transverse mass. As discussed above, this leads to a split of the six conduction band valleys. The four transverse valleys are higher in energy. Thus, the sub-bands shown in figure 4.1 are those for the two valleys with their major axis in the (001) direction. These two valleys exhibit the lighter mass in the transverse directions (along the channel), and this yields higher mobility. This effect is exaggerated when strain is introduced. In the figure, the effective potential is the quantity that is plotted and used in the simulations. This potential is discussed in chapter 10, where other forms are also discussed.

The quantization results from the self-consistent potential. In MOSFETs, however, this varies widely, so that the quantization also varies widely through the device. As the goal is to improve the mobility by seeking to use the smaller effective mass for motion along a minor axis of the ellipsoidal bands, other methods have resulted in a quantum well that is more consistent with the gate voltage quantum well and more consistent subsequent quantization. One of these is the use of strain to further separate the conduction band valleys.

4.3.2 Strain

When strain is added to a MOSFET, its band structure is modified and this changes the transport properties. This can now be termed a quantum effect [33, 34]. In the case of an n-channel MOSFET, the strain is tensile and is usually achieved by growing a Si_3N_4 layer over the gate stack. This produces a tensile strain that tries to stretch the channel between the source and drain regions. As described in the previous section, quantization in the inversion layer separates the sixfold conduction band into a twofold set of ellipsoidal energy surfaces whose major axis is normal to the interface and a fourfold set whose minor axis is normal to the interface. The use of tensile strain increases this separation, putting most of the carriers in the twofold pair of sub-bands. A 1% uniaxial strain in the [110] transport direction can lead to a split as large as 70 meV [57]. Because the transport is now dominated by the transverse mass of the twofold set of valleys, an increase in mobility of approximately 40% is achieved in Si [58]. The carrier scattering between the twofold and fourfold sets of valleys is greatly reduced due to this separation, and this also leads to higher mobility for the carriers in the inversion channel. Because these valleys are anisotropic, the actual motion in the channel can be quite complicated [59].

In figure 4.6, the effect of strain on mobility is shown for an n-channel device that is fabricated on the (001) surface of Si. As discussed above, the uniaxial strain is applied in the [110] direction [34]. The strain deforms the normally circularly symmetric shape of the twofold valleys (normal to the interface) and creates the change in mobility. The elongation of the normally circular energy surface in the plane of the device causes an decrease in the effective mass in the channel direction.

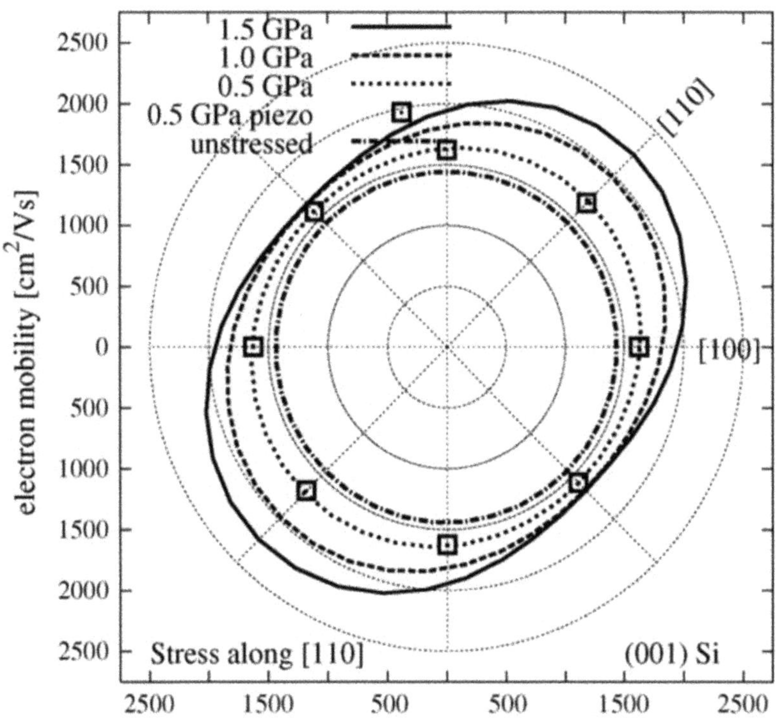

Figure 4.6. Mobility enhancement for uniaxial stress applied in the (110) direction. The piezoelectric effect is also included in this enhancement. Reprinted from [34], Copyright (2007), with permission from Elsevier.

Different levels of strain are shown by the different curves, and one can see that the mobility, though larger, becomes anisotropic in the valleys. These effects of the strain on the bands were calculated using an empirical pseudopotential method (chapter 6), and the transport was simulated using an ensemble Monte Carlo technique (chapter 9). The results are comparable to what is seen experimentally in such devices. Strain in the [110] direction elongates the crystal in that direction. Since momentum is proportional to the inverse of the lattice constant, this compresses the normal energy circle, so that the long axis of the constant energy line would lie along [110], which is normal to the strain. The resulting constant energy line is then written as

$$E = \frac{\hbar^2}{2}\left(\frac{k_{[\bar{1}10]}^2}{m_{[110]}} + \frac{k_{[\bar{1}10]}^2}{m_{[\bar{1}10]}}\right). \tag{4.2}$$

The fact that the major axis in the ellipse is normal to the strain direction means that the masses are different in different crystal directions. Here, the masses in the two normal directions are related as follows:

$$m_{[\bar{1}10]} > m_{[110]}. \tag{4.3}$$

Hence, the mass is reduced in the strain direction, a result in keeping with the theoretical calculations described above [34]. This creates greater mobility in the direction parallel to the strain.

While the two valleys that contribute to the lowest sub-band are normally degenerate, there can be many-body quantum interactions between carriers in these two valleys that produce a split between them. This has been observed experimentally [60]. Strain applied in the [110] direction leads to the warping of the energy surfaces in the plane (figure 4.6) and a significant split between these two valleys. These effects can become stronger in ultrathin layers of Si, such as silicon-on-insulator (SOI) sheets or nanosheets [61, 62].

The holes in a p-channel device require compressive strain in order to provide a lower effective mass. When a strained SiGe layer is grown on a Si substrate, this layer is subjected to compressive stress. A subsequent Si layer grown on top would bury the SiGe layer, and it would appear as a quantum well [63]. The top Si cap layer moves the holes away from the surface as a means to reduce the surface-roughness scattering, while the strain in the SiGe layer splits the light- and heavy-hole bands at Γ, reducing the transport effective mass. A drawback of this approach is that the SiGe layer has significant alloy scattering which tends to lower the mobility, so that there are competing effects. The valence band in such a structure exhibits quantum effects, but a detailed analysis requires a full-band calculation for the complicated splits, warps, and crossings of the various valence bands [64].

The approach in which p-channel devices use SiGe layers mainly in the source and drain regions has become a common approach. However, another approach is to use SiGe grown only in in the source and drain regions. The larger lattice constant of the SiGe then squeezes the channel between these two regions and puts the channel under compressive strain [33]. This uniaxial compression raises the light-hole band above the heavy-hole band at the zone center, so that the effective mass of holes becomes smaller and the mobility is raised. This subsequently creates another problem, in that the two bands tend to cross at a momentum value away from the zone center, as the light-hole band has a much higher curvature than the heavy-hole band. In actuality, this becomes an avoided crossing because interaction between the two bands leads to a hybridization between them. This provides the avoided crossing. The result gives a very non-parabolic behavior for the initial light mass band. To help the reader to understand this better, the normal unstrained valence bands in Si are shown in figure 4.7 along with equal-energy surfaces for the light- and heavy-hole bands [65]. The phrase 'warping' can be understood from these two constant energy surfaces. The maximal extension of the heavy-hole surface is in the (111) directions, while that of the light-hole surface is in the (100) directions. These energy surfaces are shown in 3D in figure 4.8 at low energy. As the energy increases, the warping becomes much more pronounced, as can be seen in figure 4.8 for the unstressed bands (the upper row shows the light-hole band after hybridization, while the lower row shows the heavy-hole band). The holes see a much reduced effective mass and therefore enhanced mobility.

The effect of the strain in the presence of the gate field is shown in figure 4.9, where it can be seen that this field has little effect on the constant energy surface for the dominant band [66]. The lowest energy (light-hole) band is shown for both the strained and unstrained energy surfaces. These surfaces were calculated with a $\boldsymbol{k} \cdot \boldsymbol{p}$ method (see chapter 6). As the surfaces are unaffected by the gate field, the masses are similarly unaffected. Similar calculations for the valence band structure in the presence of the uniaxial strain along [110] have been performed by Shifren *et al* [67], Wang *et al* [68], and Kotlyar *et al* [69], among others.

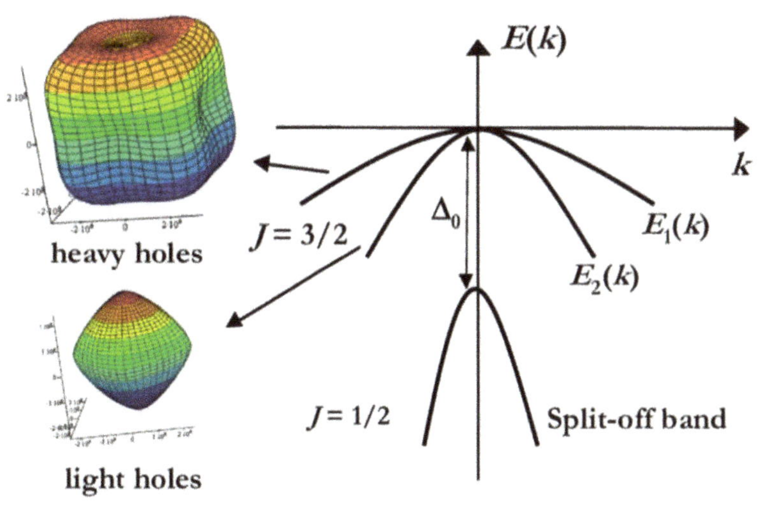

Figure 4.7. A schematic view of the valence bands in unstrained Si. E_1 is the heavy-hole band, while E_2 is the light-hole band. Δ_0 is the spin–orbit split-off band. The constant energy surfaces near $k = 0$ are shown pictorially on the left. Reprinted from [65], Copyright (2006), with permission from Elsevier.

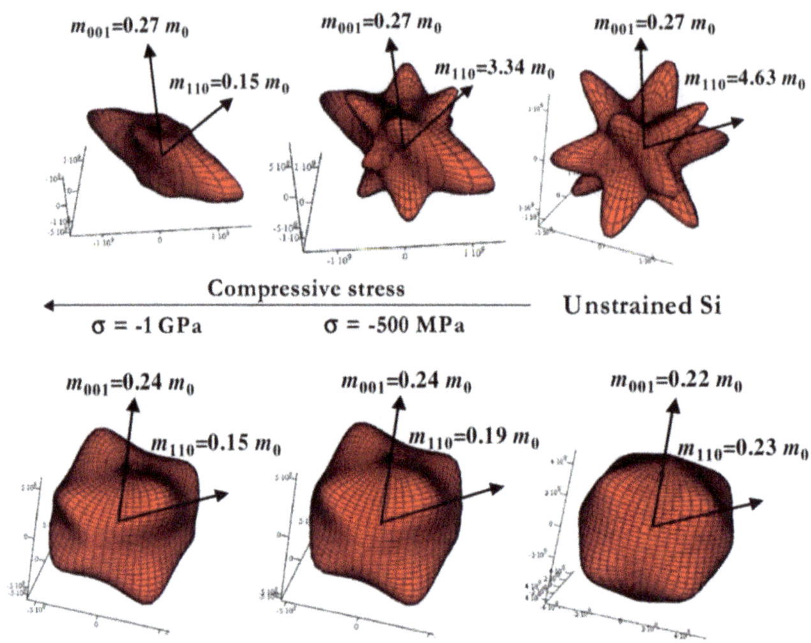

Figure 4.8. The constant energy surfaces at 39 meV in the strained and unstrained cases. It can be seen that the mass in the [110] direction is dramatically reduced as the strain is increased. Reprinted from [65], Copyright (2006), with permission from Elsevier.

4.3.3 Discrete impurities

Keyes [39] noted half a century ago that when the number of discrete impurities within a semiconductor device became small, there would be a problem with the size of the fluctuations in this number: when the number of actual impurities N was small, then the fluctuation $\sqrt{N}$ would be comparable to this number. In the small device imaged earlier, with a semiconductor region 10 nm long, 40 nm wide, and 5 nm thick, doped to 10^{18} cm^{-3}, there were, on average, two dopants in the volume. The relative fluctuation is given by $\sqrt{N}/N \sim 0.7$. Thus, there are perhaps zero, one, two, or three impurities in the entire device. This represents an uncontrollable variation in the density, which means there is an uncontrollable variation in the threshold voltage and current as well. In large devices, such fluctuations are small, but in small devices, they become almost unimaginable.

The fluctuation problem was highlighted when Wong and Taur first included the atomistic nature of these impurities in their MOSFET simulations [70]. Subsequently, random doping distributions became widely adopted in simulations through the inclusion of the actual positions of the dopants and the incorporation of molecular dynamics to treat the interactions between the dopants and the carriers [71–74]. In figure 4.10, variations caused by atomistic doping are illustrated by plotting the drain current as a function of the gate voltage for a great many different implementations. This is done for a MOSFET with a 0.1 μm gate length and the channel doped to 8×10^{17} [72]. It may be seen in the figure that the current varies by

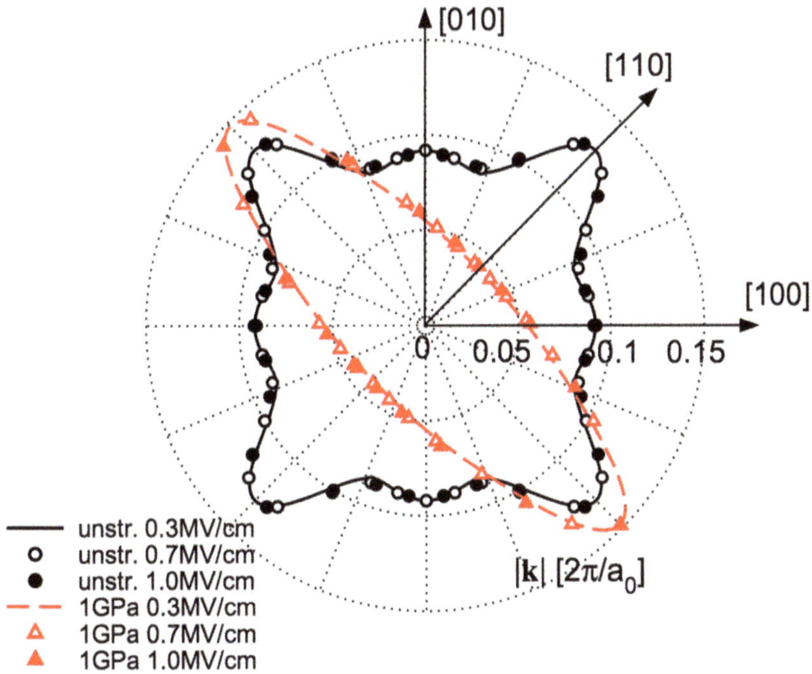

Figure 4.9. The constant energy surface of the lowest hole sub-band for Si (001) surface devices with compressive strain along [110] for several different gate fields. The black curves and data points represent unstrained channels, while the red curves and data points represent the strained channels. Reprinted from [66], Copyright (2009), with permission from Elsevier.

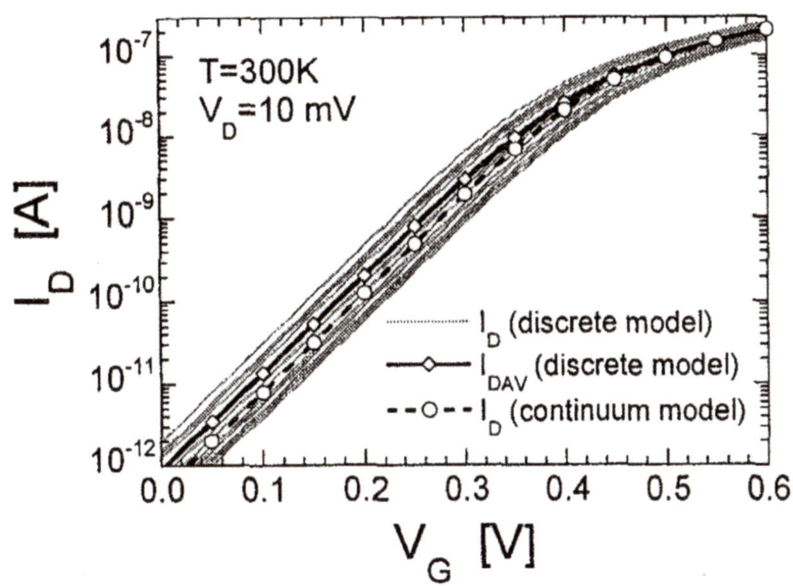

Figure 4.10. Drain currents for a great many different simulations (differing by the random positions of the dopants) for a 0.1 μm MOSFET. © 1998 IEEE. Reprinted, with permission, from [72].

more than an order of magnitude in the subthreshold region. This is an effect that is not small and not ignorable. This study also demonstrated that impurities near the source–channel transition affected the current far more than impurities far from this region [75]. The increase in the fluctuations as the devices are scaled down suggests that the best course may well be to remove doping from the substrate. For operation at less than 1 V, adequate voltage separation from the substrate still can be achieved, especially by double-gate FinFET-type devices. The name *junctionless device* was coined some time ago [76].

While the fluctuations in the impurity number are important, there is another aspect that becomes important for quantum transport. Impurity scattering is coulombic in nature, which means that it is anisotropic and long range. This has important consequences for coherence and interference. The reader should be familiar with the so-called two-slit optical experiment of Young [77]. Such an experiment can be replicated with electrons. It has been shown that a charged wire causes electrons to pass on either side of it. In a transmission electron microscope (TEM), this is known as a biprism. The two paths of electrons then form an interference pattern [78], just as with photons in Young's experiment. Importantly, a single electron in the TEM or a single photon [79] leads to the interference pattern. Since the single electron (or photon) must have both wave and particle properties [80], it creates interference in its wave properties. Further, this behavior means that a single electron passing a single impurity can create the same interference pattern due to its wave-like self-interactions in diffracting around the impurity. The existence of this interference in the wave properties of electrons is illustrated in figure 5.1 of chapter 5.

When there are a few impurities in the device, the interference becomes much more complex than that of a single impurity [52]. The impurities lead to an inhomogeneous potential given by Poisson's equation. The randomness in the impurity density gives a random nature to this potential. If this randomness is sufficiently large, it can localize a number of the band states [81], which means these carriers no longer contribute to conduction. But the random potential has another effect. Transport through such a potential is quite sensitive to the position of the Fermi level and the presence of any magnetic field, even the self-magnetic field of the current in the FET channel. Small variations in these quantities lead to large variations in the conductance through the channel and even chaotic behavior. These fluctuations are typically referred to as 'universal conductance fluctuations,' and have been experimentally observed in a large Si MOSFET at low temperatures [82]. In small modern devices, they are observed as significant drain current fluctuations [83]. In addition, one has to worry about transmission mismatches when moving from a three-dimensional source or drain to a quasi two-dimensional channel. When one of these latter interfaces is crossed, a discontinuity occurs in the carrier momentum. This creates the possibility of quasi-bound states, often called *resonances*, in the wave function for carriers in the channel. Such resonances are characterized by large peaks in the drain current. Both of these effects may be seen in figure 4.11, which shows the results of a simulation of quantum ballistic transport in a Si MOSFET [84]. The channel has a length of 9.8 nm, and the cross-section is

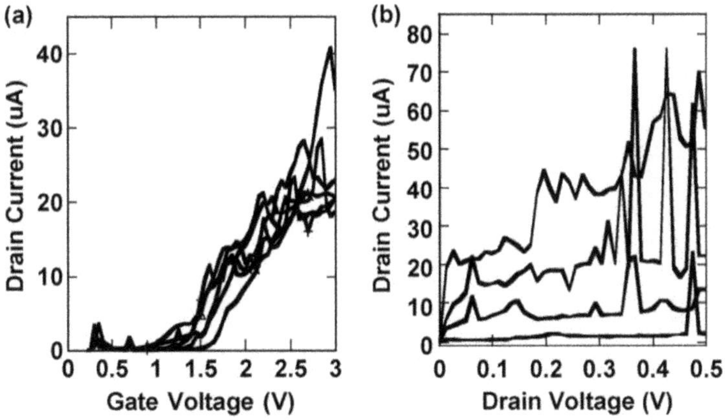

Figure 4.11. (a) The gate characteristics for six devices. The fluctuations are caused by the interferences due to the random positions of the dopants, as discussed in section 2.4. (b) The drain characteristics for one device. The smaller fluctuations are due to the dopants, but the large peaks are due to longitudinal resonances. © 2005 IEEE. Reprinted, with permission, from [84].

18.5 nm wide and 6.5 nm in depth. The channel is doped *p*-type at a concentration of 5×10^{18} cm^{-3}. Panel (a) shows the results are shown for the input characteristics of six different devices, in which the spatial dopant configuration is different in each case (the dopants are randomly placed according to the doping concentration and the grid size in the simulation). The peaks in the current arise from the fluctuations discussed above when only a few dopants are present. There are, on average, only about six dopants in the channel, so that their exact positions introduce considerable variability. In panel (b), the output characteristics are shown for a single device at different values of the gate voltage. There are still some small fluctuations due to the random nature of the dopant, but the large peaks arise from the longitudinal resonances caused by the mismatches at the source and drain transitions.

Even when adequate scattering exists in the channel, the random nature of the impurity distributions produced by atomistic doping can lead to the appearance of current filaments in the channel [85]. Such filaments are observable in the simulations depicted in figure 4.12. For this simulation, an *n*-channel, 50 nm MOSFET was considered. In this figure, the channel runs from the positions corresponding to 50–100 nm and the current is shown in A cm^{-2}. The doping in the channel was 5×10^{17} cm^{-2}, while the source and drain regions were doped to 2×10^{19} cm^{-2}. The individual dopants are depicted as individual dots or small circles. In the channel region, the size of the circle corresponds to the depth of the impurity below the surface. These impurities interact with the carriers in real space through a molecular dynamics simulation of the corresponding forces. However, the impurity potential is split into a long-range part, which appears in the Poisson equation, and a short-range part that leads to the molecular dynamics forces [86]. An interesting effect is that the current filaments are quite stable and do not disappear until they have extended almost 40 nm into the drain region!

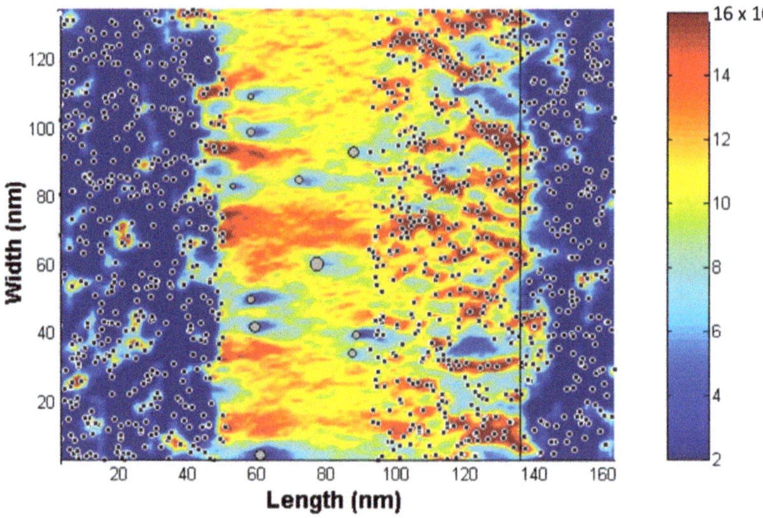

Figure 4.12. The current inhomogeneity in an *n*-channel MOSFET with a gate 50 nm long, the details of which are given in the text. The current scale on the right is in A cm^{-2}. Reproduced from [85].

4.4 MESFETs

The metal–semiconductor FET (MESFET) is one of the older semiconductor devices, having been conceived in the 1960s [87]. The MESFET differs from the MOSFET in some important ways. First, the MESFET is a natural majority carrier device, as opposed to having to create an inversion layer to host the majority carriers. Second, it is a depletion device rather than an accumulation device, so that it does not need to create an inversion layer. Control of the device is obtained via the gate electrode, just as in the MOSFET, but here the gate electrode is a metal that creates a metal–semiconductor junction, often called a Schottky barrier. The metal–semiconductor junction is much older than the MESFET, dating back to the late 19th century [88]. An understanding of this device and its rectification was provided by Schottky [89]. This junction is then used to deplete the carriers in the semiconductor layer under the gate. Naturally, this gate could take the form of a *p*–*n* junction, and, in fact, it did so in earlier concepts of such a depletion device. The metal gate has advantages, however, such as a lower gate resistance and the fact that the entire depletion region is in the semiconductor layer, which are both favored for microwave applications.

4.4.1 The Schottky barrier

The operation of a metal–semiconductor barrier is generally believed to depend upon the work function difference between the metal gate and the semiconductor material upon which the gate is deposited. When the metal is placed upon the semiconductor, it is thought that contact between the back surfaces of each material allows charge to move around to accommodate band bending in the semiconductor

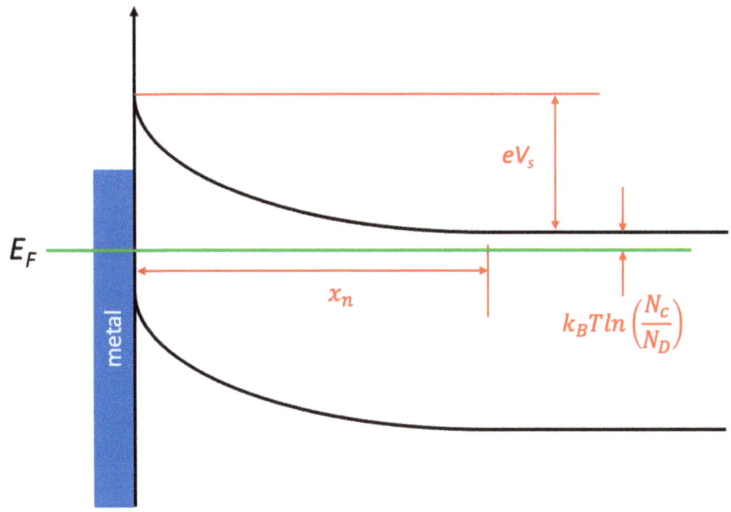

Figure 4.13. A depiction of the band structure of a metal–semiconductor contact. The built-in potential is V_s.

(there is a little such bending in the metal, but it is small due to much larger density of electrons in the metal). This band bending depletes the electrons (or holes) in the semiconductor at the interface with the metal, as shown in figure 4.13. Such band bending is required so that the Fermi energies in the two materials align with one another. A constant, spatially invariant Fermi energy is required in order for there to be no current normal to the interface.

The band bending predominately occurs in the semiconductor, as the latter is normally doped in the range of 10^{15}–10^{19} cm^{-3}. On the other hand, the electron density in the metal is typically of the order of 10^{22} cm^{-3}. Hence, the amount of charge that must be created to balance the Fermi energies is not large on the scale of the metal density. The result is that the space charge is almost entirely located in the semiconductor, as shown in figure 4.13. The charge depletion leads to a potential drop across the junction and appears as a built-in potential that has a value of

$$V_s = -\frac{eN_D x_n^2}{2\epsilon_s}, \tag{4.4}$$

where N_D is the donor's doping level (here, the material is assumed to be n-type), x_n is the depletion depth, and ϵ_s is the permittivity of the semiconductor. The net depletion sheet charge in the semiconductor is $N_D x_n$, and this produces the electric field at the interface. The built-in potential is related to the Schottky barrier height. In the case in which the built-in potential is due to the work function difference, this potential is given by

$$eV_s = \Phi_m - \Phi_s - k_B T \ln\left(\frac{N_c}{N_D}\right), \tag{4.5}$$

where the Φ's are the work functions of the metal and the semiconductor, $k_B T$ is the thermal energy, and N_c is the effective density of states in the conduction band of the semiconductor.

If the barrier height is really determined by the work functions, then the built-in potential should be linear in the difference shown in (4.5). But in many semiconductors, this has not been found to be the case. Kurtin, McGill, and Mead studied a great many semiconductors and metals to examine the role of the work function [90]. For a given semiconductor, they proposed that the built-in potential would vary with the work function of the metal as follows:

$$eV_s = S\Phi_m + \text{constant}. \tag{4.6}$$

Here, S is a parameter that is expected to vary between zero and one, the latter being its value in (4.5). Their results are shown in figure 4.14, where the measured values of S are plotted versus the electronegativity difference between the two materials making up the semiconductor material (defined to be zero in Si). In fact, S is near zero for the pure covalent semiconductors Si and Ge. In compound semiconductors, there is a small component of ionic bonding due to the unequal charges on the two atoms of the unit cell [91].

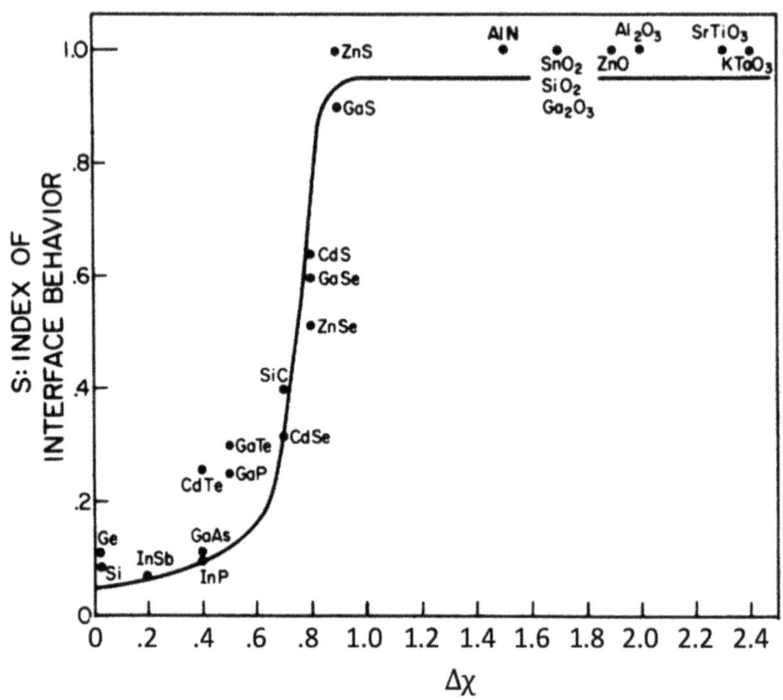

Figure 4.14. Collected data on the slopes of S obtained for different metals deposited on various semiconductors plotted against the electronegativity difference of the two atoms in the semiconductor. Reprinted with permission from [90], copyright 1969 by the American Physical Society.

An alternative theory for the built-in potential in the metal–semiconductor junction was suggested by Bardeen [92]. He pointed out that most semiconductors did not possess pristine surfaces, so that it was quite likely that defects could be found at these surfaces (and interfaces). These defects could be expected to pin the Fermi level at the semiconductor interface to their energy levels. The surface potential of the semiconductor would then be given by

$$eV_s = E_{c,\,\mathrm{surf}} - E_{\mathrm{defect}} - k_B T \ln\left(\frac{N_c}{N_D}\right). \tag{4.7}$$

Hence, the defect-dominated material would have no dependence upon the metal work function. This explained *a priori* the later results obtained through measurements. For example, for *n*-type GaAs, the Fermi level is pinned approximately 0.85 eV below the conduction band edge in all metal–semiconductor contacts, regardless of the choice of metal.

4.4.2 The MESFET

In the metal–semiconductor FET, the Schottky gate is the actual gate of the transistor. Normally, the channel is a doped layer of semiconductor grown on top of a lightly doped substrate or even a substrate of the opposite type of doping. This creates a barrier between the substrate and the active layer of the channel. The voltage on the gate then modulates the current and conductance by modulating the thickness of the active layer. The source–drain potential creates a voltage along the channel, just as in a MOSFET. If the channel voltage along the channel is $V_a(x)$, where x is the distance along the channel (see figure 4.15), the width of the depletion region is then

$$W(x) = \sqrt{\frac{2\epsilon_s[V_s + V_a(x)]}{eN_D}}. \tag{4.8}$$

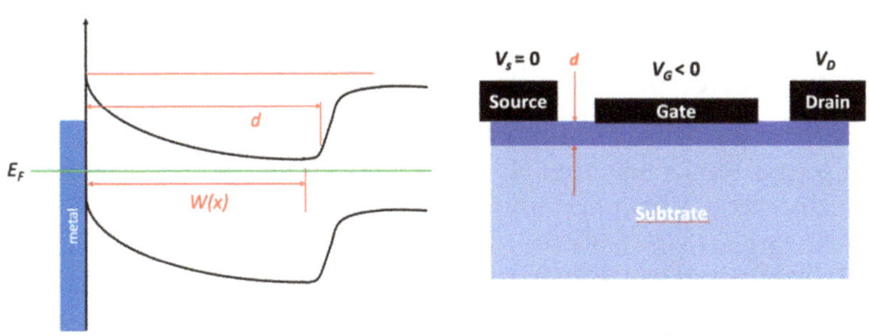

Figure 4.15. (a) The shapes of the energy bands for an MESFET under bias. (b) An illustration of the structure of the MESFET. The dark blue region represents metal (on the left) and heavily doped semiconductor (on the right).

Once this width W equals the thickness of the channel layer, the conduction is pinched off.

Just as in the MOSFET, the current throughout the structure is constant, and at any point x along the channel, this current may be written as

$$I\delta x = ZeN_D\mu[d - W(x)]dV(x). \\ = ZeN_D\mu\left[d - \sqrt{\frac{2\epsilon_s[V_s + V_G + V_a(x)]}{eN_D}}\right]\delta V(x) \cdot \tag{4.9}$$

Here, V_G is the voltage applied to the gate (<0), which further depletes the channel layer, and Z is the width (into the page) of the device. The source–drain voltage is taken to be zero at the source end of the gate ($x = 0$) and V_D at the drain end of the gate ($x = L$). This omits any potential drop in the region between the ends of the gate and the physical positions of the source and drain. Any potential drop in these latter regions is detrimental to the operation of the device, but the idea is that the resistance under the gate dominates the total device resistance. Using these boundary conditions, the differential equation (4.9) may be easily integrated to give

$$I = \frac{ZeN_D\mu}{L}\left\{dV_D - \frac{2}{3}\sqrt{\frac{2\epsilon_s}{eN_D}}[(V_s + V_G + V_D)^{3/2} - (V_s + V_G)^{3/2}]\right\}. \tag{4.10}$$

The MESFET is used in a number of applications that require strict control of the current. However, its use is limited in applications in which linearity is required because of the power law in the current of (4.10). That is, the current does not vary particularly linearly with the gate voltage, as may be seen from an examination of this latter equation. In fact, this may also be seen in the transconductance

$$g_m = \frac{\partial I}{\partial V_G} = \frac{Z\mu\sqrt{2\epsilon_s eN_D}}{L}[\sqrt{V_s + V_G + V_D} - \sqrt{V_s + V_G}]. \tag{4.11}$$

It is clear that this is not linear with respect to the gate voltage except in the case of a very small gate voltage $|V_G| \ll V_s, V_D$. The transconductance and gain are then relatively constant.

4.5 The high-electron-mobility transistor

As in the case of the MESFET described above, the HEMT is nominally fabricated from compound semiconductors. It has always been recognized that compound semiconductors would be more desirable materials than Si, because their mobility can be much larger than that of Si. As can be seen from (4.10), this gives a much larger current in the MESFET, and the same is true for the HEMT. However, the problem is that the oxides of most compound semiconductors are ill-suited for device application. The very properties that yield the Fermi level pinning in the metal–semiconductor junction are those which preclude good MOS behavior. One cannot avoid the fact that nature has blessed Si devices with an almost perfect oxide and interface. For this reason, HEMTs have always been heterostructure devices

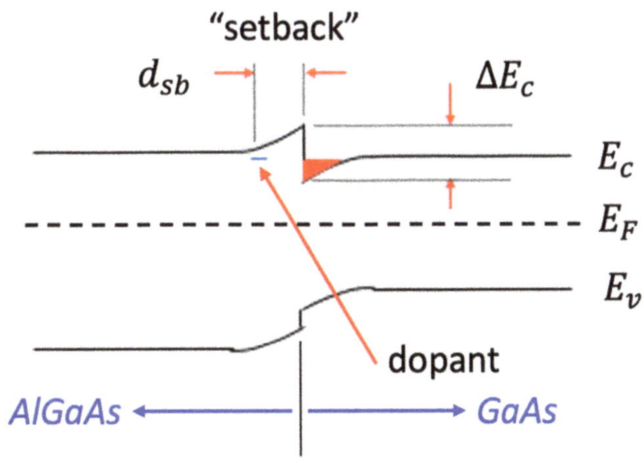

Figure 4.16. The typical structure of a modulation-doped heterostructure in AlGaAs/GaAs. The dopant is set back from the interface to reduce the impurity scattering and raise the mobility.

[93]. In modern versions, the mobility is further enhanced by moving the dopants out of the active channel using a technique known as *modulation doping* [94]. Examples of the heterostructure and the doping technique are shown in figure 4.16. Due to the difference in the bandgaps of GaAs and AlGaAs, band bending takes place at the interface. The dopant atoms are placed in the AlGaAs at a position that is set back from the interface. The electrons from these donor atoms are transferred to the quantum well on the GaAs side of the interface. The closer the donors are to the interface, the larger the fraction of electrons that transfers (determined by the self-consistent band bending at the interface), while the further they are from the interface, the more they are screened by the inversion-layer electrons and the higher the mobility of these electrons. So, the choice of setback distance is one optimization of the device. The undoped AlGaAs now acts as an 'insulator,' taking the place of the SiO_2 in the MOSFET. Nevertheless, the HEMT is usually a depletion-mode device.

The structure of a typical HEMT is shown in figure 4.17, this time for a GaN/AlGaN device [95]. This differs from the AlGaAs/GaAs HEMT in several ways, as this is a microwave power device. First, the structure is nominally inverted, with the barrier layer below the active GaN layer. In the GaN system, the würtzite structure is ferroelectric and has a built-in electric field. In a heterostructure, there is a discontinuity of the electric which is enhanced by a strain between the two layers that leads to an additional piezoelectric field. The direction of this field can be changed by growing the device on different surfaces of the crystal. In the present case, a HEMT grown on the N surface is considered, and the inversion-layer electrons (which are induced by the discontinuity in the built-in electric field) now lie at the bottom of the GaN layer, adjacent to the buried AlN layer. This moves them back from the surface so that there is little surface-roughness scattering. In this case, there still is a modulation doping layer shown as the thin Si-doped GaN layer, with an undoped GaN spacer layer to create the setback. The SiN layer serves to protect the top GaN,

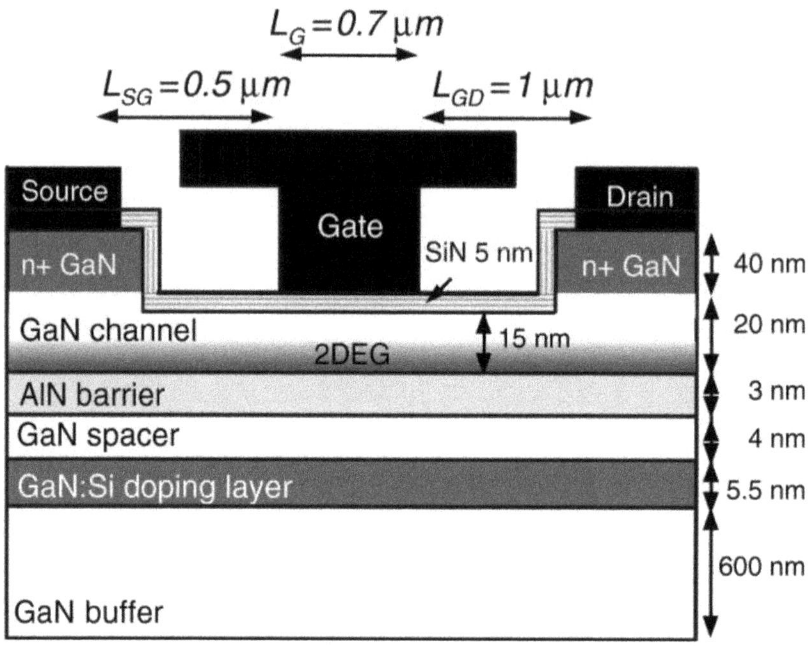

Figure 4.17. The structure used in the simulation of a N-face AlGaN/GaN HEMT with a gate length of 700 nm. Here, the barrier layer is below the active layer. © 2010 IEEE. Reprinted, with permission, from Guerra *et al* [95].

as the bare surface is likely to break down under the high bias fields of a power HEMT. The T-shaped gate provides a lower gate resistance without compromising the gate length.

The current through the HEMT is much closer to that of a MOSFET than that of a MESFET, as the lack of carriers in the top layer (under the gate) acts like the insulator of the MOSFET. However, the device is a 'normally on' device in contrast to the 'normally off' nature of the MOSFET. This means that, like the MESFET, the gate voltage is usually negative.

4.6 Other interesting devices

While the great majority of semiconductor devices have been covered in the preceding three sections, there are variations on these themes that may be found in the literature. In many cases, these are modifications of one of the above structures. But other basic concepts exist, such as tunnel diodes, that occur on their own or in combination with one of the above structures, such as the tunnel FET. A few of these are discussed in this section.

4.6.1 Tunnel diodes

Tunneling is often thought of as a pure quantum phenomenon that does not occur in classical mechanics. However, this is wrong. The barrier penetration problem is well known in classical electromagnetics and optics. The quantum mechanical properties of particles become fully apparent when we consider them to be waves. Even in a classical setting, waves can exhibit barrier penetration. In electromagnetic waves (and optics), a decaying wave is termed an *evanescent* wave. The tunneling effect between two optical media is called *frustrated total internal reflection.* A familiar version is found in the case of internal reflection from below the surface in a swimming pool. Water has a higher dielectric constant than free space, so when a person is under water, there is a critical angle beyond which one can no longer see out of the pool. A similar effect can be generated in optics using a dielectric prism. At a particular angle of incidence, the wave is always reflected from the back surface. Even when the wave is reflected, an evanescent wave remains at the back that rapidly decays away from the prism. But if one places another material that has a high dielectric constant close to the prism, a propagating wave can be generated in this new material, and this wave is excited by the evanescent wave. As a result, a fraction of the wave that is incident on the prism penetrates the barrier (the space between the prism and this new dielectric material). In essence, the second dielectric material *frustrates* the internal reflection of the incident wave. This is a purely classical property of electromagnetic waves. On the other hand, if this wave were to arise from a particle, the process would be termed *tunneling*. Esaki received the Nobel prize for first recognizing that tunneling would occur in heavily doped (doped to degeneracy) *p–n* junction diodes [96]. Nevertheless, one has to recognize that this process is only quantum mechanical because the particle is being treated as a wave.

A *p–n* junction in which the doping on the two sides of the junction is heavily degenerate has a built-in potential that is larger than the bandgap. This is depicted in the diagram of figure 4.18. When a small forward bias is applied, the built-in potential is reduced by the applied bias. This raises the conduction band edge in the *n*-type region (on the right-hand side of the figure) relative to the *p*-type region. The full states (in gold) then see empty states in the valence band on the left-hand side of the figure. As a result, the electrons on the right can tunnel to the empty states on the left, producing a current through the device. As the voltage is further increased, raising the conduction band still further, the full states on the right start to see the forbidden gap on the left and the current is reduced. Hence, the forward tunneling current only exists over a small range of positive voltages. However, at some point, normal diffusion current through the *p–n* junction begins and the current rises again. This creates a region of negative differential conductance in the current–voltage curve, which has proved useful for many applications.

Reverse bias increases the built-in potential, with the result that full states of the valence band overlap empty states in the conduction band, and a large tunneling current occurs. This current always increases with further reverse bias. Consequently, the tunnel diode has a large reverse current, which is not seen in normal *p–n* junction diodes. This may be detrimental in various applications, so care must be taken in the use of these devices.

In more recent years, a single carrier type of tunnel structure has become of interest [97]. This is usually produced using only *n*-type material, and heterostructures are used to delineate the barriers. Such a structure is depicted in figure 4.19(a),

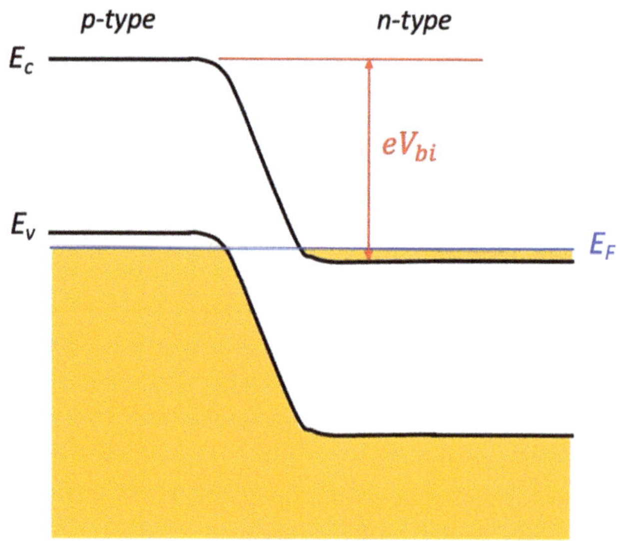

Figure 4.18. Degenerate doping on both sides of a *p–n* junction leads to an overlap between the conduction band on the right and the valence band on the left. This produces a tunnel diode. The full states on each side of the junction are indicated in gold.

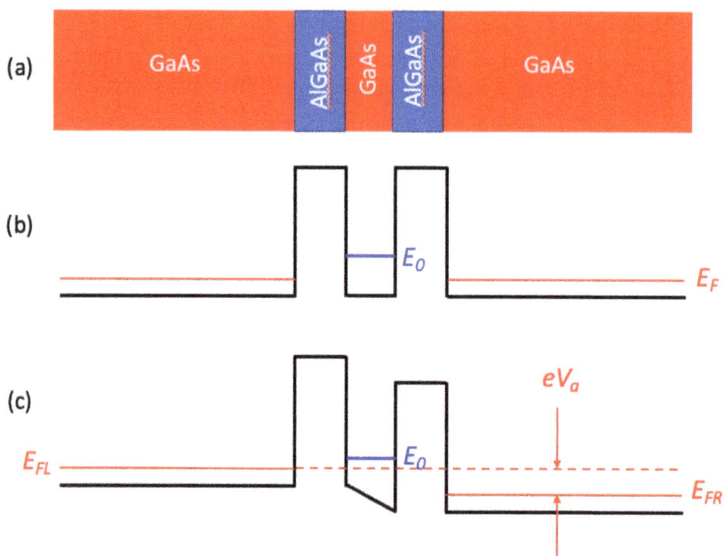

Figure 4.19. (a) The material structure of a resonant-tunneling diode created using the AlGaAs/GaAs system. (b) The conduction band energies for the structure in panel (a). The two barriers form a quantum well in which there is a resonant level E_0. (c) The shift of the bands when a forward bias is applied.

in which AlGaAs barriers are used in a GaAs base structure. The conduction band variation through this double barrier structure is indicated in panel (b) of the figure (this is an extreme case, as thermal rounding of the corners of the potential step is not shown). The two barriers create a quantum well between them, and this always has at least one bound state, indicated by the energy E_0. When the energy of an incident electron hits this bound-state energy, the tunneling probability rises to unity, provided that the two barriers are equal in shape and amplitude. Normally, this bound state sits several k_BT above the Fermi energy, so that no current flows through the structure. When a bias is applied to the structure (a positive voltage on the right-hand side of the figure), the energy of the bias pulls the conduction band on the right down relative to that on the left. Similarly, the bound state is also pulled downward by this bias, as shown in panel (c) of the figure. When the bound state gets within k_BT of the Fermi level E_{FL} on the left, current begins to flow through the device (due to the thermal spread of occupied states around the Fermi energy). This current increases with further increases in the applied voltage until the bound state passes below the conduction band edge on the left. Beyond this point, further increases in bias produces no further current until it can flow via single barrier tunneling or by thermal excitation over the barriers. This again gives a region of negative differential conductance in the current–voltage characteristics.

The resonant-tunneling diode, such as that shown in figure 4.19, has been the so-called 'fruit fly' for studies of quantum transport since at least the 1980s. It has been studied in detail with density matrices [98], Wigner functions [99], and nonequilibrium Green's functions [100]. One reason for the high level of interest in this device is its ability to reach terahertz frequencies [99, 101].

4.6.2 Tunnel FETs

As remarked above, the concept of tunneling in semiconductors is quite old. The idea of mixing the negative differential conductance of the tunnel device with the normal operation of an FET has also attracted considerable attention for a few decades and seems to have first appeared around 2007 [102]. The rationale for this idea originates from the poor subthreshold slope and the leakage current that arises as the channel length gets shorter. This becomes an acute problem in heavily scaled devices. The rise of FinFETs and nanowire FETs, as discussed previously, does not alleviate this problem, although they do allow more control over the 'off' current, which is part of the subthreshold current. But the idea of greatly improving the subthreshold slope using tunneling from the source contact into the channel appeared to be quite promising and was quickly pursued by many [103, 104]. Of course, those working with alternative materials immediately concluded that their material, such as graphene, would be a suitable material for this device [105].

It was also found that interband tunneling from a *p*-type source into an *n*-type inversion channel improved the subthreshold slope. However, there was a correlated problem with the device. While the tunneling barrier did improve the subthreshold slope, it also lowered the available 'on' current in the device, which is more or less unacceptable behavior. This problem is intrinsic to the tunneling concept [106], and this fact seems to have been overlooked in the enthusiasm to lower the subthreshold slope.

The current that passes through the constriction at the gate end of the channel defines the saturation current of the MOSFET. This current depends upon the inversion charge at this point, and this latter quantity depends upon the potential $V_G - V_S$, where V_S is the potential on the source side of this constriction. In normal MOSFETs, $V_S = 0$ because the source is grounded. However, when a tunneling structure is introduced in the source lead or at the source, this has the effect of introducing a source resistance with its concomitant voltage drop. That is, $V_S > 0$. This provides negative feedback, which reduces the gain and the saturation current in the device. If one tries to increase the current by increasing the gate voltage, the resulting increasing current raises the potential drop across the source resistance and thus works to lower the current. Thus, the MOSFET never carries as much current as it could have done without the tunnel structure.

The presence of the tunnel structure, or source resistance, also is detrimental to the use of the MOSFET for logic operations. Normally, one of the logic levels is connected to ground, and when the transistor is fully on, it normally pulls the potential to this ground level. However, when there is a source resistance, such as the tunneling structure, the lowest potential that can be reached when the device is on is this source potential V_S. Consequently, this lowers the noise margin available for this logic level and thus makes the gate much more susceptible to noise in the system (switching noise is always a concern). One may conclude, therefore, that the tunnel FET is not a meaningful approach to solving the subthreshold problem. Other approaches, some of which have been mentioned earlier, are more profitable approaches to the problem.

4.6.3 SpinFETs

A field-effect transistor based upon the spin of the carriers (Spin-FET) was theoretically predicted more than three decades ago [107]. To make such a device, one has to depend upon the blocking of one spin through a contact; i.e. the contact passes only one spin mode, either 'up' or 'down.' This normally requires a magnetic field. The prototypical Spin-FET is thus based on the use of two ferromagnetic contacts (at the source and drain) that connect to the semiconductor channel. Spin-polarized electrons are injected via the source contact, and these then flow through the channel region. As usual, the channel current is modulated by the gate voltage [108], but this results in electron spin precession (due to the spin–orbit interaction inherent in the band structure, see chapter 5). The magnetization of the drain contact acts as a filter, since only spin-aligned electrons (selected by the magnetic state of the ferromagnetic drain contact) can pass through it. This device has some attraction, but one must also realize that half of the possible current is being discarded, and this provides a perhaps unacceptable reduction of the device current.

The strength of the spin–orbit interaction dictates the minimum required length of the semiconductor channel (in order to attain sufficient spin precession). In principle, two dominant spin–orbit interaction mechanisms are known: the Rashba effect (geometrically induced structural asymmetry [108]) and the Dresselhaus effect (bulk inversion symmetry breaking [109]). In silicon thin films, the spin–orbit interaction is dominated by the Dresselhaus terms [110, 111], which arise from the breaking of inversion symmetry at the channel–oxide interface. Among the resulting challenges is the fact that a channel length of the order of microns is necessary to obtain this spin precession. This requirement is contradictory to the ongoing device downsizing.

In 2015, a Spin-MOSFET was first demonstrated; it exhibited a high on/off ratio while operating at room temperature [112]. As described above, two metallic ferromagnetic contacts were connected by a nonmagnetic semiconductor channel. The actual device allowed for both charge and spin transport. Parallel/antiparallel magnetization alignment between source and drain led to a current increase/decrease at the drain contact, respectively, as predicted by theory. The ability to change the contact magnetization of the contacts was provided either by an external magnetic field and/or by the current through the contact (the switching was due to spin-transfer torque, a process utilized in magnetic read-only memories (MRAMs)) [113]. An important feature of Spin-MOSFETs is the fact that the contact magnetization is preserved without external power, thus enabling the production of nonvolatile logic devices. However, the presence of an external magnetic field is certainly undesirable. As a concluding remark, it should be pointed out that these devices only have advantages over conventional transistor designs when no current flow is required for the fundamental transistor switching mechanism.

4.6.4 Single-atom transistors

The ultimate endpoint of scaling is a transistor that uses a single atom as the active region. A few research devices have reached this ultimate limit. One of these is a device in which the 'channel' region is composed of the states of a single impurity atom. In a sense, this device is actually closer to a set of devices known as single-electron transistors. There are a couple of methods of placing a single atom, such as a single P atom, in a desired location in a silicon structure. The heart of such a device lies in the placement of this single P atom in such a way as to to create a potential well at the site (P is a normal donor in Si, with an energy level for the bound electron a few tens of meV below the conduction band; this provides the potential well). There are at least two methods of putting single atoms into controlled positions within a Si substrate,

The first method is to use the ion implantation of a single P atom [114]. The major issues here are the generation of a 'beam' of a single atom that is to be implanted and the detection of the actual implantation of this atom. However, this is possible and is a current technology. Early work, for example, implanted a linear array of Si atoms into SiO_2 in order to create a Si nanowire of ultimately small dimensions [115]. Rather than a basic ion implantation machine, a focused ion beam microscope is usually used in this technology. This approach has been used to implant single P atoms into an MOSFET to control the flat-band voltage (and threshold voltage) in a device [116]. This use of this process has also been suggested for the implantation of P atoms to serve as qubits in quantum computers [117].

The second method for putting single atoms into place is to use chemical means and the properties of the bare Si (100) surface. Under ultrahigh vacuum conditions, this bare surface will reconstruct into a 2 × 1 (atomic) configuration. This structure has rows of dimerized dangling bonds that lower the overall energy of the surface [91]. However, this surface can be passivated by treatment with hydrogen, so that each dangling bond is attached to a single hydrogen atom. Using a scanning tunneling microscope, an individual bond can be broken [118], and this region is then exposed to other molecules, such as phosphine, which will bond to the dangling bond. Subsequent thermal processing moves the phosphorous atom into the Si lattice [119].

The single P atom for the transistor was placed by this latter process [120]. A set of source and drain metal layers was then used for contacts, and a pair of side gates were fabricated, also by metal deposition. According to measurements and simulation, the transport of a single electron from source to drain can be controllably accomplished. The single P atom produces the potential that serves as the channel in the device. Of course, the movement of more electrons is possible, but the current passed always remains lower than that of a regular MOSFET.

There is another approach to such a single-atom transistor, but this time the 'atom' is one of the dangling surface bonds itself. This approach to the structure uses the chemical fabrication method described above, but no P atom is attached. The dangling bond itself has an energy structure that contains two levels, and these can provide switching due to charge transitions from one level to the other [121]. The bond itself thus acts as a two-level atom. The atomic spacing of atoms is a small fraction of a nanometer, so that a single bond provides an active region of subnanometer dimensions. It has been demonstrated that single charges can be controllably moved between two bond 'atoms' [122]. Such switching can be considered to represent the qubits used for quantum computing. External gates located around the bond are used to move the charge states in such a manner as to manipulate the qubits. The amount of energy level movement that can be achieved using these gates is much larger than the thermal energy, which means that room-temperature operation is possible. Nearly all current embodiments of quantum computers operate at extremely low temperatures, so that the possibility of room-temperature operation is very attractive.

4.7 Ballistic transport

Ballistic transport in its truest meaning is transport without scattering. Such transport has been discussed for semiconductor devices for some years [123, 124]. In fact, ballistic transport exists in nearly all devices to some extent and began with vacuum tubes [125, 126]. In general, when transport is ballistic, the resistance of a device is independent of its length, as described by the Landauer formula (section 1.5); all the essential voltage drop occurs in the contacts. On the other hand, when scattering dominates and the transport is diffusive, the resistance linearly depends upon the length, according to Ohm's law. Naturally, in most devices, there is a characteristic length at which the transport changes from ballistic to diffusive [127]. This transition has been seen to be as small as 2 nm in Si quantum wires [127] and perhaps as long as 200 nm in InAs quantum wires [128] (both at room temperature).

A more common indicator of at least quasi-ballistic behavior is the onset of drain-induced barrier lowering (DIBL) [129]. It is the presence of scattering in the channel that isolates the gate voltage from the drain voltage. If there is no scattering, then the simple solution given by Laplace's equation tells one that the drain voltage affects the potential drop between source and gate, as the positive drain voltage pulls down the effective barrier at the source end of the gate.

However, the shape of the drain current characteristic curves does not confirm in any way that ballistic transport is present. A comparison with and without scattering must be done in order to provide some assurance [130]. The impact of ballistic transport modifies the current characteristics. In a normal MOSFET, the derivation of the current arises from the relation

$$\int_0^L I dy = ZC_{\text{ox}} \int_0^L v(V_G - V_T - V_a) dy, \tag{4.12}$$

where C_{ox} is the gate oxide capacitance, L is the channel length, V_G and V_D are the gate and drain potentials, Z is the gate width, v is the velocity along the channel, and V_a is the surface/channel voltage which varies along the channel. While one would now introduce the mobility, this has no meaning in the case of ballistic transport. Rather, if the transport is ballistic, the velocity is a function of the local potential, as $v = \sqrt{2eV_a/m^*}$. In the ballistic case, the potential can be described through [126]

$$\frac{dV_a}{dy} = \alpha\sqrt{I}\, V_a^{1/4}, \tag{4.13}$$

where

$$\alpha^2 = 8\pi\sqrt{\frac{2e}{m^*}}. \tag{4.14}$$

Introducing these last two equations into (4.12) shows that saturation still occurs in the normal manner, but that

$$I = K(V_G - V_T)^{3/2}, \quad K = \left[\frac{4\alpha W C_{\text{ox}}}{90\pi L}\right]^{2/3}. \tag{4.15}$$

Hence, the appearance of saturation cannot tell us whether the transport in the channel is ballistic or diffusive. The voltage dependence of the current does change between these two cases [131]. *However, saturation is the result of pinch-off in the channel, and is not a property of the transport. The current depends upon the injection at the source–gate region. Result (4.15) arises entirely because it is assumed that pinch-off still occurs in the channel.* The Langmuir–Child law does not impose any restrictions on the carrier motion, such as those that occur with pinch-off. If the pinch-off restriction is removed, then one might well expect diode- or triode-like behavior in the MOSFET, and even with pinch-off, (4.15) indicates some triode-like behavior. One should point out that the difference in the two approaches lies in the exponent on the voltage term in (4.15). For normal MOSFET behavior, the exponent is one, while here it is 3/2 for the ballistic transport case.

In simulations of InAs MOSFETs, one can attempt to determine the channel length at which the transport switches from ballistic to diffusive. This length can be longer than the channel length at room temperature [128, 132]. Indeed, this length is sufficiently long to affect the device characteristics of an InAs nanowire MOSFET with a 30 nm gate length [133]. The cross-section of the channel in the simulated device was 9 × 8.5 nm, and the channel region was undoped. It was found that this device exhibits strange behavior in the gate characteristics near turn-on. It also does not saturate in the normal manner. Careful studies varying the nature of the transport have confirmed that it is almost completely ballistic in nature. This explains the diode- or triode-like behavior of the drain current. More recently, studies of microwave HEMTs based upon an InP substrate with a strained $In_{0.75}Ga_{0.25}As$ channel have allowed one to carefully determine that the probability for ballistic transport between two points x_a and x_b, both measured along the channel, is [134]

$$P_{\text{ball}}(x_b - x_a) = \prod_{x_i = x_a}^{x_b} \exp\left(-\frac{\Delta x}{L(x_i)}\right), \tag{4.16}$$

where Δx is the grid spacing in the simulation (2 nm in this simulation) and $L(x_i)$ is the *effective* gate length of the device (the effective gate length is discussed in detail in section 8.2 of chapter 8). This simulation suggests that the transport is ~55% ballistic under these conditions.

4.8 Optical devices

The study of photoexcited carriers in semiconductors is one of the more interesting aspects of nanoelectronics, but the related devices seem to be limited to lasers, LEDs, solar cells, and other detectors. The role of the excess carriers is crucial to the operation of such devices. In the various detectors, the hot carriers created optically are quite different than those created by an electric field. These photoexcited carriers create a hot distribution that is no longer in thermal equilibrium, especially with the lattice, and one has already made an assumption that a temperature can describe the electron system. However, this distribution is still largely symmetric in momentum space, as the photon momentum is quite small compared to electron momenta. Nevertheless, there some asymmetry in the distribution may arise from the photons themselves. In the light-emitting devices, the nonequilibrium carriers are injected due to the nature of the *p–n* junctions. As interesting as these devices are, quantum effects mainly arise from the presence of quantum wells within the device.

The study of quantum transport has been somewhat limited as well, but there are significant examples that are of interest. Wigner functions (see chapter 11), for example, have been used to study quantum well lasers [135]. Here, the quantum approach was used to describe the lateral confinement, the quantum well energy structure, and the carrier dynamics in quantum transport. Wigner functions have also been used to describe the transport in cascade laser devices [136], which are based entirely on a single carrier type in a controlled superlattice structure [137].

The quantum effects and transport in both Si [138] and quantum well [139] photocells have been studied using nonequilibrium Green's functions (NEGFs) for some years. In such work, particularly in concentrator solar cells, the self-energy of the optically created carriers within the quantum wells can be an important aspect of the cell and needs to be incorporated into the quantum mechanical simulations [140].

References

[1] Kennard E H 1928 *Phys. Rev.* **31** 876
[2] Thompson S E *et al* 2004 *IEEE Trans. Electron Devices* **51** 1790
[3] Lilienfeld J E 1930 *US Patent* 1,745,175
[4] Heil O 1935 *UK Patent* 439,457
[5] Hoddeson L 1994 *Hist. Technol* **11** 121
[6] Bardeen J B and Brattain W H 1950 *US Patent* 2,524,035EJ
[7] Bardeen J B and Brattain W H 1948 *Phys. Rev.* **74** 230
[8] Shockley W 1950 *US Patent* 2,502,488
[9] Atalla M M, Tannenbaum E and Scheibner 1959 *Bell Syst. Tech. J.* **38** 749
[10] Kilby J S 1964 *US Patent* 3,138,743
[11] Noyce R N 1961 *US Patent* 2,981,877
[12] Wanlass F M and Sah C T 1963 *Solid-State Circuits Conf* (New York: IEEE) 32–3
[13] Wanlass F M 1967 *US Patent* 3,356,858
[14] Schwierz F and Liou J J 2002 *Modern Microwave Transistors* (New York: Wiley Interscience)
[15] Mei X *et al* 2015 *IEEE Electron Device Lett.* **36** 327
[16] Khan M A *et al* 1993 *Appl. Phys. Lett.* **63** 1214
[17] Ferry D K 2000 *Superlattices Microstruct* **27** 61
[18] Ando T, Fowler A B and Stern F 1982 *Rev. Mod. Phys.* **54** 437
[19] Fowler R H and Nordheim L 1928 *Proc. Roy. Soc.* A **119** 173–81
[20] Lenslinger M and Snow E H 1969 *J. Appl. Phys.* **40** 278
[21] Ferry D K 2019 *The Copenhagen Conspiracy* (Singapore: Jenny Stanford Publishing)
[22] Oriols X and Ferry D K 2021 *Proc. IEEE* **109** 955
[23] DiVencenzo D P 2000 *Fort. Phys.* **48** 771
[24] Weinbub J, Ballacchia M and Nedjalkov M 2018 *Phys. Stat. Sol. RRL* **12** 18000111
[25] Thornber K K 1971 *Phys. Rev.* B **3** 1929
[26] Barker J R 1973 *J. Phys.* C **6** 2663
[27] Moore G 1965 *Electronics* **38** 114
[28] Ferry D K 2008 *Science* **319** 579
[29] Rupp K and Selberherr S 2011 *IEEE Trans. Semicon. Manuf* **24** 1
[30] Ferry D K 2021 *Hot Carriers in Semiconductors* (Bristol: IOP Publishing)
[31] Ferry D K 2020 *Semiconductors: Bonds and Bands* (Bristol: IOP Publishing)
[32] Prange R E and Nee T-W 1968 *Phys. Rev.* **168** 779
[33] Thompson S E *et al* 2004 *IEEE Electron Device Lett.* **25** 191
[34] Ungersboeck E *et al* 2008 *Math. Comput. Simul* **79** 1071
[35] Mistry K *et al* 2007 *Proc. Int. Electron Device Meeting* (New York: IEEE Press) 247
[36] Hisamoto D *et al* 2000 *IEEE Trans. Electron Devices* **47** 2320

[37] Barraud S *et al* 2017 *Proc. Int. Electron Device Meeting* (New York: IEEE Press) pp 29.2
[38] Song T *et al* 2021 *2017 IEEE International Solid-State Circuits Conference (ISSCC)* (New York: IEEE) 208–9
[39] Keyes R W 1977 *Science* **195** 1230
[40] Tsutsumi T 2022 *Jpn. J. Appl. Phys.* **61** SD1043
[41] Li Y *et al* 2021 *Semicond. Sci. Technol.* **36** 125001
[42] Wong H and Kakushima K 2022 *Nanomaterials* **12** 1739
[43] Grasser T and Selberherr S 2000 *Microelectron. J.* **31** 873
[44] Dennard R *et al* 1974 *IEEE J. Solid-State Circuits* **9** 256
[45] Barker J R and Ferry D K 1980 *Solid-State Electron* **23** 531
[46] Goodnick S M *et al* 1985 *Phys. Rev.* **B32** 8171
[47] Sellier J M, Nedjalkov M, Dimov I and Selberherr S 2013 *J. Appl. Phys.* **114** 174902
[48] Hess K and Vogl P 1979 *Solid State Commun* **30** 807
[49] Ferry D K, Grubin H L and Barker J R 1982 *Molecular Electronic Devices* ed F Carter (New York: Dekker) pp 195–204
[50] Joshi R P and Ferry D K 1991 *Phys. Rev.* B **43** 9734
[51] Joshi R P and Ferry D K 1992 *Semicond. Sci. Technol.* **7** B319
[52] von Klitzing K, Dorda G and Pepper M 1980 *Phys. Rev. Lett.* **45** 494
[53] Ferry D K, Goodnick S M and Bird J P 2009 *Transport in Nanostructures* (Cambridge: Cambridge University Press) 2nd
[54] Vasileska D *et al* 1995 *J. Vac. Sci. Technol.* B **13** 1841
[55] Hedin L and Lundqvist B I 1971 *J. Phys.* C **4** 2064
[56] Vasileska D, Schroder D K and Ferry D K 1997 *IEEE Trans. Electron Devices* **44** 584
[57] Dar S *et al* 2007 *IEEE Trans. Nanotechnol.* **6** 97
[58] Ungersböck E *et al* 2007 *IEEE Trans. Electron Devices* **54** 2183
[59] Dar S *et al* 2006 *IEEE Trans. Electron Devices* **53** 3054
[60] Goswami S *et al* 2007 *Nature Phys.* **3** 41
[61] Windbacher T *et al* 2010 *Solid-State Electron.* **54** 137
[62] Osintsev D, Sverdlov V and Selberherr S 2015 *Solid-State Electron.* **112** 46
[63] Zorman B *et al* 2004 *J. Comput. Electron.* **3** 351
[64] Krishnan S, Fischetti M and Vasileska D 2006 *J. Vac. Sci. Technol.* B **24** 1997
[65] Guillame T and Mouis M 2006 *Solid-State Electron* **50** 701
[66] Conzatti F *et al* 2009 *Solid-State Electron* **53** 706
[67] Shifren L *et al* 2004 *Appl. Phys. Lett.* **25** 6188
[68] Wang E X *et al* 2006 *IEEE Trans. Electron Devices* **53** 1840
[69] Kotlyar R *et al* 2009 *J. Comput. Electron* **8** 110
[70] Wong H-S and Taur Y 1993 *Proceedings of IEEE International Electron Devices Meeting, Washington, DC, USA, 1993* (New York: IEEE) 705–8
[71] Arokianathan C R, Asenov A and Davies J H 1996 *J. Appl. Phys.* **80** 226
[72] Vasileska D, Gross W J and Ferry D K 1998 *Sixth International Workshop on Computational Electronics* (New York: IEEE) 259–62
[73] Arokianathan C R, Davies J H and Asenov A 1998 *VLSI Des.* **8** 331
[74] Wordelman C J and Ravaioli U 1999 *Physica* B **272** 568
[75] Khan H R *et al* 2004 *J. Comput. Electron* **3** 337
[76] Colinge J P *et al* 1990 *International Technical Digest on Electron Devices* (New York: IEEE) 595–8

[77] Young T 1807 *Course of Lectures on Natural Philosopy and Mechanical Arts* (London: J. Johnson) pp L. XXXIX
[78] Akashi T *et al* 2002 *Appl. Phys. Lett.* **81** 1922
[79] Taylor G I 1909 *Proc. Cambridge Phil. Soc.* **14** 417
[80] Einstein A 1909 *Phys. Z.* **10** 817
[81] Anderson P W 1958 *Phys. Rev.* **109** 1492
[82] Skocpol W J *et al* 1986 *Phys. Rev. Lett.* **56** 2865
[83] Gilbert M J, Akis R and Ferry D K 2003 *J. Comput. Electron* **2** 329
[84] Gilbert M J and Ferry D K 2005 *IEEE Trans. Nanotechnol.* **4** 355
[85] Ramey S M and Ferry D K 2004 *Semicond. Sci. Technol.* **19** S238
[86] Gross W J, Vasileska D and Ferry D K 2000 *IEEE Trans. Electron Devices* **47** 1831
[87] Mead C A 1966 *Proc. IEEE* **54** 307
[88] Braun F 1874 *Ann. Phys. Pogg* **153** 556
[89] Schottky W 1938 *Naturwiss* **26** 843
[90] Kurtin S, McGill T C and Mead C A 1969 *Phys. Rev. Lett.* **22** 143
[91] Ferry D K 1991 *Semiconductors* (New York: Macmillan)
[92] Bardeen J 1947 *Phys. Rev.* **71** 717
[93] Mimura T *et al* 1980 *Jpn. J. Appl. Phys.* **19** L225
[94] Dingle R *et al* 1978 *Phys. Rev. Lett.* **33** 827
[95] Guerra D *et al* 2010 *IEEE Trans. Electron Devices* **57** 3348
[96] Esaki L 1958 *Phys. Rev.* **109** 603
[97] Tsu R and Esaki L 1973 *Appl. Phys. Lett.* **22** 562
[98] Mizuta H and Goodings C J 1991 *J. Phys.: Condens. Matter* **3** 3739
[99] Kluksdahl N C *et al* 1989 *Phys. Rev.* B **39** 7720
[100] Lake R, Klimeck G and Datta S 1993 *Phys. Rev.* B **47** 6427
[101] Sollner T C L G *et al* 1983 *Appl. Phys. Lett.* **43** 588
[102] Knoch J, Mantl S and Appenzeller J 2007 *Solid-State Electron* **51** 572
[103] Venkatagirish N *et al* 2009 *ECS Trans.* **22** 273
[104] Verhulst A S *et al* 2009 *ECS Trans.* **25** 455
[105] Fiori G and Iannaccone G 2009 *IEEE Electron Device Lett.* **30** 1096
[106] Ionescu A M and Riel H 2011 *Nature* **479** 329
[107] Datta S and Das B 1990 *Appl. Phys. Lett.* **56** 665
[108] Bychkov Y A and Rashba É I 1984 *Sov. J. Theor. Phys. Lett.* **39** 78
[109] Dresselhaus G 1955 *Phys. Rev.* **100** 580
[110] Nestoklon M O *et al* 2008 *Phys. Rev.* B **77** 155328
[111] Prada M, Klimeck G and Joynt R 2011 *New J. Phys.* **13** 13009
[112] Tahara T *et al* 2015 *Appl. Phys. Exp* **8** 113004
[113] Sugahara S and Nitta J 2010 *Proc. IEEE* **98** 2124
[114] Tanada T *et al* 1998 *J. Vac. Sci. Technol.* B **16** 2489
[115] Tanada T *et al* 1997 *Appl. Surf. Sci.* **117/118** 677
[116] Shinada T *et al* 2000 *Appl. Surf. Sci.* **162/163** 499
[117] Schenkel T *et al* 2002 *J. Vac. Sci. Technol.* B **20** 2819
[118] Lyding J W *et al* 1994 *Appl. Phys. Lett.* **64** 2010
[119] Weber B S *et al* 2012 *Science* **335** 64
[120] Fuechsle M *et al* 2012 *Nat. Nanotechnol* **7** 242
[121] Livadaru L *et al* 2010 *New J. Phys.* **12** 083018

[122] Rashidi M *et al* 2018 *Phys. Rev. Lett.* **121** 166801
[123] Shur M S and Eastman L F 1979 *IEEE Trans. Electron Devices* **26** 1677
[124] Rahman A *et al* 2003 *IEEE Trans. Electron Devices* **50** 1853
[125] Child C D 1911 *Phys. Rev. (Ser. 1)* **32** 492
[126] Langmuir I 1913 *Phys. Rev.* **2** 450
[127] Akis R, Gilbert M and Ferry D K 2006 *J. Phys.: Conf. Ser.* **38** 87
[128] Zhou X *et al* 2006 *Appl. Phys. Lett.* **89** 053113
[129] Ferry D K, Akis R and Gilbert M J 2007 *J. Comput. Theor. Nanosci.* **4** 1149
[130] Saint Martin J, Bournel A and Dollfus P 2004 *IEEE Trans. Electron Devices* **51** 1148
[131] Ferry D K, Gilbert M J and Akis R 2008 *IEEE Trans. Electron Devices* **55** 2820
[132] Gilbert M J and Banerjee S K 2007 *IEEE Trans. Electron Devices* **54** 645
[133] Gilbert M J and Ferry D K 2006 *J. Appl. Phys.* **99** 054503
[134] Akis R *et al* 2009 *IEEE Trans. Electron Devices* **56** 2935
[135] Weetman P and Wartak M S 2007 *Phys. Rev.* B **76** 035332
[136] Jonasson O and Knezevic I 2015 *J. Comput. Electron* **15** 879
[137] Capasso F, Mohammed K and Cho A Y 1986 *Appl. Phys. Lett.* **48** 478
[138] Aeberhard U 2011 *Phys. Rev.* B **84** 035454
[139] Aeberhard U 2008 *Phys. Rev.* B **77** 125343
[140] Aeberhard U 2016 *J. Comput. Electron* **15** 1219

IOP Publishing

Quantum Transport in Semiconductor Devices
Simulation using particles
David K Ferry, Xavier Oriols and Josef Weinbub

Chapter 5

What is needed from quantum mechanics

While they are not generally discussed, quantum effects are integral to the operation of today's semiconductor devices. Quantization has been observed in the channels of both metal–oxide–semiconductor field-effect transistors (MOSFETs) and high-electron-mobility transistors (HEMTs) for many years, but this quantization is normal to the direction of current flow, so is not often observed. Quantum effects are also important in the use of strain in modern devices to enhance the conductance of both electrons and holes. Quantum effects are becoming even more important as devices move from three-dimensional transport to quasi one-dimensional transport in nanosheet MOSFETs. It was an understanding of this quantization that led to the use of strain to modify the band structure in both the conduction and valence bands and thus an enhancement in device performance. The controlled introduction of quantum modifications has been a part of device fabrication for nearly the last two decades. This has been coupled to efforts to develop simulation tools for the semiconductor world that incorporate these quantum effects as part of the (new) standard tools.

For an even longer time, people have studied quantum transport (and written books on the subject) in metals and semiconductors at low temperatures. This is the world of mesoscopic devices [1, 2]. As semiconductor devices evolve into smaller, faster, and higher-frequency applications, the world of mesoscopics moves rapidly into the world of our modern devices at room temperature (for a recent review of quantum transport in field-effect transistors (FETs), see [3]). Consider, for example, an electron in a Si quantized inversion layer (which is quasi two-dimensional in planar devices and even in FinFET devices) with a Fermi energy of some 20 meV, corresponding to an inversion density of some 1.6×10^{12} cm^{-2}, which is relatively high for Si. This electron has a momentum wave vector of some 2.5×10^{6} cm^{-1}, which leads to a de Broglie wavelength of 25 nm! Compare this with the critical dimension of a nanosheet in the neighborhood of 3 nm. It has become impossible to design the transistor without quantum mechanics. Notice that temperature does not enter this calculation, so that mesoscopics has come to the modern world.

doi:10.1088/978-0-7503-5237-6ch5

Size quantization is only the first quantum effect that is usually considered. It is, perhaps, not even the most important one. As size scales continue to be reduced, there are limits that suggest many other quantum effects will become important. These could be entanglement, tunneling, correlated scattering, and so on. In this chapter, we want to discuss some of these effects and the roles that they play in transport. To this end, it is important to first discuss some space- and timescales that become essential.

5.1 Space- and timescales

Our discussion of appropriate length scales begins with the device's characteristic length L and the mean free path $v\tau$, where τ is the mean free time between collisions and v is the mean velocity of the carriers, be they electrons or holes. These are connected with the concepts of mobility $\mu = e\tau/m^*$ and diffusivity $D(\approx\mu k_B T/e)$ (here, m^* is the effective mass of the carrier in the semiconductor, the diffusivity is given in terms of the mean velocity by $v^2\tau/d$, and d is the dimensionality—for example, two dimensions in a typical MOSFET or HEMT). In quantum transport, there is also the idea of a coherence time, or phase-breaking time τ_φ which describes the time during which the wave function retains its coherence. While this is a vague definition and description, a better understanding can only arise as the effect of this phase coherence is observed in experiments [2]. This phase-breaking time is connected to a phase-breaking length through the diffusivity via $l_\varphi = D\tau_\varphi$, which is the average distance over which the electrons diffuse before their phase is disrupted by various scattering events. Naturally, when this length becomes larger than, or comparable to, the size of the device under investigation, one can expect quantum effects to dominate. Other important lengths are the Fermi wavelength, which was introduced above, and the thermal length. This latter is a length that is a little more difficult to grasp, as it connects the diffusivity with a time defined by the thermal broadening of a typical energy level. This broadening is used to define a timescale via $\hbar/k_B T$, and this connects the diffusivity to the thermal length via $l_T = \sqrt{\hbar D/k_B T}$.

An important time for nonequilibrium systems is the energy-relaxation time τ_E, which describes the timescale over which the energy per carrier in the system returns to its equilibrium value. Semiconductor devices are far-from-equilibrium devices during operation, and the carriers gain substantial energy from the applied biases (which produce what are called 'hot' carriers [4]). But this is only one of the many characteristic timescales that are important in far-from-equilibrium transport. The importance of these scales is due to the necessity of evaluating the time dependence of the various measurable properties. The nonequilibrium distribution function also evolves over various timescales, and this affects the transport. Numerous collisions occur in semiconductors, and it is these collisions that provide the mechanism responsible for the exchanges of energy and momentum (to name just a few of the quantities) and relax these toward their equilibrium values. There are collisions between the carriers which randomize the energy and momentum *within* an ensemble but do not relax either of these quantities for the ensemble as a whole. There are also elastic collisions between the carriers and impurities or acoustic phonons which relax the momentum but do little to relax the energy. Finally, there are inelastic collisions between the carriers and lattice vibrations which relax both the energy and the momentum. In general, four generic timescales can be identified [5, 6]:

$$\tau_c < \tau < \tau_R < \tau_H. \tag{5.1}$$

The shortest time is the average *duration* of a collision and is denoted by τ_c. Generally, this timescale is quite short (perhaps on the femtosecond scale for optical phonon scattering [7]). While this is unimportant in many situations, it can affect coherence on the tens to hundreds of femtoseconds timescale, especially in very-high-frequency devices [8]. This collision duration is considered to be the time required to establish the energy-conserving delta function in the Fermi golden rule for scattering. In contrast, the average time *between* collisions, the mean free time, is denoted by the simple τ. For timescales such that $t \leqslant \tau$, the evolution of the system depends strongly upon the details of the initial state. Generally, $\tau \gg \tau_c$, but this is not always the case within high electric fields, and the breakdown of this inequality can lead to new transport effects which must be treated in a quantum mechanical manner.

The establishment of a *nonequilibrium steady state* can be achieved within a few or a few tens of τ, and the characteristic time associated with this process is the *relaxation time* τ_R. Typical situations described by a relaxation time are the momentum relaxation and, in high fields, the energy relaxation. As we will see in later chapters, the momentum relaxation time is often associated with a shift in the distribution function which gives the average velocity or momentum of the carrier system. The corresponding energy-relaxation time is associated with a broadening of the distribution, which leads to an elevated electron temperature. The phrase '*nonequilibrium steady state*' may not be familiar to the average reader, but the presence of a current flow in a semiconductor means that the system is *not* in equilibrium (there is no current flow in equilibrium). The steady state should not be confused with equilibrium, as they are quite different. In addition, it is also possible for a system to be in a non-steady state while in its nonequilibrium state. One only has to think of an oscillator. It certainly is out of equilibrium in that it is converting energy (supplied by the bias) from DC to some frequency (of the oscillator) while also dissipating energy. The fact that it is oscillating means that the total current is not constant.

Finally, consider the final timescale, τ_H. If configuration-space gradients exist, the situation becomes more complex. Relaxation in momentum space proceeds on the scale of τ_R and establishes a 'local' equilibrium over regions smaller than a macroscopic scale, perhaps only a few mean free paths in extent. The achievement of a uniform steady state of nonequilibrium requires a longer time, the hydrodynamic time $\tau_H > \tau_R$. The use of this hydrodynamic description comes from the study of current flow as an analog to fluid flow that began in the distant past [9, 10]. Only for times that are large compared to this hydrodynamic time can the ensemble truly be said to be stationary. Examples of such hydrodynamic times arise when the actual temperature of the lattice is inhomogeneous (and certainly may be different than that of any heat sink). Examples of the latter are thermoelectric effects, one example of which is the temperature differential that arises from applying an electric field and current to a semiconductor.

While this range of various timescales is quite suitable for large devices, modern devices can operate on timescales for which the set of times above is insufficient. For example, high fields can accelerate electrons (which are the primary carrier in microwave devices) to achieve an average velocity of $\sim$3–6 $\times$ 10^7 cm s^{-1}. Hence, if the average length of the active region is only $\sim$30 nm, then the carriers only spend an average of 50–100 fs in the device. This is quite short and indicates that the transport can be quasi ballistic (that is, no, or a few, scattering events take place in this time) [11, 12]. Ballistic transport can occur in very-high-frequency devices [13] as well as in high-speed logic devices [14]. This can be quite important for the observation of quantum effects in these devices, because if there is little scattering, there is little to break up quantum coherence within the device. The presence of quantum coherence then leads to transport that is definitely nonclassical in behavior.

To illustrate the appearance of quantum coherence, consider scattering when only a few impurities exist in the 'device.' For example, consider a small FinFET or nanosheet device whose active region is $\sim$5 $\times$ 20 $\times$ 20 nm^3. If this region is doped to 10^{18} cm^{-3}, on average there will only be two impurities in the device! With impurity scattering, it is necessary to understand that the coulomb force only deflects the carrier without altering its energy or the magnitude of its momentum. The carrier retains nearly all of its coherence throughout the scattering. That is, impurity scattering provides no decoherence or phase-breaking to the quantum transport. This implies that scattering by an impurity is not unlike the two-slit experiment described by Young [15]. While Young used optical waves, a similar behavior is observed in electrons. In a transmission electron microscope (TEM), a charged wire creates a biprism which splits the electron beam as it passes to either side of the wire [16]. Importantly, a single electron in the TEM, much like a single photon [17], still creates the interference pattern. That is, the single electron (or photon) must have both wave and particle properties, as recognized by Einstein [18]. For our purposes here, these observations mean that a single charged impurity behaves much like the biprism in a TEM, and a single electron passing by this impurity can create an interference pattern that represents its wavelike self-interactions in diffracting around the impurity.

A study has shown that this coherence can easily be demonstrated using impurities. Consider a simulation involving two impurities which provide a complicated wave interaction process that illustrates the same interference principle. This study used a simple nanowire structure with two impurities embedded in it, as shown in figure 5.1 [19]. The nanowire shown in this figure is 40 nm wide, and the area of interest is some 60 nm long. The impurities are modelled by their Coulomb potential that generates a double-well (potential) energy profile. The minimum energy of the two wells (−0.365 eV) is reached at the position of the two impurities, i. e. in the center of the green circles, which are constant potential energy lines at −0.175 eV. Electrons are continually injected at y = 0, x = 20 nm and absorbed completely at y = 60 nm. The electrons do not interact with each other and, as such, the study provides a single-electron picture. These electrons are each simulated by a

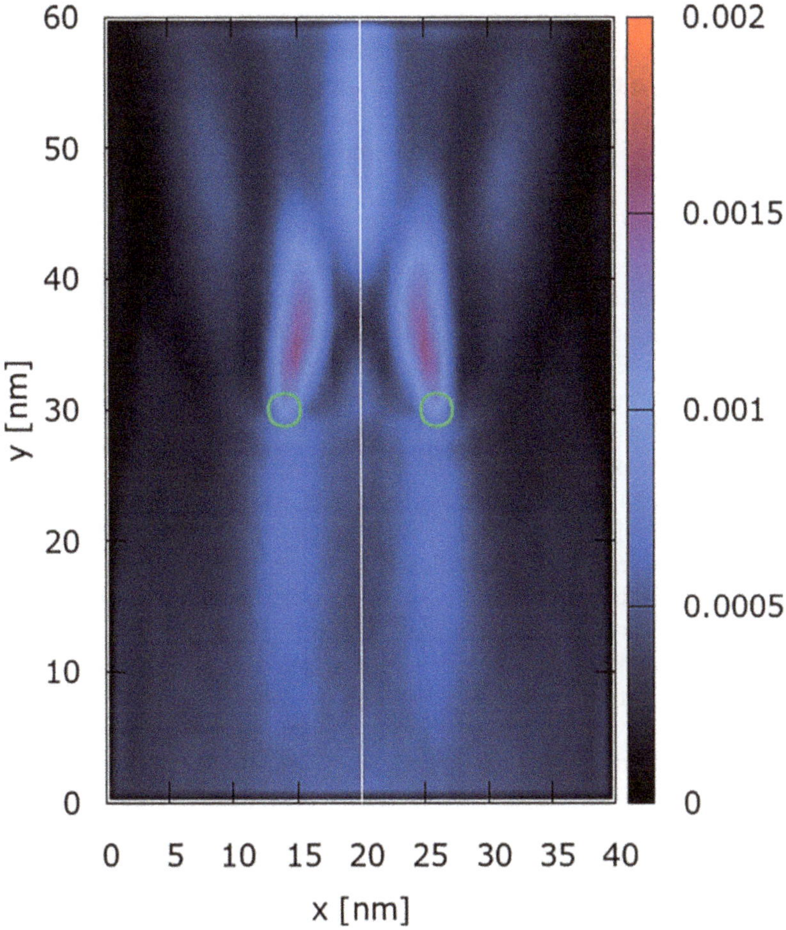

Figure 5.1. Quantum electron density after 200 fs of the initial minimum uncertainty condition. The green circles are equal energy lines at 0.175 eV of the impurity potential. Reproduced from [19]. CC BY 4.0.

Wigner function wave packet (Wigner functions are covered in chapter 11). The continual injection allows the steady-state behaviors of the electrons and impurities to be studied. The electrons are injected at an energy of 0.141 eV. This, plus the nonlocal interaction with the impurities, leads to the injected carriers forming the two beams apparent in the figure in the region before the impurities (the carriers travel from bottom to top in this figure). The carrier 'density' is plotted in the figure. The individual carrier waves are diffracted by the impurities and then interfere with themselves downstream from the impurities. This interference would be present even if only a single impurity were present, as was shown by Barker using nonequilibrium Green's functions (NEGFs) [20]. The interference peak evolves all the way to the exit of the nanowire. These interference peaks are certainly reminiscent of the two-slit experiment.

But the Wigner experiment of figure 5.1 demonstrates even more. Before the scattering takes place, the beam of individual electrons approaching the impurities is already nonlocally interacting with the impurities. Nonlocality is a pure quantum mechanical effect and cannot be reproduced with classical mechanics. One might say that it is the wavelike behavior of the electrons which leads to the interference.

5.2 Entanglement

Early in quantum mechanics, it was usually assumed that if two bodies interacted, they would be linked via a correlation that built up between two bodies that collided, even if the description of the collision was treated classically. The appearance of questions raised by Einstein, Podolsky, and Rosen (EPR) [21] brought this view sharply into question. These latter authors explicitly considered the interaction between two particles that took place at a point in time and space. Initially, the two particles were assumed to have the same, large value of the magnitude of momentum, although directed in such a manner as to interact at a specific time. They then raised the question as to how subsequent measurement of the momentum of one of the particles could affect the other; that is, how could the state of the second particle be influenced by a measurement performed upon the first? After the interaction, there was no further physical interaction between the two particles, so they suggested that knowledge of the second particle had to proceed by some 'spooky interaction at a distance.'

It was Schrödinger who defined the process of entanglement in terms of the wave function [22, 23], and, at the same time, noted that entanglement was the most important aspect of quantum mechanics. In this sense, the two particles of EPR are quantum mechanically defined by two separate single-particle wave functions before the interaction. But after the interaction, they have a single two-particle wave function, as they are entangled particles. Until some phase-breaking interaction occurs, this entanglement persists. Because this entangled wave function contains information about both particles, the measurement of one implies values for the second. This view of quantum entanglement requires no strange communication over a distance. Needless to say, controversy followed both the EPR question and the responses from a variety of others, not least of whom was Bohr [24]. The reader will have already recognized the debate about the interpretation of quantum mechanics, since it was discussed previously in several preceding chapters. The EPR problem highlights this debate in a more explicit form. However, Schrödinger raised the bar with his cat paradox [22, 23].

5.2.1 Schrödinger's cat

It was well-known that Schrödinger was no friend of the Bohr interpretation of quantum mechanics [22]. In order to illustrate the paradox raised by EPR, he proposed his own version in terms of a cat confined to a box with a radioactive element. The amount of the radioactive substance was small, so that there was a reasonable chance of decay within a given period of time. This decay would be measured by a Geiger counter, whose signal would cause a hammer to break a vial containing a poisonous gas. At a time corresponding to the half-life, it could be assumed that either the decay had occurred and the cat was dead, or it had not yet occurred and the cat was alive. And this is the paradox. According to the orthodox view, reality occurred only upon measurement, and one could not say whether the cat was alive or dead before this measurement took place. Nevertheless, the experiment is composed of a variety of purely classical objects: the box, the hammer, the cat, and the poison. So, part of the paradox lies in just how the quantum mechanics appears at the macroscopic level. If it is asserted that a quantum mechanical result is to be obtained from the measurement of this system, then the possible outcomes must be expressed in the wave function adopted for their description.

In the orthodox theory, Schrödinger tells us that [22] '...there is intrinsically only awareness, observation, measurement.' If these properties are used to determine the state of the system at a given time, then that is the best one can do, and any further questioning is meaningless. Hence, it must be assumed that the microscopic world made the jump to the macroscopic world through the supreme arbiter of truth—the measurement [25]! And this is the orthodox view. The measurement now determines whether the cat is dead or alive. But the wave function adopted must contain both of these possibilities, and because quantum mechanics is a probabilistic theory (or possibilistic, perhaps), these two results should appear with some *smearing* between them.

The cat experiment is almost entirely and completely local. The radioactive substance, the vial, and the cat are all collocated. However, in the EPR problem, the two particles are well separated in space. So, it is somewhat difficult to understand how the wave functions can be compared. Bohm addressed this in later years through a variation on the experiment, in which particle spins were used [26]. The creation of two identical particles from photon absorption would ensure that these two particles had equal and opposite spins. Either particle one would have up spin or particle two would have up spin, and the other particle would have the opposite spin. In this case, one could write the entangled wave function as

$$\psi(1, 2) = \frac{1}{2}(|\uparrow\rangle_1 |\downarrow\rangle_2 + |\downarrow\rangle_1 |\uparrow\rangle_2), \tag{5.2}$$

where the arrows denote the spin and the subscript defines to which particle the partial wave function refers. This is the type of wave function with which the cat paradox can be described (but questions can be raised, as will be discussed below).

Here, the radioactive material has either decayed ($| d\rangle_{\rm r}$) or it has not decayed ($| nd\rangle_{\rm r}$), where the subscript r designates the radioactive material. In the first case, the cat is dead ($| d\rangle_{\rm c}$), while in the second case, the cat is alive ($| a\rangle_{\rm c}$), where the subscript c denotes the cat. The most descriptive entangled wave function according to (5.2) is given by

$$\psi(r, c) = \frac{1}{2}(| d\rangle_{\rm r} | d\rangle_{\rm c} + | nd\rangle_{\rm r} | a\rangle_{\rm c}). \tag{5.3}$$

Of course, for the local problem of the cat, one does not need to know about the radioactive material, only about the state of the cat. But, to keep it within the nonlocal problem of quantum mechanics and the EPR paradox, both must be described. So, the opening of the box by the observer, as required in the orthodox Bohr interpretation, now gives us an answer about the state of the cat. It has to be remembered, as pointed out in chapter 2, that the observer is required for this interpretation, and the measurement is not complete until the observer has communicated his observation! That is, the observer has to not only perform the measurement but must also be completely certain that his observation has been sufficiently 'powerful' to alter the reality of the laboratory. The orthodox view asserts that the act of observation not only disturbs what is to be measured, but the act of observation itself is what brings about the (reality of the) outcome [22]. However, as any experimenter knows, the measurement may give the truth but not the whole truth.

The experimenter knows there are other results that can arise in an experiment such as the cat in a box described above [25]. First, there is the smearing mentioned above, which does not occur for the spin experiment. Radioactive decay is assumed to follow an exponential time variation, with the half-life being the characteristic decay constant. Thus, one would like to choose a particular time when the probability of no decay has fallen to 0.5, so that there is an equal likelihood of decay and no decay. But, there is a further delay for the Geiger counter to respond to the decay product and to then trigger the release of the hammer. How long does it take the hammer to fall and break the vial? All of this is part of the design and setup of the experiment. For the orthodox theory, this setup process is unimportant—reality occurs with the measurement! For the real experimenter, this setup is all-important. It is the initial state of the system and determines much of the outcome. Here, reality exists long before the measurement takes place. As a principle, the time chosen as the 'particular time' mentioned here cannot be so long that the cat either starves to death or dies of old age. There is then Murphy's Law, well-known to experimenters, where things just do not go according to plan. Suppose the hammer fails to break the vial? We then have a situation where decay has occurred, yet the cat still lives. In addition, since the decay process is probabilistic, there is a chance that the decay will not occur until the moment that the observer opens the box. Because of the delay in the counter and hammer fall, there is a chance that the cat will jump out of the box and the observer will be recipient of the poisonous gas, dying himself. Again, the decay has occurred, but the cat still lives. Here, there arises the conundrum of the orthodox view: *has a measurement really been made if the observer is dead and cannot communicate the result*? There remain a great many more questions. It is clear, even to a casual observer, that there are more than just two possible outcomes. In fact, the above discussion has already led to four possible outcomes. A competent experimenter is going to be aware that there are even more possibilities. So, a wave function designed to express only the naïve expected outcomes is, by nature, incomplete. And if the outcome is unexpected, a dead observer cannot transmit this information to anyone. Are we then to conclude that the entire experiment is not real? This is certainly not the case for those who prepared the experiment. Schrödinger was known to be a strong opponent of the orthodox view in quantum mechanics, and probably best expressed this point in saying [22]: '...one *cannot* put the ψ-function directly in place of the model or physical thing ... because in the realism point of view observation is a natural process like any other and cannot *per se* bring about an interruption of the orderly flow of events.' (Here, realism refers to reality before the measurement.) There is no magic in the measurement in this view, and an orderly flow of events allows one to infer that causality exists even in quantum mechanics.

5.2.2 Further thoughts

The situation here again brings the concept of observability to the forefront. This was discussed in chapter 1 with both the Heisenberg approach to quantum mechanics and the introduction of complex system in section 1.3. Consider, for example, the simple hammer in the cat paradox above. How does the experimenter measure the inherent time delay between the time when the Geiger counter signals the decay and the time at which the hammer breaks (or misses) the vial? Can the friction in the joint about which the hammer pivots be measured? As was pointed out in section 1.3, a modern complex system has a great many intrinsic variables, and many of these are either unmeasurable and/or uncontrollable. The friction in the hammer's motion is one of these.

In quantum mechanics, the *state* of a system is assumed to be definable. It is described by a wave function. Even Schrödinger understood that carrying over the concept of state to a classical system involved considerable problems. He pointed out [22] that: 'The classical concept of *state* becomes lost, in that at most a well-chosen *half* of a complete set of variables, can be assigned definite numerical values.' While perhaps underestimating the number of states of a complex system that may be observable, he nonetheless realizes that there are a great number of variables that have unassigned values in any real experiment. Nevertheless, the orthodox theory assumes that reality and complete understanding of the system arrive with the measurement.

Returning to the EPR experiment, the two particles are represented by a two-particle wave function. Each of the two particles has a position and a momentum (or a spin in the Bohm version). The entangled wave function contains this full information. If it is considered that the positions and momentum are both referenced to the point of interaction, these reduced coordinates can be described as if the two momenta are equal and opposite in vector notation and the positions similarly may be so described. Thus, a measurement of the position and momentum of one particle then completely determines the values for the second particle. But a quantum measurement does not provide this form of reality. This is because the position and momentum cannot be measured simultaneously with any accuracy. This is true classically as well as quantum mechanically. In the latter case, the uncertainty principle interferes with the measurement. While a position or a momentum may be measured for the first particle, this represents the maximum information that can be obtained about the entangled system in a single measurement. It is the entanglement that allows us to assume the values of the properties of the second particle. And, as Schrödinger pointed out [22], measurements of separated systems cannot influence each other—the measurement of such an entangled system is necessary. As a result of these considerations, it is possible to understand why EPR considered quantum mechanics to be incomplete, at least within the orthodox view. There was no specification of how to measure the unmeasurable.

In the larger, many-body world of charged particles necessary to simulate semiconductor devices, one may have N particles interacting through central forces as well as responding to external potentials. This leads to $6N$ variables—three momentum and three position variables for each particle—in the extended phase space. As described in the preceding chapters, it is possible to sum over $6(N - 1)$ of these variables and obtain a probability distribution for a single 'effective' particle, although this distribution is coupled to a two-particle distribution function. In Boltzmann's time, the two-particle distribution was dispatched through his *stosszahlansatz* (molecular chaos) assumption, which meant that the correlation between particles died out in a time much shorter than the timescale for which solutions were desired.

In quantum mechanics, this many-body world is more complicated. For a quantum description, we are stymied by the fact that if we describe the $3N$ positions perfectly, we can say nothing about the momenta. Normally, one would describe the initial conditions in solving for the motion of the particles. But the initial conditions also include the initial momenta. In some sense, the momenta are now hidden from view. This is apparent in the Schrödinger equation from the fact that the momentum is a differential operator in the coordinate representation. If one Fourier transforms the equation into momentum space, then the position becomes a differential operator. One cannot have both views. Here, it may now be seen why Schrödinger decided that only half of the variables could be assigned definite numerical values. This half was either positions or momenta. The fact that half of the variables may not be assigned definite values is due to the 'hidden' description of this half of the set [25]. Because of this huge uncertainty in the values of half of the expected variables, one can clearly understand how quantum mechanics may be considered to be incomplete, as described by EPR [21]. In fact, entanglement and the arguments of EPR and Schrödinger forced Bohr to admit [27]: '…the present formulation of quantum mechanics in spite of its great fruitfulness would seem to be no more than a first step in the necessary generalization of the classical mode of description, … we must be prepared for a more comprehensive generalization.' Nothing highlights this more explicitly than the various alternative interpretations of quantum mechanics that have arisen in the ensuing years, most of which have abandoned Bohr's view of reality via measurement.

5.3 Particles in quantum transport

It may be recalled that, in the classical world, particles and waves exist in separate universes and are not considered to have any connection. It was Planck's considerations in explaining black-body radiation that forced particles into the world of waves [28]. But subsequent investigations have examined particles themselves, beginning with Bohr's atomic theory [29]. Even Heisenberg's new quantum mechanics primarily focused upon the properties of particles [30], as particles were normal in the realm of classical mechanics. The coupling of particle and wave arrived with de Broglie [31], although the necessity for both appeared earlier in Taylor's experiments [17]. These beginnings of particles in quantum mechanics were discussed in chapter 1. Nevertheless, the continued development of atomic and molecular physics followed particles into the subsequent years, whether these were electrons, protons, or neutrons or eventually a veritable zoo of elementary particles. Indeed, interactions in high-energy physics are studied using the travel of particles through cloud chambers, or in high-energy machines such as cyclotrons. It was always the particles and their interactions that became the foci. In semiconductor devices, it is also particles which are the subject of interest—how do these electrons and holes lead to controllable currents under the action of various potentials? However, in many approaches, the study of these devices has devolved back to the study of quantum waves and their confinement. This is true whether direct solutions of the Schrödinger equation or fancier Green's functions are utilized.

In the early days of semiconductor devices, there were far too many electrons and/or holes for researchers to actually follow the motion of each particle. Thus, the focus turned to the distribution function and the Boltzmann transport equation. With the arrival of the Monte Carlo method in semiconductors [32], particles began to be used to simulate the distribution function. But this single-particle Monte Carlo approach could only give steady-state values for the velocity, as there is no ergodic theorem, nor an Einstein relation, for the nonequilibrium state [33]. It remained for the ensemble Monte Carlo (EMC), see chapter 9, to use a sufficient number of particles to allow the temporal evolution of the distribution function to be found [34]. This is because the ensemble average is always a proper way to approach statistical physics. Nevertheless, the number of particles which could be used in these simulations remained much smaller than the actual number of particles in a real device. This changed only when computers allowed the number of simulated particles to grow, while the actual device size shrunk.

Deviation from the simple forms of the Boltzmann transport equation began when interparticle forces started to be simulated via molecular dynamics treatments of the coulomb interaction [35, 36]. This once again reduced the number of particles that could be simulated due to the computational difficulty of the molecular dynamics evaluations. Incorporating this into device simulations was complicated by the additional need to solve Poisson's equation, which led to a double counting of the interparticle forces. In general, though, the time-consuming process in device simulation remains the Poisson solver. Both problems, i.e. the compute time for molecular dynamics forces and the double counting of these forces, were solved by a force-splitting approach [37]. That is, the interparticle coulomb force was split into a long-range part and a short-range part. The long-range part was included in the Poisson solver, while the short-range part was treated using molecular dynamics. One could think of the split distance as being related to the grid spacing. However, this could be formalized for a general variation, for example by writing the coulomb force as [38–40]

$$\frac{1}{r} \to [1 - \mathrm{erf}(r/r_0)]\frac{1}{r} + \mathrm{erf}\,(r/r_0)\frac{1}{r}, \tag{5.4}$$

where r_0 is a convenient cutoff (which may be related to the grid size). The first term on the right-hand side is the short-range part of the coulomb potential and is used for the molecular dynamics calculation. This limits the forces to a few nearby particles. The second term is the long-range part, which yields the charge used to compute the mesh forces from Poisson's equation. This coulomb split is applied to both the carriers and the charged impurities, so that the full coulomb interaction is treated completely in real space and not by any perturbation theory. While the processing of the Poisson equation is slowed by the additional term on the coulomb force, this can be compensated by using a larger grid size at high carrier densities and suitably adjusting the molecular dynamics forces to ensure an accurate solution is still obtained.

These days, it is not unusual for the EMC calculation to include far more particles than actually exist in a small device. Care has to be taken to ensure that the interparticle forces do not oversmooth the actual potentials that exist within the device. There is an additional factor that is important in using the molecular dynamics forces. The real-space treatment of carriers and impurities automatically accomplishes the screening function. This is important because these coulombic forces are descreened at high velocities, and this descreening is also incorporated within the force calculations [36].

As we move into quantum treatments of semiconductor devices, it is clear that we can continue to consider particles as a representation of the distribution function, which is now a quantum distribution and can incorporate spatial confinement directly. However, the application of this principle has mainly occurred with Wigner functions (due to the similarities between the Boltzmann transport equation and the Wigner function equation of motion [41]) and with Bohmian transport based on the quantum potential [42]. In the next section, a brief review of the various approaches to quantum transport is given.

5.4 Current approaches

The simplest approach to quantum transport uses the Schrödinger equation, which is straight from the textbook. This involves matching the wave function and its derivative at the interface between each pair of slices of the structure. However, this is generally unstable when a large number of such interfaces are used, due to the presence of evanescent modes that must be accurately treated. This leads to both increasing and decreasing exponentials, which are the source of the instability. Stability can be restored by modifying the recursion to one based upon the scattering matrix, as discussed in section 3.3, which has long been a staple of microwave systems; it entered quantum mechanics through the Lippmann–Schwinger equation [43]. The stability of this approach lies in the fact that modal solutions maintain their orthogonality through the scattering process. In small devices, the connection to microwave theory becomes stronger as the transport becomes dominated by the modes introduced by lateral confinement, much as in a microwave waveguide. The use of scattering states provides a method of building up an orthogonal ensemble, even with weighting of the states by e.g. a Fermi–Dirac distribution [44]. This is particularly useful for the initial conditions of the wave function in a particular semiconductor device. In the end, this approach allows one to determine the transmission from one end to the other and evaluate the conductance via the Landauer formula, also discussed in chapters 1 and 2.

To our knowledge, there has been very little work on using particles to simulate the quantum wave function in open systems. But there is no reason not to use particle approaches, as they may be faster than even the recursion approaches mentioned above. Here, the particles would be distributed in phase space in a manner representative of the confined wave function and the local quantum distribution function for the system. Another approach is to use a particle method that follows the hydrodynamic, or Bohmian, approach of section 1.2.3. In this approach, particles follow their natural trajectories under the constraints of both the classical potentials and the quantum potential (1.34). This means that the wave function must be found in order to determine the quantum potential, so this implies a particle method which is related to the wave function. However, it is a little more complicated than simply using particles to simulate the wave function. It also differs in that the hydrodynamic particles are real particles that are accompanied by the wave function, whereas the particles simulating solutions to the Schrödinger equation are thought to be wave packets. This Bohmian approach has been shown to be quite effective [42]. The quantum potential is a form of an effective potential, and this (and the Bohmian simulation approach) will be discussed in detail in chapter 10. The reader is referred to [45] for a recent review on computational methods used in nanoelectronics where quantum effects play a major role.

Both the Wigner function and nonequilibrium Green's functions were also discussed in chapter 1. Here, we give them another introduction, but this time through the development of the density matrix that is connected to both. The density matrix itself has been used for device simulations, so it has an important connection to the general development of quantum transport that goes beyond just the wave function solutions.

5.4.1 The density matrix

Generally, the Schrödinger equation is solved by assuming an expansion of the wave function in a suitable basis set so that each basis function is an energy eigenfunction according to $H\varphi_n(\boldsymbol{r}) = E_n\varphi_n(\boldsymbol{r})$. Here, E_n is the energy level corresponding to the particular basis function φ_n. For example, in a quantum well, the basis functions are the set of wave functions corresponding to the bound states of the quantum well. In a quantum wire, they are the bound states arising from the quantization resulting from the wire's width and height. In both of these cases, the basis functions are the *mode* wave functions. They have values on the grid sites used in the computation. The total wave function can be written as

$$\psi(\boldsymbol{r}, t) = \sum_n c_n \varphi_n(\boldsymbol{r}) e^{-iE_n t/\hbar}. \tag{5.5}$$

Using wave functions that have the form of (5.5), one may then define the density matrix, as a single time function, to be

$$\rho(\boldsymbol{r}, \boldsymbol{r}', t) = \sum_{m,n} c_m^* c_n \varphi_m^\dagger(\boldsymbol{r}') \varphi_n(\boldsymbol{r}) e^{-i(E_n - E_m)t/\hbar}. \tag{5.6}$$

Normally, the pair of coefficients are simply written as c_{mn}. As shown here, the off-diagonal elements are rapidly oscillating in the difference frequency, so that they are usually ignored in transport theory. However, these off-diagonal elements are generated in the complex device and represent the entanglement of the various modes, such as that occurring in figure 5.1 [46]. Because of the two wave functions, the equation of motion embodies a Schrödinger equation for each of these two wave functions and leads to the Liouville–von Neumann equation

$$i\hbar \frac{\partial \rho(\boldsymbol{r}, \boldsymbol{r}', t)}{\partial t} = \hat{H}\rho = [H, \rho]. \tag{5.7}$$

In the first term on the right (of the first equality), a commutator-generating superoperator has been introduced. The last terms allow the equation to be written in expanded form as

$$i\hbar \frac{\partial \rho(\boldsymbol{r}, \boldsymbol{r}', t)}{\partial t} = \left\{ -\frac{\hbar^2}{2m^*} \left[\nabla_r^2 - \nabla_{r'}^2 \right] + V(\boldsymbol{r}) - V(\boldsymbol{r}') \right\} \rho(\boldsymbol{r}, \boldsymbol{r}', t). \tag{5.8}$$

While no dissipation is indicated here, it is quite easy to incorporate a dissipative term [47–49]. There are many different definitions of the density matrix, which differ only in the fine detail. One can also introduce the imaginary time $t \to -i\hbar\beta$, where $\beta = 1/k_B T$ is the inverse temperature (thermal energy). The density matrix can then be written in the thermal equilibrium form

$$\rho(\boldsymbol{r}, \boldsymbol{r}', T) = \sum_n e^{-\beta E} \varphi_n^\dagger(\boldsymbol{r}) \varphi_n(\boldsymbol{r}'). \tag{5.9}$$

If the transformation leading to (5.9) is introduced to (5.8), then the left-hand derivative becomes a derivative with respect to β (the inverse temperature), and this version of the equation is often called the Bloch equation.

Generally, the superoperator of (5.7) is a fourth-rank tensor operating in the Liouvillian space [46], which is a Hilbert space of operators. Here, however, we do not need the full complexity, and it is sufficient to examine the matrix elements of the first-term on the right-hand side of (5.8). These matrix elements are

$$(\hat{H}\rho)_{kn} = \sum_{r,s} \hat{H}_{kn,\,rs} \rho_{rs} = \sum_{rs} \left(H_{kr} \rho_{rn} - \rho_{ks} H_{sn} \right). \tag{5.10}$$

If the Hamiltonian is diagonal, then

$$\hat{H}_{kn,\,rs} = (H_{kk} - H_{nn})\delta_{kr}\delta_{ns}. \tag{5.11}$$

Some other properties can be found in [50].

5.4.2 Quantum kinetic equations

Equation (5.7) gives the equation of motion for the density matrix for an arbitrary potential imposed upon the system, either internally or externally. But the equation does not contain any dissipation. On the other hand, the Boltzmann equation is a kinetic equation which describes the dynamic evolution of a distribution function under the influence of external fields and collisional forces. In this light, the resulting transport is a balance between the driving forces generated by the fields and the relaxation processes arising from the collisions. From (5.7) above, a similar kinetic equation may be derived for the density matrix. For this, we assume a semiconductor system composed of electrons and phonons and the interaction between these two subsystems. The system is assumed to be described by the Hamiltonian

$$H = H_0 + H_{\mathrm{L}} + H_{\mathrm{F}} + H_{\mathrm{eL}}, \tag{5.12}$$

where the terms on the right-hand side describe the electrons, the lattice, the external fields, and the electron–lattice interaction, respectively. The external field is taken in the scalar potential gauge $H_{\mathrm{F}} = -e\boldsymbol{F}\cdot\boldsymbol{r}$, where F is the electric field, and the electron interaction contains all of the appropriate scattering processes, typically evaluated using the semiclassical Fermi golden rule approach for perturbation theory. In modern approaches, however, the impurities are treated in real space via the local potentials they generate via the Poisson equation.

The density matrix ρ of (5.7) is that for the entire system. If there were no interactions between the electrons and the lattice, then the density matrix could be written as the tensor product of the density matrix for the electrons and the density matrix for the lattice. In this situation, a trace Tr_{L} over the lattice coordinates would yield the reduced density matrix for the electrons. Similarly, a trace Tr_{e} over the electron coordinates would yield just the density matrix for the lattice. When the interactions are present, then the simple view of two distinct density matrices is not possible. However, the relevant density matrix for the electrons may be obtained by the trace over the lattice variables, as follows:

$$\rho_{\mathrm{e}} = \mathrm{Tr}_{\mathrm{L}}\{\rho\}, \tag{5.13}$$

but this equation also yields the interactions between the electrons and the lattice, so it is no longer simple.

To proceed, the Hamiltonian (5.12) is introduced into (5.7), and this is followed by the trace (5.13). This produces an equation for the electronic density matrix, as follows:

$$i\hbar\frac{\partial\rho_{\mathrm{e}}}{\partial t} = \left[H_0 + H_{\mathrm{F}}, \rho_{\mathrm{e}}\right] + \mathrm{Tr}_{\mathrm{L}}\{[H_{\mathrm{eL}}, \rho]\}. \tag{5.14}$$

Notice that the lattice term has dropped from the equation; it is only the interaction with the lattice that is important to the carriers. The first term on the right-hand side is the electronic motion within the effective mass approximation and includes the applied field. This first term represents simple ballistic transport. The second term represents the electron–phonon interaction and the effect that this has upon the resulting electronic density matrix. Commonly, this trace over the lattice variables corresponds, in the Fourier-transformed version, to the summation over the phonon momentum vector q that appears in classical transport. This term is more complicated in the quantum case. The shorthand notation $H_e = H_0 + H_F$ will be used for the first term. But, to proceed, another shorthand approach will be introduced, namely the use of commutator-generating superoperators, with which (5.14) becomes

$$i\hbar\frac{\partial\rho_e}{\partial t} = \hat{H}_e\rho_e + \mathrm{Tr}_L\left\{\hat{H}_{eL}, \rho\right\}. \tag{5.15}$$

In the following, nested commutators are going to be found, and these are easier to follow (and to understand) in the superoperator notation. This last equation is Laplace transformed in the temporal variable to give

$$\left(s + \frac{i}{\hbar}\hat{H}_e\right)\tilde{\rho}_e = -\frac{i}{\hbar}\mathrm{Tr}_L\left\{\hat{H}_{eL}\tilde{\rho}\right\} + \rho_e(0). \tag{5.16}$$

The tilde over the density matrix refers to the Laplace transformed form. The Laplace transform works only for positive time, so that $t = 0$ denotes the initial condition for the electronic density matrix. All forces are considered to be 'turned on' at this initial time. The last term on the right-hand side is this initial condition. It is a useful result of the Laplace transform that this initial condition always appears. At the same time, this one-sided time transform ensures causality for the results of this approach.

Projection superoperators were already introduced in section 3.2.1. These operators may now be used to reduce the many-body density matrix to a one-electron density matrix similar to the Boltzmann transport equation. These projection superoperators and their complement are introduced below:

$$\tilde{\rho}_1 = \hat{P}\tilde{\rho}_e, \quad \hat{P}^2 = \hat{P}, \quad \hat{Q} = \hat{1} - \hat{P}. \tag{5.17}$$

The last two expressions define the idempotent property of the projection superoperator and the existence of a projection operator onto the complement space ($\hat{Q}$). The use of these operators provides the reduction that is usually achieved via the Bogoliubov–Born–Green–Kirkwood–Yvon (BBGKY) hierarchy. Just as with this latter hierarchy, one complication is that the equation for the one-electron density matrix results in an additional term that involves a two-electron function, and in this case, the problem lies within the scattering process—the first term on the right-hand side of (5.16). The projection superoperator commutes with the trace operation in (5.15), so that the scattering superoperator may be introduced through the definition

$$\hat{\Sigma}\tilde{\rho}_1 = \mathrm{Tr}_\mathrm{L}\left\{\hat{P}\hat{H}_{\mathrm{eL}}\tilde{\rho}\right\}. \tag{5.18}$$

We will return to this term below. First, however, using the projection operators, (5.16) may be rearranged to yield

$$\tilde{\rho}_1 = i\hbar\hat{P}\frac{1}{i\hbar s - \hat{H}_\mathrm{e} - \hat{\Sigma}}\left[\hat{P}\rho_\mathrm{e}(0) + \hat{Q}\rho_\mathrm{e}(0)\right]. \tag{5.19}$$

The initial condition has been split into the projected part and its complement for convenience in the following development.

In (5.19), there is a central term, which is termed the resolvent operator. It actually introduces much of the time variation in the equation of motion that will be obtained below. This resolvent is just

$$\hat{R}(s) = \frac{1}{i\hbar s - \hat{H}_\mathrm{e} - \hat{\Sigma}} \equiv \frac{1}{i\hbar s - \hat{H}'} = \frac{1}{i\hbar s - \left(\hat{P} + \hat{Q}\right)\hat{H}'\left(\hat{P} + \hat{Q}\right)}, \tag{5.20}$$

where the second term has introduced another shorthand notation. This form can be expanded into two different versions that will useful to the development. These are

$$\begin{aligned}\hat{R}(s) &= \frac{1}{i\hbar s - \hat{P}\hat{H}'\hat{P}}\left\{1 - \left[\left(\hat{P} + \hat{Q}\right)\hat{H}'Q + \hat{Q}\hat{H}'\hat{P}\right]\hat{R}(s)\right\} \\ &= \frac{1}{i\hbar s - \hat{Q}\hat{H}'\hat{Q}}\left\{1 - \left[\left(\hat{P} + \hat{Q}\right)\hat{H}'\hat{P} + \hat{P}\hat{H}'\hat{Q}\right]\hat{R}(s)\right\}\end{aligned}. \tag{5.21}$$

The first line is premultiplied by the projection operator and the second line is premultiplied by the complement operator, and these replace the last terms in each line. The resolvent can then be rewritten in the rather complex form

$$\begin{aligned}\hat{R}(s) = &\left(\hat{P} + \hat{Q}\frac{1}{i\hbar s - \hat{Q}\hat{H}'\hat{Q}}\hat{Q}\hat{H}'\hat{P}\right)\frac{1}{i\hbar s - \hat{P}\hat{H}'\hat{P} - \hat{C}}\left(\hat{P} + \hat{P}\hat{H}'\hat{Q}\frac{1}{i\hbar s - \hat{Q}\hat{H}'\hat{Q}}\hat{Q}\right) \\ &+ \hat{Q}\frac{1}{i\hbar s - \hat{Q}\hat{H}'\hat{Q}}\hat{Q}\end{aligned}. \tag{5.22}$$

Here, the bare collision operator $\hat{C}$ has been introduced as a shorthand:

$$\hat{C} = \hat{P}\hat{H}'\hat{Q}\frac{1}{i\hbar s - \hat{Q}\hat{H}'\hat{Q}}\hat{Q}\hat{H}'\hat{P}. \tag{5.23}$$

This expression has a strong connection to the normal Fermi golden rule, as it involves the terms that connect a 'diagonal' element (in the first set of projection operators) to an off-diagonal element via the central term and reconnects them by the conjugate operation (the complement terms in the last set of operators). This is just the type of term which leads to the Fermi golden rule in the long-time limit. The central term leads ultimately to the energy-conserving delta function in the semi-classical situation. Moreover, the Hamiltonian in the central term also involves the electronic Hamiltonian, so that the electron energy is incorporated as required to fit the Fermi golden rule. Recognizing all of this, the collision operator (5.18) can now be written as

$$\hat{\Sigma}\tilde{\rho}_1 = \mathrm{Tr}_{\mathrm{L}}\left\{\hat{P}(\hat{H}_{\mathrm{eL}} + \hat{H}_{\mathrm{e}})\hat{Q}\frac{1}{i\hbar s - \hat{Q}(\hat{H}_{\mathrm{eL}} + \hat{H}_{\mathrm{e}})\hat{Q}}\hat{Q}\tilde{\rho}\right\}. \tag{5.24}$$

This complicated structure gives much information. First, the electron–electron interactions should be included in the Hamiltonian although the single-electron density matrix is being extracted. This is because the electron–electron interaction provides the screening of the appropriate interactions. Second, the electron–phonon interactions can lead to real energy shifts, which are the self-energy corrections to the single-particle energies. This is in addition to the normal dissipative scattering processes. So, this particular form has the full off-shell self-energy contributions as well as the screening processes.

Using the above developments, the Laplace transform for the one-particle density matrix (5.16) can now be written as

$$\begin{aligned} i\hbar\tilde{\rho}_1(s) &= i\hbar\rho_1(0) + \hat{P}\hat{H}_{\mathrm{e}}\hat{P}\tilde{\rho}_1 + \hat{\Sigma}\tilde{\rho}_1 \\ &+ \hat{P}(\hat{H}_{\mathrm{eL}} + \hat{H}_{\mathrm{e}})\hat{Q}\frac{1}{i\hbar s - \hat{Q}(\hat{H}_{\mathrm{eL}} + \hat{H}_{\mathrm{e}})\hat{Q}}\hat{Q}\tilde{\rho}_{\mathrm{e}}(0)^{\cdot} \end{aligned} \tag{5.25}$$

The last term involves the entire many-electron initial density matrix, while the first three terms on the right involve only the single-electron density matrix. The last term on the right has been thought to produce a number of effects, including the random force used for the Langevin equation at the lowest order of the electron–phonon interaction [51] and a screening of the driving field in higher orders [52]. While these suggestions have been made, much more work is necessary to study the effects of this last term. The temporal equation can now be obtained by retransforming this last equation, which results in the following for the force due to the electric field:

$$\frac{\partial\rho_1}{\partial t} - \frac{e}{\hbar}\boldsymbol{F}\cdot\frac{\partial\rho_1}{\partial\boldsymbol{k}} = -\frac{1}{\hbar^2}\int_0^t \hat{\Sigma}(t - t')\rho_1(t')dt'. \tag{5.26}$$

The initial state decay, i.e. the last term of (5.25), has been neglected. This form of the quantum kinetic equation was first derived by Barker [53], but its form is quite analogous to the Prigogine–Resibois equation [54], for which the EMC approach is quite well suited. However, it may be observed that an assumption of a homogeneous semiconductor has been made. In the case of an inhomogeneous semiconductor, the second term on the right-hand side, which gives the field term, also yields a spatial gradient term.

Except for the convolution form of the collision term, the above equation is essentially the same as the Boltzmann transport equation, which means that some quantum effects, such as quantization of the states, have been left out or are buried in the form of the one-electron density matrix. The initial condition has to contain the physical confinement effects, and these carry through into the final one-electron density matrix. Temporal correlation effects are naturally included through the convolution integral for the scattering. Spatial correlations are included within the quantum nature of the self-energy term in the integral. An extensive investigation into the full spectrum of the density matrix in the Liouvillian space for non-equilibrium carriers also has been undertaken [55]. This has been applied to the subfemtosecond initial state of transport [56], and for connecting to the correct Green's functions [57]. Because of the direct analogy between (5.26) and the Boltzmann equation, it can be used in the drifted Maxwellian approximation to generate a set of balance equations in the hydrodynamic view. This set has been generated and used to study transient transport in high electric fields [58].

The EMC approach to solutions of the Boltzmann transport equation, based upon converting the equation into a path integral, may be carried over to (5.26). The difference between the Boltzmann transport equation and its quantum equivalent (5.26) lies in the convolution integral for the scattering function on the right-hand side of the latter equation. However, this should not be a hindrance to the development of a path integral formulation that would allow for an EMC treatment of quantum transport. First, (5.26) is rewritten using a more common form of the retarded collision term, as follows:

$$\begin{aligned}\frac{\partial \rho_1(\boldsymbol{k}, t)}{\partial t} - \frac{e}{\hbar}\boldsymbol{F} \cdot \frac{\partial \rho_1(\boldsymbol{k}, t)}{\partial \boldsymbol{k}} = \int_0^t dt' \sum_{\boldsymbol{k}'} \{ & S(\boldsymbol{k}, \boldsymbol{k}'; t, t')\rho_1(\boldsymbol{k}', t') \\ & - S(\boldsymbol{k}', \boldsymbol{k}; t, t')\rho_1(\boldsymbol{k}, t')\}\end{aligned} \quad (5.27)$$

The two momenta variables in (5.27) are explicit functions of the retarded time t' on the right-hand side. This is dramatically different than what is found in the classical Boltzmann equation. This is because the scattering process in quantum mechanics is not instantaneous, as discussed in section 3.1.3.

Quantum mechanically, the problem is expressed primarily by asking just how long it takes to establish the Fermi golden rule. This is quite similar to asking how long it takes for an atom to emit an electron under optical illumination. The Fermi golden rules for these two situations are quite similar and contain the same physics. The first to ask this question seems to have been Geltman [59], who was followed by Lane in analyzing the decay of a field-created state [60]. In both cases, an incident electromagnetic field creates a correlation between the initial state and the final state [61]. At some small time later, the conjugate field then breaks up this correlation, leading either to the completed process or to a virtual process in which the phonon/photon is not created or absorbed and the initial state is recovered. In this process, both Geltman [59, 61] and later Lipavsky *et al* [62] found that weak oscillations about the asymptotic result were observed. The collision duration for optical phonons in GaAs is shown in figure 5.2 [7], as computed using the appropriate two-particle Green's functions for the interactions described above [63]. In this figure, the polar mode energy is 36 meV, and the coupling constant is 10^{14} s^{-1}. For the zone edge phonon, the energy is 31 meV and

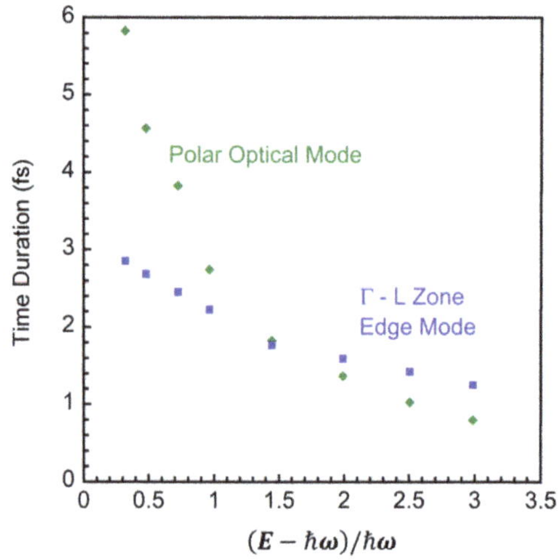

Figure 5.2. The average time taken by the zone center LO modes (green) and the zone edge intervalley modes (blue) to complete a phonon emission process for transfer from Γ to L. Adapted from [7].

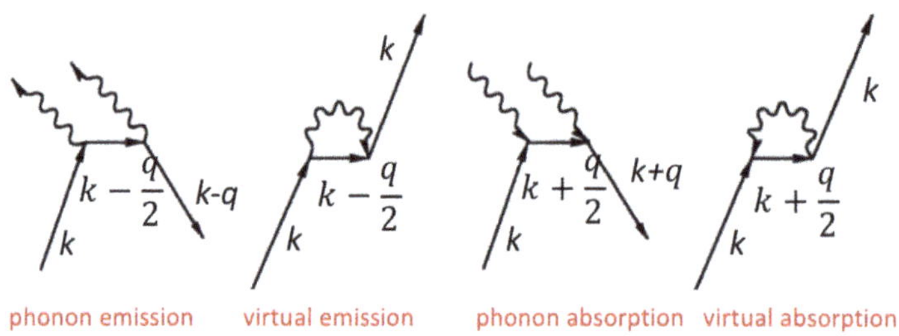

Figure 5.3. Diagrams representing the various types of interactions (second-order terms) that lead to the total scattering in the Fermi golden rule. These terms are labeled for both real (completed) and virtual processes. In virtual processes, no real scattering event occurs and coherence is not broken.

the coupling constant is $7 \times 10^{12}\ \mathrm{s}^{-1}$. Depending upon the excess energy of the carriers, the duration varies from about 1 fs to several times this value.

Dealing with scattering in a quantum system evolution is far more complicated because of this collision duration. Yet this can be included in a more complicated diagram than a simple Feynman diagram, and this guides us through the full collision process. There are some caveats that must be satisfied [64]: (1) For electron–phonon interactions or potential interactions (such as from impurities), the number of vertices must be even. Some of these diagrams are shown in figure 5.3. Both real (completed) collisions and virtual interactions are shown. In the virtual interaction, the collision is not completed, the particle returns to its initial state prior to the interaction, and the phonon is neither emitted or absorbed. The diagrams in this figure are not Feynman diagrams. For the latter, the two vertices shown in the figure are collapsed into a single vertex, which is only a portion of the Feynman diagram. (2) The times at which each of the interaction vertices occurs differ by the duration of the collision, which is determined by figures such as 5.2. (3) The momentum change at each vertex is determined by the momentum change expected for the interaction. For the scattering interaction, one half of the phonon momentum modifies the wave vector at each vertex. (4) It is critical to carefully consider the phase of the wave function that is the basis of the path. It is the phase interference that gives rise to effects such as those of figure 5.1, and this phase affects the wave function. In figure 5.3, a real collision has both phonon lines either leaving (emission) or arriving (absorption) at both vertices. On the other hand, a virtual emission combines these two types of lines. Similar behavior is found in optical absorption, in which the ratio of the number of real absorptions to total (real plus virtual) interactions gives the quantum efficiency for optical absorption. One could define a similar parameter here.

In addition to the time variation throughout the collision, one also has to address the fact that the momenta are likely to change through the collision. The two momenta variables in (5.27) are explicit functions of the retarded time t' through the accelerative expressions

$$\begin{aligned} \boldsymbol{k}(t, t') &= \boldsymbol{k}(t) - \frac{e}{\hbar}\int_{t'}^{t} \boldsymbol{F}(t'')dt'' \\ \boldsymbol{k}'(t, t') &= \boldsymbol{k}'(t) - \frac{e}{\hbar}\int_{t'}^{t} \boldsymbol{F}(t'')dt'' \end{aligned} . \tag{5.28}$$

This is required because of the convolution integral and represents the acceleration of the particle *through* the collision process, which also arises because of the nonzero duration of the collision itself [7]. The equivalent of the Fermi golden rule scattering process is now generalized to

$$S(\boldsymbol{k}, \boldsymbol{k}';t, t') = \mathrm{Re}\left\{\frac{2}{\hbar^2}\sum_{\boldsymbol{q}}| M_q |^2 f(N)\delta_{\boldsymbol{k}, \boldsymbol{k}'\pm \boldsymbol{q}} e^{-\frac{i}{\hbar}\int_{t}^{t'} dt''\zeta(\boldsymbol{k}, \boldsymbol{k}';t', t'')}\right\}, \tag{5.29}$$

where the factor $f(N)$ introduces the Bose–Einstein distributions for the phonons, as follows:

$$F(N) = N_q + \frac{1}{2} \pm \frac{1}{2}, \tag{5.30}$$

where the upper sign denotes the emission and the lower sign denotes the absorption of a phonon. The factor ζ in the exponential term of (5.29) gives the evolution of the phase:

$$\zeta(\boldsymbol{k}, \boldsymbol{k}';t', t'') = E(\boldsymbol{k}(t'')) - E(\boldsymbol{k}'(t'')) \pm \hbar\omega_{\boldsymbol{q}}. \tag{5.31}$$

This term also accounts for the collision duration and the variation of the momenta described in (5.28).

In the semiclassical approach to transport, the last exponential in (5.29) represents the joint spectral densities of the two states. This would become the energy-conserving delta function if there were a restriction to remaining on the energy shells. The first term in the curly brackets of (5.27) refers to scattering *in* to the state with momentum $\boldsymbol{k}$ from all the possible states with momentum $\boldsymbol{k}'$. This process is weighted by the occupancy of the initial momentum through the density matrix at this momentum. On the other hand, the second term corresponds to scattering *out* of the state $\boldsymbol{k}$ to all the possible states with momentum $\boldsymbol{k}'$ and is weighted by the density matrix of this initial state. The temporal evolution of the scattering functions creates a barrier to generating an integral equation approach, but this can be overcome by the introduction of a self-scattering process, just as is done in normal EMC to simplify an equivalent integral. To do this, a new scattering function is introduced via

$$S^*(\boldsymbol{k}, \boldsymbol{k}'; t, t') = S(\boldsymbol{k}, \boldsymbol{k}'; t, t') + [\Gamma(t, t') - \Gamma_{\rm out}(t, t')\delta(\boldsymbol{k} \cdot \boldsymbol{k}')], \tag{5.32}$$

with

$$\begin{aligned} \Gamma_{\rm out}(t, t') &= \sum_{\boldsymbol{k}'} S(\boldsymbol{k}, \boldsymbol{k}'; t, t') \\ \Gamma(t, t') &= \Gamma_0 \delta(t - t') \end{aligned} . \tag{5.33}$$

The terms in the square bracket of (5.32) make no contribution to the scattering integral, but the term in $\Gamma_{\rm out}$ plays a very important role. This term projects out the contribution due to uncompleted out-scattering processes, and the retardation in the out-scattering term is accounted for here rather than in the equivalent term in (5.27). With these specific terms, (5.27) can be written as [65]

$$\left[\frac{\partial}{\partial t} - \frac{e}{\hbar}\boldsymbol{F} \cdot \frac{\partial}{\partial \boldsymbol{k}} + \Gamma_0\right]\rho_1(\boldsymbol{k}, t) = \int_0^t dt' \sum_{\boldsymbol{k}'} S^*(\boldsymbol{k}, \boldsymbol{k}'; t, t')\rho_1(\boldsymbol{k}', t'). \tag{5.34}$$

This equation is now in the form of the Boltzmann transport equation and can be approached in precisely the same way as the Boltzmann equation to yield a path integral [66]. The result is

$$\begin{aligned} \rho_1(\boldsymbol{k}, t) &= \int_0^t dt' e^{-\Gamma_0(t-t')} \\ &\times \int_0^{t'} dt'' \sum_{\boldsymbol{k}'} S^*(\boldsymbol{k}, \boldsymbol{k}'; t', t'')\rho_1(\boldsymbol{k}', t'') \end{aligned} , \tag{5.35}$$

where the momenta are functions of t'' through (5.28). This last equation is sometimes known as the Barker–Ferry equation.

The evolutionary properties that are inherent in (5.35) lead to what is known as an intracollisional field effect [67, 68]. This effect is very important when the field is large or when the collision duration is a significant fraction of the mean time between collisions; it has previously been analyzed for steady-state transport in uniform, time independent fields [53, 69]. In the steady-state case, the intracollisional field effect induces a broadening and skewing of the usual energy-conserving delta functions [8]. This leads, for example, to a lowering of the threshold for phonon emission due to acceleration during the collision, a lowering of the scattering strength in high fields, and early intervalley transfer and impact ionization [70] at high energies.

5.4.3 Wigner functions

The Wigner function begins with the density matrix discussed in the previous sections. For this purpose, (5.6) is rewritten as

$$\rho(\boldsymbol{r}, \boldsymbol{r}', t) = \psi^{\dagger}(\boldsymbol{r}', t)\psi(\boldsymbol{r}, \boldsymbol{t}), \tag{5.36}$$

where the total wave functions have been used instead of the basis sets. To proceed, the center-of-mass coordinates (1.53) are introduced, as follows:

$$\begin{aligned} \boldsymbol{R} &= \frac{1}{2}(\boldsymbol{r} + \boldsymbol{r}') \\ \boldsymbol{s} &= \boldsymbol{r} - \boldsymbol{r}' \end{aligned}. \tag{5.37}$$

Wigner wanted to develop a phase-space distribution that was analogous to the Boltzmann distribution function [71] and introduced these center-of-mass coordinates and subsequent Fourier transform:

$$f_{\mathrm{W}}(\boldsymbol{R}, \boldsymbol{p}, t) = \frac{1}{h^3}\int ds \rho\left(\boldsymbol{R} + \frac{\boldsymbol{s}}{2}, \boldsymbol{R} - \frac{\boldsymbol{s}}{2}, t\right)e^{i\boldsymbol{p}\cdot\boldsymbol{s}/\hbar}. \tag{5.38}$$

This yields a distribution function that is defined in position, momentum, and time. That is, the distribution is defined in a six-plus-one-dimensional coordinate system (three for momentum, three for position, and one for time). If transformation (5.38) is introduced to the Liouville–von Neumann equation (5.8), then the Wigner equation of motion (in the absence of scattering) is found to be

$$\frac{\partial f_{\mathrm{W}}}{\partial t} + \frac{\boldsymbol{p}}{m^*}\cdot\frac{\partial f_{\mathrm{W}}}{\partial \boldsymbol{R}} - \frac{1}{h^3}\int d\boldsymbol{p}' W(\boldsymbol{R}, \boldsymbol{p}')f_{\mathrm{W}}(\boldsymbol{p} + \boldsymbol{p}') = 0, \tag{5.39}$$

where

$$W(\boldsymbol{R}, \boldsymbol{p}') = \frac{1}{i}\int d\boldsymbol{s}\exp\left(\frac{\boldsymbol{s}\cdot\boldsymbol{p}'}{\hbar}\right)\left[V\left(\boldsymbol{R} + \frac{\boldsymbol{s}}{2}\right) - V\left(\boldsymbol{R} - \frac{\boldsymbol{s}}{2}\right)\right], \tag{5.40}$$

and V is the electrostatic potential that appears in the Hamiltonian in (5.8). The term $W(\boldsymbol{R}, \boldsymbol{p}')$ is termed the *Wigner potential* in most applications. This potential is nonlocal, as can be seen from the term in brackets in (5.40). Scattering may easily be included by adding a gain–loss term to the right-hand side of (5.39), which is given as the right-hand side of (5.27), although this latter term still has to be Fourier transformed with the center-of-mass coordinates introduced first [72]. Needless to say, this does not produce the classical form of scattering but includes both the collision duration and the nonlocal nature of scattering.

The Wigner equation of motion (5.39) is quite similar to the Boltzmann equation, except for the complication of the Wigner potential (5.40). Nevertheless, it is possible to transform the equation of motion into a path integral form, from which a Monte Carlo procedure can be developed. Thus, the idea of using particles in quantum mechanics discussed quite early by Kennard [73] naturally carries into particle solutions of the Wigner function. The concept of a Wigner path for the particle is thus useful, as it provides a representation of the quantum evolution and leads to numerical simulation of the Wigner equation of motion. Such a path is followed by the particles which are a sample of the Wigner function as it evolves through phase space. Moreover, this path evolves quite like a classical particle, including provisions for scattering and for rapidly varying quantum potentials. More on the particle solutions for (5.39) will be covered in chapter 11.

In general, the Wigner function is not a positive definite function. This is a consequence of the uncertainty relationship and the development of entanglement away from equilibrium. As discussed above, the off-diagonal terms of the density matrix carry the entanglement, and it is exactly these off-diagonal terms that give rise to the components of the Fourier transform momentum $\boldsymbol{p}$. If the Wigner function is integrated over either position or momentum, a positive definite function gives the probability functions (square magnitude of the wave functions) in position or momentum, just as in normal quantum mechanics where the entanglement is difficult to see. In fact, the negative excursions of the Wigner function exist over phase-space regions whose volume is of the order of $\hbar^3$, so that smoothing the function over a volume corresponding to the uncertainty principle produces a positive definite function. The Wigner function is always positive definite in the ground state appropriate to equilibrium. However, the onset of negative excursions of the Wigner function (away from equilibrium) is always viewed as the onset of correlation/entanglement in the problem [41]. In this regard, modern approaches to the Wigner function provide a distinct formulation that is recognized as equivalent to normal operator-based quantum mechanics but provides a different perspective [74]. Because of the phase-space nature of the Wigner distribution, it is possible to identify where quantum corrections enter a problem by performing a comparison with the classical version.

Wigner himself derived an effective quantum potential that can be used as a correction term for the potentials in thermodynamics [71]. This gives an effective potential that arises from the two potential terms in (5.40) via their Taylor series expansions. This correction to the natural normalization will be dealt with in chapter 10 so that it can be compared to other forms of effective potentials. But the natural form of the Wigner potential can easily be exhibited using (5.39) in the absence of any potential terms. The simplest, minimum uncertainty wave packet in quantum mechanics is a Gaussian wave packet. This can be represented as

$$\psi(\boldsymbol{r}) = \frac{1}{(2\pi\sigma^2)^{1/4}} e^{-r^2/4\sigma^2 - i\boldsymbol{k}_0 \cdot \boldsymbol{r}}, \tag{5.41}$$

in which the momentum has arisen from the assumption of a smooth propagation of the wave packet with velocity $v = \hbar \boldsymbol{k}_0/m^*$. This wave function can be used in the Wigner formulation (5.39) to give the result

$$f_W = \frac{1}{h} e^{-2\sigma^2(\boldsymbol{k}-\boldsymbol{k}_0)^2 - r^2/4\sigma^2}. \tag{5.42}$$

This is a positive definite function, which is appropriate for what is basically a ground-state function, although the motion of the packet is represented by the shift in momentum to the new center $\boldsymbol{k}_0$. This Wigner function can now evolve directly from the equation of motion (5.39), either with or without scattering.

5.4.4 Green's functions

Like the Wigner function, the Green's functions were already introduced in section 1.5.4. There, it was shown that there are several types of Green's functions—the equilibrium functions which give rise to the retarded and advanced functions. In the usual case, these functions operate at zero temperature and often refer to the full Fermi sea, in which all states up to the Fermi level are full and all states above the Fermi level are empty. At nonzero temperatures, one can then develop the thermal, or Matsubara Green's functions, which involve the Fermi–Dirac function for $T > 0$ [75]. Schwinger developed NEGFs for semiconductor devices as well as other systems which are far from equilibrium [76]. The developments of the equilibrium Green's functions and the NEGFs follow somewhat different paths. Here, we describe the basics of the equilibrium Green's functions. NEGFs will be left to chapter 12, where their use and limitations in devices will be discussed.

The propagator (1.60) may be generally related to the density matrix by extending the latter to two times. Thus, it is related to (5.36) as follows:

$$K(\boldsymbol{r}, \boldsymbol{r}'; t, t') = \psi^\dagger(\boldsymbol{r}', t')\psi(\boldsymbol{r}, \boldsymbol{t}), \tag{5.43}$$

In this case, the wave functions are to be interpreted as field operators. This propagator is used to describe two basic equilibrium Green's functions:

$$\begin{aligned} G_r(\boldsymbol{r}, \boldsymbol{r}'; t, t') &= -i\Theta(t - t')\langle\{\hat{\psi}(\boldsymbol{r}, t), \hat{\psi}^\dagger(\boldsymbol{r}', t')\}\rangle \\ G_a(\boldsymbol{r}, \boldsymbol{r}'; t, t') &= i\Theta(t_0 - t)\langle\{\hat{\psi}^\dagger(\boldsymbol{r}', t'), \hat{\psi}(\boldsymbol{r}, t)\}\rangle \end{aligned}. \tag{5.44}$$

If we consider the full Fermi sea, the retarded function creates an electron above the Fermi level, leaving a hole below the Fermi level. The electron then propagates to $\boldsymbol{r}$, t, where it is annihilated. The advanced function runs backward in time (for an electron) and thus correlates with the motion of the hole that accompanies the electron. The $\Theta(x)$ is the Heavyside step function, which is one for $x > 0$ and zero otherwise. The angle brackets have been added to indicate an ensemble average, which corresponds to a sum over the appropriate basis states, which here is the equilibrium ground state (of zero excitations). The curly brackets denote an *anti*commutator relation in which the normal negative sign is replaced by a positive sign, which is the normal case for fermions, and may be written as

$$\{c, c^{\dagger}\} = cc^{\dagger} + c^{\dagger}c = 1. \tag{5.45}$$

For the field operators, the value of unity is actually $\delta(\boldsymbol{r} - \boldsymbol{r}')\delta(t - t')$. Both of the functions in (5.37) are connected with the general zero-order, non-interacting Green's function, which must satisfy two forms of the Schrödinger equation:

$$\begin{aligned} \left[i\hbar\frac{\partial}{\partial t} - H_0(\boldsymbol{r}) - V(\boldsymbol{r})\right]G^0(\boldsymbol{r}, \boldsymbol{r}'; t, t') &= \hbar\delta(\boldsymbol{r} - \boldsymbol{r}')\delta(t - t') \\ \left[-i\hbar\frac{\partial}{\partial t} - H_0(\boldsymbol{r}') - V(\boldsymbol{r}')\right]G^0(\boldsymbol{r}, \boldsymbol{r}'; t, t') &= \hbar\delta(\boldsymbol{r} - \boldsymbol{r}')\delta(t - t') \end{aligned}. \tag{5.46}$$

In a homogeneous system in equilibrium, it may be assumed that the Green's function is a function only of the differences in position and time and that there is no potential in the system. The two equations above can then be Fourier transformed in these difference variables and the resulting function is simply

$$G^0(\boldsymbol{k}, \omega) = \frac{\hbar}{\hbar\omega - E(\boldsymbol{k})}. \tag{5.47}$$

The bare Hamiltonian in (5.46) produces the energy as a function of momentum from the Fourier transform of the *difference* in positions, while the frequency arises from the Fourier transform of the *difference* in times. Neglecting the potential means that a freely propagating Green's function for a homogeneous system with no boundaries is being considered in (5.47). The retarded and advanced Green's functions can be produced by introducing a convergence factor described by η. Hence, the retarded and advanced bare (or ground state) Green's functions arise from (5.47) as follows:

$$G^0_{\mathrm{r,\,a}}(\boldsymbol{k}, \omega) = \frac{\hbar}{\hbar\omega - E(\boldsymbol{k}) \pm i\eta}. \tag{5.48}$$

The inverse transform in time can be written as

$$G^0_{\mathrm{r,\,a}}(\boldsymbol{k}, t) = \int \frac{d\omega}{2\pi} G^0_{\mathrm{r,\,a}}(\boldsymbol{k}, \omega)e^{-i\omega t}. \tag{5.49}$$

The integration proceeds along the real frequency axis, following a path such as that shown in figure 5.4. However, as for any integration involving a complex variable, it has to be closed to obtain a full contour integration. For $t > 0$, the contour has to close in the lower half-plane, in order to contain the proper pole. On the other hand, for $t < 0$, the contour has to close in the upper half-plane. The convergence factors are introduced into (5.48) in order to move the poles off the real axis. Hence, closure in the lower half-plane requires the upper sign. Using these conventions, the bare retarded and advanced functions become

$$G^0_{\mathrm{r}}(\boldsymbol{k}, \omega) = \frac{\hbar}{\hbar\omega - E(\boldsymbol{k}) + i\eta}, \quad G^0_{\mathrm{a}}(\boldsymbol{k}, \omega) = \frac{\hbar}{\hbar\omega - E(\boldsymbol{k}) - i\eta}. \tag{5.50}$$

Figure 5.4. The path of integration along the real frequency axis for (5.49). Note that this assumes that the two infinite limits are the same, which implies that time-reversal symmetry holds in these equilibrium Green's functions.

The integral (5.49) leads to a number of important properties that are associated with the Green's functions. First, there is the property of complex functions such as (5.50). When η is vanishingly small, as it is for these two functions when it serves only as a convergence factor, one has the limiting form

$$\frac{1}{x \pm i\eta} = P\frac{1}{x} \mp i\pi\delta(x). \tag{5.51}$$

Here, P denotes the principal part of the function (and the resulting integral). The delta function yields the on-shell relationship between the energy $\hbar\omega$ and the momentum-dependent kinetic energy $E(\boldsymbol{k})$. Hence, for the non-interacting Green's functions, the delta function arises from

$$2\pi\delta(\hbar\omega - E(\boldsymbol{k})) = -\mathrm{Im}\left\{G_\mathrm{r}^0(\boldsymbol{k}, \omega) - G_\mathrm{a}^0(\boldsymbol{k}, \omega)\right\}. \tag{5.52}$$

When interacting Green's functions are used, such as when scattering is added, the convergence factors become the self-energies, and (5.52) is replaced by the spectral density function

$$A(\boldsymbol{k}, \omega) = -\mathrm{Im}\{G_\mathrm{r}(\boldsymbol{k}, \omega) - G_\mathrm{a}(\boldsymbol{k}, \omega)\} = i[G_\mathrm{r}(\boldsymbol{k}, \omega) - G_\mathrm{a}(\boldsymbol{k}, \omega)]. \tag{5.53}$$

These are now the interacting forms of the functions and entail the full off-shell results of the interactions.

In the interaction representation, it is assumed that there is a base Hamiltonian that contains the ground-state Hamiltonian H_0 and a perturbing potential $V(t)$, from which the time variation arises. It is usually further assumed that the perturbation series converges and can be represented in the interaction picture by the unitary operator [75]

$$U(t, t_0) = T\left[\exp\left(-\frac{i}{\hbar}\int_{t_0}^{t} V(t')dt'\right)\right]. \tag{5.54}$$

An important point here is that the operator U is only unitary if the potential is real. If it contains imaginary components, then problems arise. As the perturbing potential's time dependence arises in the Heisenberg notation, this includes the Hamiltonian itself, and any self-energy can appear here. Nevertheless, this operator leads to the S-matrix, which is defined from (5.54) as follows:

$$S(t, t') = U(t, t_0)U^{\dagger}(t', t_0) = T\left[\exp\left(-\frac{i}{\hbar}\int_{t'}^{t} V(t'')dt''\right)\right]. \tag{5.55}$$

The operator T that appears in these last two equations is the time-ordering operator. In the series expansion of the exponential, there are products of the potential V at different times. The time-ordering operator arranges these terms so that the earliest time potential is on the right-hand side of the product and the latest time potential is on the left, and all times are ordered according to this requirement. In addition, the field operator terms must arrange according to Wick's theorem, in which each term has a creation operator and an annihilation operator, the latter of which is placed to the right following the so-called operator ordering [77]. Using these operators, the normalized general Green's function arising from (5.43) can be written as

$$G(\mathbf{r}, \mathbf{r}'; t, t') = \frac{\langle 0 \mid T[\psi(\mathbf{r}, t)\psi^{\dagger}(\mathbf{r}', t')S(\infty, -\infty)] \mid 0\rangle}{\langle 0 \mid S(\infty, -\infty) \mid 0\rangle}. \tag{5.56}$$

In this expression, the state $\mid 0\rangle$ is the ground state mentioned above.

There are two very important aspects of result (5.56). First, the path of integration shown in figure 5.4, along the real axis, involves the assumption that returning via $t \to \infty$ gives the same state as beginning at $t = -\infty$. This requires that time-reversal symmetry is maintained throughout the evolution of the equilibrium Green's functions. Real semiconductors, or even any current-carrying system, are not in equilibrium, and time-reversal symmetry is immediately broken. Thus, the use of equilibrium Green's functions is quite questionable in any computation of conductance, which itself involves current flow. Second, the convergence of the perturbation series and its connection to the unitary operator (5.54) have never been proven [78]. If this series does not converge, then the actual perturbation has no close connection to its use in (5.54), which in turn casts doubt on the usefulness of the unitary operator.

To illustrate the evaluation of (5.56) and the use of Feynman diagrams, the carrier–carrier interaction is considered. While complicated, the use of this interaction will be discussed further after the evaluation. In the numerator of (5.56), there are two waves plus the interaction potential. The latter is the coulomb interaction between two electrons, which means it has four more wave functions. This gives a total of six wave functions, which include three creation field operators and three annihilation field operators. While the actual interaction involves a two-particle Green's function, the expansion can be done with single-particle functions (but at the lowest order, a return to the two-particle function follows below). The expansion can be represented as one plus six diagrammatic terms, but this number can be reduced to four by a judicious interchange of variables. This means, at the lowest order, there are four irreducible diagrams, and these are depicted in figure 5.5. The first term is a bare G° term arising from the unity term in the expansion of the exponential in (5.55). The next two terms are the lowest-order terms for the direct interaction and the exchange interaction. The last two terms are known as disconnected diagrams.

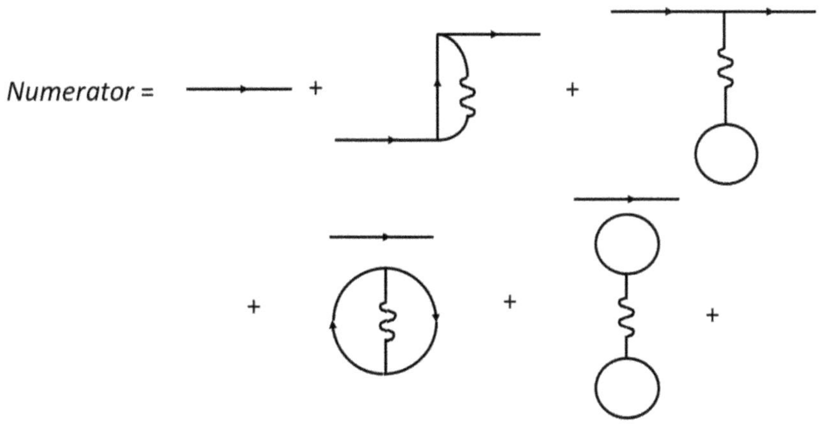

Figure 5.5. The lowest-order Feynman diagrams for the electron–electron interaction. The straight lines, curved lines, and circles are G° terms, while the wiggly lines (and their extensions) are photon lines representing the coulomb interaction between the Green's functions. The various terms are discussed in the text.

As one examines higher orders of the perturbation expansion, new terms arise in both the connected and disconnected diagrams. Notice that the denominator now has only four wave functions, all coming from the S term. The expansion here gives a unity term plus the disconnected diagrams, so that the disconnected diagrams cancel between the numerator and the denominator [75]. Hence, it is important that the result in equilibrium is that only connected diagrams appear in the Green's function expansion.

It was remarked that the proper form of the first potential term in the exponential expansion in (5.54) should involve a two-particle Green's function. In the one-particle Green's function used above, the two-particle function was expanded as a pair of one-particle functions, and the correlation was ignored, yielding a version of the Hartree approximation. To involve this correlation, the first potential term in the exponential expansion should be written as

$$1 + V\Pi, \tag{5.57}$$

where $\Pi(\boldsymbol{k}, \omega)$, in Fourier transform form, is the two-particle correlation or polarization function that appears in the Bethe–Salpeter equation [79]. This arises from the two-particle Green's function

$$\begin{aligned} G_2(\boldsymbol{r}'', t, \mathbf{r}'', t^-; \boldsymbol{r}, t, \boldsymbol{r}', t') = -\langle T[\psi^\dagger(\boldsymbol{r}', t') \\ \times \psi^\dagger(\mathbf{r}'', t^-)\psi(\boldsymbol{r}'', t)\psi(\boldsymbol{r}, t)]\rangle \end{aligned}, \tag{5.58}$$

as the solution to the (Fourier transformed) integral equation

$$\begin{aligned} \Pi(\boldsymbol{k}, \omega) = G_r(\boldsymbol{k}, \omega)G_a(\boldsymbol{k}, \omega) \\ \times \int \frac{d^3k'}{(2\pi)^3}\left\{\boldsymbol{k}\cdot\boldsymbol{k}'\partial(\boldsymbol{k}-\boldsymbol{k}') + \frac{\boldsymbol{k}\cdot\boldsymbol{k}'}{k'^2}T(\boldsymbol{k}-\boldsymbol{k}')\Pi(\boldsymbol{k}', \omega)\right\}. \end{aligned} \tag{5.59}$$

The first term in the curly brackets is just the unity term in (5.57), and the T term contains the interaction potential. Equation (5.59) gives rise to an entire set of doubled Feynman diagrams, some of which are shown in figure 5.6. As mentioned, the first term on the right-hand side is just the Hartree contribution from the lowest approximation to the two-particle Green's function, and was used in figure 5.5. The second term is the standard diagram for the carrier–carrier interaction that is found is discussions of the self-energy. The third and fourth terms are some of the second-order terms which lead to the ladder diagrams (third term) and maximally crossed diagrams (fourth term). Both of these terms appear more often in impurity scattering and lead to the normal conductance with impurity scattering and the weak localization corrections to the conductance (where the lower Green's function lines are reversed to uncross the diagrams), respectively. The fifth term is a more complicated diagram in which the interaction times are mixed. However, note that the red parts of the diagram are actually the same as the second term with one of the two Green's function lines reversed and one of the two ends pulled together. In this sense, this diagram is similar to a vertex correction to the third diagram in figure 5.5. It may be concluded that the proper inclusion of carrier–carrier scattering is quite complicated when Green's functions are adopted and that molecular dynamics with particles, in which all orders are naturally concluded in the real-space interaction, is much easier and simpler to use.

Another important point is that phonon scattering involves a gain–loss formulation that vanishes in equilibrium. Thus, the phonons contribute no perturbation terms to this formulation. The only scattering mechanisms are therefore those produced by impurities and the carrier–carrier interaction. As the impurities are subject to important memory effects, such as those shown in figure 5.1, they should be handled in real space via a molecular dynamics approach, as discussed above. This leaves only the carrier–carrier interactions to be dealt with, and these are shown in the diagrams of figures 5.5 and 5.6, both of which also suggest the use of the molecular dynamics approach.

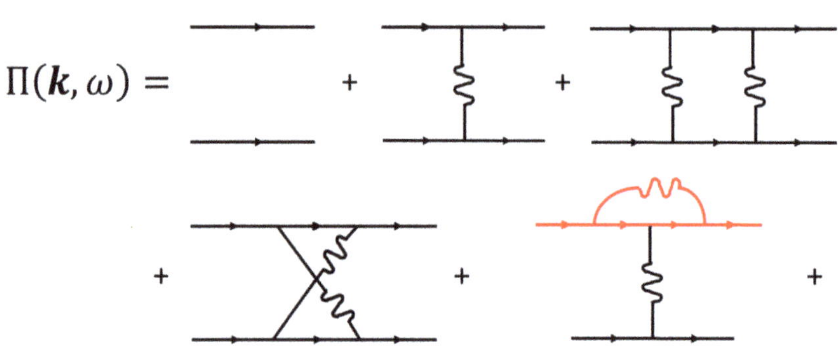

Figure 5.6. The leading terms in the carrier–carrier polarization. The second term on the right-hand side (first row) is the standard electron–electron diagram, in which the two carriers exchange a photon (the coulomb interaction). The red parts of the last diagram are discussed in the text.

Now, as attention moves away from equilibrium to current-carrying states appropriate to semiconductor devices, important differences arise. First, (5.56) is normalized via the ground state of the system. In real cases, this is no longer the situation. The Matsubara Green's functions are normalized by the Fermi–Dirac distribution. With NEGFs, one must find the distribution function from $G^{<}$, as noted in chapter 1. In both of these latter cases, an S term is involved in the normalization. This has the important result that *there is no cancellation of disconnected diagrams* in the out-of-equilibrium situation, and there may also be a failure of Wick's theorem [78, 80]. These disconnected diagrams do not usually affect the probabilities but can be important in determining the proper phase. In addition, the expansion of the two-particle Green's function makes this a drastic approximation, since there are correlations between the carriers, just like those that arise for impurity scattering. The evaluation of the two-particle Green's function, or the three-particle Green's function that is actually likely to arise with six wave functions, is far more complicated. Two-particle Green's functions are usually evaluated using the Bethe–Salpeter equation and the resulting doubled Feynman graphs. Three-particle Green's functions have certainly been dealt with in high-energy physics, and those approaches may be carried over to NEGFs.

5.5 Tunneling with particles

Tunneling is often described as a pure quantum process that has no classical counterpart. This is, of course, not the case. The classical counterpart is found in optics; in particular, in a process known as frustrated total internal reflection. A light beam may be sent into prism at such an angle that it is totally reflected at the back surface. But light waves actually decay exponentially away from that surface outside the prism. If a lens through which the waves would normally pass is brought close to the back surface of the prism, the decaying waves can excite propagating waves within the lens. That is, the light tunnels from the prism to the lens. Tunneling is a quantum phenomenon for particles only in the sense that it requires them to be described by the quantum waves that accompany the particles. Hence, most treatments of tunneling must then use a variant of the standard textbook approach [81].

So, how is one to simulate tunneling through a barrier using particles? There are many modern semiconductor devices in which tunneling is important, not the least of which is the tunnel FET [3], but tunneling is also important in cases of gate tunneling or source–drain tunneling. Since the most efficient simulation of these devices is via particle-based Monte Carlo approaches, it seems that these would benefit from treating tunneling via these particles. The first approach to solving this problem appears to be that of Barker [82]. Here, the tunneling barrier is treated as a scattering center. The knowledge required to use such an approach is a quantum analysis of the actual tunneling probability and the tunneling time, as the particle must be properly retarded as it passes through the barrier. An updated version was used to simulate a resonant-tunneling diode (RTD). In this version [83], the particle was treated as a wave packet whose centers of position, energy, and momentum were determined from the incident particle. This wave packet then remains within the structure, undergoing multiple reflections until it is either transmitted or totally reflected.

The wave packet passing through the RTD is represented entirely by knowledge of the quantum transmission and reflection coefficients, the tunneling delay time, and the packet's average position within the structure. Much of this information is determined by solving the plane wave tunneling problem known from standard textbooks but with modifications for nonparabolic bands, transverse momenta, and position-dependent effective masses. Thus, the one-dimensional Schrödinger equation may be written as follows [83]:

$$\begin{aligned}&\frac{\hbar^2}{2}\frac{d}{dx}\left\{\frac{1}{m^*(x)[1+\alpha(E-V(x))]}\frac{d\psi(x)}{dx}\right\}\\&+\left\{E-V(x)-\frac{\hbar^2k_T^2}{2m^*(x)[1+\alpha(E-V(x))]}\right\}\psi(x)=0\end{aligned}, \tag{5.60}$$

where E is the energy of the particle, including that of the transverse momentum, and α is the nonparabolicity factor. The factor in square brackets of each term accounts for the increase of mass with energy in the nonparabolic bands (this correction will be treated in some detail in section 6.4). The form of (5.60) arises from the need to have the velocity constant across the heterostructure interface [84], in order to satisfy current continuity. The spatial derivative is proportional to the momentum and must be divided by the (nonparabolic mass) to yield the velocity. Of course, for tunneling, one requires $E < V(x)$. The tunneling and reflection delay times are computed using the phases of the transmitted and reflected waves given by the solutions of the Schrödinger equation [82]. These times may be defined by

$$\tau_\mathrm{d}=\begin{cases}\dfrac{1}{v(k_0)}\left(a_\mathrm{B}+\dfrac{d\phi^+}{dk}\right), & k_x>0,\\[2ex] \dfrac{1}{v(k_0)}\dfrac{d\phi^-}{dk}, & k_x<0.\end{cases} \tag{5.61}$$

In this last equation, k_0 is the average momentum of the wave packet, a_B is the thickness of the barrier, ϕ^+ is the phase of the transmission coefficient (which itself has a complex value), and ϕ^- is the phase of the reflection coefficient. When a particle enters the tunneling structure, (5.60) is solved for k and for $k + \Delta k$, where the additive part is the momentum width of the wave packet. This yields the ranges of the transmission and reflection coefficients and the delay time (and phases). Subsequent motion depends upon the transmission and reflection coefficients and random number generation. If the packet tunnels, the particle is placed at the average position of the range of delay times.

The use of the delay times as tunneling times has been questioned over the years. As already mentioned, this choice seems to have arisen with Barker [82], but other calculations have agreed with these results [85, 86]. Actual measurements of the delay time in a microwave measurement of wave tunneling also have shown some agreement with this phase time as well as with a transversal time [87] discussed by Büttiker and Landauer [88]. In fact, over most of the frequency range above the waveguide cutoff, these two times were quite close to one another.

5.6 Spin

Spin arises in all semiconductors due to the spin split that results from the Zeeman effect [89]. Normally, the effect is most familiar from the optical spectroscopy of impurity levels in a solid, in which complex spin structures lead to a split into a family of curves. For a free carrier in a semiconductor, however, the Zeeman effect leads to just two levels, given by the additional energy

$$E_z = g\mu_B \boldsymbol{S} \cdot \boldsymbol{B} = \pm\frac{1}{2} g\mu_B B_z, \tag{5.62}$$

where the magnetic field is oriented in the z-direction in the last form. Here, $\mu_B = e\hbar/2m_0$ is the Bohr magneton, 57.94 μV T^{-1}. The factor g is referred to as the Landé g-factor, which is mainly a 'fudge' factor that has a value of two for a truly free electron. It differs greatly from this value in semiconductors and can even be negative (~ -0.43 in GaAs at low temperatures [90]), which has the effect of reversing the normal ordering of the two spin-split energy levels. The need for a spin component of the wave function and the development of the Dirac spinors may be found in many textbooks [81].

While the Zeeman effect is the most common spin effect that can occur in semiconductors, there are others that have become known in recent years as interest in spin-based devices has grown [91]. The spin of an electron can be manipulated in many ways, but it is preferable to find a purely electrical way to do this. For this reason, a great deal of interest has risen in the spin-Hall effect, in which a transverse spin current arises in response to a longitudinal charge current, without the need for external magnetic fields [92]. The spin-Hall effect most commonly arises from the Rashba form of spin–orbit interaction [93], which is present in a two-dimensional electron gas at a heterostructure interface due to the spatial variation of the confining potential. This is known as *structural inversion asymmetry*. In addition to Rashba spin–orbit interactions, a term can arise from *bulk inversion asymmetry*, and this is known as the Dresselhaus effect [94]. Only the structural inversion asymmetry will be treated here, as it is presumed to be much stronger than the bulk term.

In a single heterostructure interface, band bending appears at this interface. For example, in a GaAs–AlGaAs interface, a quantum well is formed on the GaAs side of the interface, and band bending produces an effective electric field. Through relativistic effects, currents along the interface produce a magnetic field through a $\boldsymbol{v} \times \boldsymbol{E}$ interaction. This effective magnetic field, normal to both the electric field and the velocity, is given by

$$H_{\rm SIS} = r\boldsymbol{\sigma} \cdot (\boldsymbol{k} \times \nabla V) = \alpha_z\left(k_y\sigma_x - k_x\sigma_y\right), \tag{5.63}$$

where the Dirac spinors are

$$\boldsymbol{\sigma} = \begin{bmatrix} 0 & 1 \\ 1 & 0 \end{bmatrix}\hat{a}_x + \begin{bmatrix} 0 & -i \\ i & 0 \end{bmatrix}\hat{a}_y + \begin{bmatrix} 1 & 0 \\ 0 & -1 \end{bmatrix}\hat{a}_z, \tag{5.64}$$

and

$$r \cong \frac{P^2}{3}\left[\frac{1}{E_{\rm G}^2} - \frac{1}{(E_{\rm G} + \Delta)^2}\right] \tag{5.65}$$

arises from the spin–orbit interaction (discussed in section 6.4). The normal to the interface is taken to be the z-direction. The parameter is $\alpha_z = r\langle E_z\rangle$. Near the band edge, the bands can be assumed to be near parabolic, so that the Hamiltonian is given by

$$H = \begin{bmatrix} \dfrac{\hbar^2k^2}{2m^*} & \alpha_z\left(k_x + ik_y\right) \\ \alpha_z\left(k_x - ik_y\right) & \dfrac{\hbar^2k^2}{2m^*} \end{bmatrix}, \tag{5.66}$$

and this leads to spin-split energy levels as follows:

$$E = \frac{\hbar^2k^2}{2m^*} \pm \alpha_z k. \tag{5.67}$$

Here, it may be seen that the split is isotropic in momentum, which is an important point. Thus, the energy bands are two intersecting paraboloids, and a constant energy surface is composed of a pair of circles with a common center. In (5.67), the inner circle represents the positive sign and the outer circle the negative sign. The two spin wave functions are as follows:

$$\phi_\pm = \frac{1}{\sqrt{2}}\begin{bmatrix} 1 \\ \mp e^{i(\vartheta+\pi/2)} \end{bmatrix}, \tag{5.68}$$

where ϑ is the angle that $\boldsymbol{k}$ makes with the [100] direction of the crystal. The spin direction is tangential to the two constant energy circles, and points in the negative angular momentum direction for the inner circle and the positive angular momentum direction for the outer circle.

5.7 Time dependence

As is well-known in classical simulations, time dependence in the current and voltage response of a semiconductor device is crucial to full understanding. In logic devices, an accurate determination of the delay time is required. In microwave devices, the ability to follow large signal amplitudes into the upper gigahertz and even the terahertz region is critical for determining how devices perform. This can even lead to the need to understand how the device operates in a circuit [95]. Time is a simple enough concept in classical physics. It turns out that this simple concept largely carries over to quantum physics in the nonrelativistic domain. Even in quantum mechanics, time is a simple parameter that measures the evolution of the system under applied fields and dissipative mechanisms. Time is *not* a dynamic variable that is subject to uncertainty.

The uncertainty principle governs conjugate variables, such as position and momentum. This principle ensures that these two conjugate variables cannot be simultaneously measured with any accuracy. There is a variable that is often considered to be the conjugate to time, namely energy. But energy is often an integral invariant of motion (energy is conserved for the system as a whole), and energy eigenfunctions are used as a basis set for the wave functions. A general property of eigenvalues and eigenfunctions in quantum mechanics is that they have no uncertainty. These eigenfunctions are expressed in (5.4), in which the quantum energy function is the Hamiltonian itself. Thus, it is easy to show that

$$\langle H \rangle = \int \psi^{\dagger} H \psi dx = E, \tag{5.69}$$

and

$$\langle H^2 \rangle = \int \psi^{\dagger} H^2 \psi dx = E^2, \tag{5.70}$$

so that

$$(\Delta H)^2 = \langle H^2 \rangle - \langle H \rangle^2 = 0. \tag{5.71}$$

While the integrals become somewhat more complicated with the sum over the basis sets, the result holds, as it is a basic mathematical result not constrained to quantum mechanics.

The facts that the energy has no uncertainty and the time is not a dynamic variable subject to uncertainty lead to the conclusion that there is no energy–time uncertainty principle, and this is generally true. So where does the belief in an energy–time uncertainty principle originate? Its origin is life off the shell. In the real world, interactions lead to a self-energy for the system, and the system Hamiltonian. The self-energy has an imaginary part, and when there is an imaginary part, the spectral function is no longer a simple delta function between the energy and the function of momentum. This self-energy leads to a broadening of the energy levels, which begins to violate the eigenfunction equation (5.4). Hence, the broadening gives rise to an uncertainty in the measurement of e.g. the energy. In this sense, the uncertainty in energy winds up being a measurement problem.

Appendix A Classical and quantum brackets

People often initially learn of commutators from their first introductions to quantum mechanics. This is particularly true with engineers who have only limited exposure to proper classical mechanics at the level of Goldstein [96]. In fact, commutators are well-known in classical mechanics and applied mathematics and appear to engineers at the time of their introduction to matrices in circuit theory. The brackets of most importance in mechanics are the Poisson brackets [97], named for the French mathematician and physicist of the early 19th century.

Quite generally, if a function of the trajectories of particles is defined by $f(x, p, t)$ (only a single spatial dimension is considered at this point), then the total derivative of this function is described as

$$\frac{df}{dt} = \frac{\partial f}{\partial x}\frac{dx}{dt} + \frac{\partial f}{\partial p}\frac{dp}{dt} + \frac{\partial f}{\partial t}. \tag{A.1}$$

From Hamilton's equations of mechanical motion, the time derivatives of position and momentum may be written as

$$\begin{aligned} \frac{dx}{dt} &= \frac{\partial H}{\partial p}, \\ \frac{dp}{dt} &= -\frac{\partial H}{\partial x}. \end{aligned} \tag{A.2}$$

Using (A.2) in (A.1) produces

$$\begin{aligned} \frac{df}{dt} &= \frac{\partial f}{\partial x}\frac{dH}{dp} - \frac{\partial f}{\partial p}\frac{dH}{dx} + \frac{\partial f}{\partial t} \\ &= \{f, H\} + \frac{\partial f}{\partial t}. \end{aligned} \tag{A.3}$$

The term in curly brackets defines the Poisson bracket to be (now in three dimensions)

$$\{f, H\} = \sum_i \left(\frac{\partial f}{\partial x_i}\frac{dH}{dp_i} - \frac{\partial f}{\partial p_i}\frac{dH}{dx_i} \right), \tag{A.4}$$

that holds for any arbitrary functions f and H. Of course, for mechanics, the interest is in the Hamiltonian and other operator functions or functions such as the action or Lagrangian. Using this notation, it is not difficult to extend (A.2) to the form

$$\begin{aligned} \frac{dx}{dt} &= \frac{\partial H}{\partial p} = \{x, H\}, \\ \frac{dp}{dt} &= -\frac{\partial H}{\partial x} = \{p, H\}. \end{aligned} \tag{A.5}$$

In quantum mechanics, similar results are obtained for operators and their time derivatives. It is natural to think that simply extending the Poisson brackets to quantization would lead to these quantum brackets, but this is not strictly the case. Extending the Poisson brackets to quantization actually leads to Moyal brackets that appear in treatments of the Wigner function (which is dealt with in chapter 11). A discussion of these brackets is left to that chapter.

Obtaining the most useful quantum brackets, and those usually referred to in commutators, begins with the Schrödinger equation. When the solution to the Schrödinger equation yields a wave function, the expectation value of any operator is given by

$$\langle O \rangle = \int d^3 r \psi^\dagger(\mathbf{r}, t) O(\mathbf{r}, t) \psi(\mathbf{r}, t). \tag{A.6}$$

If the time derivative, or more explicitly, the expectation value of the time derivative of the operator is computed from the normal chain rule as follows:

$$\begin{aligned}\left\langle \frac{\partial O}{\partial t} \right\rangle = & \int d^3 r \frac{\partial \psi^\dagger}{\partial t} O(\mathbf{r}, t) \psi \\ & + \int d^3 r \psi^\dagger \frac{\partial O(\mathbf{r}, t)}{\partial t} \psi \\ & + \int d^3 r \psi^\dagger O(\mathbf{r}, t) \frac{\partial \psi}{\partial t},\end{aligned} \tag{A.7}$$

each of the time derivatives of the wave function is now replaced by the Schrödinger equation itself. This leads to

$$\begin{aligned}\left\langle \frac{\partial O}{\partial t} \right\rangle = & \frac{i}{\hbar} \int d^3 r H \psi^\dagger O(\mathbf{r}, t) \psi \\ & + \int d^3 r \psi^\dagger \frac{\partial O(\mathbf{r}, t)}{\partial t} \psi \\ & - \frac{i}{\hbar} \int d^3 r \psi^\dagger O(\mathbf{r}, t) H \psi.\end{aligned} \tag{A.8}$$

A principal tenet of quantum mechanics is that the Hamiltonian is usually Hermitian for any observable ($H^\dagger = H$), but this is not always the case. There are many situations that involve non-Hermitian Hamiltonians. Here, the Hermitian nature is assumed. Therefore, the first term can be rearranged and (A.8) becomes

$$\left\langle \frac{\partial O}{\partial t} \right\rangle = \frac{i}{\hbar} \int d^3 r \psi^\dagger (HO - OH) \psi + \int d^3 r \psi^\dagger \frac{\partial O(\mathbf{r}, t)}{\partial t} \psi. \tag{A.9}$$

Thus, the quantum equivalent of (A.3) for operators becomes

$$\frac{\partial O}{\partial t} = \frac{i}{\hbar} [H, O] + \frac{\partial O(\mathbf{r}, t)}{\partial t}. \tag{A.10}$$

An important aspect of this relationship is that if the operator does not have an intrinsic time dependence, then it obtains one through (A.10), as it is relatively easy to show that

$$O(\mathbf{r}, t) = e^{iHt/\hbar} O(\mathbf{r}) e^{-iHt/\hbar}. \tag{A.11}$$

This expression holds in the Heisenberg representation, in which the basis functions are not functions of time, but the operators become functions of time through (A.11).

In the Schrödinger representation, the basis functions are time dependent and the operators are not. Generally, the basis functions define a Hilbert space, that is a multidimensional space in which the basis functions are like 'unit' vectors along each of the various axes of the space. In the Schrödinger representation, the space itself rotates around the operators, while in the Heisenberg representation, the space stays in place while the operators rotate through the space.

Appendix B Spin and second quantization for fermions

The idea of spin of the particle was introduced in section 5.6, where the effect of a magnetic field on the quantum dynamics was discussed. So far, the distinction between electrons and photons (or phonons) has only been discussed in the sense that electrons are charged particles. More importantly, photons and phonons are bosons with integer spin. As a result, these latter particles do not obey the Pauli exclusion principle. Fermions, on the other hand, must obey the Pauli exclusion principle, and these are particles with half-integer spin. In this case, only a single particle may reside in a quantum state once the spin is considered. The two particles that occupy the normal momentum state must have opposite spin. These include electrons, holes, and a number of other fundamental particles.

One of the peculiarities is that the distinction between fermions and bosons can be confused in the presence of many-body interactions. For example, two electrons with opposite spins can *condense* into a new quasi particle that is a boson, termed a Cooper pair [98]. This new quasi particle has charge of $2e$, a mass of $2m^*$, and a spin of zero. These Cooper pairs are essential to the Bardeen–Cooper–Schrieffer (BCS) theory of superconductivity [99]. Another example arises in the quantum Hall effect, in which, at a filling factor of around $\nu = 1/2$, an electron can condense with a pair of flux quanta (the quantum of flux is h/e, and these lead to quantum vortices) to create a *composite fermion* [100]. These composite fermions have a mass that is quite different from the GaAs host material [101]. It seems that composite fermions that exhibit the integer quantum Hall effect [102] are strongly related to the normal electrons that exhibit the fractional quantum Hall effect [103]. So, while one normally thinks of simple electrons (or holes) wandering around in the semiconductor device, it must be recalled that electrons are not simple and can undergo some marvelous effects when interactions appear (the quantum Hall effect was, after all, discovered in a Si MOSFET [102]).

B.1 Spin and antisymmetry

As was pointed out in section 5.6, the spin of a particle can have two possible states, which are commonly referred to as the 'spin up' and 'spin down' states. As was pointed out, the basis function must now be augmented by a spin wave function that describes the state of the electron (or hole) spin. This means that the normal Hilbert space is augmented by a two-dimensional spin space. In this latter space, two spin wave functions can be defined as follows:

$$\phi\left(\frac{1}{2}\right) = \begin{pmatrix}1\\0\end{pmatrix}, \phi\left(-\frac{1}{2}\right) = \begin{pmatrix}0\\1\end{pmatrix}. \tag{B.1}$$

In this equation, the first row of the matrix on the right-hand side refers to the 'up' state, while the second row refers to the 'down' state. That is, the spin up state has the spin eigenvalue of 1/2, while the spin down state has the eigenvalue of −1/2. In this description, the spin has been assumed to be directed along the z-axis in real space. Thus, we expect the z-component of the spin angular momentum results from the spin operator

$$\boldsymbol{S}_z = \frac{\hbar}{2}\begin{pmatrix}1 & 0\\0 & -1\end{pmatrix}, \tag{B.2}$$

so that using this operator on the spin up wave function of (B.1) produces a spin angular momentum of $\hbar/2$ (times the up wave function), while operating on the spin down wave function produces a spin angular momentum of $-\hbar/2$ (times the down wave function).

There are, of course, three components of the overall spin that point in the three coordinate directions. These angular momentum vectors have the same properties of normal angular momentum in quantum mechanics [81]. Thus, the square magnitude of the total angular momentum can be written as

$$|\boldsymbol{S}|^2 = s(s+1)\hbar\begin{pmatrix}1 & 0\\0 & 1\end{pmatrix} = \frac{3}{4}\hbar\begin{pmatrix}1 & 0\\0 & 1\end{pmatrix}. \tag{B.3}$$

From this and the normal angular momentum states, one can easily find that

$$\begin{aligned}\boldsymbol{S}_x &= \frac{\hbar}{2}\begin{pmatrix}0 & 1\\1 & 0\end{pmatrix},\\ \boldsymbol{S}_y &= \frac{\hbar}{2}\begin{pmatrix}0 & -i\\i & 0\end{pmatrix}.\end{aligned} \tag{B.4}$$

The three matrices in (B.2) and (B.4) are usually referred to as the Pauli spin matrices.

Once the spin wave functions are used to augment the basis functions, the overall wave function must be antisymmetric. To illustrate this more fully, we introduce the composite vectors $\boldsymbol{\xi}_i = (\boldsymbol{r}_i, \sigma_i)$, where the subscript refers to the ith particle (not a coordinate) and σ_i is the spin orientation of that particle. This vector may be thought of as a four-vector in which the spin axis corresponds to the fourth direction. Now consider a wave function corresponding to two particles, which is written as $\psi(\boldsymbol{\xi}_1, \boldsymbol{\xi}_2)$. When these two particles are exchanged with one another (both their positions and spins are interchanged), the wave function must undergo a phase change, according to

$$\psi(\boldsymbol{\xi}_1, \boldsymbol{\xi}_2) = e^{i\vartheta}\psi(\boldsymbol{\xi}_2, \boldsymbol{\xi}_1). \tag{B.5}$$

The phase ϑ is a real constant phase shift which occurs due to the particle exchange. If we repeat this exchange,

$$\psi(\xi_1, \xi_2) = e^{i\vartheta}\psi(\xi_2, \xi_1) = e^{i2\vartheta}\psi(\xi_1, \xi_2). \tag{B.6}$$

Of course, the magnitude squared of the wave function must be the normalization of unity, so the result of (B.5) is that

$$e^{i2\vartheta} = \pm 1. \tag{B.7}$$

In classical mechanics, one expects only the positive sign. And, indeed, the positive sign arises in quantum mechanics *when the spin is not important*; e.g. with integer spin. With fermions, however, the wave function must be antisymmetric, which requires the negative sign in (B.7). Just what does this antisymmetric wave function mean? The presence of the negative sign in (B.7) requires certain constraints to be placed upon the wave functions. For example, if the two-particle wave function is composed of single-particle basis functions, certain constraints are placed upon the manner in which the two-particle function is described. Consider the Schrödinger equation for a two-particle wave function. This may be written as

$$[H(\xi_1, \xi_2) - E]\psi(\xi_1, \xi_2) = 0. \tag{B.8}$$

The Hamiltonian may be assumed to be invariant under the exchange of the two particles, but there is often an energy cost, which appears in density functional theory via the exchange interaction [104]. The presence of the magnetic field, however, introduces a spin-dependent interaction that changes this assumption, but this will be ignored here. So, whether the wave function is symmetric or antisymmetric has no effect on (B.8). Thus, one can often separate the equation to yield two equations, one for each particle, but the exchange interaction couples them. One can begin with the single-particle basis functions that are augmented by the spin functions. Thus, it may be asserted that

$$\psi(\xi_1, \xi_2) \sim \psi_1(\xi_1)\psi_2(\xi_2). \tag{B.9}$$

However, (B.9) does not possess the required antisymmetry of (B.7). This can be restored by writing (B.9) in a different form, such as

$$\psi(\xi_1, \xi_2) = \frac{1}{\sqrt{2}}[\psi_1(\xi_1)\psi_2(\xi_2) - \psi_2(\xi_1)\psi_1(\xi_2)]. \tag{B.10}$$

It may now be observed that interchanging the two particles does yield the desired negative sign and is thus antisymmetric.

The wave function (B.10) is, in fact, a form of a generalization that can be applied to any many-body wave function for an arbitrary number of particles. To see this, (B.10) is first rewritten in a different form as follows:

$$\psi(\xi_1, \xi_2) = \frac{1}{\sqrt{2}}\begin{vmatrix} \psi_1(\xi_1) & \psi_1(\xi_2) \\ \psi_2(\xi_1) & \psi_2(\xi_2) \end{vmatrix}. \tag{B.11}$$

The right-hand side has been written in terms of a determinant, rather than a matrix. This can be extended to an arbitrary number of particles, for which the prefactor becomes $\sqrt{2} \to \sqrt{N!}$, where N is the number of particles and the order of the determinant. These determinants are known as *Slater determinants* [105].

B.2 Second quantization for fermions

In appendix A.2, creation and annihilation operators for the harmonic oscillator were introduced. In this previous treatment, spin was not considered, so these operators were considered to be primarily boson operators. This process of second quantization can also be applied to the spin fermions. This involves a notation change, however. In appendix A.2, the operators were denoted by $a^\dagger$, a. Sometimes when both carriers and bosons are present (and with no consideration of spin), the bosons are denoted by $b^\dagger$, b. In contrast to these, the spin fermions are usually denoted by the symbols $c^\dagger$, c. These latter operators have very different properties than the ones introduced in appendix A.2.

It still remains the case that $c^\dagger$ is the creation operator. If the state in which no electrons are present is denoted by $|\,0\rangle$, then the creation operator excites an electron into this state as follows:

$$c^\dagger\,|\,0\rangle = |\,1\rangle. \tag{B.12}$$

In the case of bosons, a second particle could be excited into the new state. But in the case of spin fermions, this is not an allowed operation, so that we require

$$(c^\dagger)^2\,|\,0\rangle = c^\dagger\,|\,1\rangle = 0. \tag{B.13}$$

Since only a single electron can populate a single spin state, the use of the creation operator on an occupied state is forbidden by the Pauli exclusion principle. Similarly, annihilation can operate on a full state as follows:

$$c\,|\,1\rangle = |\,0\rangle. \tag{B.14}$$

If we try to use the annihilation operator again, we obtain the same result that we obtained by doubling the creation operator:

$$(c)^2\,|\,1\rangle = c\,|\,0\rangle = 0. \tag{B.15}$$

Since the first operator empties the state, there is nothing to remove from that state, hence the zero result. The product $c^\dagger c$ remains the number operator, since it produces

$$\begin{aligned} c^\dagger c\,|\,1\rangle &= c^\dagger\,|\,0\rangle = 1\,|\,1\rangle, \\ c^\dagger c\,|\,0\rangle &= c^\dagger 0 = 0\,|\,0\rangle. \end{aligned} \tag{B.16}$$

Hence, the occupation number is limited to zero or one, which is appropriate for the spin states. The reverse operation produces

$$\begin{aligned} cc^{\dagger} \mid 1\rangle &= c0 = 0 \mid 1\rangle, \\ cc^{\dagger} \mid 0\rangle &= c^{\dagger} \mid 1\rangle = 1 \mid 1\rangle. \end{aligned} \tag{B.17}$$

Now, (B.16) and (B.17) produce the *anticommutator* for these spin operators, namely

$$\{c^{\dagger}, c\} = c^{\dagger}c + cc^{\dagger} = 1. \tag{B.18}$$

The curly brackets are used to denote the anticommutator, which utilizes the positive sign, while the normal commutator utilizes the negative sign.

These operators also have limitations regarding the state to which they are applied, just as for the operators in appendix A.2. This means that if we consider two states designated by λ and μ, (A.18) becomes

$$\left\{c_{\lambda}^{\dagger}, c_{\mu}\right\} = c_{\lambda}^{\dagger}c_{\mu} + c_{\mu}c_{\lambda}^{\dagger} = \delta_{\lambda\mu}. \tag{B.19}$$

(The use of curly brackets here should not be confused with the Poisson brackets of the previous appendix. This anticommutator involves operators, while the Poisson brackets involve functions of phase space.) Contrary to the number operator for bosons, the number operator for these spin fermions is idempotent. That is, they satisfy

$$\begin{aligned} (c_{\lambda}^{\dagger}c_{\lambda})^2 &= c_{\lambda}^{\dagger}c_{\lambda}(1 - c_{\lambda}c_{\lambda}^{\dagger}) \\ &= c_{\lambda}^{\dagger}c_{\lambda} - c_{\lambda}^{\dagger}c_{\lambda}c_{\lambda}c_{\lambda}^{\dagger} = c_{\lambda}^{\dagger}c_{\lambda} \end{aligned}. \tag{B.20}$$

This result should have been obvious, since squaring zero or one still produces zero or one. In the last line, the result of (B.15) was used to show that the second term is zero.

The generation of a two-electron state follows from the normal product order of the operators but must follow the antisymmetry requirements. Thus, for example, the two-electron state can be generated as follows:

$$\mid \lambda\mu\rangle = c_{\lambda}^{\dagger}c_{\mu}^{\dagger} \mid 0\rangle = -\mid \mu\lambda\rangle. \tag{B.21}$$

This means that the order of the creation operators is important in defining the created state. This can be arbitrarily extended to the N-particle state. Interchanging K pairs of particles modifies the wave function by the factor $(-1)^K$, which assures the antisymmetry.

References

[1] Ferry D K 2020 *Transport in Mesoscopic Semiconductor Devices* (Bristol: IOP Publishing) 2nd

[2] Ferry D K, Goodnick S M and Bird J P 2009 *Transport in Nanostructures* (Cambridge: Cambridge University Press) 2nd

[3] Ferry D K *et al* 2022 *Semicond. Sci. Technol.* **37** 043001

[4] Ferry D K 2021 *Hot Carriers in Semiconductors* (Bristol: IOP Publishing)

[5] Mori H, Oppenheim I and Ross J 1962 *Studies in Statistical Mechanics* ed J de Boer and G E Uhlenbeck (Amsterdam: North Holland)
[6] Chester G V 1963 *Rep. Prog. Phys.* **26** 411
[7] Bordone P, Vasileska D and Ferry D K 1996 *Phys. Rev.* B **53** 3846
[8] Ferry D K *et al* 1991 *Phys. Rev. Lett.* **67** 633
[9] Maxwell J C 1855 *Trans. Camb. Phil. Soc* **10** 155–229
[10] Fröhlich H and Paranjape B V 1955 *Proc. Phys. Soc.* B **69** 21
[11] Shur M S and Eastman L F 1979 *IEEE Trans. Electron Devices* **26** 1677
[12] Rahman A *et al* 2003 *IEEE Trans. Electron Devices* **50** 1853
[13] Akis R *et al* 2009 *IEEE Trans. Electron Devices* **56** 2935
[14] Bude J 2000 *Proc. Int. Conf. SISPAD* (New York: IEEE Press) 23–6
[15] Young T 1807 *A Course of Lectures of the Natural Philosophy and Mechanical Arts* 1 Lect. (London: J Johnson)
[16] Akashi T *et al* 2002 *Appl. Phys. Lett.* **81** 1922
[17] Taylor G I 1909 *Proc. Camb. Phil. Soc.* **14** 417
[18] Einstein A 1909 *Phys. Z.* **10** 817
[19] Weinbub J, Ballicchia M and Nedjalkov M 2018 *Phys. Status Solidi RRL* **12** 1800111
[20] Barker J R 2003 *J. Comput. Electron* **2** 153
[21] Einstein A, Podolsky B and Rosen N 1935 *Phys. Rev.* **47** 777
[22] Schrödinger E 1935 *Naturwissenschaften* **23** 807 823 844 transl.
Trimmer J D 1980 *Proc. Am. Phil. Soc.* **124** 323
[23] Schrödinger E 1935 *Proc. Camb. Phil. Soc.* **31** 555
Schrödinger E 1935 *Proc. Camb. Phil. Soc.* **32** 446
[24] Bohr N 1935 *Phys. Rev.* **48** 696
[25] Ferry D K 2019 *The Copenhagen Conspiracy* (Singapore: Pan Stanford Publishing)
[26] Bohm D 1952 *Phys. Rev.* **85** 166 180
[27] Bohr N 1937 *Phil. Sci.* **4** 289
[28] Planck M 1900 *Verh. Dtsch. Phys. Ges. Berlin* **2** 237
[29] Bohr N 1913 *Phil. Mag., Ser. 6* **26** 1
[30] Heisenberg W 1925 *Z. Phys.* **33** 879
[31] De Broglie L 1924 *Phil. Mag., Ser. 6* **47** 446
[32] Kurosawa T 1966 *J. Phys. Soc. Jpn. Suppl.* **21** 424
[33] Price P 1979 *Semiconductors and Semimetals* ed R K Williamson and A C Beer 14 (New York: Academic) pp 249
[34] Jacoboni C and Reggiani L 1983 *Rev. Mod. Phys.* **65** 645
[35] Montefusco L and Jacoboni C 1972 *Solid State Commun* **10** 71
[36] Lugli P and Ferry D K 1986 *Phys. Rev. Lett.* **56** 1295
[37] Gross W J, Vasileska D and Ferry D K 1999 *IEEE Electron Device Lett.* **20** 463
[38] Kelbg G 1964 *Ann. Phys.* **468** 385
[39] Kohn W, Meir Y and Makarov D E 1998 *Phys. Rev. Lett.* **80** 4153
[40] Morawetz K 2002 *Phys. Rev.* E **66** 022103
[41] Ferry D K and Nedjalkov M 2018 *The Wigner Function in Science and Technology* (Bristol: IOP Publishing)
[42] Oriols X and Mompart J 2019 *Applied Bohmian Mechanics: From Nanoscale Systems to Cosmology* (Singapore: Pan Stanford)
[43] Merzbacher E 1970 *Quantum Mechanics* 2nd (New York: Wiley)

[44] Kriman A M, Kluksdahl N C and Ferry D K 1987 *Phys. Rev.* B **36** 5953
[45] Weinbub J and Kosik R 2022 *J. Phys.: Condens. Matter* **34** 163001
[46] Knezevic I and Ferry D K 2002 *Phys. Rev.* E **66** 016131
[47] Kohn W and Luttinger J M 1957 *Phys. Rev.* **108** 590
[48] Siegel J L and Argyres P N 1969 *Phys. Rev.* **178** 1016
[49] Argyres P N and Siegel J L 1973 *Phys. Rev. Lett.* **31** 1397
[50] Ferry D K 1991 *Semiconductors* (New York: Macmillan) pp 528
[51] Pottier N 1983 *Physica* **A117** 243
[52] Barker J R 1980 *Physics of Nonlinear Transport in Semiconductors* ed D K Ferry, J R Barker and C Jacoboni (New York: Plenum) 127–51 pp
[53] Barker J R 1978 *Solid-State Electron* **21** 197
[54] Balescu R 1975 *Equilibrium and Nonequilibrium Statistical Mechanics* (New York: Wiley)
[55] Knezevic I and Ferry D K 2003 *Phys. Rev.* E **67** 066122
[56] Knezevic I and Ferry D K 2005 *J. Comput. Electron* **3** 359
[57] Knezevic I and Ferry D K 2004 *Semicond. Sci. Technol.* **19** S220
[58] Ferry D K, Barker J R and Grubin H L 1981 *IEEE Trans. Electron Devices* **28** 905
[59] Geltman S 1977 *J. Phys.* B **10** 831
[60] Lane A M 1983 *Phys. Lett.* **99A** 359
[61] Haan S L and Geltman S 1982 *J. Phys.* B **15** 1229
[62] Lipavsky P *et al* 1991 *Phys. Rev.* B **43** 6650
[63] Ferry D K 2017 *An Introduction to Quantum Transport in Semiconductors* (Singapore: Pan Stanford)
[64] Jacoboni C *et al* 2001 *Math. Comput. Simul* **55** 67
[65] Barker J R and Ferry D K 1979 *Phys. Rev. Lett.* **42** 1779
[66] Budd H F 1967 *Phys. Rev.* **158** 798
[67] Thornber K K 1971 *Phys. Rev.* B **3** 1929
[68] Barker J R 1973 *J. Phys.* C **6** 2663
[69] Ferry D K 1980 *Physics of Nonlinear Transport in Semiconductors* ed D K Ferry, J R Barker and C Jacoboni (New York: Plenum) 577–88 pp
[70] Bude J, Hess K and Iafrate G J 1992 *Phys. Rev.* B **45** 10958
[71] Wigner E 1932 *Phys. Rev.* **40** 749
[72] Levinson I B 1970 *Sov. Phys. JETP* **30** 362 http://jetp.ras.ru/cgi-bin/e/index/e/30/2/p362?a=list
[73] Kennard E H 1928 *Phys. Rev.* **31** 876
[74] Dias N C and Prata J N 2004 *Ann. Phys.* **313** 110
[75] Fetter A L and Walecka J D 1971 *Quantum Theory of Many-Particle Systems* (New York: McGraw-Hill)
[76] Martin P C and Schwinger J 1959 *Phys. Rev.* **115** 1342
[77] Wick G C 1950 *Phys. Rev.* **80** 268
[78] Dyson F 1951 *Phys. Rev.* **82** 428
[79] Salpeter E E and Bethe H A 1951 *Phys. Rev.* **84** 1232
[80] Dyson F 1951 *Phys. Rev.* **83** 608
[81] Ferry D K 2021 *Quantum Mechanics* (Boca Raton, FL: CRC Press) 3rd
[82] Barker J R 1985 *Physica B+C* **134** 22
[83] Baba T, Al-Mudares M and Barker J R 1989 *Jpn. J. Appl. Phys.* **28** L1682
[84] Bastard G 1981 *Phys. Rev.* B **24** 5693

[85] Jauho A-P and Nieto M M 1986 *Superlattices Microstruct* **2** 407
[86] Collins S, Lowe D and Barker J R 1987 *J. Phys. C: Solid-State Phys.* **20** 6213
[87] Ranfangni A *et al* 1991 *Appl. Phys. Lett.* **58** 774
[88] Büttiker M and Landauer R 1982 *Phys. Rev. Lett.* **49** 1739
[89] Zeeman P 1897 *Phil. Mag.* **43** 226
[90] Oestreich M and Rühle W W 1995 *Phys. Rev. Lett.* **74** 2315
[91] Datta S and Das B 1990 *Appl. Phys. Lett.* **58** 665
[92] Murakami S, Nagaosa N and Zhang S 2003 *Science* **301** 1348
[93] Bychov Y A and Rashba E I 1984 *J. Phys.* C **17** 6039
[94] Dresselhaus G 1955 *Phys. Rev.* **100** 580
[95] Guerra D *et al* 2012 *2012 IEEE/MTT-S International Microwave Symp.* 6258430 (Piscataway, NJ: IEEE) 1–3
[96] Goldstein H 1951 *Classical Mechanics* (Cambridge, MA: Addison-Wesley)
[97] Arnold D H 1981 *Siméon Denis Poisson Et La Science De Son Temps* (Paris: Diffusion)
[98] Cooper L 1956 *Phys. Rev.* **104** 1189
[99] Bardeen J, Cooper L N and Schrieffer J R 1957 *Phys. Rev.* **106** 162
[100] Jain J K 1989 *Phys. Rev. Lett.* **63** 199
[101] Smet J H *et al* 1993 *Phys. Rev. Lett.* **77** 2272
[102] von Klitzing K, Dorda G and Pepper M 1980 *Phys. Rev. Lett.* **45** 494
[103] Tsui D, Stormer H L and Gossard A C 1982 *Phys. Rev. Lett.* **48** 1559
[104] Ferry D K 2020 *Semiconductors: Bonds and Bands* (Bristol: IOP Publishing) 2nd
[105] Slater J C 1929 *Phys. Rev.* **34** 1293

IOP Publishing

Quantum Transport in Semiconductor Devices
Simulation using particles
David K Ferry, Xavier Oriols and Josef Weinbub

Chapter 6

Electron–atom interaction: band structure

Particles in semiconductor devices are not like free particles. They move in an array of potentials that arise from the atoms of the crystal structure. This means that the particles (the electrons and holes in which the current interest is centered) are subject to a great many quantum mechanical forces and potentials. Thus, they are rattled around within the bulk of the semiconductor and they possess properties that differ from those of free electrons or holes. But the goal in life is to simplify things and particularly the behavior of these electrons within the bulk material. One direct simplification is to recognize that, quantum mechanically, this may be achieved by understanding the electronic band structure that arises from these atomic potentials. In section 3.1.1, it was pointed out that the total Hamiltonian may be separated into electronic and atomic parts, and that the electronic part may be solved for the band structure with the atoms in fixed positions in the lattice. Then, within a single one of the available energy bands (which we will describe below), the electron takes on the role of a *quasi particle*, which may be either electron-like or hole-like, depending upon where its particular kinetic energy lies in the range of that single band. This quasi particle is then considered to have an effective mass which differs from that of a free electron and describes the motion of the quasi particle within that band.

The above simple viewpoint of the quasi particle arises from developing an understanding of the electronic band structure. The goal is to simplify the multitude of forces and potentials into a simpler and more usable form. The electron (or hole) is then replaced by what is termed a quasi particle. In this procedure, the effective mass is introduced in a manner such that the totality of the quantum forces to which this quasi particle responds is reduced to only the additional external forces. To understand how this transition occurs, one first needs to understand the electronic band structure in the semiconductor. Here, we deal first with the energy bands, determined with the atoms in their equilibrium positions. The extra motion of the atoms away from their equilibrium positions then introduces the role of the lattice vibrations—the phonons—into the problem.

doi:10.1088/978-0-7503-5237-6ch6

The calculation of the electronic structure of an atom is quite old. Although the first apparent comments on the properties of Si predate even Faraday's discovery of semiconductors [1], the first studies of carrier transport in Si predate even quantum mechanics [2]. Modern approaches for condensed-matter systems began in 1964 with the Hohenberg–Kohn theorems [3]. The first of these pointed out that the ground state of a many-electron system was uniquely determined by the three-dimensional electron density (that is, the average density). The second theorem showed that an energy functional could be written in terms of the local density, giving the local-density approximation. The correct ground-state electron density then minimized this functional. In later work, Kohn and Sham showed that the many-electron problem could be reduced to noninteracting electrons moving in an effective potential [4]. Finding the correct effective potential allowed the electronic structure to be determined by solving the one-electron Schrödinger equation. This is the basis upon which all modern band structure theory is based. At this point, it is important to note that the electronic structure depends critically upon the crystal lattice and its various periodicities. Differences between semiconductors arise primarily from differences in the lattice constant, although the chemical properties of the atoms themselves have an effect.

Today, density-functional theory relies upon two potentials: (1) the sum of all external potentials, which is constructed from the crystal structure and the atomic compositions of the materials (e.g. the atomic potentials), and (2) the effective potential that arises from the interelectronic interactions. At this point, it must be pointed out that the 'electrons' include all of the electrons, bonding electrons, and inner-shell electrons which are strongly bound to the individual nuclei. Since these strongly bound inner-shell electron wave functions do not greatly affect the computation, it was apparently first suggested by Hellman that they be removed from the calculation in a manner that led to a modified atomic potential, called the pseudopotential [5].

However, there are many ways to proceed from these early determinations. One can perform *ab initio* calculations, in which the pseudopotential, the wave functions, and the density are all computed self-consistently. These so-called 'first-principles' calculations begin with the atomic wave functions and supposedly yielded the 'correct' wave functions, potentials, and energy levels for the electrons. In fact, these approaches do yield the so-called ground state quite well. But in semiconductors, the ground state means only the valence bands. For many years, these were limited by the so-called bandgap problem, in that these methods did not work well for the conduction bands and often suggested a bandgap for Si that was as little as 10% of the actual value. In recent times, corrections have been developed, and these will be discussed later in the chapter.

The presence of the bandgap problem led to another approach, which is the empirical approach. In most semiconductors, optical as well as electrical studies have given values for the principal energy at many points of the conduction and valence bands. Knowing this, why spend a lot of time and effort to find the full spectrum by *ab initio* methods, when the various values for the interaction parameters can be fit to the experimental values? These empirical approaches are used fairly extensively in semiconductor simulations and are designated by the name *empirical.*

In both approaches, there are real-space and momentum-space methods. In a real-space approach, one begins with the nature of the outer shell orbitals that provide the bonding—typically the atomic *s*- and *p*-orbitals in semiconductors. In chemistry, this is termed the linear combination of atomic orbitals (LCAO), while in physics this is usually termed the tight-binding method (TBM) [6, 7]. In empirical approaches, the values of the atomic energies for these orbitals and the values of the overlap integrals between them (on neighboring atoms) are treated as adjustable parameters and used to match the computed bands to the experimental data. The corresponding momentum-space method is termed the empirical pseudopotential method (EPM) [8]. Its main adjustable parameters are the Fourier coefficients for the Fourier transform of the pseudopotential itself. These approaches will be discussed later in this chapter.

First, it is necessary to discuss how the atomic lattice and its periodicity determine the nature of the electronic structure. This leads to the Bloch functions that are themselves combinations of a wave and a term that is periodic in the lattice. We will use this to create a simple lattice and band structure to introduce the real bands and the complex bands. Following that, we will discuss a variety of real-space and momentum-space situations that illustrate the various methods for computing the actual energy bands in the semiconductor. We will then turn to the role played by defects and impurities that change the band structure somewhat. The perturbative spin–orbit interaction and the effective-mass approximation will be discussed in turn. Then, we will discuss how scaling between various semiconductors can be used. Finally, we will turn to the role of phonons and their effect upon the carriers.

6.1 The basics of energy bands

It was mentioned above that the most important part of determining the energy structure is the crystal lattice and its periodicities. The lattice defines certain symmetries, and the energy structure must also possess these symmetries. The mathematical approach must ensure that this periodicity is imposed upon the wave functions (solutions) that arise from the Schrödinger equation (one dimension is considered here for simplicity):

$$-\frac{\hbar^2}{2m_0}\frac{\partial^2\psi(x)}{\partial x^2} + U(x)\psi(x) = E\psi(x). \tag{6.1}$$

Here, m_0 is the free electron mass. If the potential is weak, one obtains almost free electron solutions. However, if the potential has a periodicity, the wave function solutions $\psi(x)$ must exhibit behavior that is consistent with this periodicity. The periodicity is defined by the potential $U(x)$. Suppose that this periodicity is indicated by a displacement a. This appears in the potential as the requirement

$$U(x + a) = U(x). \tag{6.2}$$

That is, one cannot determine where the position is within the crystal merely by looking at the potential. While the resulting wave function may itself be complex, the real probability that arises from this wave function must have the same periodicity. This means that

$$|\psi(x+a)|^2 = |\psi(x)|^2, \tag{6.3}$$

and this must hold for each and every value of a (by which this means it holds for a, $2a$, $3a$, …, Na). Thus, the wave function itself can differ by at most a phase factor, or

$$\psi(x+a) = e^{i\varphi}\psi(x). \tag{6.4}$$

In a real system, even a one-dimensional one, the line of atoms is not infinite, rather it has a finite length, say $L = Na$. That is, the semiconductor is not physically infinite in extent but has a finite size. In order to ensure that the results do not depend upon the ends of this chain of atoms, periodic boundary conditions are assumed. If there are N unit cells (each unit cell has length a) in the chain, as mentioned, then

$$e^{iN\varphi} = 1, \quad \varphi = \frac{2n\pi}{N}, \tag{6.5}$$

where n is an integer. The smallest value of φ (other than zero) is $2\pi/Na$. The largest value is $2\pi/a$, at which point the values begin to repeat. The invocation of this longer periodicity means that the Nth atom is actually also the zeroth atom and the length of the crystal is $L = Na$.

The value $2\pi/a$ is important, as it is the basic length in the Brillouin zone—the Fourier transform of the lattice into momentum space. To see this, one writes the wave function in terms of its Fourier transform through the summation

$$\psi(x) = \sum_k C(k)e^{ikx}. \tag{6.6}$$

At the same time, the Fourier transform of the potential may be written in terms of the basic lattice constant $2\pi/a$, as

$$U(x) = \sum_G U_G e^{iGx}, \quad G = n'\frac{2\pi}{a}, \tag{6.7}$$

where n' is an arbitrary integer. Hence, the quantities G are the harmonics of the basic quantity $2\pi/a$. These two Fourier transforms are inserted into the Schrödinger equation (6.1), so that

$$\sum_k \left[\frac{\hbar^2k^2}{2m_0}C(k) + \sum_G U_G C(k)e^{iG} - EC(k)\right]e^{ikx} = 0. \tag{6.8}$$

While it may not be immediately apparent, one can assume that the analogy to (6.4) leads to a displacement operator in momentum space, as follows:

$$C(k+\lambda) = e^{i\lambda x}C(k). \tag{6.9}$$

Here is an important point: the periodicity of the potential given by (6.2) enforces periodicity upon the wave function given by (6.3) but also forces periodicity in the Fourier space given by (6.9). The momentum shift inherent in the second term of the square brackets of (6.8) ensures that the exponential term in (6.9) is a displacement operator. This reflects the fact that, in quantum mechanics, x is a differential operator in momentum space [9, 10]. The role of this displacement operator is specifically to shift the position (in momentum space) of the wave function-like quantity $C(k)$ to $C(k - G)$. The relation (6.9) establishes this periodicity in momentum space, and the Fourier transforms of the wave function must correspond exactly to the periodicity (6.4). From (6.7), this periodicity in momentum space has the length $2\pi/a$, as determined by the periodicity a in real space.

To satisfy (6.8), it is sufficient for the square brackets to vanish, and this leads to

$$\left(\frac{\hbar^2 k^2}{2m_0} - E\right)C(k) + \sum_G U_G C(k - G) = 0. \tag{6.10}$$

This result actually represents an entire set of equations, one for each value of k. The entire set of equations must be solved to find the family of Fourier coefficients $C(k)$. The second term in (6.10) represents a convolution summation of these coefficients with the Fourier coefficients of the potential. This equation will appear in a variety of different forms throughout this chapter. It is fundamental to the determination of the electronic band structure, regardless of which solution approach is followed. All results depend upon the Fourier transform of the potentials, or pseudopotentials, through the coefficients U_G. In the case in which the so-called 'first-principles' calculations are used, one must also determine the Fourier coefficients self-consistently with the wave functions, represented by the $C(k)$'s. This is the heart of the momentum-space approaches.

It is apparent that the spectrum of Fourier coefficients do not consist of a continuous spectrum. Only a discrete number of values of the vector k are allowed by the discretization (6.5) introduced by the periodic boundary conditions. This number is N, which is the number of unit cells (each of length a) in the crystal. This is often thought to be the number of atoms, but this is true only for systems with a single atom per unit cell. The values of k are selected by the values of G. These latter values form the *reciprocal lattice* in momentum space, and the set of values k, formed via (6.7), span one unit cell of this reciprocal lattice. This cell is called the (first) Brillouin zone of the reciprocal lattice. (Normally, a value of k which runs through $-\pi/a < k < \pi/a$ is used in order to provide a centered cell.) The wave function (6.6) may be written in terms of the shifted vector of the second term of (6.10) as

$$\psi(x) = \left[\sum_G C(k - G)e^{-iGx}\right]e^{ikx}. \tag{6.11}$$

The term in the square brackets is a function that is periodic in the lattice and in the reciprocal lattice ($Ga = 2\pi$). This term is the Fourier transform of a real-space function $u_k(x)$ which is periodic in the unit cell, so that the *Bloch function* is defined to be

$$\psi(x) = e^{ikx}u_k(x). \tag{6.12}$$

The Bloch functions (6.12) are the general solutions of the Schrödinger equation in a periodic potential. These functions combine the properties of a wave in a periodic structure (the exponential factor) and a cell periodic function (the second term). These solutions are not unique to quantum mechanics but arise in any periodic system [11].

An interesting point has been reached. The wave functions for the periodic crystal are Bloch functions which represent the presence of the periodic potential. This dramatically confines the basic nature of the propagating waves that characterize the electrons. They are also periodic in momentum space (within the Brillouin zone). As a result, the energy bands are also periodic in momentum space. For all practical purposes, therefore, only the first Brillouin zone needs to be studied in order to understand all of the energy bands and the nature of particles within the solid.

6.1.1 Potentials and pseudopotentials

In the above discussion, it was remarked that the atomic potentials could be modified to remove the effect of the core–shell electrons. In tetrahedrally coordinated semiconductors as well as the lower-dimensional layered semiconductors, the bonds are actually only formed by the outer shell electrons. These are the so-called valence electrons. The electrons in the inner core shells play very little role in the bonding which occurs between different atoms in the crystal. For example, the Si bonds are composed of the $3s$ and $3p$ electrons in the outer shell. Other atoms have different levels, but it is still mainly the outer shell that provides the electrons for the bonding process. Hence, one treats only the four outer shell electrons from each of the atoms in group IVB of the periodic table.

The Hamiltonian from section 3.1.1 can be simplified when we are treating only the outer shell bonding electrons. In this case, the Hamiltonian can be rearranged into the form

$$H_e = \sum_{i,\mathrm{val}}\left\{\frac{p_i^2}{2m_0} + e_i\left[V(\mathbf{r}_i) + \sum\nolimits_{j,\,\mathrm{core}}\frac{e_j}{4\pi\varepsilon_0 r_{ij}}\right] + \sum_{j,\mathrm{val}}\frac{e_ie_j}{4\pi\varepsilon_0 r_{ij}}\right\} + e_ie_jE_{\mathrm{core}}. \tag{6.13}$$

Here, the coulomb interaction of (3.2) has been divided into three terms. The first term (the second term in the square brackets) represents the interaction between the outer shell (subscript 'val' for valence) electrons and the inner core (subscript 'core') electrons. This term actually represents a potential through Coulomb's law, and this potential modifies the total potential, which is the first term in the square brackets.

The second of the three terms represents the interactions between just the outer shell electrons, and this is the last term in the curly brackets. This term creates an interaction that leads to the Hartree and higher-order corrections to the one-electron approximation of the Hohenberg–Kohn (H–K) theorems.

Finally, the third of the three terms represents the interactions among just the inner-shell electrons, and this term is represented by the last term of (6.13). This term can be treated as a rigid energy shift of the band structure. While important for many applications, such as photoemission, it is not particularly important in our discussion of the electronic structure, since the energies are usually referenced to either the bottom of the conduction band or the top of the valence band.

The second sum in the square brackets represents another potential, so that one can combine it with the first term in the square brackets and rewrite both terms in square brackets as a new potential, which is termed a *pseudopotential.* This can be written as

$$V_P(x_i) = V(x_i) + \sum\nolimits_{j,\,\mathrm{core}} \frac{e_j}{4\pi\varepsilon_0 x_{ij}}. \tag{6.14}$$

The difference between the actual potential and the pseudopotential is illustrated in figure 6.1. The pseudopotential is smoother and does not have the strong singularity at zero, and these two factors make computation of the necessary overlap integrals (discussed in the next section) considerably easier. However, the true shape of the pseudopotential cannot be determined until the atomic structure is known. Thus, one needs to know the atomic nature of the wave functions, charge density, and the self-consistent potential in order to determine the pseudopotential for a particular atom. Thus, in density functional calculations, the shape and amplitude of the pseudopotential are among the things that must be determined in the self-consistent solution. In empirical methods in momentum space, as will be seen below, only a few lower-order Fourier coefficients are needed.

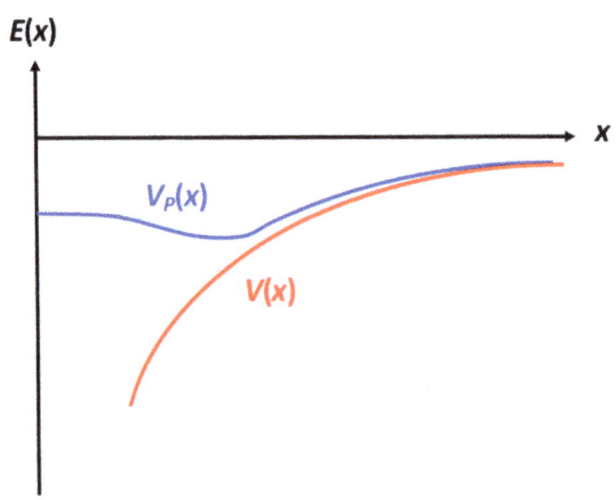

Figure 6.1. When the screening of the total potential by the inner-shell electrons is taken into account, the external potential (red) is simplified to the pseudopotential (blue).

6.1.2 A diatomic lattice

As an example of the atomic potential approach to band structure, a linear chain of two distinct atoms is adopted, as shown in figure 6.2. The two distinct atoms provide a diatomic basis, in which each unit cell contains two unequal atoms (one blue and one red in this example). The lattice can be defined using either the blue atoms or the red atoms, but each lattice site contains a *basis* of two atoms. However, these two atoms do not have to be different chemical elements (such as the case of Si, which still has two atoms per three-dimensional unit cell). The diatomic basis can arise from the different spacings shown in the figure.

As described in section 6.1, the structure is assumed to have periodic boundary conditions, although these only appear specifically in the use of the Brillouin zone and its properties. With the periodic boundary conditions, the Nth atom folds back onto atom zero, so that these are the same atom. An index j is adopted to indicate which atom in the chain is being considered. The basis set for determining the band structure is one in which each individual atomic wave function is localized upon a single atom so that othonormality for basis functions r and s on the atom at point j appears as

$$\langle r|s\rangle_j = \delta_{rs}, \tag{6.15}$$

where Dirac notation has been adopted. The right-hand side is the Kronecker delta function, which vanishes unless $r = s$ and is unity when the two indices are the same. It is assumed that the use of Dirac notation simplifies the equations and reduces confusion and clutter, and it will be used here as much as possible. The basis functions in (6.15) are energy eigenfunctions and the diagonal energies may be different for each of the two atoms of the basis, so that

$$H_0\,|r\rangle = E_r\,|r\rangle = E_{1,2}\,|r\rangle, \tag{6.16}$$

where the subscript on the last energy designates which of the two atoms of the basis is treated with this equation. Here, we designate the blue atom by the subscript 1 and the red atom by 2. In this last step, it is assumed that only a single atomic basis

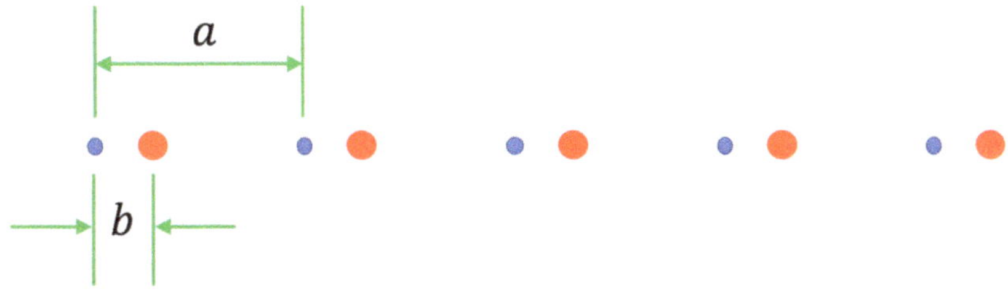

Figure 6.2. A diatomic lattice formed from two distinct atoms. Each unit cell contains one blue atom and one red atom, and this changes the electronic structure even though it remains a one-dimensional lattice. The two atoms do not have to be different chemical elements, but since the spacings are different, the unit cell requires two atoms per cell to achieve the periodicity.

function is involved, and the energy is labeled accordingly. Equivalently, this means that each atom has a single electron in the outermost, bonding shell.

The approach that has been initialized here is that in which a single chemical state provides the basis function and can be, for example, the *s*-state of each atom. Thus, the approach is a real-space use of atomic orbitals and connects with the LCAO or ETB models of chemistry. In keeping with the simple approach, the nearest-neighbor interaction is assumed to dominate the electronic structure. There are two Bloch functions, one arising from each of the two atoms in the basis. These have to be formed from the atomic orbitals. This, in turn, means that two equations are needed to account for the two atoms. From figure 6.2, it is apparent that there are two distances involved in the interactions between the *blue* and *red* atoms. In one interaction, the two atoms are separated by b. In the second interaction, the two atoms are separated by a–b. The lattice constant, however, remains a. Because of the two interactions, a slightly more complicated notation has to be adopted. The blue atoms are indexed by the site j when j is an even number (including zero). Similarly, the red atoms are indexed by the site j when j is an odd number. Two equations are required, one when the central site is a blue atom and the other when the central site is a red atom. It really does not matter which sites are picked, but the adjacent sites are chosen so that these two equations may be written as

$$\begin{aligned} E_1 \,|\, j\rangle + V_{\mathrm{P}} \,|\, j+1\rangle + V_{\mathrm{P}} \,|\, j-1\rangle &= E \,|\, j\rangle \\ E_2 \,|\, j+1\rangle + V_{\mathrm{P}} \,|\, j+2\rangle + V_{\mathrm{P}} \,|\, j\rangle &= E \,|\, j+1\rangle \end{aligned}. \tag{6.17}$$

Note that the atomic orbital energy (E_1, E_2) is taken to be different for the two different atoms. The integer j is taken to be an even integer (blue atom) in order to evaluate the integrals. The quantity V_{P} is the atomic pseudopotential of the atoms. The first of these equations is multiplied by the complex conjugate basis function of the site j and the second by the complex conjugate basis function of the site j+1, and the implied integration is performed. Equation (6.17) then becomes

$$\begin{aligned} E_1 + \langle j|V_{\mathrm{P}}|j+1\rangle + \langle j|V_{\mathrm{P}}|j-1\rangle &= E \\ E_2 + \langle j+1|V_{\mathrm{P}}|j+2\rangle + \langle j+1|V_{\mathrm{P}}|j\rangle &= E \end{aligned}. \tag{6.18}$$

There are now four so-called overlap integrals to evaluate, and these become

$$\begin{aligned} \langle j|V_{\mathrm{P}}|j+1\rangle &= e^{ikb}\langle j|V_{\mathrm{P}}|j\rangle \equiv e^{ikb}A_1 \\ \langle j|V_{\mathrm{P}}|j-1\rangle &= e^{ik(b-a)}\langle j|V_{\mathrm{P}}|j\rangle \equiv -e^{ik(b-a)}A_2 \\ \langle j+1|V_{\mathrm{P}}|j+2\rangle &= e^{-ik(b-a)}\langle j+1|V_{\mathrm{P}}|j+1\rangle \equiv -e^{ik(a-b)}A_2 \\ \langle j+1|V_{\mathrm{P}}|j\rangle &= e^{-ikb}\langle j|V_{\mathrm{P}}|j\rangle \equiv e^{-ikb}A_1 \end{aligned}. \tag{6.19}$$

In the above definitions, the overlap integrals are different for the two different atoms. The actual shape of the pseudopotential in the region between the two atoms appears in the overlap integrals, so the various integrals are mostly distinct. There are only two due to the two distances a and b. The choice of the sign on the A_2 terms ensures that the gap occurs at $k = 0$. The choice of the translation direction in (6.18) ensures that the Hamiltonian is Hermitian. The two equations now give a secular determinant which must be solved:

$$\begin{vmatrix} (E_1 - E) & (A_1 e^{ikb} - A_2 e^{ik(b-a)}) \\ (A_1 e^{-ikb} - A_2 e^{-ik(b-a)}) & (E_2 - E) \end{vmatrix} = 0. \tag{6.20}$$

The two off-diagonal elements are complex conjugates of one another, which ensures that the Hamiltonian is Hermitian and the energy solutions are real. This is a basic property of quantum mechanics that is required in order to obtain real eigenvalues. Two bands are formed from this diatomic lattice, and these are mirror images around an energy midway between the lowest energy of the upper band and the highest energy of the lower band. The energy is given by

$$E = \frac{E_1 + E_2}{2} \pm \sqrt{\left(\frac{E_1 - E_2}{2}\right)^2 + A_1^2 + A_2^2 - 2A_1A_2\cos(ka)}\,. \tag{6.21}$$

The two bands are shown in figure 6.3 for the values $E_1 = 5$ eV, $E_2 = 7$ eV, $A_1 = 2$ eV, and $A_2 = 0.5$ eV. The two bands are mirrored around the value 6 eV, and each has a band width of ~1.89 eV. It may be noted that only the last term in the square root bracket is a function of the wave momentum. Hence, it is the two overlap integrals yielding A_1 and A_2 that determine the shape of the bands. The local energies E_1 and E_2 determine the value of the energy and contribute to the size of the gap. Since the energies can be considered to be relative, most treatments of the band structure usually shift the energy so that the top of the valence band or the bottom of the conduction band is taken to be $E = 0$.

As required by section 6.1, each band contains N momentum states, as there are N unit cells in the lattice. Each band can hold $2N$ electrons when one considers the spin degeneracy of each momentum state. The two atoms per unit cell provide exactly $2N$ electrons, since each atom contributes a single outer shell electron to the bonding. These electrons completely fill the lower band of figure 6.3. This lower band is the valence band, as it contains the valence electrons contributed by the atoms. Hence, this diatomic lattice represents a semiconductor or an insulator, depending upon the width of the bandgap. Such a material has a full valence band and an empty conduction band (the upper band in figure 6.3). Here, the gap is ~2.24 eV wide, which would normally be construed as a very wide-bandgap semiconductor or a poor insulator.

While it is easy to consider this discussion of the one-dimensional material as mainly a tutorial of limited interest, one-dimensional (or quasi one-dimensional) materials have appeared in recent years. Foremost among these are the molecules bonded to metals or semiconductors for studies of their conduction properties [12]. But one can also have pseudo one-dimensional conduction in biological ion channels which translate potassium and sodium into and out of individual cells [13]. Quasi one-dimensional structures have also been created by putting single phosphorous atoms into a chain embedded in undoped silicon to produce an n^+ silicon nanowire [14]. One also can make atomic chains which create nanowires of e.g. gold [15] on an

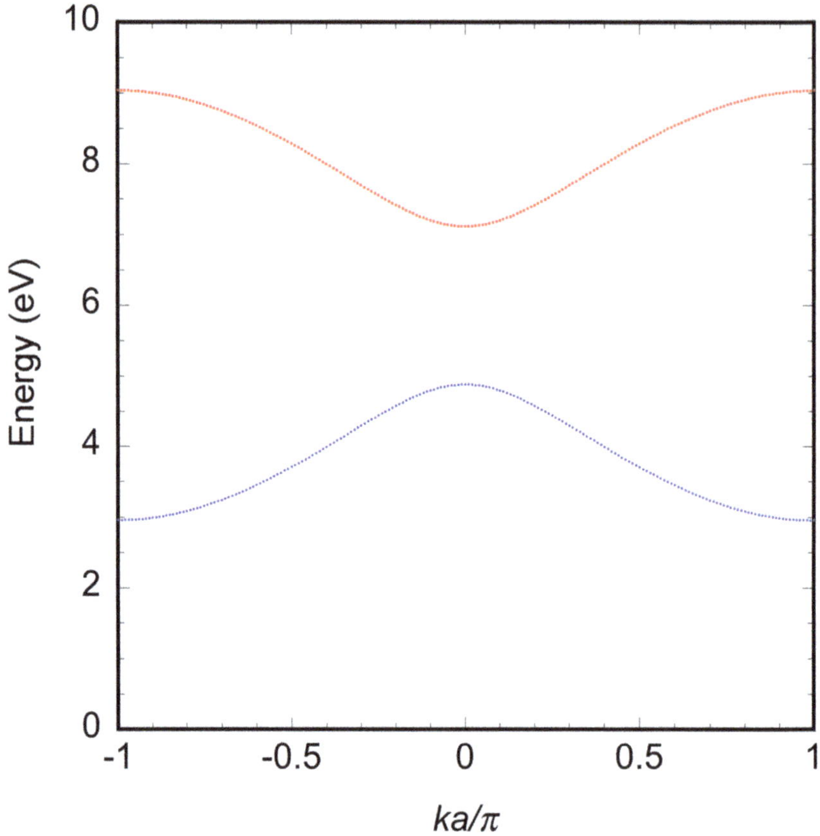

Figure 6.3. The two bands that arise for a diatomic lattice are plotted here. The values of the various parameters are shown on the figure. The parameters used for the plot are $E_1 = 5$ eV, $E_2 = 7$ eV, $A_1 = 2$ eV, and $A_2 = 0.5$ eV, and the bands follow (6.21).

insulating substrate. More recent is the growth of nearly vertical nanowires on a silicon substrate [16]. While these latter nanowires are not atomically thin, they remain quasi one-dimensional in nature due to the strong quantization arising from their small lateral dimensions [17]. The one-dimensional band structure provides a first estimate of the nanowire bands, while the transverse quantization provides more detail.

6.1.3 The complex band structure

Most studies of band structure reach this point and immediately begin to study more complex structures such as real semiconductors. But there is more to be said about figure 6.3. The band structure of figure 6.3 describes propagating states in which wave functions exist that have a propagating nature; i.e. they describe particles traveling through the crystal. However, there is more to band structure than merely the propagation states. A common semiconductor device which is often taken to be the ‘fruit fly’ for quantum transport is the resonant tunneling device [18, 19]. The prototypical device is a GaAs structure with two embedded AlGaAs barriers (one such was discussed in section 3.1.3). Between the two barriers is a thin GaAs quantum well, which has a bound state lying within the well. As the energy of electrons incident from either side of the barrier and well and the barrier structure hits the energy of the bound state, the transmission becomes unity in the simplest circumstance [20]. Since the transport takes place via tunneling, there are no propagating waves in the energy range of the AlGaAs barriers. Thus, the wave functions in the AlGaAs are evanescent waves with imaginary values of k. How does one determine these imaginary values?

In simple quantum mechanics, this value of k is determined by the energy difference between the conduction band edge of the AlGaAs and the energy of interest through

$$k = \sqrt{\frac{2m^*}{\hbar^2}(U - E)}\,. \tag{6.22}$$

But it is a leap of faith to believe that this is the correct approach in semiconductors. Near the conduction band (or valence band) edge, it is a reasonable approach. However, this latter formula is limited in its application. Which mass should be used? Is it the GaAs effective mass, or that of AlGaAs? However, the effective mass in (6.22) is defined only in the bands, and not in the gaps (as will be discussed further in a later section of this chapter). How do we know that the evanescent states have an imaginary value of k that follows this formula? In fact, it is known that this formula is *not* generally correct for the evanescent states.

When we discussed the band structure above, only propagating waves were discussed, and no consideration was given to the possibility of evanescent waves. It is well-known that impurities and defects can introduce states in the gap regions, but there are other states that exist in this region as well. These states actually come from the same band structure but are not propagating states. This can be illustrated with the above simple two-band model of (6.21) for the two mirror-image (conduction and valence) bands. An evanescent state which decays away from its principal site in position space has an imaginary value of k. If a real value of energy is still required, then these states are also stationary states that do not decay in time. These states add to the total knowledge of the band structure, and the latter is known as the *complex band structure.* The first important region to study is this structure around the minimum of the conduction band (and the maximum of the valence band). With an imaginary value for k (which is often designated by κ), the cosine term becomes a hyperbolic cosine. We plot the same bands in figure 6.4 as in figure 6.3, except that both real and imaginary k are included and the real k bands are only in the positive k direction. (Since the hyperbolic cosine is symmetric in the sign of its argument, we can use the negative sign on κ here without any difficulty.) Here, it may be seen that the bands curl back toward each other (they actually join one another, although the accuracy of the grid selected here does not show that). The bands for imaginary k are shown in purple, but the same parameters are used for the simulation.

What is clear from figure 6.4 is that (6.22) is not a good approximation away from the extrema of the conduction and valence bands. In fact, below 6 eV, the tunneling

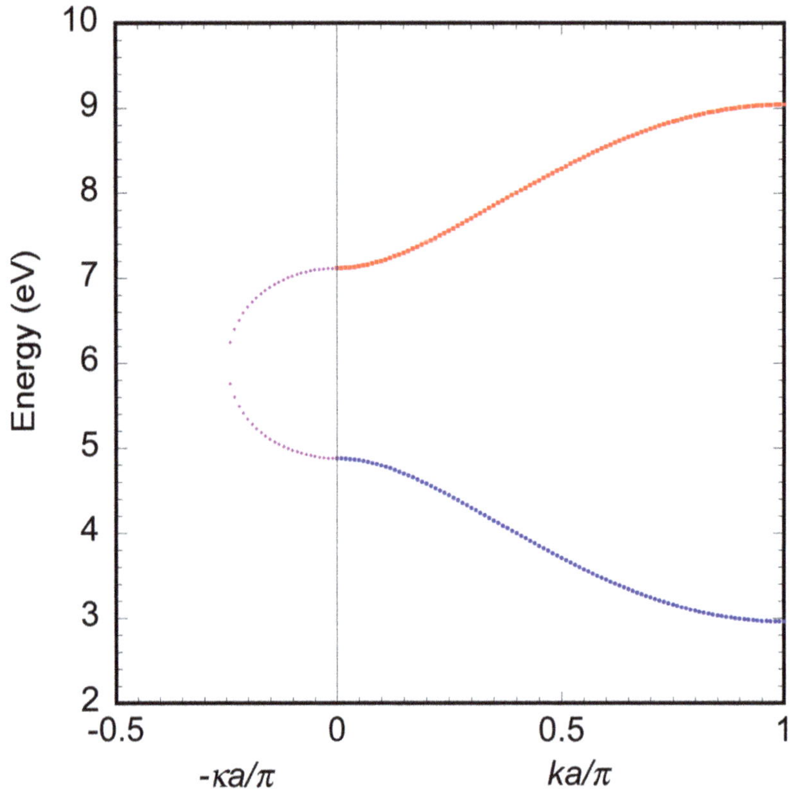

Figure 6.4. The bands shown in figure 6.3 are extended to allow for the imaginary values of k (κ) around the $k = 0$ minimum of the conduction band and the maximum of the valence band. The same parameters are used as in figure 6.3. Only those values of real and imaginary k that give a real energy are considered.

actually increases as the valence band is approached. The two evanescent bands meet at the point where the square root in (6.21) vanishes. For the parameters used in the plots, this occurs at

$$\kappa a \sim 0.\,244\pi, \tag{6.23}$$

as determined from figure 6.4. This point is called the branch point. But the complex band structure is much more complicated than this single extension. A more complete structure (with different values) is shown in figure 6.5 [21]. Several critical points are designated in both panels. In panel (a) the complete three-dimensional dispersion relation (albeit for a one-dimensional crystal) is plotted. This shows the values of complex k for real values of the energy. The real axis is labeled to run from Γ to X, although these do not correspond to such values in a typical band structure, where they are related to points in the full Brillouin zone. Comparing panel (b) to figure 6.4, it may be seen that $-a/\pi \leqslant \mathrm{Re}\{k\} \leqslant a/\pi$ and $-1 \leqslant \kappa a \leqslant 1$. The curves of figure 6.3 run from B to C to L for the valence band and K to F to G for the conduction band. The branch point lies at D and at its reflection through $\kappa = 0$ to the

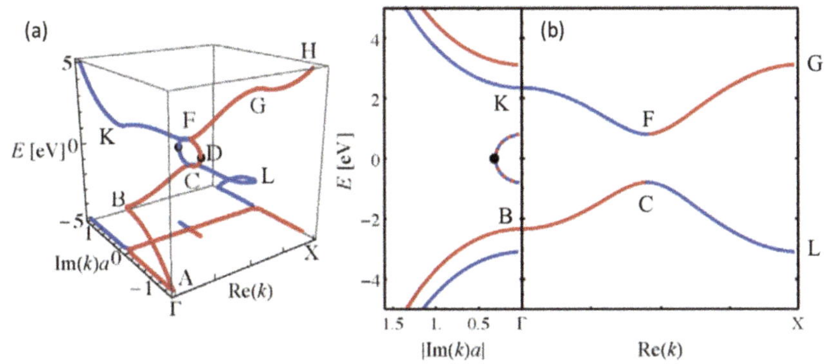

Figure 6.5. (a) A three-dimensional view of the full dispersion relation for real values of the energy in complex k. Several points have been labeled for discussion. (b) A standard two-dimensional view of the dispersion relation, which may be compared to figure 6.4. The letters A–H denote the full spread of one of the roots of the eikonal equation for the energy. The two roots are shown in red and blue. Reproduced from [21].

equivalent point shown in the blue curve. The uppermost red curve (for $\kappa \neq 0$) extends from the G point, and the lowermost blue curve (for $\kappa \neq 0$) extends from the L point, as may be determined from panel (a). It may be concluded from this figure that an extension of the bands' imaginary values of the wave momentum k begins and ends at the extreme points of the real bands. It might also be supposed that the red line ending at H actually connects to the end of the red line near X with a flip of the sign of the energy (and similarly with the blue lines ending at ± 5 at the far left of panel (a)), although not with real energy. With complex or imaginary energy, the states decay temporally, so are not stationary states.

The full dispersion relation is a central object in complex band theory, and the extensions to complex k provide important information [21]. For example, at surfaces and interfaces, the periodicity of the crystal is broken in the direction normal to the surface or interface. Nevertheless, there remains some repeat unit, particularly in superlattices, which may be artificially introduced to compute a periodic 'super cell' structure. The complex band structure is important in understanding how this impacts surface and interface physics. The full complex band structure describes in detail both the static and the dynamic properties of the material(s). Reuters points out that, in fact, it may well be the minimal amount of information needed for this purpose [21].

6.2 Real-space approaches

In real-space methods, the electronic structure is determined by writing the Hamiltonian and the wave functions as real-space quantities. The complement of this, namely momentum-space approaches, will be discussed in the next section. In the real-space approach, the pseudopotential version of the Schrödinger equation written as

$$H(r)\psi(r) = H_0(r)\psi(r) + V_P(r)\psi(r) \tag{6.24}$$

still needs to be solved. It should be noticed that there is a remaining interaction among the electrons, which appears in (3.2). This latter term leads to a nonlinear behavior in the Schrödinger equation. But in calculating the one-electron band structure, this term may be ignored for the moment. Nevertheless, various approximations to this term have been pursued through the years in order to determine a usable form that adds to the band structure yet avoids the complication of the full nonlinear behavior. One approach is to simply assume that the bonding electrons lead to a smooth additional potential determined from the average local density of charge, and this is known as the Hartree approximation. This treats the interaction as that of a single electron interacting with a mean-field background potential. A somewhat better approximation is to explicitly include the exchange energy—an energy correction that arises from the interchange in the position of any two electrons (on average), which leads to the Hartree–Fock approximation. A still more general approach is to follow the H–K theorems and adopt an energy functional term for both the exchange and the correlation energies. In this case, the energy correction is a function of the local density that is then included within the self-consistent solution for the wave functions and the energies. This last approach is known as the local-density approximation (LDA) [22].

In spite of this range of approximations, all first-principles calculations still have a problem in achieving values for the bandgaps that are correct for semiconductors. The energy gaps can be as much as an order of magnitude too small. Even though a number of corrections have been suggested for LDA, none of these has totally solved the bandgap problem. Humorously, this approach has led to the use of the 'scissors operator,' in which a plot of the band structure is cut through the gap with scissors and the two parts moved apart to give the right bandgap! The modern approaches that are most successful in solving this bandgap problem are based upon determining the Green's function for the electrons and convolving this with the interelectronic coulomb forces (labeled as a W potential) to give the GW approximation.

In (6.24), H_0 is only the kinetic energy of the electron and V_P describes the pseudopotential at the particular site. Any multielectron effects are then included through a term added to H_0 that is described through H–K theorems, as described in the previous paragraphs. The pseudopotential V_P is just (6.14), depicted by the blue curve in figure 6.1. Whether or not the terms for the many-body interactions are used depends upon the exact approach taken to find the bands. Approaches which include these interactions are often called *ab initio* approaches. On the other hand, approaches which fit the overlap integrals so that bandgaps agree with experiment are called *empirical* approaches. To proceed, a lattice must be specified, as in section 6.1.2, and a basis set of the wave functions properly chosen. In a real-space approach, the basis set is composed of one or more atomic orbitals that are localized on a particular lattice site. The basis functions at each site are assumed to satisfy orthonormality, and this is imposed on each different lattice site. One very important point is that only two-point integrals and their interactions are considered. That is, integrals in which the wave functions are on atoms one and two, while the potential may arise from atom three, are ignored. While these may be important

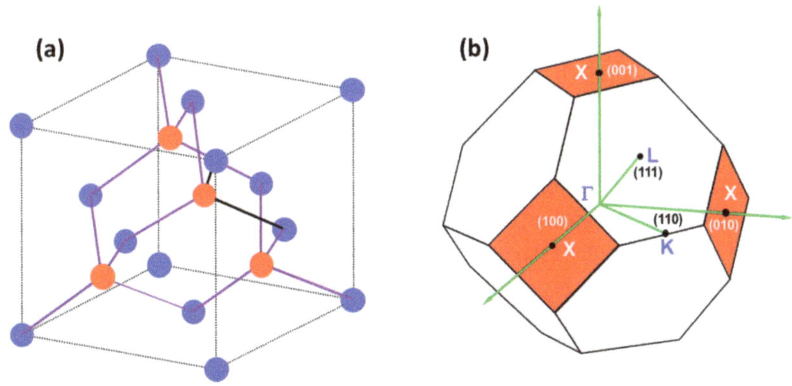

Figure 6.6. (a) The zinc-blende basic face-centered cubic cell. The blue and red atoms are the pair that form the basic unit cell. These could be Ga and As, for example. In diamond, for example, they are the same atom. (b) The momentum-space Brillouin zone for the face-centered cubic cell. The blue and white letters denote special points in the Brillouin zone that lie at the zone boundaries in various crystal directions.

in some cases of first-principles calculations, they are not usually considered necessary for empirical approaches.

Most of the tetrahedral semiconductors have either the zinc-blende lattice or or the diamond lattice. On average, each atom has four outer shell electrons (in III–V or II–VI semiconductors, the two atoms of the basis total to eight outer shell electrons, so they fit this rule). These are characterized as a single *s* state and three *p* states. These four orbitals hybridize into four directional bonds (connecting a red atom and a blue atom), each of which points toward one of the four nearest neighbors. As these four neighbors correspond to the vertices of a regular tetrahedron, these materials are usually referred to as tetrahedral materials. In figure 6.6, the lattice is shown in panel (a) of the figure. The basic cell is a face-centered cube (FCC). The lattice is defined by e.g. the blue atoms at the eight corners and those at the centers of each face. Each lattice site has a basis of two atoms, one red and one blue, as shown in the figure. This may be either two Si atoms or, for example, a Ga atom and an As atom in III–V materials. The tetrahedral coordination is indicated by lines connecting atoms—these point toward the nearest neighbors in each case. Only four of the second basis atoms (red) lie within this FCC cell. The FCC cell is not the primitive unit cell, but is the shape most commonly used to describe this lattice. The presence of the four nearest neighbors means that there are four exponentials that contribute to the Bloch sums; these are equivalent to (6.11) and (6.12).

The Brillouin zone for the FCC lattice is shown in panel (b) of figure 6.6. This is a truncated octahedron. It may also be viewed as a cube that has eight removed corners. In either case, there are important points in this zone that define important values of the wave vector. Critical directions have been indicated in the figure by the blue and white letters. The important momentum points are the Γ point, located at the center of the zone ($k = 0$), the X points at the centers of the square faces in the three main directions ((001), (010), and (100)), and the L points at the centers of the hexagonal faces (in the various (111) directions). The shapes of 'truncated octahedrons' can be stacked neatly upon one another by just shifting the second cell along any two of the other axes. Hence, if one moves from Γ in the (110) direction, one passes the point K and enters the second zone's top square face there. Thus, one arrives at the X point in the (001) direction of that second Brillouin zone. This becomes useful when the energy bands are plotted.

To summarize the previous paragraphs, each atom has four basis functions, and there are two atoms per primitive unit cell. Therefore, the Hamiltonian matrix is an 8×8 matrix, and this matrix can be decomposed into four diagonal 4×4 blocks. When only interactions between the nearest neighbors are considered, the form of these blocks becomes quite simple. The two diagonal blocks are themselves diagonal in nature, and the atomic s and three atomic p energies lie along this diagonal. One of these two blocks represents the A atom (blue in figure 6.6), and the other represents the B atom (red in figure 6.6). The other two blocks—the upper-right block and the lower-left block—are full rank 4×4 matrices. However, the lower-left block is the Hermitian conjugate (complex conjugate and transpose) of the upper-right block, as required for the Hamiltonian to be Hermitian. These two blocks represent the interactions of an orbital on the A atom with an orbital on the B atom. The Bloch sums are important in these blocks, as they define the various overlap integrals. An A atom has interactions with its four B neighbors and vice versa. The Bloch sums contain four exponential terms that are the four translation operators from one atom to its four neighbors. If atom A is the origin of the reference coordinates, then the four nearest neighbors are positioned according to the four vectors

$$\begin{aligned} \boldsymbol{r}_1 &= \frac{a}{4}(\boldsymbol{a}_x + \boldsymbol{a}_y + \boldsymbol{a}_z), \quad \boldsymbol{r}_2 = \frac{a}{4}(\boldsymbol{a}_x - \boldsymbol{a}_y - \boldsymbol{a}_z), \\ \boldsymbol{r}_3 &= \frac{a}{4}(-\boldsymbol{a}_x + \boldsymbol{a}_y - \boldsymbol{a}_z), \quad \boldsymbol{r}_4 = \frac{a}{4}(-\boldsymbol{a}_x - \boldsymbol{a}_y + \boldsymbol{a}_z). \end{aligned} \tag{6.25}$$

The first vector, for example, points from the lower-left blue atom to the red atom shown in the left panel of figure 6.6, along with the corresponding bond. The full 8×8 matrix has the general form

$$\begin{bmatrix} E_s^A & 0 & 0 & 0 & H_{ss}^{AB} & H_{sx}^{AB} & H_{sy}^{AB} & H_{sz}^{AB} \\ 0 & E_p^A & 0 & 0 & H_{xs}^{AB} & H_{xx}^{AB} & H_{xy}^{AB} & H_{xz}^{AB} \\ 0 & 0 & E_p^A & 0 & H_{ys}^{AB} & H_{yx}^{AB} & H_{yy}^{AB} & H_{yz}^{AB} \\ 0 & 0 & 0 & E_p^A & H_{zs}^{AB} & H_{zx}^{AB} & H_{zy}^{AB} & H_{zz}^{AB} \\ & & & & E_s^B & 0 & 0 & 0 \\ & & & & 0 & E_p^B & 0 & 0 \\ & & & & 0 & 0 & E_p^B & 0 \\ & & & & 0 & 0 & 0 & E_p^B \end{bmatrix}, \tag{6.26}$$

where the lower-left block is filled in by the appropriate transpose complex conjugate of the upper-right block (which also reverses the AB to BA). A shorthand notation of x, y, z has been adopted for the orbitals p_x, p_y, p_z [6]. If the two subscripts are reversed (when they are different), this leads to the conjugate of the appropriate complex energy. Interchanging the A and B atoms inverts the coordinate system. These simple operations have the result that the number of different Bloch sums is reduced to only four. The five principal interactions that appear in the upper-right block of (6.26) may be written as the Bloch sums

$$\begin{aligned} H_{ss}^{AB} &= \langle s^A|H|s^B\rangle(e^{i\mathbf{k}\cdot\mathbf{r}_1} + e^{i\mathbf{k}\cdot\mathbf{r}_2} + e^{i\mathbf{k}\cdot\mathbf{r}_3} + e^{i\mathbf{k}\cdot\mathbf{r}_4}) = E_{ss}B_0, \\ H_{xx}^{AB} &= \left\langle p_x^A\middle|H\middle|p_x^B\right\rangle(e^{i\mathbf{k}\cdot\mathbf{r}_1} + e^{i\mathbf{k}\cdot\mathbf{r}_2} + e^{i\mathbf{k}\cdot\mathbf{r}_3} + e^{i\mathbf{k}\cdot\mathbf{r}_4}) = E_{xx}B_0(\mathbf{k}), \\ H_{sx}^{AB} &= \left\langle s^A\middle|H\middle|p_x^B\right\rangle(e^{i\mathbf{k}\cdot\mathbf{r}_1} + e^{i\mathbf{k}\cdot\mathbf{r}_2} - e^{i\mathbf{k}\cdot\mathbf{r}_3} - e^{i\mathbf{k}\cdot\mathbf{r}_4}) = E_{sx}^{AB}B_1(\mathbf{k}), \\ H_{sy}^{AB} &= \left\langle s^A\middle|H\middle|p_y^B\right\rangle(e^{i\mathbf{k}\cdot\mathbf{r}_1} - e^{i\mathbf{k}\cdot\mathbf{r}_2} + e^{i\mathbf{k}\cdot\mathbf{r}_3} - e^{i\mathbf{k}\cdot\mathbf{r}_4}) = E_{sx}^{AB}B_2(\mathbf{k}), \\ H_{sz}^{AB} &= \left\langle s^A\middle|H\middle|p_z^B\right\rangle(-e^{i\mathbf{k}\cdot\mathbf{r}_1} - e^{i\mathbf{k}\cdot\mathbf{r}_2} + e^{i\mathbf{k}\cdot\mathbf{r}_3} + e^{i\mathbf{k}\cdot\mathbf{r}_4}) = E_{sx}^{AB}B_3(\mathbf{k}). \end{aligned} \tag{6.27}$$

In a similar manner, the other matrix elements can be written as

$$\begin{aligned} H_{xy}^{AB} &= E_{xy}B_3(\mathbf{k}), \quad H_{xz}^{AB} = E_{xy}B_2(\mathbf{k}), \\ H_{xy}^{AB} &= E_{yz}B_1(\mathbf{k}), H_{xx}^{AB} = H_{yy}^{AB} = H_{zz}^{AB}. \end{aligned} \tag{6.28}$$

It should be noted that there is an inherent symmetry among k_x, k_y, and k_z in the Bloch sums that arises from the crystal symmetry. The coordinate axes are quite easily interchanged due to the spherical symmetry of these orbitals. This leads to the obvious $k_x \to k_y \to k_z \to k_x$.

When we interchange the two atoms of the basis, the tetrahedron is inverted, and the four vectors (6.25) are reversed. This reverses the directions of the momentum in the Bloch sums. This takes each B into its complex conjugate. Reversing the order of the wave functions in (6.27) also introduces a complex conjugate into the energies, but since these energies are taken to be real, this operation does not change anything. However, one has to keep track of which atom has the s orbital in the s–p matrix elements, as this mixes different orbitals from the two atoms. The off-diagonal block —the upper, right block of (6.26)—may be written as

$$\begin{bmatrix} E_{ss}B_0 & E_{sx}^{AB}B_1 & E_{sx}^{AB}B_2 & E_{sx}^{AB}B_3 \\ E_{sx}^{BA}B_1^* & E_{xx}B_0 & E_{xy}B_3 & E_{xy}B_2 \\ E_{sx}^{BA}B_2^* & E_{xy}B_3^* & E_{xx}B_0 & E_{xy}B_1 \\ E_{sx}^{BA}B_3^* & E_{xy}B_2^* & E_{xy}B_1^* & E_{xx}B_0 \end{bmatrix}. \tag{6.29}$$

As discussed above, the rest of the 8×8 matrix is filled in to make the final matrix Hermitian. The total 8×8 matrix is now diagonalized to find the energy levels, and repeating this for each value of the momentum k gives the bands.

As mentioned above, one can use actual wave functions and the pseudopotentials to perform a first-principles, or *ab initio*, calculation. One of the better approaches to first-principles tight binding was formulated by Sankey [23]. A set of localized atomic orbitals, called *fireballs*, are coupled to the pseudopotentials along with an efficient exchange and correlation functional. This approach was extended to three-center integrals as well. Including molecular dynamics—allowing the lattice to relax while computing the inter-atomic forces—also leads to optimized structures. A thorough review of the fireball approach has recently appeared [24]. Another current first-principles approach to tight-binding calculations is commercially provided through the Spanish Initiative for Electronic Simulations with Thousands of Atoms (SIESTA) code [25].

In contrast, many people prefer the simpler and more computationally efficient route of the semiempirical tight-binding approach [6]. In this approach, the parameters that appear in the overlap integrals of (6.27) are *adjusted* in such a manner that the bands fit to experimentally determined gaps. This usually adjusts gaps, instead of adjusting actual points in the band. There always remains a question as to *which* experiments should be used for these measurements, and *whose* results should be accepted. The only answer to such questions is in the eyes of the one doing the *adjusting*. For semiconductor devices, it seems much better to fit the extrema of the conduction band and the principal energy gap at Γ. This is because the positions of the X and L points (relative to the bottom of the conduction band at Γ) are more important for transport in semiconductor devices. Most of these devices utilize n-type material in which the transport is provided by electrons in the conduction band, and much of the conduction band will be sampled.

As a first example, the band structures of GaAs and AlAs are shown in figure 6.7 [26]. Here, the tight-binding interactions have been extended to the second-nearest neighbors. The second-neighbor interactions appear as off-diagonal terms in the diagonal blocks of (6.26). There are second-neighbor interactions between the two s-states, which are between the same type of atom, as well as between the p-states

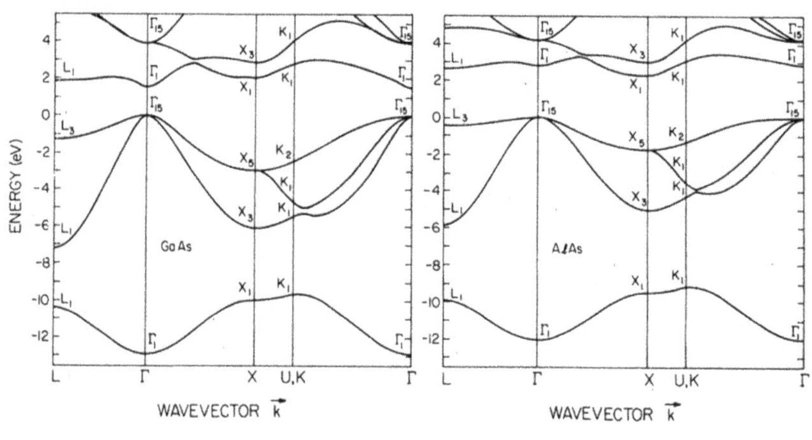

Figure 6.7. The band structures of GaAs (left) and AlAs (right). These are computed by empirical tight binding that includes second-neighbor interactions, as discussed in the text. Reprinted with permission from [26], Copyright (1979) by the American Physical Society.

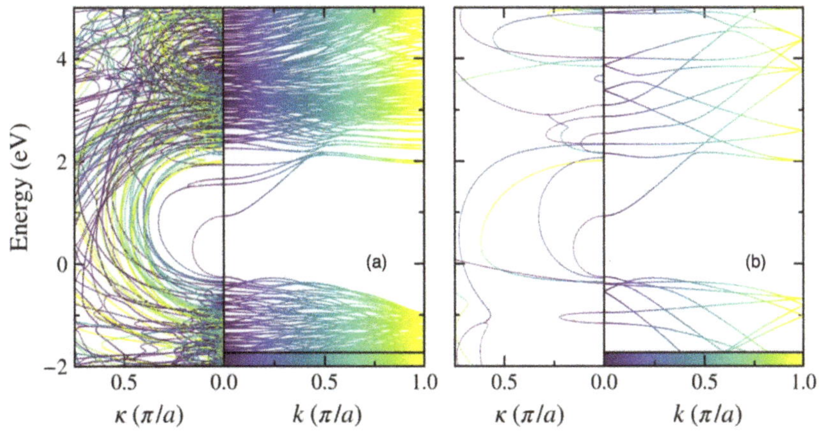

Figure 6.8. The complex band structure for an InAs nanowire that has a cross-section of 2.73×2.73 nm^2 is plotted for (a) $N = 1800$ and (b) $N = 48$. In each case, the right-hand portion depicts the bands for real k, and the left-hand portion depicts the bands for imaginary k. Reproduced from [27].

and between the *s*- and *p*-states. With this extension, there are some 19 different interaction energies, although the matrix remains 8×8. These calculations can be extended to interfacial states for a heterojunction between these two materials and to evanescent waves in the AlAs for tunneling structures.

The second example is considerably more complicated. This is an InAs nanowire, so that there is no transverse periodicity. As a result, the only periodicity is along the wire, and the equivalent unit cell must include all the atoms in the transverse plane as well as the multiple layers of the cubic unit cell of figure 6.6(a). Hence, the size of the equivalent cell depends upon how many layers are treated before repetition is invoked. The general matrix is $N \times N$, but the number of atoms in the cell is less than N due to the number of orbitals taken at each atom. One can either describe the matrix by its size or by the number of orbitals times the number of atoms considered. In the calculations considered [27], a wire grown in the (100) direction is considered. The number of orbitals has been extended to include the highest *d*-states from the next lowest shell, as these have energies approaching (or even exceeding) those of the *s*-states in the shell considered. The *s*-states of the next highest (empty) shell are also considered, as these have energies near the considered *p*-states. This gives a description of the entire set of orbitals, namely $sp^3d^5s^*$. In figure 6.8, the band structure for an InAs nanowire of 2.73×2.73 nm^2 cross-section is plotted for $N = 1800$ and $N = 48$. The bands for both real and imaginary k are plotted.

6.3 Momentum-space approaches

In section 6.1, the Fourier transforms of both the potential and the wave functions were adopted. Moreover, in both transforms, various exponentials were encountered. As a result, it is generally the case that momentum-space approaches utilize exponential wave functions to build the energy structure. If the potentials due to the atoms are ignored, this plane-wave approach is exact, but it is not of much interest in semiconductors. It is the perturbation of these so-called free electron bands by the atomic potentials that is important. When the atomic potentials have relatively smooth variations with position, as with pseudopotentials, the resulting plane-wave approach can be accurate. It is important to recognize that this approach is *three-dimensional*. It involves the Fourier transform of three-dimensional quantities—the pseudopotential as well as the wave functions. If we are going to use the momentum-space approach for a one-dimensional or two-dimensional crystal, then we have to generate a full three-dimensional structure. For the one-dimensional material, we create a two-dimensional lattice of one-dimensional wires in the plane perpendicular to the one dimension of the material. For the two-dimensional layered material, we use multiple layers to generate the three-dimensional structure. In each of these cases, however, the desired one-dimensional or two-dimensional structures must be sufficiently separated in space that the replicas do not interact! That is, for e.g. layers of graphene, we need to space the layers further apart than they would be in graphite so that the individual layers do not interact [28]. Similarly, for the transition-metal dichalcogenides (TMDCs), multiple layers must be used, and by varying the separation, one can study the transition from the bulk to a monolayer. An example of this is a study of the adsorption of a methane molecule on the surface of Au [29]. Six layers of Au were used, and the methane molecule was adsorbed on the top layer of this structure. Then the structure was assumed to repeat at some distance above the methane molecule, and this distance was sufficiently large that the molecule did not interact with the upper Au (the vacuum distance above the molecule was 1.0–1.5 nm in this study). The simulations were then performed using *ab initio* density-functional theory (DFT) with a variety of density functionals.

Since we are working in momentum space, it is actually the Fourier transform (6.7) of the pseudopotential which is of interest. For a particular atom located at site $\boldsymbol{r}_1$, the pseudopotential may be expressed as

$$V_P(\boldsymbol{r} - \boldsymbol{r}_1) = \sum_{\boldsymbol{G}} \widetilde{V}_1(\boldsymbol{G}) e^{i\boldsymbol{G}\cdot(\boldsymbol{r}-\boldsymbol{r}_1)}. \tag{6.30}$$

As in (6.7), $\boldsymbol{G}$ is the set of reciprocal lattice vectors. In practice, only a limited set is introduced (to reduce the computational complexity), and this sets a cutoff on the maximum energy of the calculation. The accuracy is improved as more reciprocal lattice vectors are used. The inverse of (6.30) now becomes

$$\widetilde{V}_1(\boldsymbol{G}) = \frac{1}{\Omega} \int d^3r e^{-i\boldsymbol{G}\cdot(\boldsymbol{r}-\boldsymbol{r}_1)} V_P(\boldsymbol{r} - \boldsymbol{r}_1). \tag{6.31}$$

The quantity Ω is the volume of the unit cell in the crystal. It becomes convenient to work with both of the two atoms in the primitive unit cells of the zinc-blende or diamond crystals [30]. For convenience, the lattice vectors are referenced to the midpoint between the two atoms. Then, $\boldsymbol{r}_a - \boldsymbol{r}_1 = \boldsymbol{t}$, $\boldsymbol{r}_a - \boldsymbol{r}_2 = -\boldsymbol{t}$, and (6.30) becomes

$$\begin{aligned} V_P(\boldsymbol{r}) &= \sum_{\alpha=1,2} V_P(\boldsymbol{r} - \boldsymbol{r}_\alpha) \\ &= \sum_G [\widetilde{V}_1(\boldsymbol{G})e^{i\boldsymbol{G}\cdot\boldsymbol{t}} + \widetilde{V}_2(\boldsymbol{G})e^{-i\boldsymbol{G}\cdot\boldsymbol{t}}]e^{i\boldsymbol{G}\cdot\boldsymbol{r}}. \end{aligned} \tag{6.32}$$

In the zinc-blende lattice or the diamond lattice, these two vectors refer to the positions of the two atoms of the basis set. In this case, $\boldsymbol{t} = (a/8)(111)$ or one-eighth of the body diagonal across the face-centered cube (this is not the unit cell), while a is the length of the edge of this face-centered cube. At this point, it is convenient to introduce symmetric and antisymmetric potentials, as follows:

$$\begin{aligned} V_S &= \frac{1}{2}[\widetilde{V}_1(\boldsymbol{G}) + \widetilde{V}_2(\boldsymbol{G})] \\ V_A &= \frac{1}{2}[\widetilde{V}_1(\boldsymbol{G}) - \widetilde{V}_2(\boldsymbol{G})] \end{aligned}, \tag{6.33}$$

so that

$$\begin{aligned} \widetilde{V}_1(\boldsymbol{G}) &= V_S(\boldsymbol{G}) + V_A(\boldsymbol{G}) \\ \widetilde{V}_2(\boldsymbol{G}) &= V_S(\boldsymbol{G}) - V_A(\boldsymbol{G}) \end{aligned}. \tag{6.34}$$

Using the definitions (6.34) in (6.32), the *cell* pseudopotential becomes

$$\begin{aligned} V_c(\boldsymbol{G}) &= \widetilde{V}_1(\boldsymbol{G})e^{i\boldsymbol{G}\cdot\boldsymbol{t}} + \widetilde{V}_2(\boldsymbol{G})e^{-i\boldsymbol{G}\cdot\boldsymbol{t}} \\ &= 2V_S(\boldsymbol{G})\cos(\boldsymbol{G}\cdot\boldsymbol{t}) + 2iV_A(\boldsymbol{G})\sin(\boldsymbol{G}\cdot\boldsymbol{t}) \end{aligned}. \tag{6.35}$$

Using the Schrödinger equation for a 'pseudo wave function,' the Hamiltonian matrix form can be developed. A plane-wave function defined at a particular momentum **k** and lying within the first Brillouin zone may be written as a sum over reciprocal lattice vectors:

$$|\boldsymbol{k}\rangle = \sum_G C_G\,|\boldsymbol{k} + \boldsymbol{G}\rangle. \tag{6.36}$$

By introducing (6.35) and (6.36) into the Schrödinger equation for this wave function, one obtains the principal equation for the momentum-space approach, as follows:

$$\sum_G C_G\left[\frac{\hbar^2}{2m_0}(\boldsymbol{k} + \boldsymbol{G})^2 + V_c(\boldsymbol{r}) - E\right]|\boldsymbol{k} + \boldsymbol{G}\rangle = 0. \tag{6.37}$$

Just as in the tight-binding approach, this equation is premultiplied by the adjoint state $\langle \boldsymbol{k} + \boldsymbol{G}'|$, and the inferred integration is performed. The resulting matrix elements between the reciprocal lattice vectors $\boldsymbol{G}$ and $\boldsymbol{G}'$ yield the equation:

$$\sum_G C_G\left\{\left[\frac{\hbar^2}{2m_0}(\boldsymbol{k} + \boldsymbol{G})^2 - E\right]\delta_{G,G'} + \langle \boldsymbol{k} + \boldsymbol{G}'|\,V_c(\boldsymbol{r})\,|\boldsymbol{k} + \boldsymbol{G}\rangle\right\} = 0. \tag{6.38}$$

Equation (6.38) gives a different equation for each value of $\boldsymbol{G}'$. Each equation within curly brackets produces one row of the resulting eigenvalue matrix (which corresponds to (6.26) for the tight-binding approach). That is, there is a single row for each value of $\boldsymbol{G}$ chosen for the basis set. If nine basis vectors are chosen, the matrix will be 9×9. If 137 basis vectors are chosen, the matrix will be 137×137. The choice of size is exactly the limit on the number of reciprocal lattice vectors mentioned above, and therefore relates to an energy cutoff. This choice of vectors will be discussed further below. The important interaction matrix element is the last term in the curly brackets, which may be expressed as

$$\begin{aligned} \langle \boldsymbol{k} + \boldsymbol{G}' | \, V_c(\boldsymbol{r}) \, | \, \boldsymbol{k} + \boldsymbol{G} \rangle &= \frac{1}{\Omega} \int d^3 r \sum_{\boldsymbol{G}''} e^{-i(\boldsymbol{k}+\boldsymbol{G}')\cdot \boldsymbol{r}} V_c(\boldsymbol{G}'') e^{i(\boldsymbol{G}''+\boldsymbol{k}+\boldsymbol{G}')\cdot \boldsymbol{r}} \\ &= \sum_{\boldsymbol{G}''} V_c(\boldsymbol{G}'') \delta(\boldsymbol{G}'' + \boldsymbol{G} - \boldsymbol{G}') = V_c(\boldsymbol{G} - \boldsymbol{G}') \end{aligned} . \qquad (6.39)$$

Here is the important point. The necessary Fourier coefficients of the pseudopotential are given by the differences between the vectors $\boldsymbol{G}$ and $\boldsymbol{G}'$. The diagonal contribution ($\boldsymbol{G} = \boldsymbol{G}'$) to this is $V_c(0)$, which only provides an energy shift of the overall spectrum. Generally, this allows one to align the energy scale so that $E = 0$ corresponds to the top of the valence band, although this reference energy can be placed anywhere. Although (6.39) could be prepared with an infinite number of reciprocal lattice vectors, a finite number is typically used that is large enough to allow some degree of convergence in the calculation.

The difference between *ab initio* calculations and empirical calculations now lies in the method used to determine these Fourier coefficients. For *ab initio* calculations, (6.35) and (6.36) define the Fourier transforms of the pseudopotential and the wave functions. Using these, the matrix elements can be evaluated from the initial values of these two quantities. These are then used to diagonalize the energy matrix (6.38). In doing so, we obtain a square transformation matrix that defines the similarity transform that diagonalizes the matrix. The rows and columns of this transform matrix define the coefficients of the wave functions at each energy and for the chosen value of $\boldsymbol{k}$. These three-dimensional coefficients are then used to find the real-space wave functions and the corresponding real-space charge density. This is used via Poisson's equation to find a new pseudopotential, and this is Fourier transformed to momentum space for the next iteration. In doing this, the exchange-correlation function (the density functional) is evaluated from the local density and is added to the pseudopotential prior to the Fourier transform, and the lattice relaxation can be included from molecular dynamics forces at each iteration. Good software packages are available that may be used for this process, two of which are VASP [31] and Quantum Espresso [32].

In the empirical pseudopotential approach, the matrix elements (6.39) are directly used to adjust the energy bands throughout the Brillouin to match experimentally determined values of critical points [8], just as with the empirical tight-binding approach. Usually, a range of $\boldsymbol{k}$ values is used, and one can then see all the principal values in the zone. This is, of course, a major difference from the *ab initio* approach.

The key element is the choice of the basis set. The obvious first member is the plane wave with $\boldsymbol{G} = 0$, which defines the Γ point of the first Brillouin zone. This may be taken as $\boldsymbol{G}_0$. The next-nearest-neighbor points arise from translations of the Γ point to the nearest-neighbor Brillouin zones. These are the set of {111} translations, and there are eight of these vectors. So, these become the plane waves for $\boldsymbol{G}_1 - \boldsymbol{G}_8$. The second-neighbor cells of the reciprocal lattice lie in the set of directions {200}, and there are six of these, forming the set $\boldsymbol{G}_9 - \boldsymbol{G}_{14}$. At this point, it is important to return to the matrix element to recall that the actual Fourier coefficients needed arise from the difference $\boldsymbol{Q} = \boldsymbol{G} - \boldsymbol{G}'$. The difference between $\boldsymbol{G}_0$ and the nearest neighbors is just $|\,Q\,| = 3$ (the vector designation arises from the squared magnitude of the vector $1^2+1^2+1^2$) and this is usually denoted as the Q_3 coefficient. Between the various second neighbors, one has the sets {200}, {220}, and {222}. So, one needs the Q_4, Q_8, and Q_{12} Fourier coefficients for these interactions. One can continue this consideration of larger values of both the reciprocal lattice vectors and the Fourier coefficients. By the time one reaches the ninth-nearest neighbors, the set of {422} vectors, the basis set has grown to 137 plane waves. Primarily, the smooth pseudopotential leads to the fact that the various coefficients tend to become small after $\sim Q_{11}$. The coefficients that appear depend upon the types of atoms in the unit cell. For Si or Ge, (6.35) produces only the symmetric values, and Q_4 vanishes. Similarly, compound semiconductors have both real and imaginary values for most coefficients, although Q_8 is entirely real. The vanishment of some real and some imaginary values is a result of the periodicity of the lattice.

Often, it is found that the local pseudopotential may have too few parameters to fit the available data. In tight binding, one may extend the interactions to second-neighbor interactions, or one can add chemical levels such as *d*-levels or unfilled *s*-levels to create more parameters. In this momentum-space approach, additional parameters are added by extending the basis set to more reciprocal lattice vectors and to longer-range interactions (higher Fourier coefficients). Some have suggested that the free electron mass should be replaced by an adjusted value [33]. Pandey and Phillips [34] pointed out that there is no physical justification for this, and the sign is often wrong and thus does not provide the proper correction. Instead, they suggested the use of a nonlocal pseudopotential. These authors pointed out that, in Ge, this provides an additional parameter which can represent repulsive effects from the 3*d* core states. This provides some additional angular behavior and an additional set of parameters with which to improve the fit between the computed band structure and the experimental data that is used to achieve the fit.

To introduce a nonlocal (or angular variation) effect into the pseudopotential, the local pseudopotential is itself expanded into spherical harmonics. Even with a spherically symmetric potential, it is known that the solution to the Schrödinger equation involves angular variables and that the angular solutions are spherical harmonics. Here, the nonlocal part of the pseudopotential for atom one may be written in terms of a projection operator P_l onto the subspace Y_{lm} (m runs from $-l$ to l) of spherical harmonics with the same value of l as [34]

$$V_{\mathrm{P}}^{\mathrm{NL}}(\boldsymbol{r}) = \sum_{l} V_l(r) P_l, \tag{6.40}$$

where

$$V_l(r) = \begin{cases} A_i, & r < r_{0l} \\ 0, & r > r_{0l} \end{cases}. \tag{6.41}$$

Here, r_{0l} is an effective radius in the spherical harmonic. Each of the two atoms contributes its own nonlocal correction term to $V(\boldsymbol{r})$, as in (6.35), so that each of the two terms is handled separately. The combined results use (6.35), so that the matrix element (6.39) now becomes

$$\langle \boldsymbol{k} + \boldsymbol{G}' |\, V_{\mathrm{c}}(\boldsymbol{r}) \,| \boldsymbol{k} + \boldsymbol{G} \rangle = \frac{1}{\Omega} \int d^3\boldsymbol{r} \sum_{l} e^{-i(\boldsymbol{k}+\boldsymbol{G}')\cdot\boldsymbol{r}} V_l(r) P_l e^{i(\boldsymbol{k}+\boldsymbol{G})\cdot\boldsymbol{r}}. \tag{6.42}$$

To proceed, we use the common form of expanding exponentials in terms of Legendre polynomials, as follows:

$$e^{i\boldsymbol{K}\cdot\boldsymbol{r}} = \sum_{l} (i)^l (2l + 1) j_l(Kr) P_l(\cos\vartheta). \tag{6.43}$$

The angle in the Legendre polynomial is that between the radius vector **r** and the momentum vector $\boldsymbol{K}$. With two exponentials, primed and unprimed indices are used to correspond to the primed and unprimed $\boldsymbol{G}$ vectors that are in (6.42). The projection operator in (6.42) selects the appropriate Legendre polynomial and the corresponding spherical Bessel function j_l from the second exponential upon which it operates. The matrix element is rewritten using the reduced notation $\boldsymbol{K} = \boldsymbol{k} + \boldsymbol{G}$ and $\boldsymbol{K}' = \boldsymbol{k} + \boldsymbol{G}'$, and after some manipulation of the angular integrals, the matrix element becomes [34]

$$\langle \boldsymbol{K}' | V_{\mathrm{P}}^{\mathrm{NL}} | \boldsymbol{K} \rangle = \frac{4\pi}{\Omega} \sum_{l} A_l P_l(\cos\vartheta_{KK'}) F(Kr, K'r), \tag{6.44}$$

with

$$F(x, x') = \begin{cases} \dfrac{r_{0l}^2}{x^2 - x'^2}\left[x j_{l+1}(x) j_l(x') - x' j_l(x) j_{l+1}(x')\right], & x \neq x' \\ \dfrac{r_{0l}^2}{2}\left[j_l^2(x) - j_{l+1}(x) j_{l-1}(x)\right], & x = x' \end{cases}. \tag{6.45}$$

This result has also been found by Chelikowsky and Cohen [30], although they differ on the definition of the volume element discussed above. It is important to point out that the magnitudes of this latter function can be equal even in the off-diagonal elements, but the nonlocal correction is applied only to the off-diagonal elements of the matrix.

An additional inclusion in the momentum-space approach is the spin–orbit interaction. While the energy bands are composed of both the *s*- and *p*-orbitals of the individual atoms in the semiconductors, it has been found that the spin–orbit interaction affects atomic energy levels, so likely affects the energy bands. The spin–orbit interaction is a relativistic effect coupling the angular motion of the electron (around the atom) with the gradient of the confining potential. This produces an effective magnetic field. This field couples to the spin just as in the Zeeman effect. The most noticeable effect is a split of the threefold bands at the top of the valence band, in which one band is lowered in energy and referred to as the *split-off band*. This latter band lies from a few meV to a significant fraction of an eV below the top of the valence band in various semiconductors. Usually, the effect in the conduction band is smaller, but it can still be significant. The first inclusion of the spin–orbit interactions in pseudopotential calculations for semiconductors is thought to be due to Bloom and Bergstresser [35]. The interaction is known to be stronger in heavier atoms, so they considered the heavier materials InSb and CdTe. Subsequently, this interaction has been applied to most of the major semiconductors. In general, the formulation is written as [36]

$$\langle \boldsymbol{K}', s'|H_{SO}|\boldsymbol{K}, s\rangle = (\boldsymbol{K}' \times \boldsymbol{K}) \cdot \langle s'|\boldsymbol{\sigma}|s\rangle \sum_l \lambda_l P'_l(\cos\vartheta_{KK'}) S(\boldsymbol{K}' - \boldsymbol{K}), \tag{6.46}$$

where $\boldsymbol{K}$ and $\boldsymbol{K}'$ are the shifted wave vectors defined above (6.44), P'_l is the derivative of the Legendre polynomial, and S is the structure factor (the sine and cosine terms used earlier in (6.35)). The term $\boldsymbol{\sigma}$ is the vector of 2×2 Pauli spin matrices, and the direction of the leading cross-product term picks out one Pauli term. The wave function has just become more complicated. If 137 plane waves are used in the basis discussed earlier, then there will be 137 basis states with spin up and another 137 basis states with spin down. So, each plane wave is now associated with a spin wave function and the matrix has doubled in rank. Since the bonding electrons only represent the s- and p-states, only the $l = 1$ term needs to be kept in (6.46). There are also two atoms in the basis, and this leads to even and odd values for the parameter λ_l. This version of the spin–orbit interaction is a *full-zone* correction. It is basically different than the perturbative spin–orbit interaction that will be treated in the next section. The matrix element now becomes

$$\begin{aligned} \langle \boldsymbol{K}', s'|H_{SO}|\boldsymbol{K}, s\rangle &= -i(\boldsymbol{K}' \times \boldsymbol{K}) \cdot \langle s'|\boldsymbol{\sigma}|s\rangle \\ &\times \{\lambda_S \cos[(\boldsymbol{K}' - \boldsymbol{K}) \cdot \boldsymbol{t}] + i\lambda_A \sin[(\boldsymbol{K}' - \boldsymbol{K}) \cdot \boldsymbol{t}]\}, \end{aligned} \tag{6.47}$$

with

$$\lambda_S = \frac{1}{2}(\lambda_1 + \lambda_2), \quad \lambda_A = \frac{1}{2}(\lambda_1 - \lambda_2). \tag{6.48}$$

The subscripts 1 and 2 refer to atoms A and B in the unit cell, with

$$\begin{aligned} \lambda_1(K, K') &= \mu B_n(K) B_n(K') \\ \lambda_2(K, K') &= \alpha\mu B_m(K) B_m(K'). \end{aligned} \tag{6.49}$$

The subscripts n and m correspond to the row of the periodic table in which the atom resides. The parameters μ and α give two new fitting parameters. The functions B in (6.49) are given by spherical Bessel functions, as follows:

$$B_n(K) = i\sqrt{12\pi}\,C\int_0^{\infty} j_n(Kr)R_n(r)dr. \tag{6.50}$$

Pötz and Vogl [37] have shown that the functions (6.50) can be approximated by the relations

$$B_2 = \frac{1}{\left(1+\kappa_2^2\right)^3},\quad B_3 = \frac{5-\kappa_3^2}{5\left(1+\kappa_3^2\right)^4},\quad B_4 = \frac{5-3\kappa_4^2}{5\left(1+\kappa_4^2\right)^5}, \tag{6.51}$$

where

$$\kappa_n = \frac{Ka_{\mathrm{B}}}{\zeta_n}. \tag{6.52}$$

Here, a_{B} is the Bohr radius and ζ_n is the normalized (to the Bohr radius) radial extent of the core wave functions.

The first example we consider is the band structure of Si; we use only the local pseudopotential and the empirical approach. This is shown in figure 6.9. The dots represent the values of $\boldsymbol{k}$ that are used in the simulation. Here, the plot runs from the L point at the center of a (111) face to the Γ point at the zone center, then along (100) to the X point. It then shifts to an X point in the second zone that sits along the (110) line beyond the K point. The path then runs from this point back through K to Γ. The lack of inclusion of the spin–orbit interaction is apparent from the threefold set of bands at the top of the valence band at Γ. Nevertheless, this approach gets the conduction band minimum at Δ (along the (100) axis about 85% of the way to X) correct, since this is one of the principal points to which the bands have been fit in the empirical approach. This view may be compared with the Si band structure, again computed using a local pseudopotential approach, which also includes some of the complex bands, shown in figure 6.10 [38]. Here, k_y and k_z are considered to be fixed parameters and the bands are plotted along the Γ to X line for real k, designated by $k_{X\Gamma}$. The bands for imaginary k are extended from the two ends of this line with $\kappa = k_{X1}$. Here, the energy bands may be considered to be different sheets of the entire complex function $E(k_x)$, in which the energy bands for real k are connected by lines of complex k [38].

In figure 6.11, the band structure is shown for $In_{0.65}Al_{0.35}As$, a compound semiconductor that is lattice matched to InP and is used heavily in high-electron-mobility transistors (HEMTs) and optical devices [39]. Here, the full nonlocal pseudopotential, including the spin–orbit interaction, was used for the simulation, and 137 plane waves were used for each spin state. The valence bands near the Γ point at the zone center are extremely anisotropic, which is usual in group IV and III–V semiconductors. While the material is a direct bandgap semiconductor, the upper valleys along Δ are very Si-like in that the minima are located away from the X points. The expected gap between the split-off valence band and the top of the

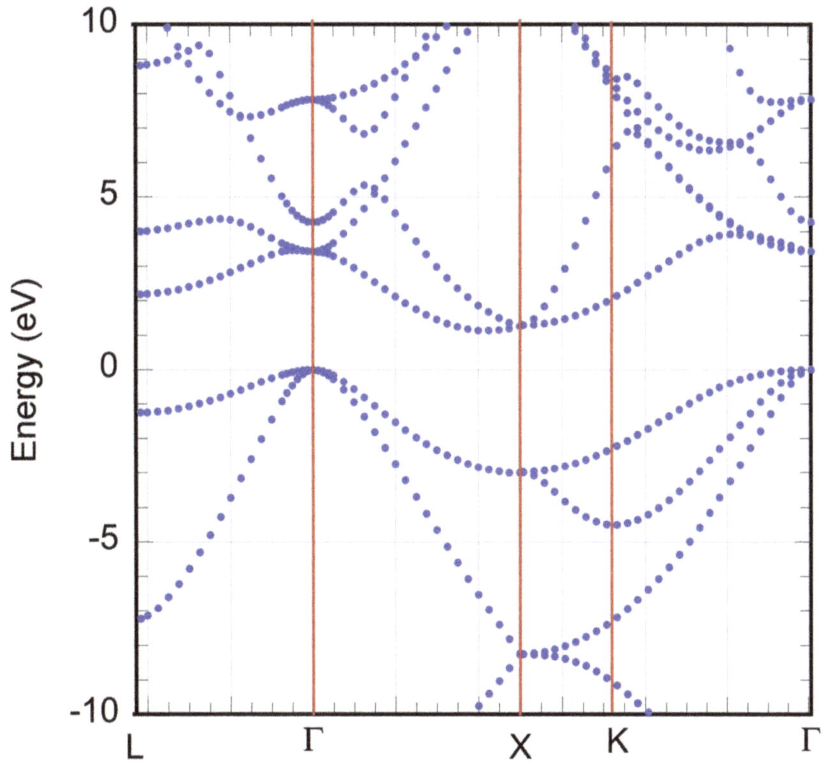

Figure 6.9. The band structure for Si using the local empirical pseudopotential approach. The dots represent the values of $\boldsymbol{k}$ for which the energy levels are calculated. The principal points of the Brillouin zone are identified along the lower axis.

valence band is apparent at Γ. This value is ~340 meV in $In_{0.65}Al_{0.35}As$. There is also a spin split in the conduction band at the minimum at Γ, but this is only about 1 meV, so it is hardly noticeable.

6.4 The $k \cdot p$ approximation

It was pointed out above that the spin–orbit interaction is easily incorporated within the pseudopotential method. However, it is often treated as a perturbational approach to correct the bands produced by other methods. In this case, it is usually coupled with the $\boldsymbol{k} \cdot \boldsymbol{p}$ terms of the Hamiltonian. One reason for this approach is to find an analytical formula for the energy bands that also expresses the non-parabolic behavior of the bands. Away from the actual minima or maxima of the bands, the dispersion (the variation of the energy bands with momentum) becomes non-parabolic and thus nonclassical in the dependence of energy on momentum. This arises from the interaction between the wave functions of the conduction band and the valence bands. While the approach adopted here uses just these four bands, higher-order perturbation theories often involve more bands. The $\boldsymbol{k} \cdot \boldsymbol{p}$ interaction and the spin–orbit interaction usually are considered to be relativistic effects. As discussed in section 6.3, the

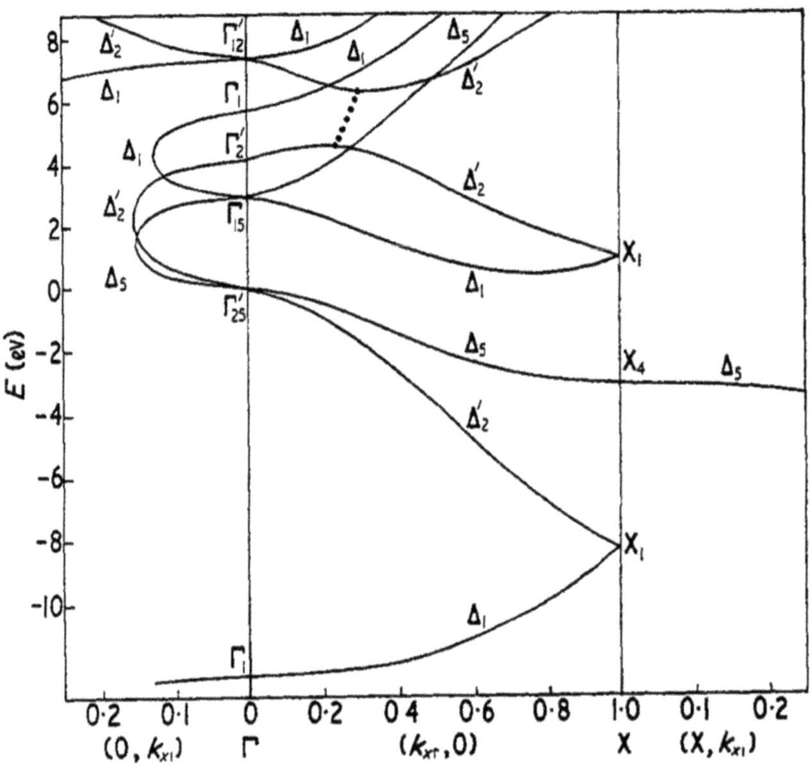

Figure 6.10. Energy bands for Si for real and imaginary k along the (100) axis. The bands for real k are in the central panel. Reproduced from [38].

spin–orbit interaction is a of the basic s- and p-states, which alters the various admixtures of the wave vector in the different wave functions.

The general solutions of the Schrödinger equation in a periodic potential were shown above to be Bloch wave functions. It is these Bloch functions that will now be used to define the Hamiltonian for the $\boldsymbol{k} \cdot \boldsymbol{p}$ method. Here, $\boldsymbol{k}$ remains the wave vector describing the propagation of the wave in a particular band. We now introduce the quantum mechanical momentum operator $\boldsymbol{p} = -i\hbar\nabla$ in a real-space formulation. Using the vector form of the Bloch function (6.12) in the Schrödinger equation leads to

$$\left[-\frac{\hbar^2}{2m_0}\nabla^2 - \frac{i\hbar^2}{m_0}\boldsymbol{k} \cdot \nabla + V(\boldsymbol{r})\right]u_k(\boldsymbol{r}) = \left(E - \frac{\hbar^2 k^2}{2m_0}\right)u_k(\boldsymbol{r}). \tag{6.53}$$

Here, the exponential term has been eliminated, since it is common to all terms in (6.53). The second term in the square brackets of (6.53) is known as the $\boldsymbol{k} \cdot \boldsymbol{p}$ term (the spin–orbit interaction has not yet been introduced). The latter term is often treated by perturbation theory and yields a correction to the effective mass. This is not necessary in the present formulation. Using a limited basis set, a diagonalization of the resulting matrix incorporates these terms exactly. The approach here follows

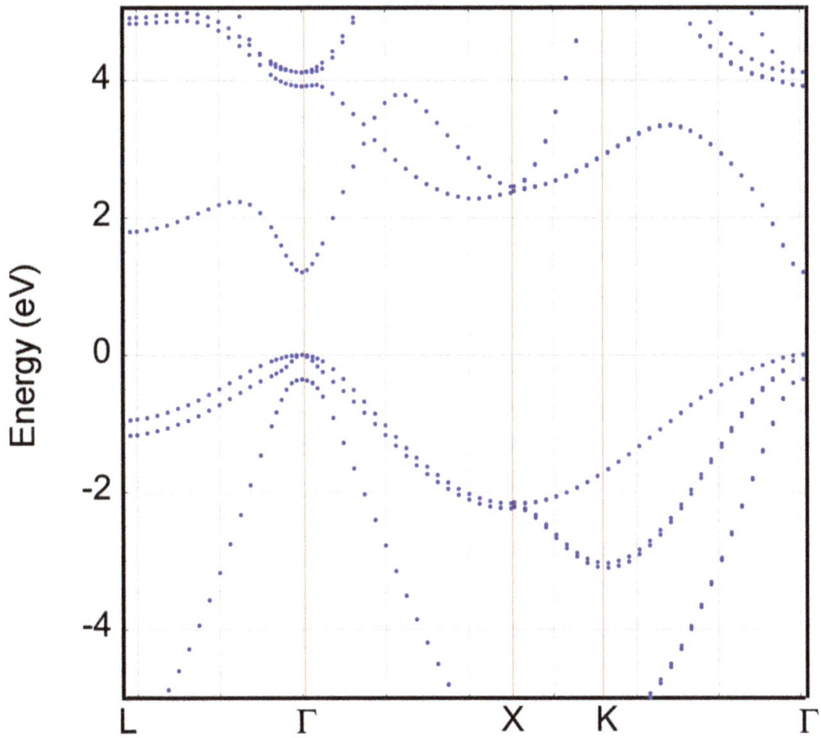

Figure 6.11. Energy bands for $In_{0.65}Al_{0.35}As$ determined using the empirical nonlocal pseudopotential method, including the spin–orbit interaction.

that of Kane [40, 41], and the spin–orbit interaction is also included. The basis states for the Hamiltonian (the $u_k(\boldsymbol{r})$ in the above equation) are the four hybridized orbitals arising from the single s and three p states. But these are doubled to add the spin wave function to each orbital as an eight-band model. A 16-band version has been given by Yu and Cardona [42].

The spin–orbit interaction terms must now be added to the Hamiltonian in (6.53). This interaction splits the triply degenerate (sixfold degenerate when spin is taken into account) valence bands at the band maximum (usually at the zone center, Γ), as discussed in the last section. This split is common in all tetrahedrally coordinated semiconductors. Two terms arise from the coupling of the orbital angular momentum (as well as the free electron momentum) and the spin angular momentum (the motion of the electron spinning on its own axis) of the electrons. These are generally expressed as

$$\frac{\hbar}{4m_0}\{[\nabla V(\boldsymbol{r}) \times \boldsymbol{p}] \cdot \boldsymbol{\sigma} + [\nabla V(\boldsymbol{r}) \times \boldsymbol{k}] \cdot \boldsymbol{\sigma}\}, \tag{6.54}$$

where $\boldsymbol{\sigma}$ is the vector of Pauli spin matrices already introduced in section 6.3. The second of these terms gives a term that is linear in k at the zone center. This tends to shift the maxima of the valence band a negligibly small amount away from Γ in compound semiconductors, in which the crystal lacks inversion symmetry. This is a small effect. Indeed, the following treatment does not produce the proper behavior of the heavy-hole bands, and a further correction is often applied. This is discussed below. Because of this correction, this second term can be neglected in the current calculations. The first term of (6.54) is k-independent and is the major term to be included in (6.53).

While one could use simply the s- and p-state atomic functions, the resulting Hamiltonian would be considerably more complex. For this reason, a different combination of s- and p-state atomic functions is used [40, 41]. This leads to a simpler form of the Hamiltonian. The basic sp^3 hybridization of the conduction and valence bands suggests the use of the s (denoted by S in the spin-split set of functions), p_x (denoted by X), p_y (denoted by Y), and p_z (denoted by Z) wave functions. While these are not proper Bloch functions, they have the correct symmetry for the designated band states and are the local cell part of the Bloch states. When the new interaction is added to (6.53), the resulting bands are doubly (spin) degenerate, which eases the problem and size of the Hamiltonian matrix to be diagonalized. For the eight (spin-dependent) basis states, the wave functions are taken to be the eight functions (the arrow denotes the direction of the electron spin in the particular state) [40, 41]

$$\begin{aligned} &| iS\downarrow\rangle, \left| \frac{X - iY}{\sqrt{2}} \uparrow \right\rangle, | Z\downarrow\rangle, \left| \frac{X + iY}{\sqrt{2}} \uparrow \right\rangle, \\ &| iS\uparrow\rangle, \left| \frac{X + iY}{\sqrt{2}} \downarrow \right\rangle, | Z\uparrow\rangle, \left| \frac{X - iY}{\sqrt{2}} \downarrow \right\rangle. \end{aligned} \tag{6.55}$$

The four states along the top row of (6.55) are degenerate with the set of states along the bottom row. The Hamiltonian matrix is 8×8, and the wave vector k is assumed to point in the z-direction for convenience (it is the direction normal to the assumed form of the X and Y functions). Choosing an arbitrary direction for the momentum would cause the matrix to be more complicated. Using standard coordinate rotations, the z-axis can be rotated to an arbitrary direction, and this changes the Hamiltonian matrix appropriately.

Given the choice of the basis functions in (6.55), the 8×8 matrix separates into a simpler block diagonal form that contains two 4×4 matrices, one for each spin direction. These are the diagonal blocks of the larger matrix. The off-diagonal blocks are 4×4 as well, but these blocks are empty (all elements are zero). Hence, each of the diagonal blocks can be diagonalized on its own. The form of these diagonal blocks is

$$\begin{bmatrix} E_s & 0 & kP & 0 \\ 0 & E_p - \frac{\Delta}{3} & \sqrt{2}\frac{\Delta}{3} & 0 \\ kP & \sqrt{2}\frac{\Delta}{3} & E_p & 0 \\ 0 & 0 & 0 & E_p + \frac{\Delta}{3} \end{bmatrix}. \tag{6.56}$$

The parameter Δ is given by

$$\Delta = \frac{3i\hbar}{4m_0^2}\left\langle X\left|\left(p_y\frac{\partial}{\partial x} - p_x\frac{\partial}{\partial y}\right)\right|Y\right\rangle > 0. \tag{6.57}$$

This quantity gives a simple value for matrix elements (2,3) and (3,2) in (6.56) and arise from one of the p wave functions (the p direction normal to k). Empirically, Δ is related to the spin–orbit split energy that separates the split-off valence band from the other valence bands. This is a measured quantity for most materials. The fourth line in the matrix (6.56) is an isolated level and is the heavy-hole band. This isolated level is given by $E_p = -\Delta/3$, so that the spin–orbit band is split away from the remaining valence bands by this amount. The momentum matrix element P in (6.56) arises from the $\boldsymbol{k} \cdot \boldsymbol{p}$ term in (6.53) and is defined as

$$P = -\frac{i\hbar}{m_0}\langle S|p_z|Z\rangle. \tag{6.58}$$

The quantities E_s and E_p are averages of the corresponding atomic energy levels (these were used earlier in the tight-binding formulation of section 6.2). The top of the valence band is the reference zero, as has been assumed in the previous sections. As mentioned, the heavy-hole band has an energy given by the free electron curvature of the second term in (6.53), and this curvature has the wrong sign, a point to which we return later. Once this energy level choice is made, the characteristic equation for the determinant of the remaining 3×3 matrix is

$$E'(E' - E_G)(E' + \Delta) - k^2P^2\left(E' + \frac{2\Delta}{3}\right) = 0, \tag{6.59}$$

where the reduced energy E' represents the energy term in parentheses on the right-hand side of (6.53), and $E_G = E_s$ in recognition that the bottom of the conduction band is primarily determined by the s states. There are several approximations that can now be made in order to solve (6.59).

6.4.1 The energy bands

Small *kP*. If the last term in (6.59) is taken to be small (e.g. the energy is near to the band extremum), the solutions are basically just those arising from the products of the first term with a slight shift arising from the kP term. When this assumption is used, the three bands arise from the three zeroes of the first term as follows:

$$\begin{aligned} E_C &= E_G + \frac{k^2P^2}{3}\left(\frac{2}{E_G} + \frac{1}{E_G + \Delta}\right) + \frac{\hbar^2k^2}{2m_0} \\ E_{lh} &= -\frac{2k^2P^2}{3E_G} + \frac{\hbar^2k^2}{2m_0} \\ E_\Delta &= -\Delta - \frac{k^2P^2}{3(E_G + \Delta)} + \frac{\hbar^2k^2}{2m_0} \end{aligned} \quad . \tag{6.60}$$

These three results give the dispersion for the conduction band, the light-hole band, and the split-off (spin–orbit) band, respectively. The free electron contribution (the correction of the parentheses on the right-hand side of (6.53)) has been restored in each line of (6.60). Here, each of these bands is parabolic for small values of k. In each case, the effective mass is inversely proportional to the square of the momentum matrix element. For example, for the conduction band (the first line of (6.60)), the effective mass is given by

$$\frac{1}{m_c} = \frac{1}{m_0} + \frac{2P^2}{3\hbar^2}\left(\frac{2}{E_G} + \frac{1}{E_G + \Delta}\right). \tag{6.61}$$

This mass is considerably smaller than the free electron mass m_0 due to the size of the last term. Similarly, masses can be inferred for the other two bands. These values are all slightly different, although they depend on both the energy gap and the momentum matrix element. Near $k = 0$, the effects of the higher bands are relatively minor.

$\Delta = 0$. The two-band model. If the spin–orbit energy Δ is set to zero, the split-off band becomes degenerate with the light-hole and the heavy-hole bands at $k = 0$. One energy level can then be factored out of the resulting equation (this is the heavy-hole band). The remaining two bands are almost mirror images of one another and yield the conduction and light-hole bands. These are described by

$$\begin{aligned} E_c &= \frac{E_G}{2}\left[1 + \sqrt{1 + \frac{4k^2P^2}{E_G^2}}\right] + \frac{\hbar^2k^2}{2m_0} \\ E_{lh} &= \frac{E_G}{2}\left[1 - \sqrt{1 + \frac{4k^2P^2}{E_G^2}}\right] + \frac{\hbar^2k^2}{2m_0} \end{aligned}. \tag{6.62}$$

This form here is a typical hyperbolic description used to describe non-parabolic bands in a relativistic manner (these two equations are quite close to Einstein's equation when one sets $E_G = mc^2$). The interaction energy depends upon the momentum matrix element and the energy gap. As above, the two effective masses at the band edge (in the $k \to 0$ limit) may be found to be

$$\begin{aligned} \frac{1}{m_c} &= \frac{1}{m_0} + \frac{2P^2}{\hbar^2 E_G} \\ \frac{1}{m_{lh}} &= -\frac{1}{m_0} + \frac{2P^2}{\hbar^2 E_G} \end{aligned}. \tag{6.63}$$

The opposite signs on the free electron mass term prevent these two bands from being pure mirror images about the center of the energy gap.

$\Delta \gg E_G, kP$. When the spin–orbit split Δ is large, a somewhat different expansion of (6.59) is obtained. The resulting equation can be solved to yield three energy relations, as follows:

$$E_c = \frac{E_G}{2}\left[1 + \sqrt{1 + \frac{8k^2P^2}{3E_G^2}}\right] + \frac{\hbar^2k^2}{2m_0}$$
$$E_{lh} = \frac{E_G}{2}\left[1 - \sqrt{1 + \frac{8k^2P^2}{3E_G^2}}\right] + \frac{\hbar^2k^2}{2m_0}. \tag{6.64}$$
$$E_\Delta = -\Delta - \frac{k^2P^2}{3(E_G + \Delta)} + \frac{\hbar^2k^2}{2m_0}$$

The last line may be recognized as the spin–orbit split-off. The two leading equations represent the conduction band and light-hole band, just as in (6.62), but the terms in the square root are different than those found earlier. This difference is just the numerical factor of 2/3, and these two bands remain almost mirror images.

The two approximations (6.62) and (6.64) show that the conduction band and the light-hole band are both hyperbolic in shape when Δ is nonzero. Of course, this is expected, as this is primarily a result of the relativistic corrections that arise when this parameter is nonzero. The major cause of the relativistic band shape is actually introduced by the *kp* interaction, while the spin–orbit interaction leads primarily to the split-off band separation from the light- and heavy-hole bands. It is common practice to introduce the band-edge masses directly into a hyperbolic energy relationship without worrying about the size of the spin–orbit energy. This gives the most common form seen for the hyperbolic bands, namely that of (6.62) with the effective masses given by (6.63). For most of the direct-gap compound semiconductors (formed from the group III and V columns of the periodic table), the conduction band and light-hole band effective masses are small and typically lie in the range of 0.01–0.1 (times, of course, m_0). One important observation is that the momentum matrix element P arises from the sp^3 hybrids, and the results of (6.63) suggest that the actual effective masses scale with the bandgap in an almost linear fashion, with only modest variations from the momentum matrix element. Materials with narrow bandgaps usually have very small values of the effective masses.

In studies of transport, it is often useful to rearrange the mirror-image bands, which are hyperbolic in nature, to express the momentum wave vector k directly in terms of the energy. This is easily done, with the results that

$$\frac{\hbar^2k^2}{2m_c} = E\left(\frac{E}{E_G} + 1\right)$$
$$\frac{\hbar^2k^2}{2m_{lh}} = -E_G - E\left(1 + \frac{E}{E_G}\right). \tag{6.65}$$

Here, it must be remembered that the energy is measured from the conduction band minima. If we shift the zero of energy to the bottom of the conduction band, then the signs are changed in the two parentheses.

6.4.2 Anisotropic valence bands

As remarked in the spin–orbit discussion of the previous section, the equation for the heavy-hole band does not, in fact, have the right curvature. This band needs to be corrected, usually by possible interactions with higher-lying bands. If the heavy-hole band is to be corrected, then more bands are required with interactions between them, and these interactions are handled by perturbation theory. In general, these lead to the linear k terms mentioned above, which are important at very small values of k in the valence band. There are also quadratic terms in k that lead to variations of the masses beyond those of the hyperbolic bands and quadratic cross-terms that lead to warping (nonspherical symmetry) of the band away from $k = 0$. All of this is traditionally handled by perturbation theory [43], typically with a 16-band formulation [42]. The dominant effect of these terms, however, mainly affects the light- and heavy-hole bands, and these bands may be written as [43]

$$E_{\mathrm{h}} = -\left[Ak^2 \pm \sqrt{B^2k^4 + C^2\left(k_x^2k_y^2 + k_y^2k_z^2 + k_z^2k_x^2\right)}\right]. \tag{6.66}$$

The upper sign gives the light-hole band and the lower sign gives the heavy-hole band. The three constants A, B, and C have been determined for many semiconductors. For Si, they take the values 4, 1.1, and 4.1, respectively. In addition, the quartic terms in the parentheses under the square root lead to significant warping of the bands, which makes them nonspherical.

In general, three parameters describe the interaction with the higher-lying conduction bands, and the A, B, and C parameters are defined in terms of these other three parameters. The three that arise from the actual perturbation theory are given below [43]:

$$L = \frac{\hbar^2}{m_0^2}\sum_{lj}\frac{\left|\left\langle \epsilon_1^+ \middle| p_x \middle| lj\right\rangle\right|^2}{E_{\mathrm{G}} - E_l}$$
$$M = \frac{\hbar^2}{m_0^2}\sum_{lj}\frac{\left|\left\langle \epsilon_2^+ \middle| p_y \middle| lj\right\rangle\right|^2}{E_{\mathrm{G}} - E_l}. \tag{6.67}$$
$$N = \frac{\hbar^2}{m_0^2}\sum_{lj}\frac{\left|\left\langle \epsilon_3^+ \middle| p_z \middle| lj\right\rangle\right|^2}{E_{\mathrm{G}} - E_l}$$

Here, l represents the band and j is a state in that band (the band here is the valence band). The energies are $\epsilon_1^+ \sim YZ$, $\epsilon_2^+ \sim ZX$, and $\epsilon_3^+ \sim XY$ in the notation of the previous section. The energy E_l represents the term E' defined by the term in the parentheses on the right-hand side of (6.53). These three matrix elements yield the three parameters in (6.66) as follows:

$$
\begin{aligned}
A &= \frac{1}{3}(L + 2M) + \frac{\hbar^2}{2m_0} \\
B &= \frac{1}{3}(L - M) \\
C^2 &= \frac{1}{3}[N^2 - (L - M)^2]
\end{aligned}
\quad . \tag{6.68}
$$

A, B, and *C* can be determined from the band-edge effective masses in the various directions, and using (6.66), these can written as [42]

$$
k\|(100): \begin{aligned}
\frac{1}{m_{hh}} &= \frac{2}{\hbar^2}(-A + B) \\
\frac{1}{m_{lh}} &= \frac{2}{\hbar^2}(-A - B)
\end{aligned}
, \tag{6.69}
$$

and

$$
k\|(111): \begin{aligned}
\frac{1}{m_{hh}} &= \frac{2}{\hbar^2}\left[-A + B\left(1 + \frac{|C|^2}{3B^2}\right)\right] \\
\frac{1}{m_{lh}} &= \frac{2}{\hbar^2}\left[-A - B\left(1 + \frac{|C|^2}{3B^2}\right)\right]
\end{aligned}
. \tag{6.70}
$$

The values of the *A*, *B*, and *C* parameters can be found in many textbooks, and they can be used to estimate the effective masses of the light- and heavy-hole bands at the extrema of these bands [42, 44]. Of course, they are determined by the actual band structures that are calculated.

6.5 Broadening of the band edges

The preceding material utilized the Bloch states of section 6.1 that exist in a regular crystal lattice and possess well-ordered parameters. In a sense, these discussions have been predicated upon the fact that the band states have long-range order. But there are many ways in which this long-range order can be disturbed. This can arise from a large density of defects in the lattice or a large concentration of impurity atoms that disturb the truly periodic nature of the lattice. The fundamental problem in weakly disordered semiconductors (those assumed to generally have long-range order) is how the spectrum of allowed energy levels and the corresponding nearest-neighbor coupling energies are modified in the presence of this disorder. When disorder is present, the existence of long-range order may or may not be retained, but even if it is retained, this is only for a reduced set of energy states. Once the long-range order is broken, one can consider that the disordered material contains many small regions in which the band properties vary but remain more or less uniform within a single region. There may be bands in a region, but these bands develop very 'fuzzy' edges and there are likely to be states within the bandgaps. The general experimental situation is that this may lead to an excitation gap which is often larger than the expected bandgap of the base semiconductor, especially in amorphous materials. This can also lead to the localization of the carriers in these fuzzy edges. In earlier days, it was assumed that this localization, which is often called strong localization in distinction from the weak localization which is a perturbative effect on the conductivity, was caused by this strong disorder. At the same time, it was believed that weak localization was caused by phase interference in the potential landscape created by the weak disorder arising from a large impurity or defect density. In fact, the two are closely related, which should have been obvious from the fact that the same theoretical model is used to simulate both effects! This model is the Anderson model [45].

The fundamental problem with disorder is to determine how it affects the allowed energy levels in the material. Normally, electron waves are free to propagate, and a bandgap occurs, in which the electron waves become localized and non-propagating (evanescent). In the presence of disorder in the band semiconductor, the edge between the band and the gap becomes a dispersed gray area (our fuzzy set of states). Usually, the width of the band increases, allowing states to broaden into the gap area, which is called band tailing. But many of the band states in this tail, as well as some in the normal band region, can become localized. Here, the long-range order of the crystal can be compromised, in the ultimate limit producing amorphous material. Because of the extension of the band region and its interaction with localized states, a new parameter arises which plays the role of the normal band edge, and this is the mobility edge. The mobility edge now separates the localized states from the conducting, or long-range ordered, states. The excitation mentioned above then occurs when carriers below the valence mobility edge are excited to states above the conduction mobility edge, and this transition is larger than the normal bandgap.

It is a remarkable idea that there is a rapid transition, which occurs when some critical energy (the mobility edge), from strongly localized states that have only short-range order to a set of extended states that have long-range coherence of their wave function. This idea of such a transition warrants further investigation. Anderson [45] discussed several possible reasons for this occurrence. In the principal approach, a set of atomic sites is assumed in the crystal (once more, only a single orbital per atom), in which the atomic energies are randomly but uniformly distributed over a fixed energy range, which is commonly denoted by W. This means that the energy is a random function with a uniform probability distribution function given by $1/W$ for energies within this given range, usually denoted by $-W/2 \leqslant E \leqslant W/2$. This has the effect of *shifting the zero of energy to the center of the band* (which is somewhat strange, since the band edges have normally been the focus up until now). The reason for this will become clear later. It will be found below that there is a critical value of the width W, such that if W is greater than this critical value, all the states in the band are localized. The value of W at this critical value is normally known as the Anderson transition. Without dwelling on the details of how this occurs, the main point is that it is found that

$$\frac{W}{2V} > e^1 Z \sim 2.\,7Z \sim 10.\,8 \tag{6.71}$$

is required to completely localize the band ($Z = 4$, the number of nearest neighbors, is taken for tetrahedrally coordinated semiconductors). In general, we can say that $W > e^1 \Delta E$, where $\Delta E = 2VZ$, totally localizes the band, so that no extended states exist. Here, V is the average nearest-neighbor interaction potential. At the other extreme, we can take $W = 0$, and we arrive at the equivalent inequality that says states are localized if $E > \Delta E/2$. As ΔE is the width of the normal band, this is just the normal requirement that localized states exist outside the normal band width. These localized states are just the bands for imaginary k.

Of interest is the point at which an energy at the edge of the distribution, $E = W/2$, begins to be localized. When $W > e^1 \Delta E/4$, localized states appear at the edge of the band. This means that there is a critical size of the disorder that must be reached before the band states near the edge begin to be localized. These different conditions are sketched in figure 6.12 [46]. The red curve is the minimum energy required before states begin to be localized. The blue curve is the band edge (the figure shows the upper half of the energy band; exactly the same behavior is seen for negative E). The blue shaded region is that of the localized states, which are pulled out of the band because the total number of states is given by the number of unit cells in the solid and does not change with the disorder.

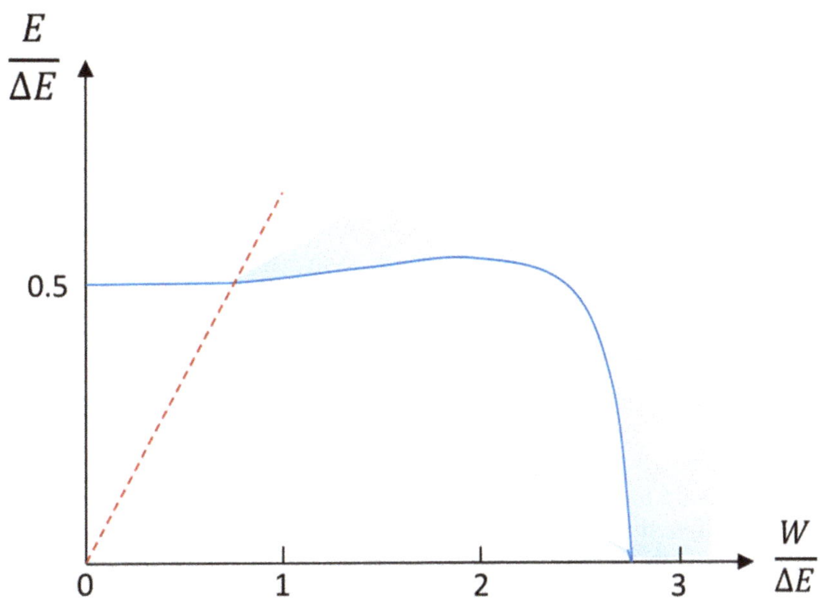

Figure 6.12. The critical energy that differentiates extended from localized states (blue curve), with the localized states (blue shaded region) outside the curve. The red curve shows the minimum value of the disorder needed to begin to localize states. States below the blue curve maintain their long-range order. Here, W is the width of the energies of the various atoms, ΔE is the energy width of the band in question (zero is the center of the band), and E is the energy of the carrier. Reprinted from [46].

A very illustrative example can be found in the TMDCs. These materials form interesting two-dimensional semiconductor layers that have been shown to be useful for electronic applications. However, they currently suffer from the presence of defects that act as both deep levels and random potentials leading to localization. In particular, it has been shown that charge impurities such as S vacancies in MoS_2 and impurities from the dielectric environment not only have a significant impact on the mobility of MoS_2 devices [47] but can also lead to the formation of an impurity band tail within the bandgap region. In addition, localized states are found above the conduction band edge [48]. The TMDCs tend to have a significant number of chalcogenide vacancies (e.g. missing S in this case). These vacancies lead to both effects. First, the vacancy itself tends to produce a deep level that lies near the mid-gap in the material. Second, the missing atomic potential produces a modulation of the crystal potential formed by the pseudopotentials of the atoms themselves. This modulation creates a random potential critical to the Anderson model and localization found in this model. So, in this case, the atomic defect plays two roles—deep level and localization. A model for this behavior was proposed by Zhu *et al* [48]. From extensive capacitance voltage measurements between a back gate on the TMDC and the layer itself (using a SiO_2 insulator grown on the Si back gate), these authors found that there is a mid-gap state that has a population of approximately 10^{12} cm^{-2}. They also found an extremely broad distribution of localized states, as in the Anderson model, and an extremely high mobility edge for the conduction band. They then adapted a simulation model to fit the data. In this model, the impurity band tail was incorporated by the following density of states distribution:

$$D_n(E) = \begin{cases} \alpha D_0 \exp\left[\frac{E}{\varphi}\right], & -\frac{E_G}{2} < E < 0, \\ D_0 - (1-\alpha)\exp\left[-\frac{E}{\varphi'}\right], & E > 0. \end{cases} \tag{6.72}$$

The energies here are referenced to the conduction band edge *in the absence of the disorder*. Here, D_0 is the normal two-dimensional density of states, using an effective mass of $0.4m_0$. The $(1 - \alpha)$ term for positive energy accounts for the states that are removed from the conduction band to form the band-tail states of the first line (the blue shaded region in figure 6.12). The parameter φ is the characteristic width of the localized states that form the band tail discussed above and is found to be ~100 meV, and φ' is chosen so that the two piecewise functions have a continuous gradient at $E = 0$. The parameter α is found to be ~0.33. From this model and the measured conductance, the mobility edge is found to lie ~10 meV above the conduction band edge.

To calculate the free carrier density, one considers only the extended states above the mobility edge energy E_M with respect to the conduction band edge:

$$n_{\text{free}} = \int_{E_M}^{\infty} D_n(E) P_{FD}(E) dE. \tag{6.73}$$

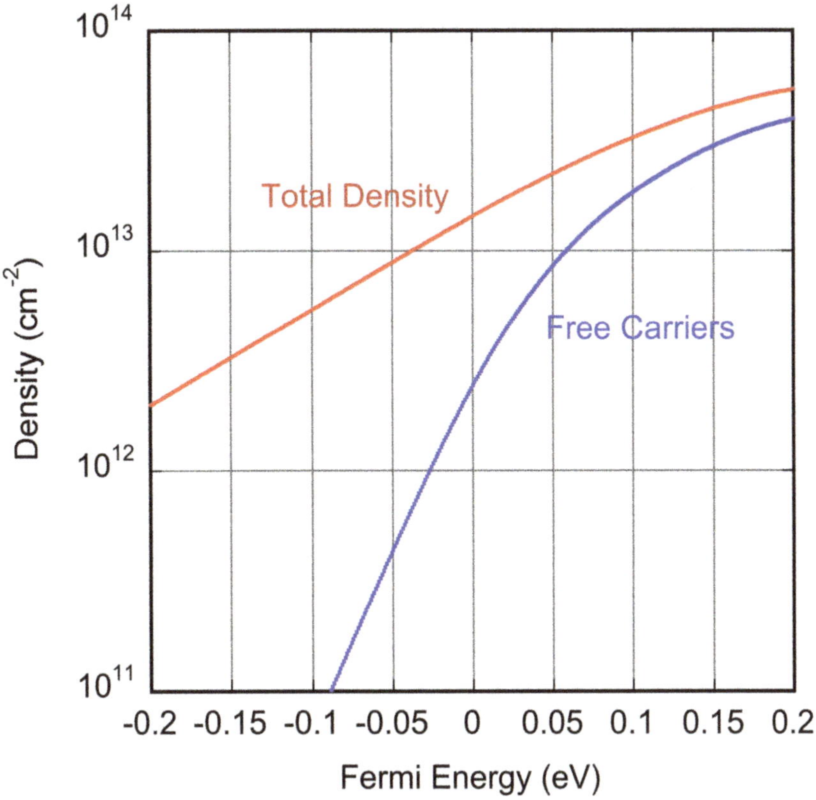

Figure 6.13. Localized states trap electrons into immobile states. This plot shows the total electron density and the free carrier density for a MoS_2 monolayer at 300 K. The parameters are discussed in the text. Reprinted from [46]. © IOP Publishing Ltd. All rights reserved.

In different experiments reported by Xiao *et al* [49], a slightly different value of the band tailing, $\varphi \sim 110$ meV, a mobility edge of 10 meV, and a positively charged oxide impurity density of 3.5×10^{12} cm^{-2} were found. As in the previous example, these were found for a layer of MoS_2 placed upon an oxidized Si wafer so that the Si could be used as a back gate to vary the electron concentration. Using this data and parameters, the total induced electron concentration as well as the free carrier (those above the mobility edge) concentration were determined at room temperature. These two quantities are plotted in figure 6.13. The difference between these two density yields demonstrates how a significant fraction of the carriers are trapped in the localized states.

6.6 The effective-mass approximation

The properties of an electron in one of the energy bands are not the same as those of a free electron. The electron is a wave inside the crystal, a fact that was established by Davisson [50] and Thompson [51], and is of course used in the Schrödinger equation in the above discussions. While the idea of an effective mass used above arose from the energies for the bands near the extrema, a more general method is desired to find the *effective* mass at an arbitrary point in the energy band. This is what the effective-mass approximation gives us. The intuitive way to achieve this is to recognize that what we call an electron in a semiconductor is not a real electron but what is called a quasi particle. This is because it has certain properties that real electrons have, but are different from the free electron inside the semiconductor. The important properties are the mass and, in the case of holes, the charge. The most intuitive way to do this says that the quasi particle must have the same momentum and energy that the real particle would have had. Thus, its kinetic energy is given by the wave frequency through the Planck relation, as follows:

$$E = \hbar\omega, \tag{6.74}$$

where the energy is measured from the convenient zero (typically at the band edge). The kinetic momentum must then be the same as the wave momentum, where the kinetic momentum is assigned to the quasi particle and the wave momentum comes from the band structure. This means that the relation

$$m^* v = \hbar k \tag{6.75}$$

must be enforced. As for any normal wave, the position of the quasi particle in this situation moves with the group velocity of the wave, or

$$v = \frac{\partial\omega}{\partial k} = \frac{1}{\hbar}\frac{\partial E}{\partial k}, \tag{6.76}$$

which leads to an effective mass of

$$\frac{1}{m^*} = \frac{v}{\hbar k} = \frac{1}{\hbar^2 k}\frac{\partial E}{\partial k}. \tag{6.77}$$

This is different from what one sees in a great many textbooks. This is because the textbooks usually only deal with the behavior at the band edges. This result is actually the correct result [52, 53].

When the electron moves in the potential that arises from the atoms (and from the varieties of electron–electron potentials), it follows the energy band and is thus nonstationary in time. On the other hand, if we connect it to the Bloch wave function, then it sits with a wave momentum and a corresponding energy at the momentum $\hbar k$ and can be thought of as having that momentum. The detailed velocity and momentum are different from the velocity and momentum that are associated with the Bloch wave function. This leads to the idea of a quasi particle that has properties different than those of a free electron. This is in spite of the fact that the free electron is the basis upon which the energy bands are constructed. Nevertheless, the quasi particle has an effective mass that is needed to complete the description of its response to external fields and potentials. This now leads to the result (6.77) [52, 53].

To connect with the effective mass defined in many textbooks, there is a problem at the bottom of the band in (6.77), where both the derivative with respect to k and k

itself vanish (k is defined from the point of the band minimum). In this situation, one must use L'Hôpital's rule and take the derivatives of the numerator and denominator with respect to k, and this leads to

$$\frac{1}{m^*} = \lim_{k \to 0}\left(\frac{1}{\hbar^2}\frac{1}{\partial k/\partial k}\frac{\partial^2 E_n(\boldsymbol{k})}{\partial k^2}\right) \sim \frac{1}{\hbar^2}\frac{\partial^2 E_n(\boldsymbol{k})}{\partial k^2}. \tag{6.78}$$

This is the result given in most textbooks, but it must be remembered that this is a limiting form and is valid only near the bottom of the conduction band (or, equivalently, near the top of the valence band). While this latter form is found in most textbooks, there is usually little discussion of its range of validity. Indeed, most, if not all, of the common derivations of this last expression have conceptual errors in them that obscure these limitations, a fact that necessitates the present approach. It is worth pointing out that the mass defined in (6.77) has long been known as the *cyclotron* mass [54], as cyclotron resonance measures a cross-sectional area of the energy surface, and this is related to the value given in (6.77).

Let us examine what happens here for the energy-dependent mass in the hyperbolic bands. The conduction band mass from (6.77) may be determined from the hyperbolic band (6.62) (the energy of the conduction band is now referenced to the minimum of this band):

$$E_c = \frac{E_G}{2}\left[\sqrt{1 + \frac{2\hbar^2 k^2}{m_c E_G}} - 1\right], \tag{6.79}$$

where m_c is the effective mass at the band minimum. Using (6.77) to find the energy-dependent mass gives

$$\frac{1}{m^*} = \frac{1}{\hbar^2 k}\frac{\partial E(\boldsymbol{k})}{\partial k} = \frac{1}{\hbar^2 k}\left(\frac{\hbar^2 k}{m_c}\right)\frac{1}{\sqrt{1 + \frac{2\hbar^2 k^2}{m_c E_G}}}, \tag{6.80}$$

and

$$m^* = m_c\left(1 + \frac{2E}{E_G}\right). \tag{6.81}$$

That is, the mass gets heavier as the energy increases away from the minimum of the conduction band. The result that the mass gets larger as we increase the momentum is often referred to as a relativistic effect, and the general $\boldsymbol{k} \cdot \boldsymbol{p}$ correction is often called the relativistic correction, due to the connection between this approach and the Dirac equation, in which the static mass term corresponds to the bandgap in semiconductors.

6.7 Strain

In modern logic devices, strain is added to modify the transport properties by modifying the band structure; this modification thus becomes a quantum effect [55, 56]. For an

n-channel metal-oxide semiconductor field-effect transistor (MOSFET), tensile strain is used to split the conduction bands and is introduced by placing a Si_3N_4 layer over the gate stack. As discussed in chapter 4, confinement in the inversion layer splits the sixfold conduction band into a twofold set of elliptical energy surfaces whose major axis is normal to the interface and a fourfold set whose minor axis is normal to the interface. The use of tensile strain increases this separation so that the conduction process is dominated by carriers in the twofold pair of sub-bands. A 1% uniaxial strain in the (110) transport direction can give as much as a 70 meV split between the two sets of valleys [57]. The effective mass for transport in the channel then becomes the transverse mass, ($\sim 0.19 m_0$ in Si). This results in an approximately 40% increase in the mobility relative to the normal inversion mobility in Si. In addition, the carrier scattering between the twofold and fourfold sets of valleys is greatly reduced, and this further increases the mobility for carriers in the inversion channel.

In figure 6.14, the effect of strain in the (110) direction is plotted for an n-channel device on the (001) surface of Si [56]. The strain deforms the normally circularly symmetric shape of the twofold valleys, as can be seen in the figure. Several different levels of strain are represented by the different curves. It is clear that the mobility, though larger, also becomes anisotropic in the valleys. The strain in the bands was calculated using an empirical pseudopotential method (section 6.3), and the

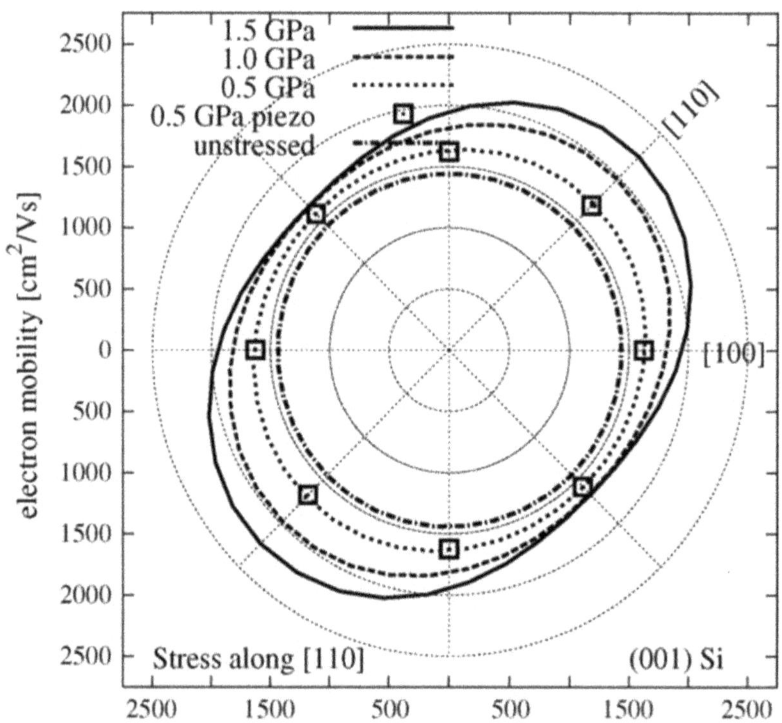

Figure 6.14. Mobility enhancement for uniaxial stress applied in the (110) direction. The piezoelectric effect is also included in this enhancement. The horizontal axis is the same as the vertical. Reprinted from [56], Copyright (2007), with permission from Elsevier.

transport was determined using an ensemble Monte Carlo method (chapter 9). The strain elongates the normal circular energy surface in the plane of the device, and this causes an decrease in the effective mass in the channel direction. Note that this figure actually plots the mobility as a function of the crystalline direction, and the plot does not represent the energy surfaces.

In *p*-channel devices, a compressive stress is introduced to vary the effective mass and mobility of the holes. This is often achieved by growing a strained SiGe layer as a quantum well [58] for the holes. The strain in the SiGe layer splits the light- and heavy-hole bands at Γ, so that the holes respond to the light-hole mass rather than an admixture of the masses of the two bands. One drawback is that the SiGe layer has significant alloy scattering, and this tends to lower the mobility. To determine the valence band structure in the presence of both the confinement effects and strain requires a form of full band calculation for the complicated splits, warps, and crossings of the various valence bands [59]. One additional problem is that the two bands now tend to cross at a momentum value away from the zone center. This crossing is actually an avoided crossing, as the two bands hybridize, but the result is nevertheless an extremely non-parabolic behavior. The effect of the gate voltage on the strain is illustrated in figure 6.15, where it can be seen that the gate electric field has only a small effect on the constant energy surface for the dominant band [60].

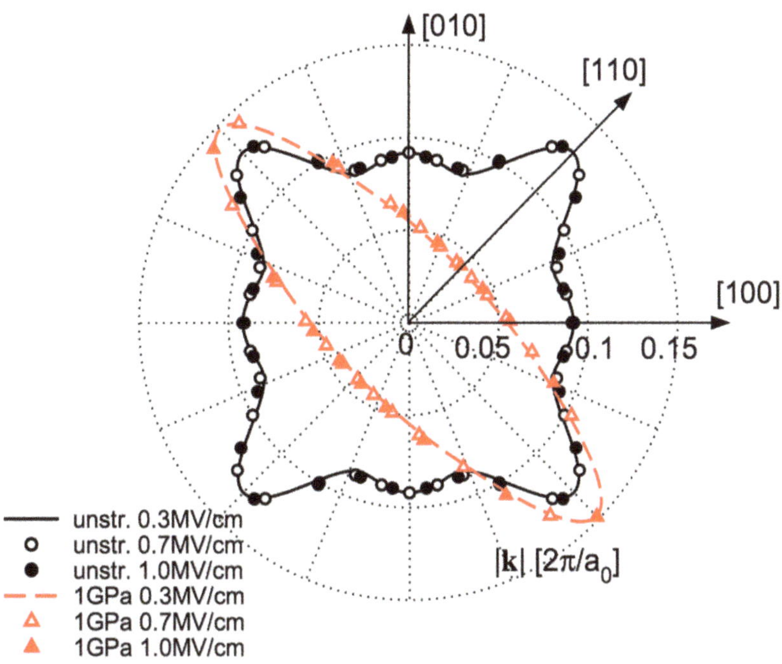

Figure 6.15. The constant energy surface of the lowest hole sub-band for Si (001) surface devices with compressive strain in the (110) direction for several different gate fields. The black curves and data points are for unstrained channels, while the red curves and data points are for strained channels. Reprinted from [60], Copyright (2009), with permission from Elsevier.

6.8 Connecting transport to the band structure

In full self-consistent calculations, particularly in devices, the electric field is determined from the applied voltage and any self-consistent potential. However, the transport calculations are still performed using the effective-mass approximation. This is most often done with parabolic bands but occasionally with the non-parabolic hyperbolic bands that arise in $\boldsymbol{k} \cdot \boldsymbol{p}$ theory. But one can go further when it is desired to completely characterize electrons (or holes) that reach very high kinetic energies. In this case, the full band structure throughout the Brillouin zone can be used. The first researchers to do this were apparently Shichijo and Hess [61]. The band structure for GaAs was determined using an empirical pseudopotential method [62]. This was a local pseudopotential that did not include the spin–orbit interaction. The idea was to study impact ionization in GaAs, and the authors wanted a better description of the bands for carriers at the high energy required for this. The rationale was that the actual band structure was quite far from an analytical band. To implement the full band structure, it must be gridded within the Brillouin zone, but due to the symmetry of the zone, only a 1/48 segment of the zone needs be used, and symmetries are used to find the other parts of the zone. Nevertheless, scattering rates between all values of energy and momentum have to be computed and stored. While the accelerating force remains that of the electric field, the velocity is no longer a simple function of momentum. The band in which the particle is located must be kept as a parameter, and the velocity at any momentum determined from (6.76). As a result, the grid chosen has to be fine enough to allow the velocity to be determined with minimum noise. The major problems are the storage required for the scattering tables and the random selection of final momenta in the 48 different segments of the band structure. However, other problems arise in which the bands split (or cross in silicon, for example), such as at X in the valence band and near Γ in the second conduction band. The usual approach is to assert that the particle will follow the band with approximately the same slope in momentum space [61]. This is shown in figure 6.16 for the band structure of Si along the (100) lines. The case is clear at X, where the particle normally just moves into the second conduction band (following the red arrow) at this point. It is more complicated at Γ, because both bands come into the point flat (at zero velocity). The motion after passing Γ can be into either of the two bands, so this may be selected by a random number, as used in Monte Carlo simulations (discussed in chapter 9).

Perhaps the most published follow-on to the introduction of full band approaches is that of Fischetti and Laux, which is associated with the development of the device analysis using Monte Carlo et Poisson solver (DAMOCLES) code at IBM [63, 64]. This effort began with the same local pseudopotential as that of Shichiho and Hess [61] without the spin–orbit correction, but was aimed at the simulation of semiconductor devices from the start, which means coupling to the Poisson equation. Here, they showed that storing the entire Brillouin zone at 41 000 points would provide a speed advantage by avoiding the symmetry operations required by using just a 1/48 wedge. In addition, the advantage of using self-scattering is lost, as the equations of motion cannot be simply integrated and must be numerically

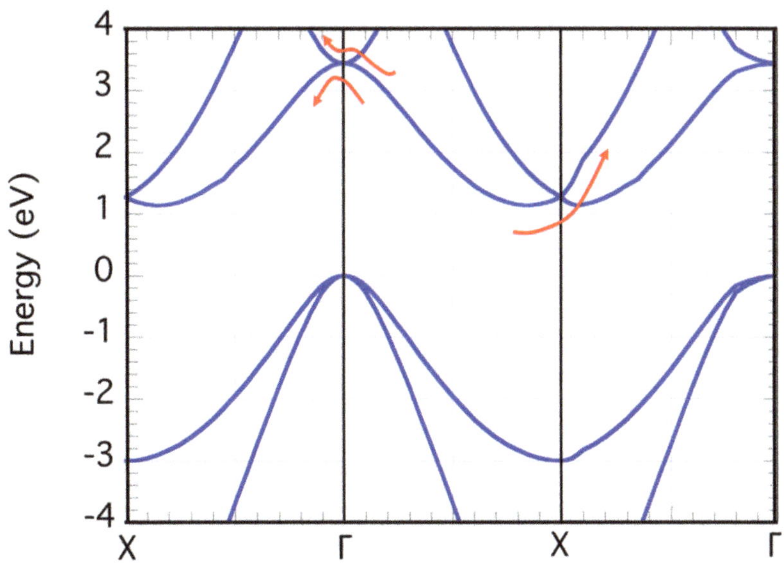

Figure 6.16. Variation of the transport across a symmetry point where multiple bands meet. In the full-zone band structure approach, it is usually assumed that the particle follows the same momentum as it crosses the meeting point. This is shown by the red arrows. At the zone center, variations may occur, as discussed in the text.

integrated. Nevertheless, the use of this code allowed the pair to improve the knowledge of band-structure parameters, masses, and phonon coupling constants in the principal valleys of the conduction band. Further work has allowed them to extend this development to a wider set of semiconductors and semiconductor devices.

Another advance in full band calculations was made in the development of what is known as cellular Monte Carlo, in which the scattering rates are computed between cells in the Brillouin zone as a function of the initial momentum, rather than being tabulated with energy cells [65]. This code has been adapted with some quantum corrections and is capable of device simulations, as it also incorporates the Poisson equation [66, 67]. This method has been used extensively to study InGaAs strained channel HEMTs [68, 69]. As the carriers tend to move all valleys of the conduction band (or all bands in the valence band), the full-zone approach is fully warranted.

Probably the first ensemble Monte Carlo (EMC) simulation of a semiconductor transistor was that of an HEMT [70]. This was performed by the same group that had verified the real-space transfer in the AlGaAs/GaAs system [71], so it was a small step to simulate the entire transistor. The authors compared both a normal single delta-doped structure with one that had delta-doping on both sides of the GaAs channel. By varying the doping level in the various layers and the setback distance (of the doping layer from the channel), they were able to vary the threshold voltage of the device by a considerable amount. As a result of these simulations, the authors were already suggesting that cutoff frequencies could be in the neighborhood of 50 GHz for a 0.5 μm gate length. Later simulations of InGaAs strained-layer HEMTs moved this cutoff frequency to more than 140 GHz [72] for a 0.12 μm gate length device, which compared well with experiments that yielded a 114 GHz cutoff [73].

The first EMC simulations for Si MOSFETs were actually done with the full band structure program at IBM [63, 64]. Later work examined EMC simulations for Si MOSFETs, with a focus upon a strained Si on SiGe MOSFET [74]. A focus upon *p*-channel MOSFETs seems to have begun at about the same time [75]. The study of a great many devices via EMC simulations has grown fairly rapidly since these early years. Such simulations have provided a great deal of insight into the physics and the circuit operation of various devices and enhanced our ability to progress to ever smaller dimensions and higher frequencies of operation.

6.9 Phonons: beyond the Born–Oppenheimer approximation

As important as the band structure is, the phonon dynamics are equally important. The motion of the atoms causes scattering of the electrons. In a one-dimensional chain of atoms, exactly as shown in figure 6.2, the motion of the atoms provides the phonon model. The two atoms of the basis, one blue and one red, are assumed to have different masses, M_1 and M_2, respectively. Each atom can move in three dimensions, and with two atoms in the basis of the unit cell, that means that there are six different modes of vibration. In three of these modes, the atoms move together; these are long-wavelength modes, also known as acoustic modes. In the other three modes, the atoms move oppositely to each other, so that the wave has a very short wavelength. These three are the optical modes, so called because of their high frequency. In each case, there is one longitudinal mode in which the displacement is parallel to the propagation vector, taken as $\boldsymbol{q}$ (fully equivalent to the electron wave number $\boldsymbol{k}$). The other two modes in each case are transverse modes, in which the vibration direction is normal to the propagation vector $\boldsymbol{q}$.

There is a basic difference between the situations in which the two atoms are the same and those in which they are different. When the two atoms are the same, all three of the optical modes are degenerate at the Γ point. However, in the III–V and II–VI materials, these two atoms are different, and this is significant. When there are two dissimilar atoms per primitive unit cell, there is a small, but opposite charge, called the *effective charge, e^**, on each of the two atoms. This leads to a dipole force between the two atoms that is extremely effective in scattering the electrons by a coulomb interaction. This dipolar field can interact with external electromagnetic waves, which provides an oscillatory polarization arising from the motion of the two atoms. This polarization contributes to the dielectric function and provides a different permittivity below the characteristic frequency of the oscillation [76]. That is, below the characteristic longitudinal optical (LO) phonon frequency, the polarization adds to the normal value contributed by the core electrons of the atoms. Above the LO frequency, the atoms cannot follow the electromagnetic wave frequency, so this polarization does not contribute. As a result, the dielectric permittivity is smaller at optical frequencies than it is at low frequencies. This is seen in the spectrum of the phonons in the form of a split between the LO mode and the transverse optical (TO) modes at the Γ point. The polarization of the atoms is a longitudinal oscillation, and this is why only the LO mode sees this effect.

The phonon spectrum of bulk GaAs is shown in figure 6.17, although this can be considered to be the form for any generic direct-gap III–V semiconductor. This set of phonon bands was calculated using a valence force field model [46] provided by the 'band structure lab' tool available at NanoHub.org. The split between the LO and TO modes at the zone center Γ is clearly evident. The LO and TO modes cross between Γ and the zone edges. Experimentally, the crossings are a little closer to Γ than in the results shown here [77]. In addition, in group IV materials such as Si or Ge, the LO and LA modes are degenerate at X. This latter degeneracy is broken by the fact that the two atoms in the III–V materials have different masses. The LA branch describes the motion of the heavier of these two atoms, while the LO branch describes the lighter atom's motion. The split in GaAs is found to be only about 1.8 meV, as determined by neutron spectrometry [77]. But the split is also a reflection of the dipolar nature of the diatomic lattice.

The wurtzite structure in both the III–V and II–VI compounds is thought to arise from a relaxation of the zinc-blende lattice. The presence of the *effective* charge

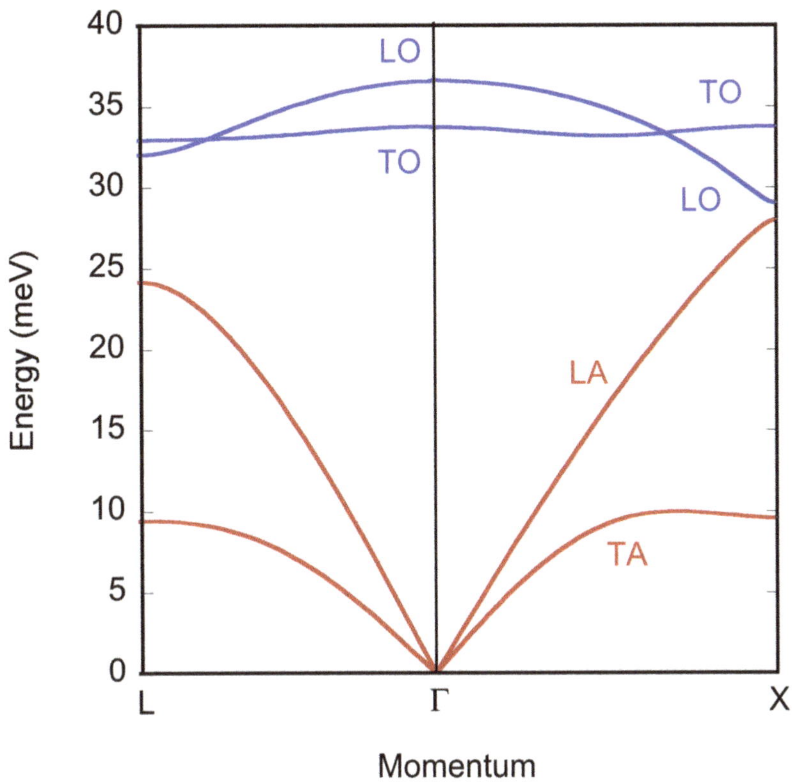

Figure 6.17. The phonon spectrum of GaAs calculated using a valence force field model. Reprinted from [46].

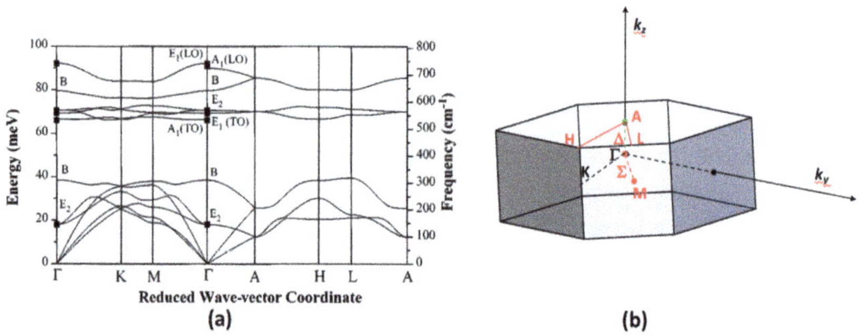

Figure 6.18. (a) The phonon spectrum for GaN in the wurtzite phase. Doubling of the lattice in the z-direction gives four atoms per unit cell and hence 12 phonon modes. Reprinted with permission from [78], copyright (1997) by the American Physical Society. (b) The Brillouin zone for GaN. Reprinted from [88], with the permission of AIP Publishing.

transfer in these compounds, and the resulting ionicity, can destablize the cubic form when this ionicity is too large. The relaxation to the wurtzite structure doubles the unit cell in the z-direction, i.e. normal to the resulting hexagonal planes. This doubling now yields four atoms per unit cell and 12 phonon modes, as shown in figure 6.18(a) [78]. There are still only three acoustic modes, and there are now nine optical modes. The various points in the dispersion curve for GaN in panel (a) can be identified from the Brillouin zone in panel (b). The phonons may be divided into modes that lie in the basal plane of the material and modes with polarizations normal to the basal plane. Since there are nine modes in figure 6.18(a), there are three pairs of LO–TO modes. The A_1 mode is characterized by the movement of the two N atoms in one direction and the movement of the two Ga atoms in the opposite direction, in which the motion is parallel to the (0001) axis (the c-axis). The E_1 mode is a similar vibration but the motion now lies in the (x,y) plane. The E_2 modes also lie in the latter plane, but now one of the atoms (e.g. Ga) is immobile (fixed) while the other atoms are moving. The B modes are similar to the E_2 modes, but the motion is now parallel to the c-axis. Normally, the low-energy B and E_2 (low) modes are known as dark modes that do not couple to light.

The phonons are always quantized. The periodicity of the lattice discretizes the wave numbers, as given in (6.5); the reasons for this are the same as the reasons for the discretization of the energy bands. Each distinct wave number and its corresponding energy (which depends upon the mode under consideration) are quantized as a harmonic oscillator. Thus, the phonon occupation number of that mode describes how many harmonic oscillator states are occupied in that particular case.

6.9.1 Nonequilibrium phonons

When a semiconductor is illuminated by intense laser radiation, nonequilibrium populations, particularly those of the longitudinal optical (LO) modes, can form in polar materials [79, 80]. This is especially the case at low lattice temperatures. The laser creates electron–hole pairs with considerable excess energy, particularly for the electrons in direct-gap semiconductors. These carriers then cool by the emission of

LO phonons, particularly the polar LO modes in compound semiconductors [81]. These phonons have extremely low group velocities, so they do not move far from where they are created. If sufficient energy is transferred from the relaxing carriers to the phonon system, the phonon distribution can be driven far out of equilibrium. The resulting population can be studied by many methods, one of which is Raman scattering. Here, the phonon population may be estimated from the relative strengths of the Stokes and anti-Stokes transitions [81].

Outside of optical studies, nonequilibrium phonons and possible so-called phonon *bottlenecks* have been found in semiconductor devices. These have been studied for some time. Hot phonon populations for the optical modes as well as the acoustic modes have been studied in Si MOSFETs [82], GaN HEMTs [83–86], as well as carbon nanotubes [87]. The relaxation or decay of these out-of-equilibrium phonons occurs through three phonon processes that primarily involve the third-order elastic constants [88]. The decay of the optical modes and the possibility of a bottleneck is the same whether the dynamic carriers are excited by photons or by an electric field in the semiconductor device. The physics of the phonon decay and lifetime are the same.

With nonequilibrium phonons, the resulting phonon distribution, normally the Bose–Einstein distribution, is driven out of equilibrium for those wave numbers that can be excited by the electrons. This nonequilibrium distribution does not have the traditional Bose–Einstein shape, as there are preferred wave numbers that are found in the scattering process. That is, there is a minimum and a maximum value of the phonon momentum q that can be excited by the electron emission. This leads to a very narrow distribution of nonequilibrium phonons. These limits may be found from the energy conservation conditions on electron–phonon scattering [46]. For the emission of a phonon, the limits are

$$k - \sqrt{k^2 - \frac{2m_c\omega_q}{\hbar}} < q < k + \sqrt{k^2 - \frac{2m_c\omega_q}{\hbar}}, \tag{6.82}$$

and for the absorption of a phonon, they are

$$\sqrt{k^2 + \frac{2m_c\omega_q}{\hbar}} - k < q < \sqrt{k^2 + \frac{2m_c\omega_q}{\hbar}} + k, \tag{6.83}$$

where ω_q is the phonon energy (of the particular harmonic oscillator for that mode and state).

The phonon population at a particular value of wave number satisfies a gain–loss equation that can be written as

$$\frac{dN(q)}{dt} = G(q) - \left.\frac{dN(q)}{dt}\right|_{\mathrm{sc}}, \tag{6.84}$$

where

$$G(q) = \Gamma_{\mathrm{em}} - \Gamma_{\mathrm{abs}} \tag{6.85}$$

is the difference between the phonon emission and phonon absorption processes [46]. The scattering rate for the emission of a phonon is usually a few times 10^{12} s^{-1}, so that it takes a few picoseconds for the hot carriers to emit 10–15 phonons. In this time, the carrier may reabsorb only about two to three phonons, so that there is a net gain in the phonon population. As the phonons are heated, these statistics change. The two processes approximately depend upon the population as follows:

$$\begin{aligned}\Gamma_{em} &= \frac{me^2}{4\pi\gamma\hbar^2\omega_{LO}}(N_{LO}+1)W_{em}(k)\\ \Gamma_{abs} &= \frac{me^2}{4\pi\gamma\hbar^2\omega_{LO}}N_{LO}W_{abs}(k)\end{aligned}, \tag{6.86}$$

where γ is given by

$$\frac{1}{\gamma} = \omega_{LO}^2\left(\frac{1}{\epsilon_\infty} - \frac{1}{\epsilon(0)}\right), \tag{6.87}$$

and describes the size of the atomic polarization. The W terms are given by

$$W = \ln\left(\frac{q_{max}^2 + q_0^2}{q_{min}^2 + q_0^2}\right) - \frac{1}{2}\frac{4q_0^2\sqrt{k^2 \mp \frac{2m\omega_q}{\hbar}}}{\left(\frac{2m\omega_q}{\hbar} \mp q_0^2\right)^2 + 4k^2q_0^2}, \tag{6.88}$$

and q_{max} and q_{min} are given by (6.82) and (6.83). The upper and lower signs arise from these latter equations as well. The factor q_0 is the Debye screening wave number for the coulomb scattering process.

Generally, the bottleneck occurs when the emission and absorption processes for the polar LO phonons become comparable. In this situation, (6.84) shows that the energy relaxation process for the excited 'hot' carriers is slowed as the right-hand side becomes small. According to (6.86), this can occur when $N_{LO} \gg 1$. In general, one can use the lifetime for the phonons given by the right-hand side of (6.84) to express the relaxation-time approximation as follows:

$$\left.\frac{dN(q)}{dt}\right|_{sc} = \frac{N_{LO} - N_{eq}}{\tau_{LO}}, \tag{6.89}$$

where N_{eq} is the population in equilibrium and τ_{LO} is the phonon lifetime. This lifetime has been measured for a great many materials [88].

6.9.2 Treating phonons as particles

The idea of treating the phonons as distinct particles seems to have arisen from a similar treatment of gas particles [89], which appeared at about the same time as the first Monte Carlo method for semiconductors [90]. This was taken further using a Monte Carlo random number process in real space for the phonons [91, 92]. Phonons are decidedly more complicated than gas particles because, first, they can multiply as the distribution goes out of equilibrium and the hot LO modes decay into a pair of lower-energy modes. Second, momentum conservation is complicated by the ability of *Umklapp* processes to affect such interactions (in periodic media, momentum is only conserved up to a reciprocal lattice vector). This means that the collision model is not a simple one, due to the fact that phonons can be created and destroyed. In the latter work, a set of real-space cells was adopted, and phonon particles were allowed to propagate from one cell to another during the various drifts and interactions [91]. Within a cell, the number and types of phonons determined the amount of energy in the cell, and this was tied to an assumed distribution with an effective temperature. This method was used to study heat propagation from a higher-temperature boundary condition.

A proper EMC approach was adopted somewhat later to study the LA and TA modes in a thin film of material [93], and the growth of the phonon populations was handled by a renormalization procedure governed by the assumed temperature of the film. Heat transfer due to a phonon Boltzmann equation for these same modes was studied for Si and Ge nanostructures with a similar EMC approach using a relaxation-time approximation for phonon relaxation [94]. Further studies of heat transfer from these modes followed [95, 96]. Such an approach was later used to study the thermoelectric properties of Si nanowires, again focusing upon the acoustic modes of the lattice [97]. The role of surface scattering in thermal conductivity was investigated in Si nanowires [98]. In more recent work, the phonon Boltzmann equation has been solved using the cellular Monte Carlo approach (a variant of EMC which is thought to be faster at the expense of more storage in the computer) [99].

Optical modes were studied via the EMC technique to examine their role in mid-infrared quantum-cascade lasers [100]. Here, optical phonon absorption is an important process in exciting carriers from the ground state of one injection layer to the lasing level in the next stage. Phonon emission is also present and is responsible for driving the phonons out of equilibrium for lattice temperatures below 200 K. The EMC approach was extended to include optical phonons in bulk Si using several models for the generation of sampling techniques [101].

While the particle technique has been shown to be quite useful and to give good results, it does not yet seem to have been extended to semiconductor devices. Considerably more work is needed in this area.

References

[1] Hare A C 1833 *J. Franklin Inst.* **15** 362
[2] Wick F G 1908 *Phys. Rev., Ser. 1* **27** 238
[3] Hohenberg H and Kohn W 1964 *Phys. Rev.* **136** B864
[4] Kohn W and Sham L J 1965 *Phys. Rev.* **140** A1113
[5] Hellmann H 1935 *J. Chem. Phys.* **3** 61
[6] Slater J C and Koster G F 1954 *Phys. Rev.* **94** 1498
[7] Papaconstantopoulos D A and Mehl M J 2003 *J. Phys.: Condens. Matter* **15** R413
[8] Phillips J C and Kleinman L 1959 *Phys. Rev.* **116** 287

[9] Dirac P A M 1967 *The Principles of Quantum Mechanics* 4th edn (Oxford: Oxford University Press)
[10] Merzbacher E 1970 *Quantum Mechanics* 2nd (New York: Wiley)
[11] Brillouin L 1953 *Wave Propagation in Periodic Structures* (Toronto: Dover)
[12] Mathew P T and Fang F 2018 *Engineering* **4** 760
[13] DeMarco K R, Bekker S and Vorobyov I 2019 *J. Physiol. Lond.* **597** 679
[14] Shamin S *et al* 2016 *Nano Lett.* **16** 5779
[15] Saffarzadeh A, Demir F and Kirczenow G 2018 *Phys. Rev.* B **98** 115436
[16] Vukajlovic-Plestina J *et al* 2019 *Nat. Commun.* **10** 869
[17] Hathwar R, Saraniti M and Goodnick S M 2016 *J. Appl. Phys.* **120** 044307
[18] Chang L L, Esaki L and Tsu L 1974 *Appl. Phys. Lett.* **24** 593
[19] Sollner T C L G *et al* 1983 *Appl. Phys. Lett.* **43** 588
[20] Ferry D K 2021 *Quantum Mechanics* (Boca Raton, FL: CRC Press) 3rd
[21] Reuter M G 2017 *J. Phys.: Condens. Matter* **29** 053001
[22] Kohanoff J 2006 *Electronic Structure Calculations for Solids and Molecules* (Cambridge: Cambridge University Press)
[23] Sankey O F and Niklewski D J 1989 *Phys. Rev.* B **40** 3979
[24] Lewis J P *et al* 2011 *Phys. Status Solidi* B **248** 1989
[25] Soler J M *et al* 2002 *J. Phys.: Condens. Matter* **14** 2745
[26] Osbourn G C and Smith D L 1979 *Phys. Rev.* B **19** 2124
[27] Okada J, Hashimoto F and Mori N 2021 *Jpn. J. Appl. Phys.* **60** 091002
[28] Fischetti M V *et al* 2013 *J. Phys.: Condens. Matter* **25** 473202
[29] Bhattacharya S *et al* 2020 *ACS Omega* **5** 20874
[30] Chelikowsky J R and Cohen M L 1976 *Phys. Rev.* B **14** 556
[31] Kresse G and Furthmüller J 1996 *Phys. Rev.* B **54** 11169
[32] Giannozzi P *et al* 2009 *J. Phys.: Condens. Matter* **21** 395502
[33] Chelikowsky J, Chadi D J and Cohen M L 1973 *Phys. Rev.* B **8** 2786
[34] Pandey K C and Phillips J C 1974 *Phys. Rev.* B **9** 1552
[35] Bloom S and Bergstresser T K 1968 *Solid State Commun.* **6** 465
[36] De A and Pryor C E 2010 *Phys. Rev.* B **81** 155210
[37] Pötz W and Vogl P 1981 *Phys. Rev.* B **24** 2025
[38] Jones R O 1966 *Proc. Phys. Soc.* **89** 443
[39] Welland I and Ferry D K 2019 *Semicond. Sci. Technol.* **34** 064003
[40] Kane E O 1957 *J. Phys. Chem. Solids* **1** 249
[41] Kane E O 1966 *Semiconductors and Semimetals* ed R K Willardson and A C Beer (New York: Academic)1 75 pp
[42] Yu P Y and Cardona M 2005 *Fundamentals of Semiconductors* 3rd edn (Heidelberg: Springer) sec. 2.6
[43] Dresselhaus G, Kip A F and Kittel C 1955 *Phys. Rev.* **98** 368
[44] Madelung O 1996 *Semiconductors—Basic Data* (Berlin: Springer)
[45] Anderson P W 1958 *Phys. Rev.* **109** 1492
[46] Ferry D K 2020 *Semiconductors: Bonds and Bands* 2nd edn (Bristol: IOP Publishing)
[47] Ma N and Jena D 2014 *Phys. Rev.* X **4** 011043
[48] Zhu W J *et al* 2014 *Nat. Commun.* **5** 3087
[49] Xiao Z, Song J, Ferry D K, Ducharme S and Hong X 2017 *Phys. Rev. Lett.* **118** 236801
[50] Davisson C and Kunsman C H 1921 *Science* **54** 522

[51] Thomson G P and Reid A 1927 *Nature* **119** 890
[52] Smith R A 1961 *Wave Mechanics of Crystalline Solids* (London: Chapman and Hall)
[53] Zawadzki W 2013 *Acta Phys. Pol.* A **123** 132
[54] Kittel C 1963 *Quantum Theory of Solids* (New York: Wiley) pp 227
[55] Thompson S E *et al* 2004 *IEEE Electron Device Lett.* **25** 191
[56] Ungersboeck E *et al* 2008 *Math. Comput. Simul.* **79** 1071
[57] Dar S *et al* 2007 *IEEE Trans. Nanotechnol.* **6** 97
[58] Zorman B *et al* 2004 *J. Comput. Electron.* **3** 351
[59] Krishnan S, Fischetti M and Vasileska D 2006 *J. Vac. Sci. Technol.* B **24** 1997
[60] Conzatti F *et al* 2009 *Solid-State Electron.* **53** 706
[61] Shichijo H and Hess K 1981 *Phys. Rev.* B **23** 4197
[62] Cohen M L and Bergstresser T K 1966 *Phys. Rev.* **141** 789
[63] Laux S E and Fischetti M V 1988 *IEEE Electron Device Lett.* **9** 467
[64] Fischetti M V and Laux S E 1988 *Phys. Rev.* B **14** 9721
[65] Rein A *et al* 1994 *Proc. Int. Electron Devices Meeting* (New York: IEEE Press) pp 351–4
[66] Wigner S J and Goodnick S M 2000 *Superlattices Microstruct.* **27** 417
[67] Yamakawa S *et al* 2004 *J. Comput. Electron.* **3** 299
[68] Ayubi-Moak J S *et al* 2007 *IEEE Trans. Electron Devices* **54** 2327
[69] Akis R *et al* 2008 *J. Phys.: Condens. Matter* **20** 384201
[70] Tian H, Kim K W and Littlejohn M A 1991 *J. Appl. Phys.* **70** 4593
[71] Glisson T H *et al* 1980 *J. Appl. Phys.* **51** 5445
[72] Babiker S *et al* 1998 *VLSI Des.* **8** 313
[73] Cameron N I *et al* 1995 *IEEE Microwave Theory and Technical Symp. Digest* (New York: IEEE Press) pp 435–8
[74] Formicone G F, Vasileska D and Ferry D K 1997 *Solid-State Electron.* **41** 879
[75] Formicone G F, Vasileska D and Ferry D K 1998 *VLSI Des.* **6** 167
[76] Fröhlich H 1954 *Adv. Phys.* **3** 325
[77] Waugh J L T and Dolling G 1963 *Phys. Rev.* **132** 2410
[78] Siegle H *et al* 1997 *Phys. Rev.* B **55** 7000
[79] Vella-Coleiro G P 1969 *Phys. Rev. Lett.* **23** 697
[80] Shah J, Leite R C C and Scott J F 1970 *Solid State Commun.* **8** 1089
[81] Shah J 1978 *Solid-State Electron.* **21** 43
[82] Aksamija Z and Ravaioli U 2006 *J. Comput. Electron.* **5** 431
[83] Barker J M, Ferry D K and Goodnick S M *et al* 2004 *J. Vac. Sci. Technol.* B **22** 2045
[84] Sadi T and Kell R W 2007 *IEEE Electron Device Lett.* **28** 787
[85] Leach J H, Zhu C Y and Wu M *et al* 2009 *J. Appl. Phys.* **95** 223504
[86] Ashok A, Vasileska D, Hartin O L and Goodnick S M 2010 *IEEE Trans. Electron Devices* **57** 562
[87] Auer C, Schürrer F and Ertler C 2007 *J. Comput. Electron.* **6** 325
[88] Ferry D K 2021 *Appl. Phys. Rev.* **8** 021324
[89] Bird G A 1963 *Phys. Fluids* **6** 1518
[90] Kurosawa T 1966 *J. Phys. Soc. Jpn. Suppl.* **21** 424
[91] Klitsner T *et al* 1988 *Phys. Rev.* B **38** 7576
[92] Peterson R B 1994 *J. Heat Transfer* **116** 815
[93] Mazumder S and Majumdar A 2001 *J. Heat Transfer* **123** 749
[94] Lacroix D, Joulian K and Lemmonier D 2005 *Phys. Rev.* B **72** 064305

[95] Randrianalisoa J and Baillis D 2008 *J. Heat Transfer* **130** 072404
[96] Mei S *et al* 2014 *J. Appl. Phys.* **116** 164307
[97] Ramayya E B *et al* 2012 *Phys. Rev.* B **86** 115328
[98] Maurer L N *et al* 2015 *Appl. Phys. Lett.* **106** 133108
[99] Sabatti F F M, Goodnick S M and Saraniti M 2017 *J. Heat Transfer* **139** 032002
[100] Shi Y B and Knezevic I 2014 *J. Appl. Phys.* **116** 123105
[101] Sabatti F F M, Goodnick S M and Saraniti M 2020 *J. Heat Transfer* **142** 012402

Chapter 7

Electron interactions with fields: the electrostatic approximation

Charged particles, such as electrons, interact with electromagnetic fields. An intuitive and prototypical example is to imagine a positively biased contact and a negatively biased contact. The electric field between the contacts attracts negatively and positively charged particles to the contacts. In chapter 8, a formal perspective on electric fields shows that the motion of these charges introduces a magnetic field, and vice versa, and that this coupling of the electric and magnetic field components, which forms an electromagnetic field, is described by Maxwell's equations. An important question, however, is: is it always necessary to model the fully coupled electromagnetic field problem found via solving Maxwell's equations? The answer is that it is not, as long as the electric/magnetic field components are constant in time (or only change slowly so that the changes are negligible). As will be seen in chapter 8, this considerable simplification is known as the electrostatic (or magnetostatic) approximation (also known as the quasi-stationary case). This quasi-stationary case significantly simplifies the modeling and solution process, as the time-dependent field terms can be set to zero. This important approximation has been used since the beginning of device simulations. Back then, operating frequencies were very limited, but to this day, a large body of semiconductor devices operate at frequencies which warrant the electrostatic view. Today, however, an ever-increasing amount of research is devoted to high-frequency problems in which semiconductor devices are applied. In such cases, the quasi-stationary assumption is obviously no longer valid, as will be discussed in detail in chapter 8. Still, the electrostatic approximation is an important special case, and some light will be shed on the consequences of this approximation when we consider the prototypical numerical solution approach via finite differences.

First, however, it is necessary to establish a rule regarding the applicability of the electrostatic approximation. In essence, one must ensure that the characteristic length of the semiconductor device is considerably smaller than the shortest electromagnetic

doi:10.1088/978-0-7503-5237-6ch7

wavelength in the device. For example, consider a silicon device (the relative permittivity of silicon is $\epsilon_r = 11.68$). In principle, an electromagnetic wave is thus able to propagate in this device at a speed of $c = 1/\sqrt{\mu_0\epsilon_0\epsilon_r} \approx 88 \times 10^6 \mathrm{m\ s^{-1}}$ (ϵ_0 and μ_0 are constants that refer to the vacuum permittivity and the vacuum permeability, respectively). Now consider an upper limit of $f = 100$ GHz for the frequency of the electromagnetic field. This would result in a wavelength of $\lambda = c/f \approx 880\ \mu\mathrm{m}$. The characteristic device length L in micro- and nanoelectronics is in the order of 1 μm to 1 nm: in the first case, L is roughly three orders of magnitude smaller than the wavelength; in the second case, it is roughly six orders of magnitude smaller. In both cases, L can thus be considered to be much smaller than λ, leading to the conclusion that the electrostatic approximation is warranted in the case considered here.

As hinted before, the electrostatic approximation allows one to set the time-dependent field terms to zero. As will be discussed in more detail in chapter 8, the focus here will be on the primary consequence of this simplification, i.e. the ability to establish an equation for the electric field strength based on introducing the electric scalar potential V: $\boldsymbol{E} = -\nabla V$.

In section 7.1, it will be seen that the link between the electric field and the scalar potential allows one to establish a convenient relation between the potential and the charge distributions within a semiconductor device by formulating Poisson's equation. The textbook approach of discretizing this equation with a finite-difference approach will then be introduced. The challenge of calculating the electric field strength using the same grid as for the electric scalar potential will be discussed. Section 7.2 discusses the linkage of Poisson's equation to a transport equation system to get a complete description of the transport processes. The continuity equations for both charge carrier types are defined and generation–recombination phenomena are discussed. The widely used drift-diffusion transport model is introduced and discretization aspects are considered. Section 7.3 investigates the challenge of apportioning the charge of particles to a grid in order to solve Poisson's equation. The approximative nature is highlighted and different methods are discussed. Section 7.4 introduces boundary conditions which are essential for modeling practically relevant, real-world electronic devices. The added challenge of handling boundaries in a particle transport picture is highlighted and some approaches are mentioned.

7.1 Poisson's equation and charge distributions

Using the previously discussed electrostatic approximation, the electric field strength $\boldsymbol{E}(\boldsymbol{r})$, which is a vector field, is expressed as the gradient of a scalar potential field $V(\boldsymbol{r})$ as follows:

$$\boldsymbol{E}(\boldsymbol{r}) = -\nabla V(\boldsymbol{r}) \tag{7.1}$$

Inserting this equation into Gauss' law (see section 8.1) and considering the relation between flux density and electric field

$$\boldsymbol{D}(\boldsymbol{r}) = \epsilon \boldsymbol{E}(\boldsymbol{r}) \tag{7.2}$$

yields Poisson's equation:

$$\nabla^2 V(\boldsymbol{r}) = -\frac{\rho(\boldsymbol{r})}{\epsilon}. \tag{7.3}$$

Here, a simplified case for the electric permittivity is considered in which $\epsilon = \epsilon_r \epsilon_0$, i.e. homogeneous, linear, and isotropic matter. Inhomogeneous, nonlinear, and anisotropic matter would transform this simplified scalar value into a significantly more complicated tensor field which could locally depend on other quantities, e.g. $\boldsymbol{E}(\boldsymbol{r})$.

The space charge density $\rho(\boldsymbol{r})$, which is a scalar field in the mathematical sense, is particularly important for modeling semiconductor devices, as it is primarily composed of the charges of electrons $n(\boldsymbol{r})$, holes $p(\boldsymbol{r})$, ionized donors $N_D(\boldsymbol{r})$, and ionized acceptors $N_A(\boldsymbol{r})$ (multiplied by the elementary charge constant e):

$$\rho(\boldsymbol{r}) = e[p(\boldsymbol{r}) - n(\boldsymbol{r}) + N_D(\boldsymbol{r}) - N_A(\boldsymbol{r})]. \tag{7.4}$$

As will be discussed in section 7.2, this equation allows one to consider: (i) self-consistently linking Poisson's equation with a transport equation system via $n(\boldsymbol{r})$ and $p(\boldsymbol{r})$, and (ii) the introduction of specific doping levels via $N_D(\boldsymbol{r})$ and $N_A(\boldsymbol{r})$, which are important engineering quantities. The doping levels are typically provided by a preceding process simulation (along with the generated device discretization, i.e. the simulation grid, often also referred to as mesh), in which specific doping models are used to predict the distributions of dopants [1, 2].

The numerical solution of Poisson's equation is then relatively simple to achieve. A particularly intuitive approach is to use a finite-difference scheme. A motivating target application would be the two-dimensional (2D) simulation of a metal–semiconductor field-effect transistor (MESFET). It is important to note that the discretization approaches shown in the following can be directly extended to problems of other dimensionality, such as three-dimensional (3D) problems. In this case, a similar z-component would have to be added to the x/y-components shown for 2D problems. Figure 7.1 shows a schematic representation of an

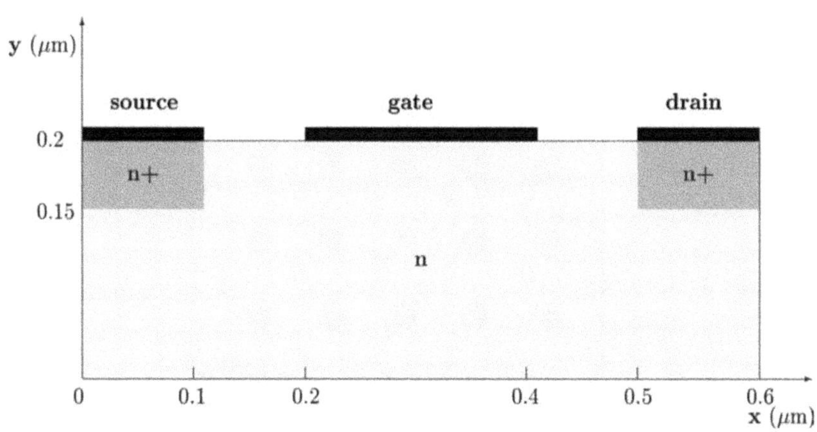

Figure 7.1. A schematic representation of a 2D MESFET. Reprinted from [3], Copyright (2005), with permission from Elsevier.

exemplary geometry [3]. The typical scenario is such that for a given space charge density ρ and a set of boundary conditions (in this case referring to the applied contact voltages at the gate, source, and/or drain), it is desired to calculate the electric field and scalar potential values according to Poisson's equation. First, however, a method of assigning these given values to every point within the virtual device is required to solve Poisson's equation. For this, the geometry needs to be discretized to be able to store the space charge density, the electric field, and the scalar potential (the latter of which is also needed for boundary conditions) in a manner that does not require infinite capacity. Therefore, the geometry shown in figure 7.1 must be decomposed into a finite set of points and connections. This discretization into a set of points is known as the (previously mentioned) *grid*. Many different methods can be used to discretize any given geometry. For finite-difference methods, regular grids in general are particularly advantageous, as they implicitly provide point-neighborhood relationships via index calculations. For the sake of simplicity, consider a Cartesian grid, which is a special case of a regular grid with equidistantly sized grid elements: the points of the grid are used to associate values with the field data. This grid type allows one to realize straightforward integer-based indexing that identifies the individual points using an equidistant point spacing of d. Figure 7.2 shows a representative schematic of a subset of grid points, allowing a focus to be placed on the approximations for the partial derivatives. In practice, a much larger number of grid points than that shown in the figure would be necessary to cover the entire 2D MESFET device geometry of figure 7.1.

Using the grid in figure 7.2, the scalar potential $V(\boldsymbol{r})$ is connected to the space charge density $\rho(\boldsymbol{r})$ at the grid points, and (7.3) can be reformulated to:

$$\frac{\partial^2 V(i, j)}{\partial x^2} + \frac{\partial^2 V(i, j)}{\partial y^2} = -\frac{\rho(i, j)}{\epsilon}. \tag{7.5}$$

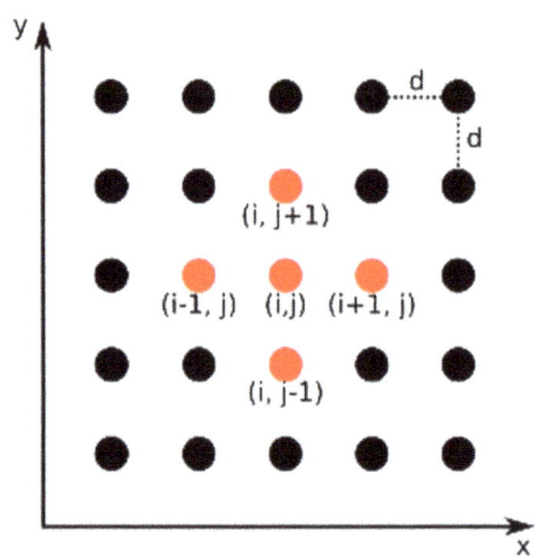

Figure 7.2. A schematic of a Cartesian grid using a five-point stencil (red points). All points are equidistantly positioned from each other at a distance of d. The axes are aligned with the axes used in figure 7.1.

The approximations used to generate the finite differences are found by expanding the scalar potential in a Taylor series. There are many ways of doing these expansions, but the most common is a normal type of second-order approximation. Second-order approximations for the second derivatives may be carried out in a forward or backward fashion. However, the centered versions are often preferred due to their inherent symmetry. The most straightforward approach to the second-order derivatives is found to be a centered three-point stencil for each dimension:

$$\frac{\partial^2 V(i,j)}{\partial x^2} \approx \frac{V(i-1,\ j) - 2V(i,j) + V(i+1,j)}{d^2} \tag{7.6}$$

and

$$\frac{\partial^2 V(i,j)}{\partial y^2} \approx \frac{V(i,\ j-1) - 2V(i,j) + V(i,j+1)}{d^2} \tag{7.7}$$

Inserting (7.6) and (7.7) into (7.5) results in the five-point stencil shown in figure 7.2 (in red) and, after rearrangement, this yields:

$$V(i,j) = \frac{1}{4}[V(i-1,j) + V(i+1,j) + V(i,j-1) + V(i,j+1)] + \frac{d^2}{4\epsilon}\rho(i,j). \tag{7.8}$$

The scalar potential at a given point thus linearly depends on the space charge density at the same point and on the scalar potentials of its surrounding points. This approach to the solution of Poisson's equation can be reformulated into a vector–matrix form $Ax = b$, where x contains the voltage values $V(i_n, j_n)$ on the grid points within the structure, A contains the linear relationships of (7.8), and b contains the boundary conditions (see section 7.4) and charge density terms. Solving for x is then realized by $x = A^{-1}b$. This would involve inverting A, which is only practical for 'small' problems. Here, the word 'small' depends upon the current state of the art of the computer used for the purpose. Near the end of the last century, a matrix of order 300 or so could be inverted in a few minutes. Today, it would be orders of magnitude faster.

For larger systems, other more efficient solution methods are readily available. In principle, there are many different methods available for solving Poisson's equation, ranging from direct methods (e.g. Gaussian elimination, lower–upper (LU) decomposition), iterative grid relaxation methods (e.g. Jacobi, Gauss–Seidel, successive over-relaxation), and iterative matrix methods (e.g. sparse matrix, conjugate gradient, multigrid). For an overview of many of these methods the reader is referred to [4]. Additionally, grid-free methods have been developed in which the particles are used as grid nodes and the solution is based on a Fourier series expansion coupled with point matching [5]. More recently, machine-learning approaches have been considered [6].

After we have solved Poisson's equation using one of the above methods, the scalar potential $V(i_n, j_n)$ is available at all points of the grid. The question now becomes: how can (7.1) be utilized to calculate the electric field strength? To achieve this, it is necessary to first separate $\boldsymbol{E}(\boldsymbol{r})$ and the gradient operator into individual components:

$$E(r) = E_x(r)a_x + E_y(r)a_y = -\frac{\partial V(r)}{\partial x}a_x - \frac{\partial V(r)}{\partial y}a_y \tag{7.9}$$

Contrary to the above approach to discretizing Poisson's equation, here, the first derivative now lies between each grid point, essentially creating a virtual second grid that can also be discretized in many ways. Again, forward, backward, and centered approximations are available. The choice of which one to use here is more subtle and requires one to look closely at the options.

First consider a forward-difference approach, which may be written in the same manner as the finite-difference forms above (e.g. it arises from formal Taylor series expansions of the scalar potential)

$$E_x(i, j) = -\frac{V(i+1,\ j) - V(i, j)}{d}, \tag{7.10}$$

and

$$E_y(i, j) = -\frac{V(i,\ j+1) - V(i, j)}{d}. \tag{7.11}$$

However, the discretization confirms that the electric field is evaluated between the grid points, as shown in figure 7.3 (the same consequence holds true for the backward finite-difference approach). In addition, the two electric field components are not defined at the same points. This can be remedied, though, by averaging the components. As was hinted before, the averaging introduces a virtual second grid on top of our original grid but shifted in both directions by $d/2$, effectively introducing a new grid interlaced with the original grid, as shown in figure 7.4.

A word of caution is needed at this point. The choice of forward or backward discretization is not particularly important for the electrostatic case. However, it is

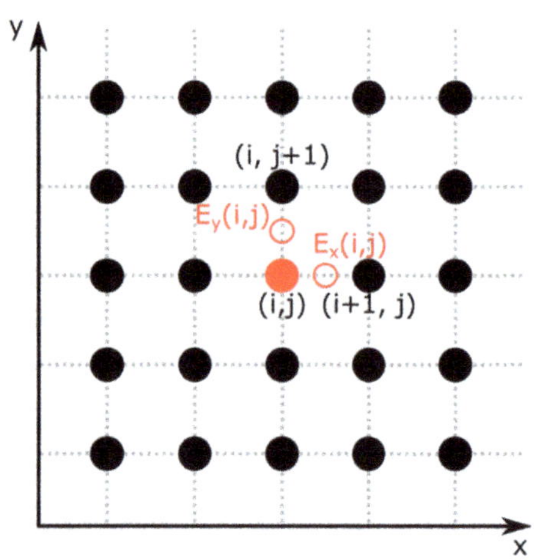

Figure 7.3. In the case of a forward-difference approximation, the electric field components (red circles) are calculated in between neighboring points based on the two scalar potential values.

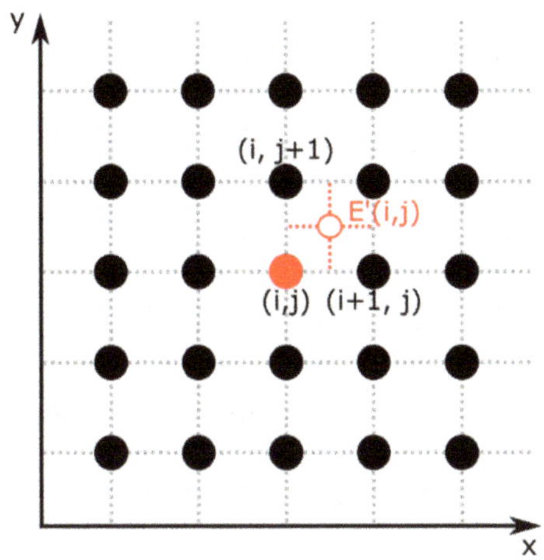

Figure 7.4. The electric field components are averaged so that they can be defined at the same points of the shifted grid.

quite important in the consideration of propagating waves or particles. In this case, the choice is based upon the direction in which the wave is propagating. That is, if the wave is moving to the right (relative to figure 7.3), then (7.10) and (7.11) would be appropriate. However, if the wave were propagating to the left, backward differencing would be preferred. This importance is reflected in quantum transport; an example is the Wigner function approach to transport (see chapter 11). Here, the phase space is usually split between upper (positive momentum) and lower (negative momentum) regions of the discretization; forward differencing is applied to the upper half, where electrons are moving to the right, and backward differencing is applied to the lower half, where electrons are moving to the left.

Finally, the electric field strength $\boldsymbol{E}'(i, j)$ at the new shifted grid is defined by:

$$\boldsymbol{E}'(i, j) = E_x'(i, j) + E_y'(i, j) = \frac{E_x(i, j) + E_x(i, j+1)}{2} + \frac{E_y(i, j) + E_y(i+1, j)}{2} \tag{7.12}$$

This shifted (or interlaced) grid approach is widely applied, particularly for general electromagnetic field solvers based on finite-difference time-domain (FDTD) methods, which were originally proposed by Yee [7] and will be covered again in chapter 8.

Alternatively, however, an approximation that evaluates the electric field components at the original grid points can be generated by using a centered finite-difference approach with twice the point spacing:

$$\begin{aligned} E_x(i, j) &= -\frac{V(i+1, j) - V(i-1, j)}{2d} \\ E_y(i, j) &= -\frac{V(i, j+1) - V(i, j-1)}{2d} \end{aligned} \tag{7.13}$$

Using a Taylor expansion, the approximations are now second order in d (the error is proportional to d^2) as compared to the first-order nature of the forward/backward scheme (in which the error is proportional to d). In other words, the centered approximation is more accurate [8].

7.2 The self-consistency of the transport equation

In principle, predicting the performance of electron devices only requires solving the many-body Schrödinger equation (or a similar transport equation) for the entire ensemble of electrons. Once such a solution is known, all properties of the device can be accessed, and there is no need for a self-consistent loop between the transport equation and the Poisson equation, as the Coulomb interaction among electrons is already accounted for in the many-body transport equation. However, in practice, implementing the many-body transport equations in real simulators is too cumbersome. They are substituted by simpler (time and/or ensemble averaged) equations of motion, where the exact Coulomb interaction among electrons disappears and it is substituted by a mean-field approximation, which is typically taken as the solution of the Poisson equation. To streamline the discussion, this section employs classical or semi-classical transport equations when delving into the self-consistent loop between transport and Poisson equations. In section 7.5, this point will be further elaborated.

Poisson's equation (7.3) needs to be solved self-consistently with a transport equation system to get a complete description of the transport process. In essence, conservation laws are needed for the charge carriers, which can be derived using Gauss' and Ampère's laws with the set of Maxwell's equations. If we consider the vector identity produced by taking the divergence of the curl of any vector field to always be zero, applying the divergence to Ampère's Law yields:

$$\nabla \cdot (\nabla \times \mathbf{H}) = 0 = \nabla \cdot \frac{\partial \boldsymbol{D}}{\partial t} + \nabla \cdot \boldsymbol{J}. \tag{7.14}$$

Rewriting this (to move the divergence inside the partial time derivative) and inserting Gauss' Law yields the continuity equation which expresses local charge conservation:

$$\nabla \cdot \boldsymbol{J} = -\frac{\partial \rho}{\partial t}. \tag{7.15}$$

As in the case of the charge density, which consists of various contributions (7.4), the current density $\boldsymbol{J}$ is composed of an electron current density $\boldsymbol{J}_n$ and a hole current density $\boldsymbol{J}_p$. Consequently, using (7.4) the charge continuity equation can be formulated as:

$$\nabla \cdot \left(\boldsymbol{J}_n + \boldsymbol{J}_p\right) + e\frac{\partial}{\partial t}(p - n) = 0 \tag{7.16}$$

Note that N_D and N_A are time independent and are thus dropped from (7.16). As a sidenote, the continuity equation can be derived using the zeroth moment of the time-dependent Boltzmann transport equation (BTE) [4]. The equation can be further split into two equations, one for each type of charge carrier, by introducing a separation parameter R:

$$\nabla \cdot \boldsymbol{J}_n - e\frac{\partial n}{\partial t} = eR, \tag{7.17}$$

and

$$\nabla \cdot \boldsymbol{J}_p + e\frac{\partial p}{\partial t} = -eR. \tag{7.18}$$

Charges are strictly speaking not generated or destroyed, which is also reflected in the fact that the right-hand side of (7.16) is zero. Every current-contributing electron in a semiconductor leaves a hole behind: since electrons and holes have opposite charges, they appear in their respective charge continuity equation with opposite signs. In n-type material, the positive charge arises from ionized donors, while in p-type material, the negative charge comes from ionized acceptors. In bipolar material, which has both electrons and holes, the separation parameter R can be interpreted as the net recombination rate, which refers to the rate at which electron–hole pairs are generated minus the rate at which they recombine. Formally, R can be defined as

$$R = R_n - G_n = R_p - G_p, \tag{7.19}$$

where R_n, G_n and R_p, G_p are the recombination and generation rates for electrons and holes, respectively. In equilibrium, $R = 0$. For nonequilibrium scenarios this is not the case, and then for accurate descriptions of the transport, specific carrier generation and recombination rate models are mandatory.

Generation–recombination phenomena are the reason for many effects affecting device performance, such as leakage current and device breakdown. As a reminder, for a semiconductor in thermal equilibrium, there is a dynamic balance between generation and recombination processes, leading to equilibrium concentrations (assuming Boltzmann statistics) for electrons n_0 and holes p_0

$$n_0 = N_{\rm C}\exp\left(\frac{E_{\rm F} - E_{\rm C}}{k_{\rm B}T}\right) = n_i\exp\left(\frac{E_{\rm F} - E_i}{k_{\rm B}T}\right), \tag{7.20}$$

and

$$p_0 = N_{\rm V}\exp\left(\frac{E_{\rm V} - E_{\rm F}}{k_{\rm B}T}\right) = n_i\exp\left(\frac{E_i - E_{\rm F}}{k_{\rm B}T}\right), \tag{7.21}$$

where $N_{\rm C}$ and $N_{\rm V}$ are the effective densities of states or band weights of the conduction and valence bands, respectively, $E_{\rm F}$, $E_{\rm C}$, and $E_{\rm V}$ are the Fermi energy, conduction band energy, and valence band energy, respectively (with $E_{\rm C} - E_{\rm V} = E_{\rm g}$

being the bandgap energy), k_B and T are the Boltzmann constant and the electron temperature in degrees Kelvin, respectively, and n_i is the intrinsic carrier concentration. The product of (7.20) and (7.21) yields the square of the intrinsic carrier concentration, allowing one to express n_i as

$$n_i = \sqrt{N_C N_V} \exp\left(-\frac{E_g}{2k_B T}\right). \tag{7.22}$$

Due to the assumed Boltzmann statistics, equations (7.20)–(7.22) are only valid for nondegenerate semiconductors. Any deviation from the equilibrium conditions also drives the relation between generation and recombination out of balance. If $np > n_i^2$, then recombination dominates, and in the case of $np < n_i^2$, then generation dominates.

Generation and recombination can take place for different physical reasons. Examples are photon/phonon absorption/emission, three-particle transitions, and transitions assisted by recombination centers. Charge carrier band transitions require energy, e.g. when an electron is elevated from the valence band to the conduction band, its energy is increased by approximately E_g. This energy must be provided, which, in essence, can have three possible sources: phonons, photons, or collisions. The relevant related processes are [4, 9, 10]: Shockley–Read–Hall generation/recombination (non-radiative recombination at traps, energy exchange via photons) [11, 12], photon recombination (radiative band-to-band, or impurity-to-band [13], leading to photon emission/absorption in direct semiconductors, such as GaAs), as well as impact ionization and Auger generation–recombination (non-radiative, three-particle processes; impact ionization and Auger processes are, in a sense, the inverse processes of one another [9]).

7.2.1 The drift-diffusion transport model

Considering the goal of a self-consistent solution of Poisson's equation (7.3) together with the charge continuity equations (7.17) and (7.18), a current transport model that defines $\boldsymbol{J}_n$, $\boldsymbol{J}_p$ is required. The simplest model is based on the 'drift-diffusion' assumption, which considers two distinct charge carrier transport mechanisms: charge carrier drift due to the presence of an electric field and charge carrier diffusion due to an inhomogeneous distribution of charge carriers, as will be discussed in the following.

7.2.1.1 The charge carrier drift component

The drift component is defined via Ohm's law and links the current density to the electric field via the electric conductivity σ:

$$\boldsymbol{J}^{\mathrm{Drift}} = \sigma \boldsymbol{E} = \sigma_n \boldsymbol{E} + \sigma_p \boldsymbol{E}. \tag{7.23}$$

The equation is also shown in its two respective parts, i.e. the electron density and the hole current density. The electrical conductivity is considered to be a scalar, which is a frequently used approximation valid for semiconductors with cubic symmetry in equilibrium, e.g. silicon. The electrical conductivity is related to the elementary charge, charge carrier density, and mobility μ as follows:

$$\boldsymbol{J}^{\text{Drift}} = en\mu_n\boldsymbol{E} + ep\mu_p\boldsymbol{E}. \tag{7.24}$$

In turn, the mobility is defined for each carrier type using the respective average carrier velocities and the electric field:

$$\overline{v}_n = -\mu_n\boldsymbol{E}; \quad \overline{v}_p = \mu_p\boldsymbol{E}. \tag{7.25}$$

As can be seen, the convention of electrons traveling against the direction of the electric field is captured by the sign of the mobility. For the sake of completeness, equation (7.24) can be reformulated to:

$$\boldsymbol{J}^{\text{Drift}} = -en\overline{v}_n + ep\overline{v}_p. \tag{7.26}$$

7.2.1.2 The charge carrier diffusion component

Minority carrier diffusion is a process in which the carriers seek to establish a thermodynamic equilibrium in an imbalanced physical system, which, in this case, is the variable distributions of charge carriers n, p. Thermodynamic equilibrium would be represented by $\nabla n = \nabla p = 0$ (in this case, charge carriers would be evenly distributed). The charge carrier diffusion process is based on carrier migration from areas of high charge concentration to areas of low charge concentration. By convention, charge carriers move in the opposite direction to the concentration gradient. The diffusion current is particularly relevant to semiconductors, in which the number of charge carriers is much smaller than in metals; in semiconductors, local variations in carrier densities are not equilibrated instantaneously (in contrast to the case of metals), making diffusion current an observable macroscopic quantity.

Formalizing the diffusion current requires the notion of a charge carrier flux density $\boldsymbol{F}$ defined for both charge carrier types, whose direction is opposite to the gradient of the carrier densities:

$$\boldsymbol{F}_n = -D_n\nabla n; \quad \boldsymbol{F}_p = -D_p\nabla p. \tag{7.27}$$

The diffusion coefficients $D_{n,\,p}$ capture the separate diffusion processes of electrons and holes. In turn, the diffusion current density links the charge carrier flux density and the elementary charge through

$$\boldsymbol{J}^{\text{Diffusion}} = -e\boldsymbol{F}_n + e\boldsymbol{F}_p = eD_n\nabla n - eD_p\nabla p. \tag{7.28}$$

The diffusion coefficient $D_{n,\,p}$ is defined by the Einstein relation for thermal equilibrium conditions and for nondegenerate carrier systems in which Boltzmann statistics are valid

$$D_{n,\,p} = \frac{k_{\text{B}}T}{e}\mu_{n,\,p} = V_{\text{T}}\mu_{n,\,p}, \tag{7.29}$$

where V_T refers to the thermal voltage ($V_T \sim 26$ mV at room temperature). In many cases, the electron temperature matches the lattice temperature. However, in nonequilibrium situations, this is not the case, i.e. for situations in which the electrons have a much higher energy (and thus temperature) than the lattice ('hot electrons') [4, 14].

Inserting the electron and hole terms of the drift (7.24) and diffusion (7.28) current densities into the continuity equations (7.17) and (7.18) and using (7.1) and (7.29) together with (7.3) and (7.4) yields a system of three coupled second-order partial differential equations (PDEs) which are the basic semiconductor equations:

$$\nabla^2 V = -\frac{e(p - n + N_D - N_A)}{\epsilon}, \tag{7.30}$$

$$\nabla \cdot \left(n\mu_n \nabla V - V_T \mu_n \nabla n\right) + \frac{\partial n}{\partial t} = -R, \tag{7.31}$$

and

$$\nabla \cdot \left(p\mu_p \nabla V + V_T \mu_p \nabla p\right) - \frac{\partial p}{\partial t} = R. \tag{7.32}$$

In general, equations (7.30)–(7.32) are convenient for numerical solution processes, as the coupled system depends on three unknowns V, n, p being present in all equations. However, a straightforward conservative discretization approach presents some limitations for (7.31) and (7.32), specifically a major stability issue [4], as summarized in the following paragraphs.

The main challenges for a suitable numerical solution procedure for (7.31) and (7.32) are to maintain: (i) charge conservation, (ii) the local positive definite nature of the carrier density, and (iii) the monotonicity of the solution. Charge conservation requires the discretization to capture the flux from one grid node to all others and vice versa. If we consider a Cartesian grid, the obvious approach is to introduce a shifted grid as previously used to evaluate the fields in section 7.1. This results in each grid point being surrounded by a centered 'box' so that the box boundaries pass through the middles of the edges (i.e. connections of the grid points) connecting all neighboring grid points. V, n, p are defined on the grid points and ∇V and ∇n, ∇p of (7.31) and (7.32) can be calculated in a straightforward manner via finite differences and are thus defined at the edge midpoints. However, the issue arises in evaluating n, p (see the first terms on the left-hand sides of (7.31) and (7.32)) at these midpoints; the question is: just how are the quantities defined there? Naively, one could use a linear interpretation

$$n(i, j) \approx \frac{n(i) + n(j)}{2}; \quad p(i, j) \approx \frac{p(i) + p(j)}{2}. \tag{7.33}$$

However, this approach introduces a very strong dependence on the grid spacing (which must be smaller than the Debye length) and would not only demand finely resolved grids for stable results but also small electric fields (strictly speaking, the linear interpretation would only be valid for the special case of $V(j) - V(i) = 0$), which renders the approach entirely impractical.

After van Roosbroeck's work on the drift-diffusion model was published [15], it took 19 years for a numerically stable, alternative discretization to be developed by Scharfetter and Gummel [16]. In essence, this highly successful approach incorporates the potential variations between the grid points involved and the thermal voltage V_T into the discretization of $n, p(i, j)$.

Finally, it has to be emphasized that the drift-diffusion model is a macroscopic transport model. Furthermore, it is the simplest such model (more accurate macroscopic models are available [17]). More specifically, it is an approximation of the BTE, which itself is a microscopic model. The drift-diffusion model can be derived from the BTE via the method of moments. Regardless of its limitations, the drift-diffusion model is one of the most widely used transport models in the device simulation community. It dates back to 1950 [15] and is still widely used today (with modifications) [17–20].

7.2.2 Monte Carlo models

Avoiding the approximations necessary for macroscopic models would require solving the BTE. In highly scaled and/or confined devices, the validity of macroscopic transport models meets its limitation: the motion and behavior of individual particles becomes increasingly important. However, solving the BTE via deterministic methods is computationally very expensive and has proved impractical in many cases. Monte Carlo solution approaches are thus widely used and successful alternatives, particularly because electron transport in semiconductors and metals is dominated by scattering events due to, e.g. impurities and lattice vibrations (phonons), which nicely fit the stochastic nature of the solution process. Ensemble Monte Carlo (EMC) approaches have proved particularly useful because they naturally map to the particle picture and thus the microscopic nature of the BTE, and will be discussed in detail in chapter 9. In EMC approaches, an *ensemble* of particles (one numerical particle essentially models one or more charge carriers) is simulated, which allows the model to describe the time-dependent evolution of the electron and hole distributions under the influence of forces such as external electromagnetic fields.

Today, EMC approaches are widely used for semiclassical device simulations. For instance, an EMC simulator for high-power devices has been developed which couples to a Maxwell solver and uses graphical processing units for the self-consistent solution process [21]. Other recent studies have used EMC approaches to study transistors, e.g. with a focus on variability effects in fin field-effect transistors (FinFETs) [22], tunneling in tunnel field-effect transistors (TFETs) [23], energy distributions of electrons [24] and confinement effects [25] in gate-all-around silicon MOSFETs, and spin transport in MOSFETs [26]. The EMC approach has also been applied in conjunction with a FDTD method (see chapter 8) to study the transport in 2D electronic systems in the presence of high-frequency electromagnetic fields [27, 28]. The EMC approach naturally also lends itself to the study of electron transport characteristics within the context of material studies in combination with density-functional theory tools, as was recently done for Al_4SiC_4 [29].

EMC approaches are also widely used in quantum transport simulations, particularly those based on a Wigner function approach (see chapter 11; for a general overview of particle Wigner methods, the reader is referred to [30, 31]). To that end, different particle models have been developed, such as the affinity [32] and signed-particle models [33]. Very recently, the EMC approach has also been considered for a particle approach to nonequilibrium Green's functions [34].

7.3 Apportioning the charge

In order to continue the discussion of section 7.2 on EMC methods for the transport problem and correctly capture locally varying changes in the electric field and externally applied electric fields, the transport problem must be linked to a grid-based solution of Poisson's equation. To optimize the solution process, the Poisson solution can be decoupled from the particle motion over a suitably small time step, which can be related to the inverse plasma frequency [4]. It is assumed that the field calculated from the solution of Poisson's equation at the previous time step is constant during any given time step and that this field can thus be used for the transport problem. At the end of a time step, Poisson's equation is solved again, and the field is updated in preparation for the subsequent time step's ensemble Monte Carlo solution of the transport problem. This coupled transport–field problem thus requires a particle–grid (also known as particle–mesh) coupling for the charge assignment and the force interpolation. The first part is necessary to couple the transport solver to the field solver by apportioning the charge to the grid points. The latter is required to couple the field solver to the transport solver by interpolating the grid-based forces back to the particles.

For the charge apportionment and the force interpolation, two widely used schemes are available [4, 35, 36]: the nearest-grid-point (NGP) scheme and the cloud-in-cell (CIC) scheme. Figure 7.5 shows an exemplary grid configuration with one grid cell highlighted. Regarding taxonomy, a grid node (shown by black/red dots in figure 7.5) is identified by (i, j), whereas a grid cell is denoted by $\langle i, j \rangle$. Consequently, a grid cell $\langle i, j \rangle$ is composed of the four grid cell nodes (i, j), $(i + 1, j)$, $(i, j + 1)$, and $(i + 1, j + 1)$. It is important to note that some studies have used different approaches to identify a grid cell: some used the cell centers (e.g. [4]) and some used the *bottom-left* node (e.g. [37]). Here, we stick to the latter (which can also be seen from the fact that the indices i, j in $\langle i, j \rangle$ and (i, j) are the same) as we consider the charge density ρ and the scalar potential V to be assigned to the grid nodes (see section 7.1). Conceptually, however, there is no difference, and the schemes can be applied to either scenario. In what follows, two essential assumptions are made: (1) the permittivity ϵ is assigned to each grid cell and considered constant within this cell; (2) the grid has a uniform spacing in each direction. If one of these two assumptions does not apply, then the NGP and CIC schemes introduce erroneous self-forces and/or the spatial accuracy of forces is reduced.

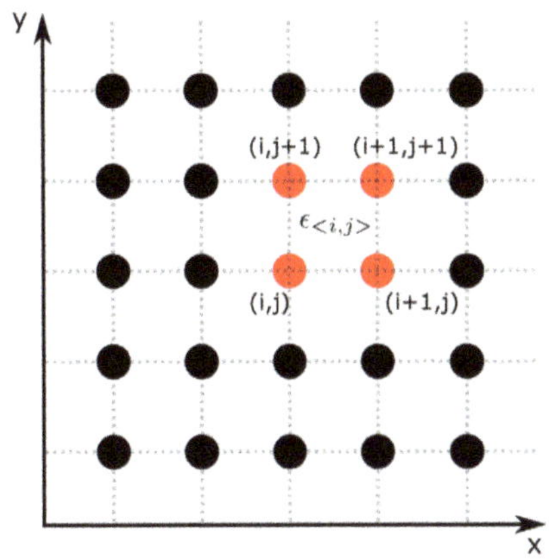

Figure 7.5. The grid node setup considered for the NGP and CIC scheme. The nodes of an exemplary grid $\langle i, j\rangle$ cell are highlighted (red). The charge density and electrostatic potential are considered to be stored on the grid points (black and red). The permittivity values are considered to be piecewise constant and are assigned to the grid cells, here only shown for the cell under consideration $\epsilon_{\langle i, j\rangle}$.

Let us assume that a particle carrying a charge ρ is located at position (x, y) which falls within the highlighted cell $\langle i, j\rangle$ in figure 7.5. The goal is to apportion this charge to the cell's nodes (red nodes). The particle's ρ is distributed to the four grid cell nodes by node-specific weights w_{kl}, where k and l allow one to identify the four grid cell nodes via $k = i, i + 1$ and $l = j, j + 1$. The weight w_{kl} sums to unity for all four nodes, ensuring charge conservation. The charge assignment schemes discussed in the following differ in their approaches to setting these weights.

Using an NGP scheme consists of picking the closest grid node to the particle and assigning the entire particle charge to this node. More formally, the closest node receives $w_{kl} = 1$, while for the other nodes, $w_{kl} = 0$. The NGP scheme offers a simple approach but introduces a noisy charge distribution, which harms numerical stability. By comparison, the CIC scheme considers a finite volume (representing the *cloud* in the scheme's name) for each particle, centered around its position. This volume consequently spans several grid cells and thus a fractional portion of the particle's charge is assigned to each grid node in dependence on the volume's share of a grid node's corresponding cell. The weights are defined as: $w_{i,j} = w_x w_y$, $w_{i+1,j} = (1 - w_x)w_y$, $w_{i,j+1} = w_x(1 - w_y)$, and $w_{i+1,j+1} = (1 - w_x)(1 - w_y)$ using $w_x = (x_{i+1} - x)/(x_{i+1} - x_i)$ and $w_y = (y_{i+1} - y)/(y_{i+1} - y_i)$. The CIC scheme is computationally more costly than NGP but provides better numerical stability due to its smoothing of the charge distribution (as compared to the NGP's discontinuous jump of the charge and thus the force at grid cell boundaries). The CIC scheme improves the situation, as the resulting force is piecewise linear and continuous, but the first derivative jumps [38]. Some improvements for the force interpolations of the NGP and CIC schemes have been suggested which focus on position-dependent permittivity but still use uniform grids [36].

Other schemes have been developed to further alleviate the limitations of NGP and CIC schemes, such as the triangular-shaped cloud (TSC) scheme [34] and the nearest-element-center (NEC) [36] scheme. Both schemes provide advantages with respect to smoothing (and thus better numerical stability) but also introduce more computational complexity into the charge assignment process. The TSC scheme offers the numerical advantage of enabling the first derivative of the force to be continuous. The NEC scheme is particularly suitable for nonuniform grids and spatially varying permittivity (both are de facto standards today). It is also possible to mix different schemes, and this allows one to further optimize the charge assignment process for a particular problem [4, 35], for example, to use different schemes for charge assignment and force interpolation algorithms or to use different schemes for different axial directions.

7.4 Boundary conditions

Finding a unique solution for the previously discussed semiconductor equations (7.30)–(7.32) requires us to define boundary conditions for the solution variables V, n, p. The need for boundary conditions (one type in particular) becomes clear if one thinks of device contacts. A typical device engineering task is to determine the currents at specific contacts for particularly biased contacts. Figure 7.6 shows exemplary output characteristics for a nanowire FET [39]. Every simulation point is the result of solving a transport–field problem, such as (7.30)–(7.32) for a specific set of boundary conditions, e.g. drain, source, and gate voltages. It is important to realize that the boundary conditions are artificial and are a necessity in order to simulate an individual device in a self-contained manner. Formally, one may consider the problem domain (the device to be simulated) as Ω, its boundary $\partial\Omega$, and, by way of example, the solution of Poisson's equation, such as (7.3). There are three types of boundary conditions to consider: Dirichlet, Neumann, and Robin.

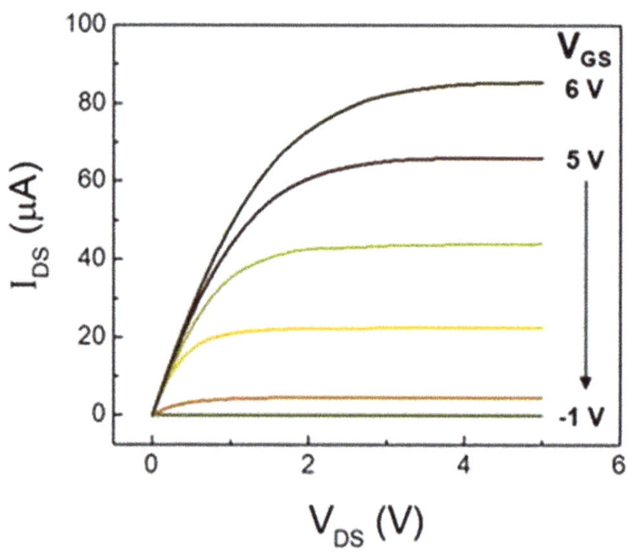

Figure 7.6. The output characteristics of an exemplary nanowire FET device. Reprinted from [39], with the permission of AIP Publishing.

Dirichlet boundary conditions allow one to explicitly set values of the solution $V(\boldsymbol{r})$ on the boundary of the simulation domain

$$V(\boldsymbol{r}) = f(\boldsymbol{r}), \quad \mathbf{r} \in \partial\Omega. \tag{7.34}$$

This directly maps to the example given above, in which a specific voltage is assigned to a device contact. Note that here the desired value $f(\boldsymbol{r})$ is assigned to the entire boundary. Of course, the boundary can be further decomposed into boundary sections $\Gamma_1 \ldots \Gamma_n$ such that the boundary condition can be applied to specific boundary parts, allowing one to mix boundary conditions. This boundary decomposition can be done for all types of boundary conditions. Dirichlet boundary conditions are particularly important for Poisson's equation; finding a unique solution to the simulation problem requires setting at least one Dirichlet boundary condition.

In contrast, Neumann boundary conditions enable one to define a change of the solution variable in the direction perpendicular to the boundary:

$$\frac{\partial V(\boldsymbol{r})}{\partial \boldsymbol{n}} = \boldsymbol{n} \cdot \nabla V(\boldsymbol{r}) = g(\boldsymbol{r}), \quad \mathbf{r} \in \partial\Omega, \tag{7.35}$$

where $\boldsymbol{n}$ is the unit vector perpendicular to the boundary. The physical meaning of (7.35) is that of fluxes coming out of Ω. Homogenous conditions are often used ($g(\boldsymbol{r}) = 0$) to ensure that there are no outgoing fluxes.

Finally, Robin boundary conditions are a linear combination of Dirichlet and Neumann boundary conditions:

$$\alpha(\boldsymbol{r})V(\boldsymbol{r}) + \beta(\boldsymbol{r})\frac{\partial V(\boldsymbol{r})}{\partial \boldsymbol{n}} = h(\boldsymbol{r}), \quad \mathbf{r} \in \partial\Omega. \tag{7.36}$$

This type of boundary condition is much more challenging to apply but essentially allows one to introduce weights that balance the boundary conditions. Robin conditions are more often found in heat transfer or electromagnetism problems.

The availability of the above three boundary conditions allows one to guide the solution process of the coupled semiconductor equations. However, the physical component remains: what values should be used for V, n, p at the boundary? The electronic connection between semiconductor devices and circuits is realized using metallic contacts. As such, the metal–semiconductor interface plays an important role in the correct modeling of the electronic behavior. In essence, the electrical behavior of the interface is determined by the Fermi levels in the metal and the semiconductor. When the two materials come into contact, a charge transfer manifests due to the different Fermi levels. This ultimately introduces a counter-field that leads to identical Fermi levels in both materials in equilibrium. Depending on which material has the higher Fermi level and on other aspects, such as doping concentrations, an ohmic or Schottky-type contact can be realized. The reader is referred to other literature for further details [3, 10].

In the context of particle simulations, boundary conditions require special attention. In principle, reflective or periodic boundary conditions are used to realize charge conservation. However, when the simulation includes ohmic contacts, particles must be allowed to exit the device via the contacts. Again, in order to enforce charge conservation, these particles must then be reintroduced. However, this is not straightforward; different models are available and must be carefully considered, as they provide varying degrees of complexity and physical accuracy [4]. One approach is to assign a doping concentration to contact grid cells and to maintain a fixed number of particles there throughout the entire simulation, as has been suggested in terahertz transistor simulation studies [40]. Carriers are then allowed to be injected/ejected, ensuring charge neutrality [35]. The doping within the contact cells creates a potential drop near the ohmic contacts, which emulates an effective source resistance. This allows one to study the effect of an internal contact resistance by varying the doping concentration.

7.5 Not always so simple

To simplify the discussion, the self-consistent treatment of the Poisson equation and transport equation, developed in the previous sections has been applied to classical or semi-classical transport models. In principle, generalizing the same ideas to quantum transport models can be done straightforwardly. There are numerous examples in literature where the Schrödinger equation is self-consistently solved together with the Poisson equation. In this process, the Schrödinger equation takes the potential profile $V(\boldsymbol{r})$ as an input, providing the wave function as an output, the modulus of which is assumed to represent the 'charge density' $\rho(\boldsymbol{r})$. This 'charge density' is then used as an input for the Poisson equation, yielding the potential $V(\boldsymbol{r})$ as an output. This iterative self-consistent loop eventually converges to the correct potential profile and charge density. However, there are scenarios, both in the quantum and classical regimes, where the application of the aforementioned loop can be misleading.

It is well-known that the Poisson equation and the charge density provide a way to determine the potential profile $V(\boldsymbol{r})$ at position $\boldsymbol{r}$ due to the presence of N electrons $\boldsymbol{r}_1, \boldsymbol{r}_2, \ldots, \boldsymbol{r}_N$ interacting among them. However, such potential profile $V(\boldsymbol{r})$ can also be derived by summing the distinct Coulomb potentials that the N particles create at position $\boldsymbol{r}$. From a computational point of view, the Poisson equation offers advantages over summing the potentials, including the ability to implement various boundary conditions beyond just those resulting in zero potential at infinity. Inaccuracies could appear when $V(\boldsymbol{r})$ is assumed to be the potential profile seen by electrons present in a device. This is not accurate because each electron experiences a distinct Coulomb interaction based on its proximity to other electrons.

To exactly address the Coulomb interaction between electrons, one must refer to the Hamiltonian equations (3.2) and (3.3) in chapter 3. As seen in these equations, including the Coulomb interaction among electrons in such Hamiltonians doesn't necessitate a discussion of the Poisson equation or charge density. The task only involves solving the many-body Schrödinger equation using the many-body Hamiltonian outlined in equation (3.1). As described in chapter 3, solving this many-body problem is exceedingly challenging in most scenarios, thus demanding approximations to simplify the computational burden. But, it is still very instructive to understand such many-body correlations because they allow us to understand the range of validity of some of the typical approximations in classical and quantum electron transport.

The assumption that a single electric field $E(\boldsymbol{r})$ or potential profile $V(\boldsymbol{r})$ suffices for discussing coulomb correlations constitutes a mean-field approximation. Yet, this approximation doesn't universally hold true. The phenomena of Coulomb blockade is an example that defies this mean-field assumption. Let us assume a scenario with two electrons, labeled by $\boldsymbol{r}_1$ and $\boldsymbol{r}_2$, are trying to enter, consecutively, into an initially empty quantum dot. The first electron encounters a specific potential profile, which is referred to here as the 'potential with no electrons' $V_0(\boldsymbol{r})$, facilitating its entry into the quantum dot. However, when a subsequent electron attempts to enter the quantum dot (now with one electron inside), it encounters a different potential, named here the 'potential with one electron', $V_1(\boldsymbol{r})$. If the presence or not of an electron in the quantum dot is electrostatically relevant enough, the second electron finds that $V_1(\boldsymbol{r}) \neq V_0(\boldsymbol{r})$ so that its path to the quantum dot is blocked by the first electron. This phenomenon is known as Coulomb blockade.

The fundamental reasons why different potential is needed for each electron is because the many-body potential profile that appears in the Hamiltonian of (3.2) is a function in the configuration space

$$V(\boldsymbol{r}_1, \boldsymbol{r}_2, \ldots, \boldsymbol{r}_N), \tag{7.37}$$

and not a function in the physical space

$$V(\boldsymbol{r}). \tag{7.38}$$

In other words, a mean field potential $V(\boldsymbol{r})$ is not able to discern which potential is 'seen' by electron $\boldsymbol{r}_1$ and which by $\boldsymbol{r}_2$, while the potential $V(\boldsymbol{r}_1, \boldsymbol{r}_2, \ldots, \boldsymbol{r}_N)$ can distinguish between different configurations of electrons.

Indeed, there exist methods to capture, at least to some extent, these correlations. For instance, in the discourse on quantum transport using Bohmian trajectories, detailed in section 10.5 of chapter 10, the potential profile perceived by each electron is computed through an individual Poisson equation (10.33), wherein the electron's own charge is excluded. Simultaneously, the overall electric field to compute displacement currents (as measured by an external electron) is computed using equation (10.35). Comparable approaches have also been incorporated in the Monte Carlo solution of the Boltzmann equation discussed in chapter 10. As detailed in section 5.3 of chapter 5, a typical approach to go beyond the limitations of the Poisson equation involves breaking down the interparticle Coulomb force for each electron into a long-range and a short-range component. The long-range aspect is integrated using the Poisson equation, as elucidated earlier in this chapter. Conversely, the short-range component is managed via molecular dynamics to prevent the possibility of an electron being influenced by its own charge.

In summary, we have realized that, at the fundamental level, electron transport is described in the 3N-dimensional configuration space $\boldsymbol{r}_1, \boldsymbol{r}_2, \ldots, \boldsymbol{r}_N$ for either classical or quantum transport. The relevance of the quantum 3N-dimensional correlations is evident in the many-body wave function $\Psi(\boldsymbol{r}_1, \boldsymbol{r}_2, \ldots, \boldsymbol{r}_N, t)$, while the classical correlations are manifested, for example, in the 3N-dimensional potential in (7.37). The simplification of discussing electron transport in the 3D physical space r is just a strategy to be able to deal with manageable equations. A byproduct of this strategy is the need for a self-consistency between 3D transport and 3D Poisson equation.

References

[1] Ferry D K 2020 *Transport in Semiconductor Mesoscopic Devices* 2nd edn (Bristol: IOP Publishing)
[2] Selberherr S 1984 *Analysis and Simulation of Semiconductor Devices* (Wien: Springer)
[3] Carrillo J A *et al* 2006 *J. Comput. Phys.* **214** 55–80
[4] Vasileska D and Goodnick S M 2006 *Computational Electronics* (Cham: Springer)
[5] Aksamija Z and Ravaioli U 2006 *J. Comput. Electron.* **5** 459–62
[6] Zhang Z *et al* 2019 Solving Poisson's equation using deep learning in particle simulation of PN junction *Joint Int. Symp. on Electromagnetic Compatibility, Sapporo and Asia-Pacific Int. Symp. on Electromagnetic Compatibility (EMC Sapporo/APEMC)* pp 305–8
[7] Yee K 1966 *IEEE Trans. Antennas Propag.* **14** 302
[8] LeVeque R J 2007 *Finite Difference Methods for Ordinary and Partial Differential Equations* (Philadelphia, PA: SIAM)
[9] Li S S 2006 *Semiconductor Physical Electronics* (New York: Springer)
[10] Jacoboni C 2010 *Theory of Electron Transport in Semiconductors* (Berlin: Springer)
[11] Shockley W and Read W T 1952 *Phys. Rev.* **87** 835–42
[12] Hall R N 1952 *Phys. Rev.* **87** 387
[13] Fröhlich H 1947 *Proc. Roy. Soc. London* A **188** 521
[14] Ferry D K 2021 *Hot Carriers in Semiconductors* (Bristol: IOP Publishing)
[15] Van R W 1950 *The Bell System Technical Journal* **29** 560–607
[16] Scharfetter D L and Gummel H K 1969 *IEEE Trans. Electron Devices* **16** 64–77
[17] Cervenka J *et al* 2022 *Springer Handbook of Semiconductor Devices* ed M Rudan *et al* (Cham: Springer)
[18] Altazin S *et al* 2018 *J. Appl. Phys.* **124** 135501
[19] Felekidis N *et al* 2016 *Phys. Rev.* B **94** 035205
[20] Saldutti M *et al* 2020 *Photon. Res* **8** 1388–97
[21] Cooke S J and Ancona M G 2023 *Solid-State Electron.* **199** 108487
[22] Furtado G F *et al* 2021 *J. Integr. Circuits Syst.* **16**
[23] Medina-Bailón C *et al* 2017 *IEEE Trans. Electron Devices* **64** 3084–91
[24] Reaz M *et al* 2021 *IEEE Trans. Electron Devices* **68** 2556–63
[25] Donetti L *et al* 2018 *Solid-State Electron* **143** 49–55
[26] Thorpe B *et al* 2022 *Semicond. Sci. Technol.* **37** 075009
[27] Sule N *et al* 2013 *J. Comput. Electron.* **12** 563–71
[28] Mitra S *et al* 2023 *J. Comput. Electron.*
[29] Forster S *et al* 2019 *ACS Appl. Energy Mater.* **2** 715–20
[30] Ferry D K and Nedjalkov M 2018 *The Wigner Function in Science and Technology* (Bristol: IOP Publishing)

[31] Weinbub J and Ferry D K 2018 *Appl. Phys. Rev.* **5** 041104
[32] Shifren L and Ferry D K 2002 *J. Comput. Electron.* **1** 55–8
[33] Nedjalkov M *et al* 2011 *Nano-Electronic Devices: Semiclassical and Quantum Transport Modeling* ed D Vasileska and S M Goodnick (New York: Springer) pp 289–358
[34] Ferry D K 2023 *Semicond. Sci. Technol.* **38** 055005
[35] Hockney R W and Eastwood J W 1988 *Computer Simulation Using Particles* (Boca Raton, FL: CRC Press)
[36] Jacoboni C 1989 *The Monte Carlo Method for Semiconductor Device Simulation* (Wien: Springer)
[37] Laux S E 1996 *IEEE Trans. Comput.-Aided Des. Integr. Circuits Syst.* **15** 1266
[38] Springel V 2015 *Star Formation in Galaxy Evolution: Connecting Numerical Models to Reality* ed Y Revaz *et al* (Heidelberg: Springer) pp 251–358
[39] Zhu H *et al* 2019 *Appl. Phys. Lett.* **115** 073107
[40] Ayubi-Moak J S *et al* 2007 *IEEE Trans. Electron Devices* **54** 2327

IOP Publishing

Quantum Transport in Semiconductor Devices
Simulation using particles
David K Ferry, Xavier Oriols and Josef Weinbub

Chapter 8

Beyond the electrostatic approximation

From the beginning of the use of device simulations, it has generally been assumed that the electrostatic approximations were the only approach that was needed. Indeed, for too many people, there is an assumption that Poisson's equation is exact. When devices are placed in circuits to obtain a more complete solution, it has also been assumed that Kirchhoff's current law is complete when applied to DC circuits. Neither of these assumptions is correct, as will be discussed in this chapter. Indeed, at the end of the previous century, and certainly within this century, the advent of high gigahertz and terahertz device operation has become a topic of major interest. It is in this frequency regime that the previous approaches begin to fail, because they do not address the complexities of the electromagnetics that affect the device (and the circuit). In optics, the equivalent is the onset of picosecond and subpicosecond operation that brings the same problems to the fore.

A part of the problem lies in the fact that electrons (and holes) are basically inductive in their transport behavior. Newton's acceleration laws themselves enforce a simple truth: the velocity response to an electric field lags behind any temporal behavior of this field. If this delay becomes a significant part of the device operation, such as in a terahertz field-effect transistor (FET), then the electrostatic approximations fail. The same is true in optical processes; the excitation of electrons and holes lags behind the temporal response of the optical fields. This fact is actually utilized to characterize logic circuits through the so-called delay time.

The performance of logic circuits is often characterized by constructing a so-called ring oscillator. This is a set of logic gates in which the transition from 'off' to 'on' of the first gate causes the second gate to transition from 'on' to 'off.' The third gate is then switched on, and so on around the circuit. The ring oscillator has a number of gates such that the last gate output is fed back to the first input and switches this first gate from 'on' to 'off.' Hence, this circuit builds up feedback in a manner that creates an oscillator. The delay time is not the switching time but is the difference between the time at which the first gate is 50% on and the time at which

doi:10.1088/978-0-7503-5237-6ch8

the second gate is 50% off. Hence, the oscillation frequency is $f = 1/2\pi N\tau_d$, where N is the number of gates in the oscillator and τ_d is the delay time. This delay time then gives information about the internal dynamics of the logic gate, and is a standard measurement [1]. These gate delays are typically in the subpicosecond range in devices whose effective gate lengths are 10 nm or less [1]. In addition to logic devices, high-frequency analog devices are characterized by this same delay time, but here it is referred to as the transit time through the gate region. The gate cutoff frequency then becomes $f_T = 1/2\pi\tau_{tr}$, where τ_{tr} is this transit time [2]. This will be illustrated further later in this chapter.

While the various times are important, there is also a significant change in design philosophy for high-frequency devices, especially analog power devices for microwave signal amplification. There are several reasons for this. The first arises from simple considerations. The amount of power that can be generated in an high-electron-mobility transistor (HEMT) (or any other power transistor) depends upon the current that can be carried through the device. The natural approach to raising current is to increase the gate width (or periphery, the dimension along the surface transverse to the current) and therefore the total current. There is a problem with this, in that one cannot increase the width when it begins to be a fraction of the wavelength of the radio-frequency (RF) signal. When this happens, there is a phase shift between different positions along the gate, as the signal has to propagate through the gate width, and this is detrimental to good operation. A 50 GHz signal has a wavelength in free space of 60 mm. The semiconductor has a high dielectric constant, so that the wavelength is only going to be about 15 mm in the device. A practical limit is then about 5% of this, or 750 μm. At higher frequencies, the effect is more pronounced and the limit smaller. The way around this is to break the device into multiple segments, which operate in parallel. This construction is shown in figure 8.1 [3]. In this figure, the black areas are metal, which provides both the

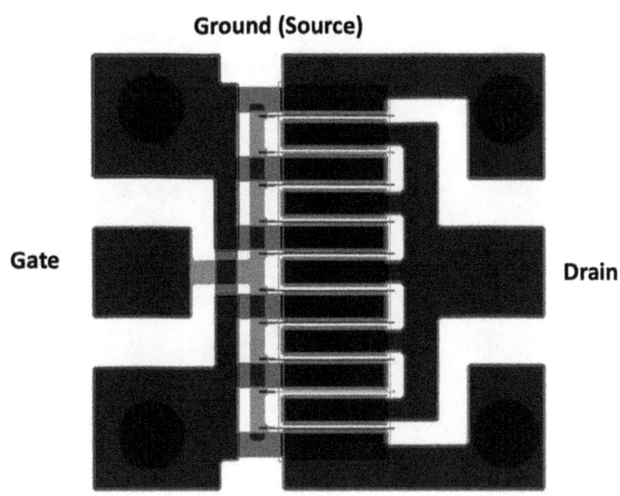

Figure 8.1. Layout for a segmented gate AlGaN/GaN HEMT. Ten gate segments of 100 μm each lead to a total gate periphery (or width) of 1 mm. This design is for an X band amplifier circuit. Reprinted from [3].

ground reference and the power/signal connections. The device is an AlGaN/GaN HEMT, with 10 gate segments (which run horizontally in the figure) of 100 μm each. The actual gates are the very thin black lines running horizontally; these terminate in the white area on the right, and the gray gate connection on the left. The structure in the center, as one goes downward from the top, is source, gate, drain, gate, source, and so on. The effective gate length is 0.3 μm, which leads to an f_T of 28.2 GHz and an f_{max} of >50 GHz. These devices are used in a two-stage amplifier designed to operate at around the 10 GHz range. The fact that the gate and source connections cross one another is addressed by air bridge technology, in which one of the connections is literally suspended (as a bridge) above the other.

The complete circuit for another power device designed to operate in the 16–20 GHz range is shown in figure 8.2 [4]. Here, an AlGaN/GaN HEMT with a gate length of 0.25 μm is used. The device shown in the figure uses a gate with eight segments of 75 μm each for a total gate width of 600 μm. This circuit provides a power output of 27 W and illustrates several other problems that have to be addressed with microwave applications. First, the strip line waveguides have an impedance of a few hundred ohms, whereas the transistor has a near-infinite input impedance due to the gate isolation (the AC impedance is less, but is dominated by the gate capacitance). This significant mismatch has to be addressed in order to actually get sufficient power into the device (and, of course, out of the device). Thus, there have to be tuning circuits in the waveguides in order to achieve this coupling. In addition, voltage has to be applied to the drain and the gates in order for the device to operate. But these voltages have to be isolated from the input and output

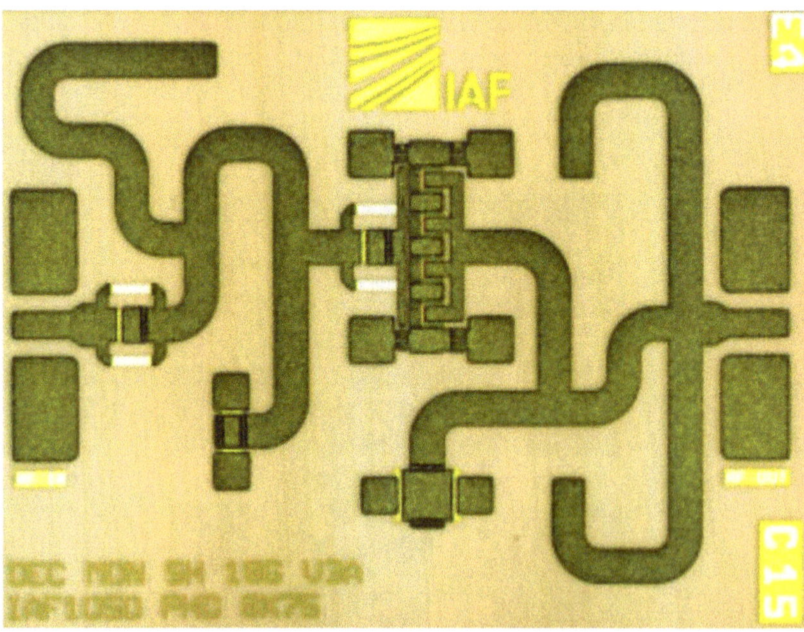

Figure 8.2. A millimeter-wave chip designed as a single-stage amplifier for the 16–20 GHz region using an AlGaN/GaN HEMT. Reproduced from [4].

microwave system; therefore, various circuits have to be used to keep the microwaves out of the bias supplies and vice versa. The design of complicated analog millimeter-wave circuits is not an easy task, but it is not one that will be addressed here to any great degree.

In this chapter, the discussion will go beyond the electrostatic approximations. It will range from the nearly classical discussion of circuit interactions to the very quantum aspect of the electron–photon interaction and optical absorption in semiconductors. To begin, a discussion of Maxwell's equations and gauge relationships is given in the next section. Explanations of the various cutoff frequencies mentioned above are then presented, and these are followed by a discussion of circuit interactions at high frequencies. This is followed by a discussion of the quantum aspects of electrons and photons.

8.1 Maxwell's equations and the gauge

The general introduction to electromagnetics given in universities almost always starts with the electrostatic world; this is the world in which most devices are treated. However, Maxwell's equations for the dynamic version of electromagnetic behavior are usually introduced eventually, albeit often quite late. It is these equations that are important for high-frequency devices and will be considered here. The first of Maxwell's equations is often referred to as Faraday's law, which is written as

$$\nabla \times \boldsymbol{E} = -\frac{\partial \boldsymbol{B}}{\partial t}, \tag{8.1}$$

where $\boldsymbol{E}$ is the electric field intensity and $\boldsymbol{B}$ is the magnetic flux density. It is through this equation that the vector potential $\boldsymbol{A}$ is introduced by relating it to the magnetic field via

$$\boldsymbol{B} = \nabla \times \boldsymbol{A}. \tag{8.2}$$

The vector potential is one of the two important potentials that arise in electromagnetics and thus govern much of the high-frequency world. It is very important in optics and the optical excitation of semiconductors, but is also found to be important in many other applications. By using (8.2) in (8.1), one obtains

$$\nabla \times \boldsymbol{E} = -\frac{\partial}{\partial t}(\nabla \times \boldsymbol{A}); \tag{8.3}$$

or

$$\boldsymbol{E} = -\frac{\partial \boldsymbol{A}}{\partial t} + \boldsymbol{C}, \tag{8.4}$$

where $\boldsymbol{C}$ is a integrating factor whose curl must vanish. Thus, $\boldsymbol{C}$ can actually be a function, so long as it satisfies the zero curl requirement. This can be satisfied by defining this function as

$$\boldsymbol{C} = -\nabla V, \tag{8.5}$$

where V is the scalar potential found in the treatment of electrostatics. Thus, (8.4) can be rewritten as

$$\boldsymbol{E} = -\frac{\partial \boldsymbol{A}}{\partial t} - \nabla V. \tag{8.6}$$

This is normally seen without the vector potential in studies of semiconductor devices, but the vector potential term is important in, e.g. the Aharonov–Bohm effect [5, 6]. The way in which one chooses to depict the electric field, and the connections between the vector and scalar potentials, is referred to as a gauge condition.

To illustrate the connection between the two potentials, the second of Maxwell's main equations, often referred to as the generalized Ampère's law, is given below using the constitutive relation $\boldsymbol{B} = \mu\boldsymbol{H}$:

$$\nabla \times \boldsymbol{B} = \mu\left(\boldsymbol{J} + \frac{\partial \boldsymbol{D}}{\partial t}\right), \tag{8.7}$$

where $\boldsymbol{H}$ is the magnetic field intensity, $\boldsymbol{J}$ the current density, $\boldsymbol{D}$ the electric flux density, and μ the permeability of the material (or $4\pi \times 10^{-7}$ newtons per ampere squared in free space and MKS units). In magnetostatics, the last term is ignored, and the divergence of this equation leads to

$$\nabla \cdot \boldsymbol{J} = 0. \tag{8.8}$$

It is this form that leads to the common DC representation of Kirchhoff's current law. If this equation is integrated over any surface shell of volume Ω, one gets

$$\int \nabla \cdot \boldsymbol{J} d\Omega = \oint \boldsymbol{J} \cdot \boldsymbol{n} dS = 0, \tag{8.9}$$

where S is the surface area of the shell and $\boldsymbol{n}$ is a unit vector normal to the surface. Hence, the second form leads to the statement that the sum of all currents flowing into a point must be zero, which is usually the expression for Kirchhoff's law [7]. The extension to the full form of (8.7) is often attributed to Maxwell in his famous 1861 paper [8], but it was actually introduced by Kirchhoff himself some years earlier [9]. This second term modifies (8.9) by the increase or decrease of charge stored at the point (the node). If another one of the constitutive relations ($\boldsymbol{D} = \epsilon\boldsymbol{E}$) is used, one can arrive at a series of expressions for (8.7), as follows:

$$\begin{aligned} \nabla \times \boldsymbol{B} &= \epsilon\mu\frac{\partial \boldsymbol{E}}{\partial t} + \mu\boldsymbol{J}, \\ \nabla \cdot \boldsymbol{J} + \frac{\partial \rho}{\partial t} &= 0, \\ \nabla \cdot \boldsymbol{E} &= \frac{\rho}{\epsilon}. \end{aligned} \tag{8.10}$$

Here, ϵ is the permittivity of the material and ρ is the electric charge density. The second line is the well-known continuity equation for charge, which was actually established by Kirchhoff [9, 10]. The last line leads to Poisson's equation. Using this last line in (8.6) yields

$$\nabla^2 V - \frac{\partial}{\partial t}(\nabla \cdot \boldsymbol{A}) = -\frac{\rho}{\epsilon}. \tag{8.11}$$

It is now clear that Poisson's equation appears if the vector potential is constant in time, which is true only in the static situation (it is what defines the static situation used for electrostatics and magnetostatics) or following the adoption of a gauge in which the divergence is zero (as discussed in the following paragraphs).

This latter point can be pursued a little further to gain additional insight into the relations between the vector and scalar potentials. For this, (8.2) and (8.6) are used in the first line of (8.10), and this leads to

$$\nabla^2 \boldsymbol{A} - \epsilon\mu\frac{\partial^2 \boldsymbol{A}}{\partial t^2} = -\mu \boldsymbol{J} + \nabla\left(\nabla \cdot \boldsymbol{A} - \epsilon\mu\frac{\partial V}{\partial t}\right). \tag{8.12}$$

If the term in parentheses could be set to zero, the inhomogeneous wave equation for the vector potential would arise with the current density as the driving term. At the same time, setting the relation to zero allows equation (8.11) to become the wave equation for the scalar potential with the charge density as the driving term. Thus, this act uncouples the two potentials, and each of these then has its own driving term. This condition is known as the *Lorentz gauge*, or sometimes simply as the *gauge equation*

$$\nabla \cdot \boldsymbol{A} - \epsilon\mu\frac{\partial V}{\partial t} = 0. \tag{8.13}$$

Remember, however, that this is a choice of convenience and is not a required condition. The fact that this choice is usually made in the study of electromagnetic fields does not make it a firm fact or theorem. Nevertheless, when this gauge equation is imposed, it is a rigid constraint upon the solutions of the two wave equations. Again, these results lead to the wave equation for the scalar potential, as follows:

$$\nabla^2 V - \epsilon\mu\frac{\partial^2 V}{\partial t^2} = -\frac{\rho}{\epsilon}. \tag{8.14}$$

From this equation, one can immediately see that Poisson's equation is an approximation that is valid only when the potential is either constant or varying very slowly with time. Neither of these conditions is valid in the very-high-frequency regime.

In the above equations, both μ and ϵ have been treated as if they are constants. They are not constants but can be complicated functions of time, position, momentum, and energy. They can even be nonlinear, in which case the polarization defined through

$$\begin{matrix} \epsilon = \epsilon_0(1 + \chi) \\ \boldsymbol{P} = \chi \boldsymbol{E} \end{matrix}, \tag{8.15}$$

is capable of being an actual power series in the electric field intensity. The latter is especially true of the nonlinear optical materials used for parametric amplifiers and frequency converters. The fact is, one must very carefully determine the nature of the material being considered to determine just what dielectric and magnetic properties will be encountered in the device of interest. The most common variation found in semiconductors is the case of compound semiconductors, such as the III–V or II–VI compounds, in which the so-called dielectric 'constant' is considerably different at optical frequencies than it is at low frequencies. This difference arises from the lattice polarization which contributes to the dielectric response at low frequencies.

In actual practice, there are further possibilities for defining a gauge that can be imposed. One approximation is to invoke the so-called coulomb gauge (or the electrostatic gauge) as it is sometimes called, in which we set

$$\nabla \cdot \boldsymbol{A} = 0, \tag{8.16}$$

for which the Lorentz gauge condition leads to

$$\frac{\partial V}{\partial t} = 0. \tag{8.17}$$

When this condition is assumed, equation (8.14) now becomes the more familiar Poisson equation in an exact form, although one must recognize that (8.17) does not always contain the exact physics. This familiar result arises from a choice of gauge; it is not automatically either true or fundamental. Assuming only the Poisson equation for a device means that only low-frequency effects are of interest, and that the potential and electric field instantaneously follow variations in charge (or that the propagation delays are much shorter than any time constant in the system), although this condition actually violates (8.17). That is, in assuming (8.17) to be strictly applicable, the charge density in the device is taken to be *quasi-static and time invariant*. The scalar potential then follows the charge density change instantaneously. So, if one wants to use the coulomb gauge to disregard the vector potential in device simulation, it must first be ascertained that any charge changes due to the imposition of self-consistency must be slow enough to validate the use of this gauge.

Let us explain this further. Charge carriers are naturally inductive. This means that, at high frequencies, the potential lags behind the charge when the latter is varies over time. The static case assumes this phase lag is irrelevant (and therefore small). But at high frequencies, it can be important, so that the complete wave equation (8.14) needs to be used. An example can be given in the form of the impact ionization avalanche transit-time (IMPATT) devices. Such devices are expected to generate a negative differential impedance at high frequencies [11, 12]. The natural delay between charge generation (avalanche) and the potential is used to give $\pi/2$ phase shift, and the delay in the ballistic transit time supposedly gives an additional $\pi/2$ phase shift, thus yielding the needed phase shift between current and voltage.

Another gauge approximation is useful when the effect of a magnetic field on various mesoscopic devices is of interest. For this situation, there are two more common gauge choices that may be made. These arise from the manner in which the magnetic field and the vector potential are connected to satisfy equation (8.2). One of these introduces the *Landau gauge*

$$\boldsymbol{A} = Bx\boldsymbol{a}_y, \tag{8.18}$$

where $\boldsymbol{a}_y$ is a unit vector in the y-direction. This yields a magnetic field that is oriented in the positive z-direction. An alternative is to orient the vector potential in the y-direction; the choice is one of convenience. It is convenient to combine these two forms into the *symmetric gauge*

$$\boldsymbol{A} = \frac{B}{2}(-y\boldsymbol{a}_x + x\boldsymbol{a}_y). \tag{8.19}$$

In both classical and quantum mechanics, the electromagnetic interactions are taken into account by a change in the momentum, which can be considered to be a gauge transformation. This uses the Peierls' substitution

$$\boldsymbol{p} \rightarrow \boldsymbol{p} - e\boldsymbol{A}. \tag{8.20}$$

This is used often for optical excitation, and it will be considered below when this is treated.

8.2 Cutoff frequencies

In the introduction to this chapter, an important device frequency was introduced, namely the 'cutoff' frequency f_T. There is another important frequency in analog power devices, f_{max}, which is the frequency at which the power gain becomes unity (0 db of gain). In this section, we discuss the manner in which these quantities may be determined using particle-based simulations. Here, the particle simulation of the devices discussed is determined by a full-band Monte Carlo simulation tool [13]. The full-band energy structure is determined using an empirical pseudopotential code that incorporates local, nonlocal, and spin–orbit interactions. The phonon characteristics are computed using a 14-parameter valence shell model [14]. Scattering rates are computed and stored for momentum-state-to-momentum-state scattering, in what is known as a cellular Monte Carlo [13]. The Monte Carlo simulations and Poisson's equation are then iteratively solved for various time steps. This is applied to an InGaAs channel HEMT [15] and a GaN HEMT [16].

The structure to be used here is a typical InGaAs channel HEMT, as shown in figure 8.3. The structure is grown on an InP substrate and the heterostructure barriers are InAlAs, lattice matched to the InP. The active channel material is not lattice matched but has a composition that is InAs-rich in order to improve the carrier mobility and velocity ($In_{0.75}Ga_{0.25}As$). An additional heavily doped and lattice-matched InGaAs layer is grown on the top of the structure to assist in contacting the device. Typical current–voltage curves for a device with a 70 nm gate length are shown in figure 8.4. Contacts are an important part of the device simulation. Here, the contacts are simulated by a set of grid cells adjacent to the contact that are appropriately doped to provide both a boundary condition for Poisson's equation and a charge-neutral boundary condition for the Boltzmann equation. The effect of the doping in these cells leads to a variation in the effective

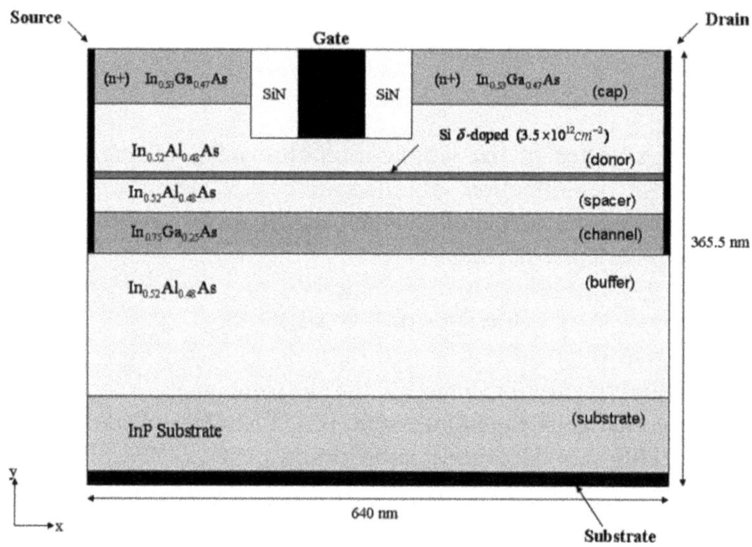

Figure 8.3. The layer structure of a strained InGaAs layer HEMT with a 70 nm gate length. The total source–drain spacing is 640 nm, and the spaces are filled with SiN. The various layers, other than the cap layer, are nominally undoped. © 2007 IEEE. Reprinted, with permission, from [15].

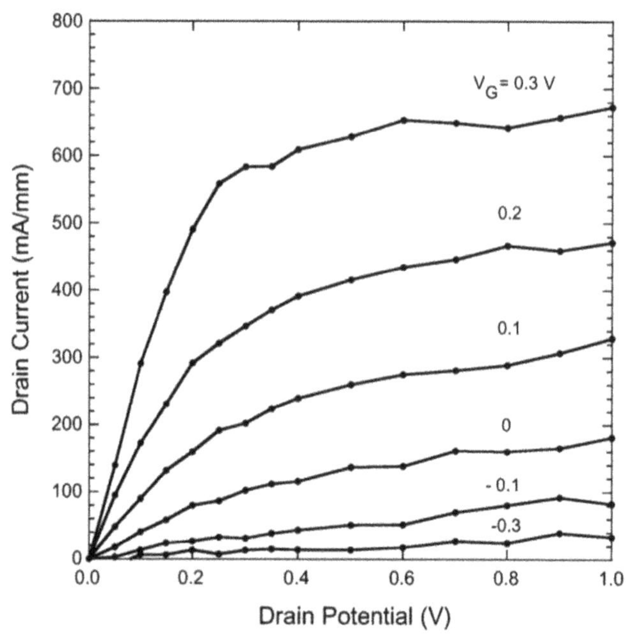

Figure 8.4. The output characteristics of the HEMT with a 70 nm gate length for a δ-doping concentration of 3.5×10^{12} cm^{-2}. The results shown here are for a 640 nm source–drain spacing and a contact doping of 10^{18} cm^{-3}. © 2007 IEEE. Reprinted, with permission, from [15].

contact resistance of the device. In figure 8.5, this is illustrated by plotting the effective contact resistance as a function of the heavily doped contacts. The variation in doping leads to a variation in the potential drop across the contact grid cell, and this changes the transconductance of the device in a manner similar to real physical contact resistance. This will be discussed further in the next section.

Over the last couple of decades, significant advances have been made in high-frequency HEMTs. As a result, many simulations of these devices have been performed to determine the upper frequencies of operation [15, 17]. Here, the InGaAs device mentioned above is used with a variety of gate lengths to illustrate the method used to compute the various cutoff frequencies. It has long been known that the longitudinal electric field in the HEMT is concentrated under the gate, due to the higher resistance in this part of the device. However, the gate itself is situated above the actual channel by some distance (d_g in the discussion here). This means that there is substantial fringing of the capacitance around the physical gate, which affects the actual velocity in the channel. Such velocities are shown in figure 8.6 for a variety of gate lengths and separations of the gate from the channel [17]. It may be noted that, in each case, the peak of the velocity lies near the drain end of the physical gate (the solid lines at the top denote the extent of the physical gate). However, the velocity begins to rise some short distance before the particles reach the actual gate. This affects the resulting cutoff frequency. In figure 8.7, the cutoff frequency at which the current gain drops to unity is plotted as a function of the inverse of the physical gate length (black data and curve) [2]. This shows saturation as the gate becomes smaller. This is the result of the acceleration of the particles

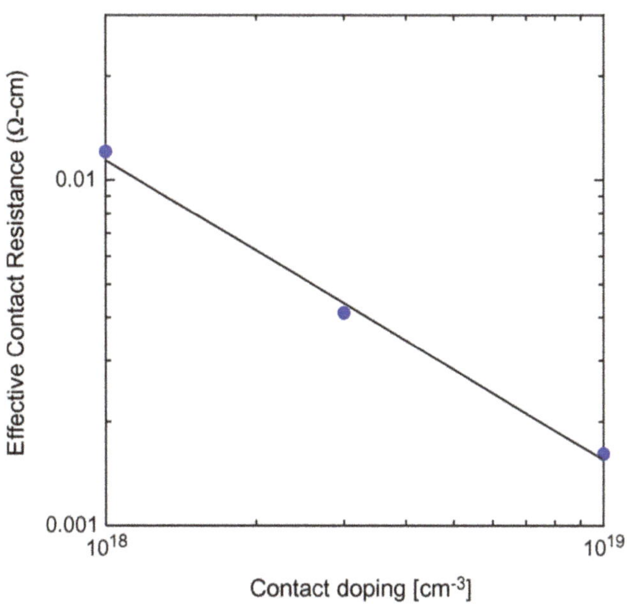

Figure 8.5. The variation in the effective contact resistance with doping. The doping modifies the potential drop across this grid cell and then changes the transconductance of the device, just as a physical contact resistance would. © 2007 IEEE. Reprinted, with permission, from [15].

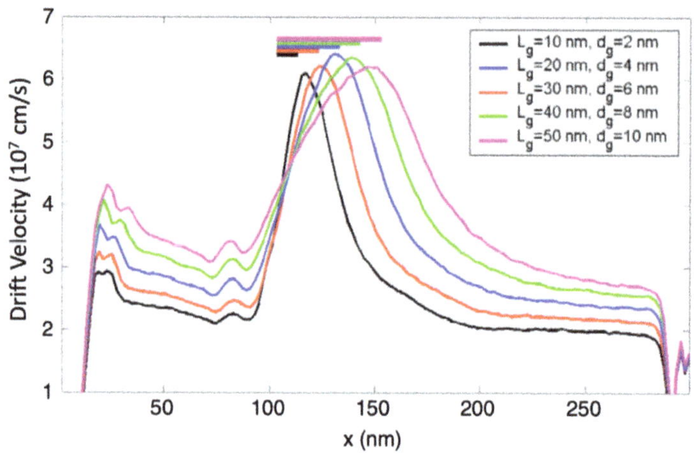

Figure 8.6. The velocity along the channel for a set of InGaAs HEMTs for different physical gate lengths (L_g) and separations (d_g) of the gate from the channel. Reproduced from [17].

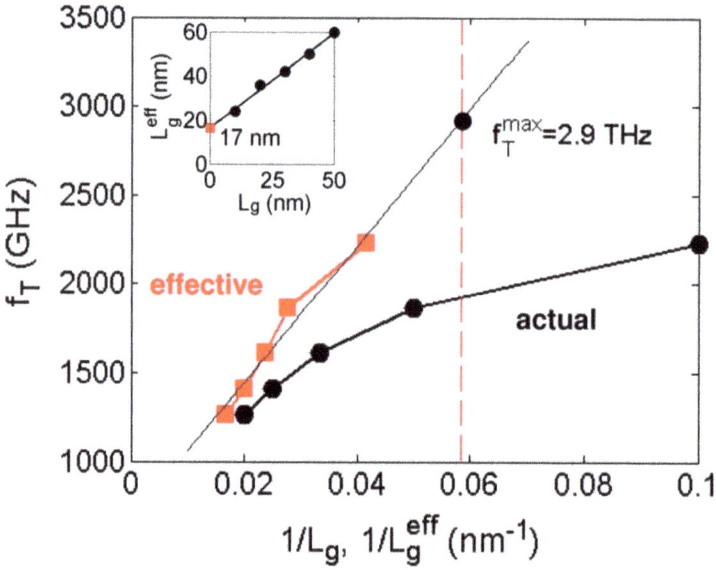

Figure 8.7. The cutoff frequency f_T as a function of the inverse of the physical gate length (L_g) and the effective gate length (L_g^{eff}, determined in the text). The inset plots the effective gate length as a function of the physical gate length. © 2008 IEEE. Reprinted, with permission, from [2].

prior to reaching the actual gate. To account for this, it is necessary to use the entire acceleration length as well as the actual position-dependent charge density in weighting the current. Here (and in many of the following figures), the x-direction is along the channel from source to drain, while the y-direction is normal to the surface of the device and measures the distance into the device from the surface. Thus, the average x-directed velocity along the channel may be defined as

$$v_x^{\text{av}}(x) = \int_{y_1}^{y_2} n(x, y)v_x(x, y)dy / \int_{y_1}^{y_2} n(x, y)dy. \tag{8.21}$$

Here, y_1 and y_2 are the upper and lower boundaries of the actual channel at position x. Once this average velocity is determined from the particle simulation, the transit time through the acceleration region may be calculated as well, giving

$$\tau_{\text{T}} = \frac{1}{2\pi f_{\text{T}}} = \int_0^{L_{\text{g}}^{\text{eff}}} \frac{dx}{v_x^{\text{av}}(x)}. \tag{8.22}$$

The actual size of the effective gate length is defined by this acceleration length, as depicted in figure 8.8 [17]. The cutoff frequency is also plotted as a function of the inverse of the effective gate length in figure 8.7 (red data and curve). A comparison of the effective gate length and the physical gate length is shown in the inset to this latter figure.

As has been described, there are two important frequencies, which are both cutoff frequencies. One is for the current gain and the other is for the power gain. These are depicted using simulation results in figure 8.9 [18]. It is important to note that determining the power gain cutoff or unity power gain (UPG) frequency is often beset with special difficulties, whether in experiment or in simulation. This is

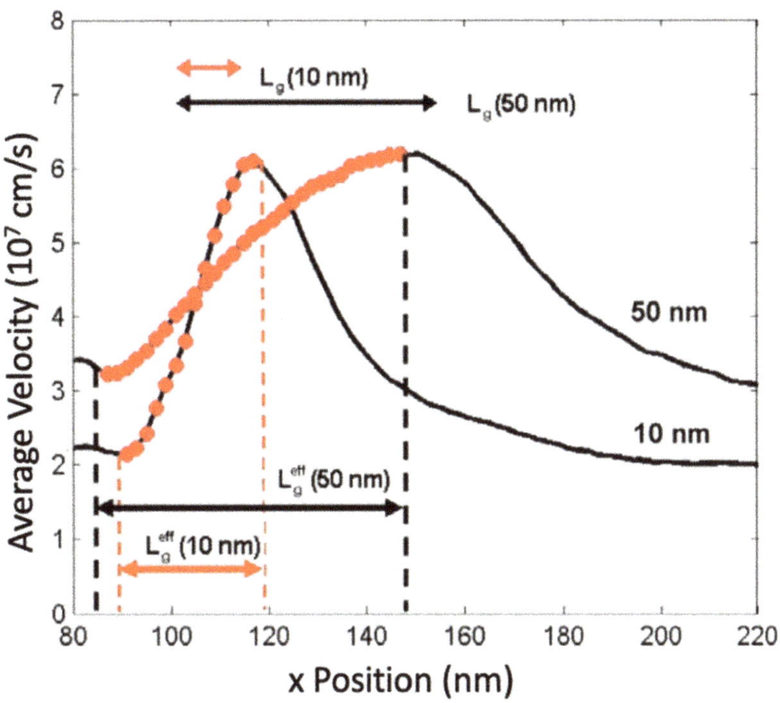

Figure 8.8. The determination of the effective gate length is depicted for devices that have gate lengths of 10 nm and 50 nm, as discussed in the text. The physical lengths are shown by the colored arrows at the top of the figure. Reproduced from [17].

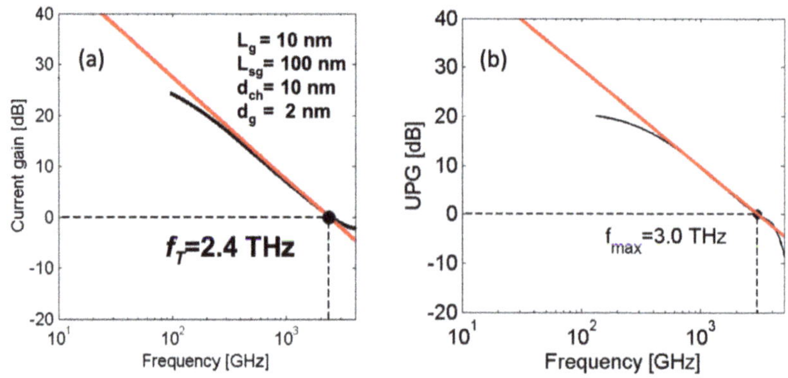

Figure 8.9. (a) The determination of f_T for an HEMT with a physical gate length of 10 nm. (b) The determination of f_{max} for the same device. [18] John Wiley & Sons. Copyright © 2010 WILEY-VCH Verlag GmbH & Co. KGaA, Weinheim.

because, in the very-high-frequency regime, one has to deal with the presence of resonances, and it can be difficult to work around these. This has been done in panel (b) of the figure, but it is common to see these resonances in other simulations and experiments. These two frequencies are for a device with a 10 nm physical gate length, a source-to-drain spacing of 100 nm, a channel lying 10 nm below the surface, and a recessed gate lying only 2 nm from the channel. This is an extreme situation in many ways, but it leads to a very high value for each of the two cutoff frequencies, strongly suggesting that this device can produce terahertz behavior. This is supported in figure 8.10, which shows the variation in f_{max} with the inverse of the effective gate length [18]. The red dots in this figure and in the inset indicate the maximum that could be expected for these devices with a zero gate length (perhaps using a vertically oriented sheet of graphene for the gate). Here, the effective gate length is still of the order of 18 nm. Nevertheless, it still holds true that these devices will be useful in the terahertz regime.

The relationship between the physical gate length and the effective length is affected by the aspect ratio of the device. Here, this ratio is defined to be L_g/d_g. It has been found from extensive simulations of the device that a minimum value of this ratio must be adhered to in order to minimize short-channel effects and achieve a better ratio of the effective gate length to the physical gate length. For example, it has been found that this minimum ratio is about 15 for GaN HEMTs but may be as small as 5–6 for InGaAs HEMTs [19]. This difference is thought to arise from the different dielectric properties which affect the fringing fields of the gates.

8.3 Circuit effects

In the field of analog microwave devices, the actual circuits have an effect on device performance. This is because mismatches between circuit impedances and device impedances directly impact the performance of the actual device. While circuit interactions are also important for logic devices, the determination of these latter effects is less difficult than in the analog case.

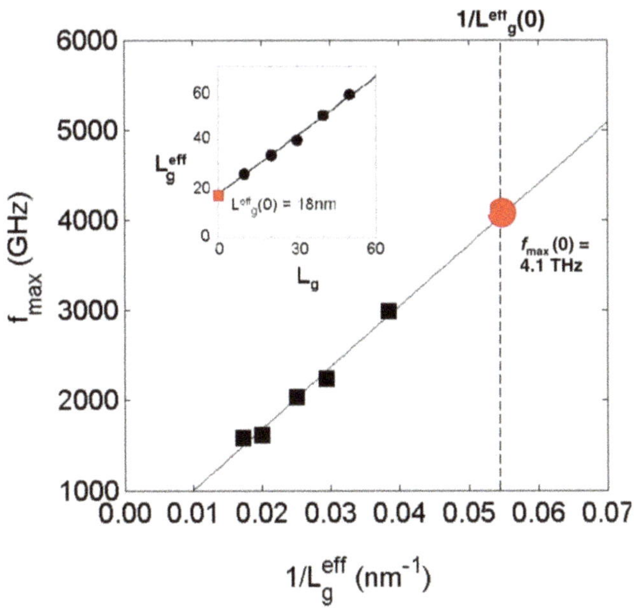

Figure 8.10. The variation of the unity power gain frequency with the inverse of the effective gate length. The red dots in both the inset and the main panel show the maximum frequency expected as the physical gate length goes to zero, as discussed in the text. [18] John Wiley & Sons. Copyright © 2010 WILEY-VCH Verlag GmbH & Co. KGaA, Weinheim.

In this section, a GaN HEMT structure will be utilized. This structure is shown in figure 8.11, which is the basic AlGaN/GaN structure [20], but with an InGaN back barrier to further confine the electrons to the channel [21]. The area around the gate is usually filled with SiN to control the surface properties of the AlGaN, but this can be detrimental to the high-frequency behavior, as it enhances the fringing fields of the gate. The structure of the gate is referred to as a 'T-gate,' and yields a larger current-carrying cross-section while retaining a short physical gate length. This device is simulated with an active load-line that represents the high Q circuit in which the device is placed. The active load line is determined in postprocessing to be [20]

$$Z_L(\omega) = \frac{v_D(\omega)}{i_D(\omega)} \equiv |Z_L(\omega)|\, e^{i\theta_{Z_L}}. \tag{8.23}$$

Here, v_D and i_D are the AC drain voltage and drain current; they are both complex phasors evaluated at the fundamental microwave frequency of operation. The magnitude and phase of the load impedance can be adjusted by varying the phase of the signal generator used to drive the circuit and device. As remarked in the last section, the effective contact resistance dramatically affects the performance of the device, and this also affects the active load line with which the device interacts. This is shown in figure 8.12 for various values of the effective contact resistance [20].

Simulating the circuit directly is a more complicated problem. The use of a time-domain analysis of the matching circuit and waveguide is a lengthy process that yields the full transient response of the reactive elements in the matching circuit. A

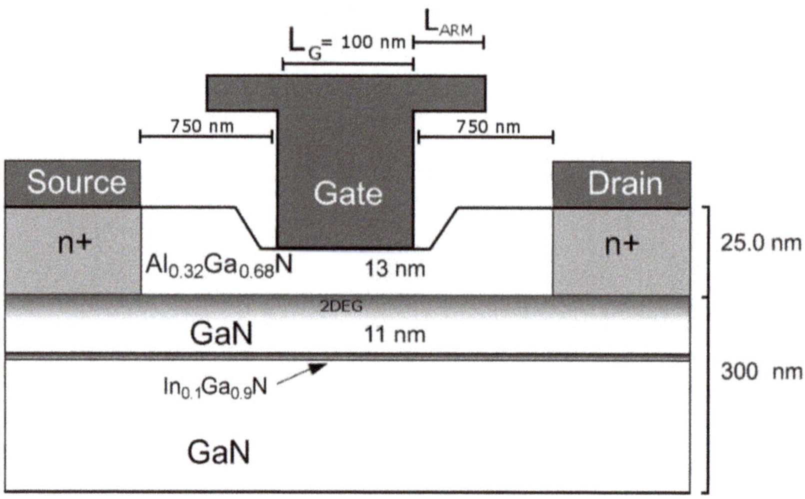

Figure 8.11. A typical AlGaN/GaN HEMT used for high-frequency power applications. The thin InGaN layer is inserted to provide additional confinement of the electrons in the channel (discussed in the text). © 2012, IEEE. Reprinted, with permission, from [20].

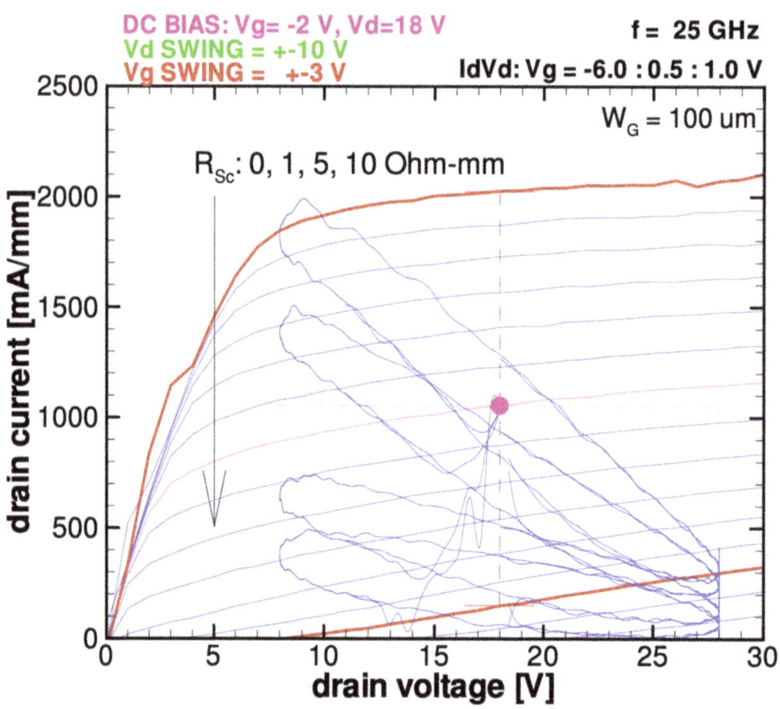

Figure 8.12. The effect of the source contact resistance is shown by the various operational loops (light blue). These affect the nominal operational point (red dot), which moves for each value of the resistance. The nominal operating conditions are shown along the top of the plot. © 2012, IEEE. Reprinted, with permission, from [20].

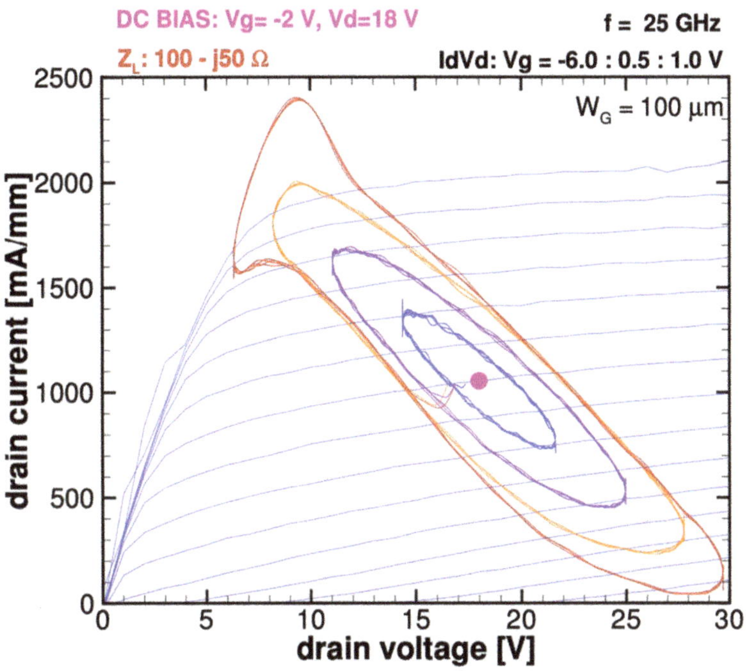

Figure 8.13. The variation of the dynamic load response for increasing input powers supplied to the HEMT at 25 GHz. The variations shown in blue, magenta, orange, and red correspond to increasing input powers under class A operation. © 2012, IEEE. Reprinted, with permission, from [24].

frequency-dependent analysis, such as the harmonic balance approach [22, 23], provides a rapid solution of the steady-state behavior of the circuit. Combining these two approaches with the full time-dependent device simulation provides a complete characterization of the overall system in a self-consistent manner [24]. An example of this is shown in figure 8.13 for a GaN HEMT at 25 GHz operating in class A mode (for signal response over the complete AC cycle). The various blue, magenta, orange, and red circles correspond to increasing amounts of input power supplied to the device. At the highest input power, the loop distortion at the highest currents is likely to lead to harmonic generation in the device. The gate resistance is an additional parasitic component that affects device performance. The effect of this is shown in figure 8.14, where the power output, the gain, and the power added efficiency are shown as a family of curves for varying gate resistances [24].

8.4 Finite-difference time-domain analysis

The use of lumped parameters for the circuit often does not work at high frequencies. As mentioned above, a time-domain analysis of the microwave waveguides and matching circuits is often applied, which may be evaluated using the finite-difference time-domain (FDTD) algorithm. This algorithm is a method of solving Maxwell's equations and has been used for some time in the evaluation of microwave systems [25]. In general, this method uses Maxwell's equations given in section 8.1 in terms of $\boldsymbol{H}$ and $\boldsymbol{E}$, along with the two constitutive relationships

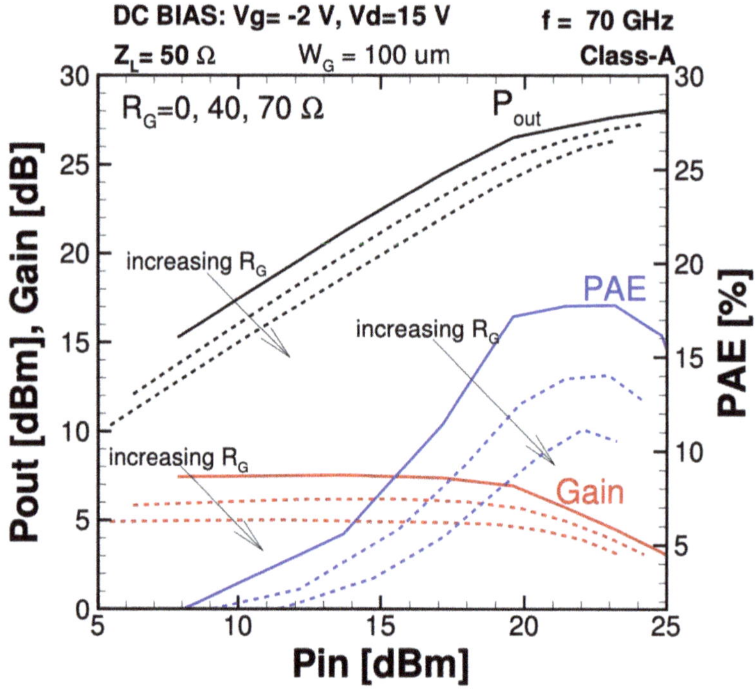

Figure 8.14. The effects of gate resistance on HEMT performance at 70 GHz under class A operation. Increasing gate resistance leads to decreasing performance. © 2012, IEEE. Reprinted, with permission, from [24].

$$\begin{aligned} \boldsymbol{D} &= \epsilon \boldsymbol{E}, \\ \boldsymbol{B} &= \mu \boldsymbol{H}. \end{aligned} \tag{8.24}$$

The permittivity and permeability are taken to be constant in microwave circuits, although this is not required and is not the case when some types of substrates are utilized. Generally, Maxwell's equations are rewritten in terms of six simpler equations:

$$\begin{aligned} -\mu\frac{\partial H_x}{\partial t} &= \frac{\partial E_z}{\partial y} - \frac{\partial E_y}{\partial z} \\ \epsilon\frac{\partial E_x}{\partial t} &= \frac{\partial H_z}{\partial y} - \frac{\partial H_y}{\partial z} - J_x \\ -\mu\frac{\partial H_y}{\partial t} &= \frac{\partial E_x}{\partial z} - \frac{\partial E_z}{\partial x} \\ \epsilon\frac{\partial E_y}{\partial t} &= \frac{\partial H_x}{\partial y} - \frac{\partial H_y}{\partial x} - J_y \\ -\mu\frac{\partial H_z}{\partial t} &= \frac{\partial E_x}{\partial y} - \frac{\partial E_y}{\partial x} \\ \epsilon\frac{\partial E_z}{\partial t} &= \frac{\partial H_y}{\partial x} - \frac{\partial H_x}{\partial y} - J_z \end{aligned}. \tag{8.25}$$

For computational purposes, these equations are re-expressed in a finite-difference form. The time derivatives are taken to be a centered difference ($t \pm \Delta t$), while the spatial differences are simple trailing differences (i, $i - 1$); a similar approach is taken for the other two axes, once a grid is generated upon which to solve the equations. This allows the user to solve for $H_x^{t+\Delta t}$, for example, in terms of $H_x^{t-\Delta t}$, E_z^t, E_y^t. Each of the other equations is handled in the same manner. In three dimensions, the grids must be interlaced so that, for example, the electric and magnetic fields are solved on separate grids at half spacings. This arrangement creates what is known as a 'Yee cell' for the grid [26]. This method provides a relatively fast and efficient approach to solving microwave problems.

8.4.1 Finite-difference time-domain and ensemble Monte Carlo methods for devices

The coupling of FDTD solutions for the electromagnetics to ensemble Monte Carlo (EMC) particle-based solutions of the device dynamics appears to have already begun in the 1990s [27, 28]. This approach was later expanded to the full-band Monte Carlo approach used for the device in section 8.2 [15, 29]. Like earlier experimental work, this approach was also applied to a coplanar waveguide-fed metal–semiconductor FET (MESFET) with a 100 nm gate length [30]. The simulation predicted an f_T of 170 GHz for this device.

Later studies that used a two-finger gate GaAs MESFET with a gate length of 200 nm focused upon the coupling problem [31]. The total device structure is shown in figure 8.15, with the active domain outlined in the light blue striped section. The microwave area is the dashed outline in black. This work simplified the microwave

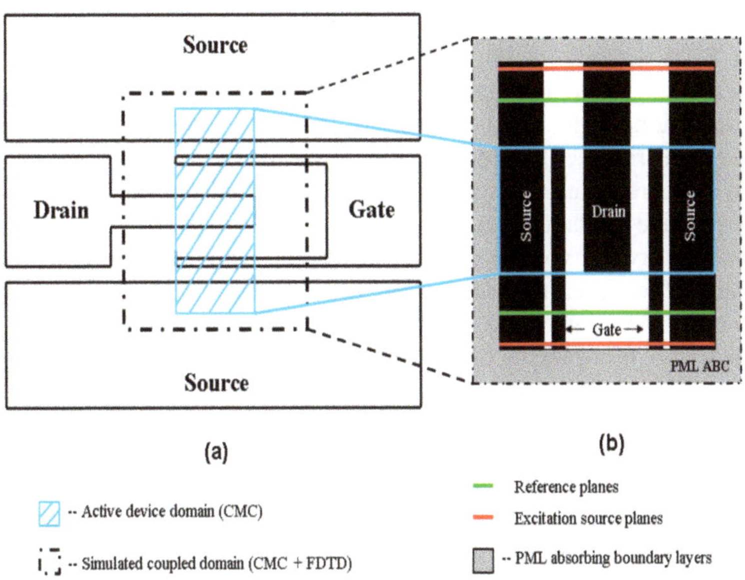

Figure 8.15. (a) A depiction of the active device area (dashed blue rectangle) that is embedded within a larger microwave structure. (b) A view of the 'coupled' microwave area used for a coplanar waveguide (ground–signal–ground metallization). © 2009, IEEE. Reprinted, with permission, from [31].

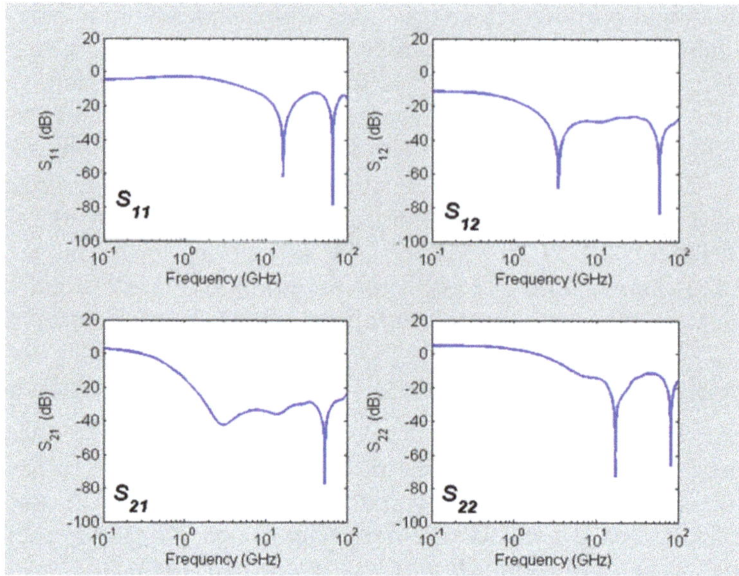

Figure 8.16. The S-parameters determined from the simulation of the structure shown in figure 8.15. These parameters are discussed in the text. © 2009, IEEE. Reprinted, with permission, from [31].

coupling structures and suffered from the resulting mismatch in impedances. The authors noted that further work was needed to more explicitly characterize the entire microwave region. Nevertheless, they were able to fully characterize the device and determine the full set of S-parameters, which are shown in figure 8.16. The parameter S_{11} is the ratio of the reflected voltage to the input voltage at the input side of the device. The parameter S_{21} is the ratio of the output voltage (traveling in the same direction as the input voltage) to this input voltage, i.e. the ratio of the transmitted (plus reflected) voltage at the output to the input voltage. Similarly, S_{22} is the ratio of the reflected voltage (plus the transmitted voltage) to the input voltage on the output side of the device. Finally, S_{12} is the reverse of S_{21}. In this sense, S_{21} determines the forward voltage gain of the device, while S_{12} determines the reverse voltage gain. The other two terms are the reflection coefficients. While the S-parameters are commonly used for microwave circuits and devices, they are often transformed into h-parameters. The important h-parameter is h_{21}, which is the forward current gain. It is this parameter that determines f_T, as this cutoff frequency is the frequency at which this forward current gain goes to zero (in dB), i.e. when the output current equals the input current. The fact that the values for S_{21} are mainly negative does not suggest that this is a very useful device for producing real gain. The conversion from S-parameters to h-parameters can be done, and is not difficult, but it usually involves various impedances for accuracy.

More recent applications of the coupled FDTD and EMC approaches have examined monolayer graphene on a SiO_2 substrate [32]. They have also been applied to the study of hole currents in Si at low temperatures [33].

8.4.2 The finite-difference time-domain method and optics

It has already been noted that the FDTD method is quite old. It has been applied to optics for about as long as it has existed. What is somewhat new is the emergence of the use of particles in the electromagnetic parts of the simulation as well as in normal device simulation. The use of 'particles,' i.e. the use of the photons themselves, is often implemented via ray-tracing algorithms. The latter often have problems with coherent effects such as interference or diffraction. Thus, the combination of FDTD and ray tracing offers a method that can address these problems rather efficiently [34]. This approach was applied to the simulation of optical illumination produced by a light-emitting diode and to a complementary metal–oxide–semiconductor (CMOS) image sensor. The FDTD approach was also applied to a CMOS image sensor in which the transport was computed by particle simulation [35].

8.5 Electromagnetic field quantization

In the previous sections, the fields were mainly treated as classical concepts. From the early days of quantum mechanics, there was an interest in quantizing the actual electromagnetic fields. This began even earlier with Planck's introduction of light packets in his treatment of blackbody radiation [36]. Planck's work had already been confirmed by experiments and was later used for a variety of new theories, such as those of Einstein [37] and Bohr [38]. As part of the rise of quantum mechanics a few years after these works, it was Dirac who began to quantize the fields using second quantization [39], which was outlined in appendix A.2 of chapter 1 and appendix A of chapter 5 for bosons and fermions, respectively. The use of these creation and annihilation operators on the field led to the actual term 'field operator' used in these appendices. Dirac used both the quantum theory and Planck's results by assuming that a field state, in Dirac notation, could be represented as $|\, \boldsymbol{k}, s\rangle$, where $\boldsymbol{k}$ is the wave number and s is the spin (of the photon). Photons are usually considered to have spin values of ± 1, which correspond to right- and left-circularly polarized waves. As these spin values are integers, the photons are bosons, as expected. From Planck's work, this field state satisfies the eigenvalue equation

$$H \,|\, \boldsymbol{k}, s\rangle = \hbar\omega \,|\, \boldsymbol{k}, s\rangle = h\nu \,|\, \boldsymbol{k}, s\rangle. \tag{8.26}$$

Since photons have zero mass, the concept of the momentum of a particle is a little more difficult, but this momentum is represented as $\hbar k$ since, in free space, the phase velocity and group velocity are equal to c, so $v = \omega/k$, and

$$\boldsymbol{p} \,|\, \boldsymbol{k}, s\rangle = \hbar k \,|\, \boldsymbol{k}, s\rangle. \tag{8.27}$$

Similarly, the eigenvalue of the z-component of spin is our ± 1. Here, the Dirac wave function notation indicates that the wave function is the product of a wave function in momentum space that depends upon $\boldsymbol{k}$ and a spin wave function (which is a simple spinor) that depends upon s.

As was shown in section 8.1, both the electric field intensity and the magnetic flux density are determined by the vector potential $\boldsymbol{A}$, especially in the coulomb gauge, which is often used for optical studies. Thus, it is the vector potential that is normally written in terms of the operators; in the case of a unbounded medium, it may be written as

$$\boldsymbol{A}(\boldsymbol{r}, t) = \sum_{k,s}\left[\boldsymbol{e}_s a_{\boldsymbol{k},s}(t)e^{i\boldsymbol{k}\cdot\boldsymbol{r}} + \boldsymbol{e}_s^* a_{\boldsymbol{k},s}^\dagger(t)e^{-i\boldsymbol{k}\cdot\boldsymbol{r}}\right], \tag{8.28}$$

where $a^\dagger$ and a are the creation and annihilation operators and $\boldsymbol{e}$ is a polarization vector that gives the vector relationship of the potential. If the fields are in a confined, bounded region, then the plane waves are replaced by the eigenfunctions defined by the nature of the boundary values and the coordinates in use.

The first quantization of the fields is normally given by developing the Fourier transform in space and time, which is already apparent in (8.28). However, it continues by assuming that each Fourier mode may be considered to be a harmonic oscillator in form. The operators in (8.28) then have the temporal form

$$\begin{aligned} a_{\boldsymbol{k},s}(t) &= \sqrt{\frac{\hbar}{2\omega\Omega\epsilon_0}}\, a_{\boldsymbol{k},s} \\ a_{\boldsymbol{k},s}^\dagger(t) &= \sqrt{\frac{\hbar}{2\omega\Omega\epsilon_0}}\, a_{\boldsymbol{k},s}^\dagger \end{aligned}. \tag{8.29}$$

This result is different from the description in (A.30), but in the case of the fields, the mass is zero and the momentum is given by $\hbar k$. These two facts change the prefactor to the current form. This obtained by considering that the plane wave (wave function) $e^{i\boldsymbol{k}\cdot\boldsymbol{r}}$ has to be normalized over its volume, which gives a term $1/\sqrt{\Omega}$, where Ω is the volume. The normal consideration of the energy density is then $n\hbar\omega = 2\epsilon_0 E^2$ (the magnetic field adds an equal energy to that of the electric field, and this gives the factor of two). Adding the volume normalization, the electric field of a single photon ($n = 1$) is then $\sqrt{\hbar\omega/2\epsilon_0\Omega}$, which arises from the fact that the normal consideration of the energy density is $n\hbar\omega = \epsilon_0 E^2$. Since this electric field is the time derivative of the vector potential, the resulting normalization inherent in the vector potential is the term in front of each line of (8.28). This implies that the plane wave also contains a factor of $e^{-i\omega t}$, which has been suppressed in the above equations of this section.

Just as in the harmonic oscillator, the quantity $a_{\boldsymbol{k},s}^\dagger(t)$ creates a photon with wave number $\boldsymbol{k}$ and spin s. Similarly, $a_{\boldsymbol{k},s}(t)$ destroys that state. Thus, the number operator may be expressed exactly as shown in (A.17):

$$n(\boldsymbol{k}, s) = a_{\boldsymbol{k},s}^\dagger a_{\boldsymbol{k},s}. \tag{8.30}$$

In this notation, a photon number state

$$N(\boldsymbol{k},s)|0\rangle = n\left(a_{\boldsymbol{k},s}^\dagger\right)^n |0\rangle \tag{8.31}$$

is termed a 'Fock' state. This is considerably different from a many-body fermion Fock state.

8.6 Electron–photon interactions

In considering the interactions between electrons and photons, it is first necessary to understand that the wave numbers for these two particles are dramatically different even if they have the same energy. For a 1 eV photon, the radian frequency is $\omega = 1.6 \times 10^{15}\ \mathrm{s}^{-1}$. This leads to a wave number of $q = 5.33 \times 10^{6}\ \mathrm{m}^{-1}$. (As in previous chapters, we distinguish between wave numbers for electrons and other excitations, such as photons, by using $\boldsymbol{k}$ for electrons and $\boldsymbol{q}$ for these other excitations.) For the electron, the wavelength is given by the de Broglie relation, and this leads to a value of $k = \sqrt{2mE}/\hbar = 3.23 \times 10^{9}\ \mathrm{m}^{-1}$. Therefore, there are almost three orders of magnitude of difference between the two wave numbers, mainly due to the much smaller wavelength of the electron. This will be important for many of the considerations of this section.

8.6.1 Electron–photon scattering

Perhaps the most common type of electron–photon scattering occurs in high-energy physics and is known as Compton scattering. It was discovered in the scattering of x-rays by relatively light elements [40], although it is often observed by sending x-rays through metallic foils. In general, this leads to a wavelength shift in the scattered x-ray of

$$\lambda' - \lambda = \frac{h}{m_0 c}(1 - \cos\vartheta), \tag{8.32}$$

where λ is the wavelength of the input photon, λ' is the wavelength after scattering, m_0 is the free electron mass, and ϑ is the scattering angle between the input and output photons. The prefactor on the right-hand side of (8.32) is commonly referred to as the Compton wavelength. While Compton thought the scattering was due to the electrons bound to the atoms, later studies tended to suggest that there was no nuclear Compton scattering; rather, the scattering came from unbound electrons such as the electron 'gas' in a metallic film. In figure 8.17, a diagrammatic interpretation of Compton scattering is shown. Here, the electron first absorbs the photon and then re-emits it at a different wavelength. This suggests that the scattering event is a two-step process, such as that shown in figure 5.3 for the emission/absorption of a phonon by an electron, in which the two steps account for the time duration and spatial shift arising from the interaction.

Another process of the electron–photon interaction arises when the electron is accelerated, particularly when the acceleration takes place at high energy in a cyclotron. In the latter, the electron is continuously accelerated inwardly as it circles around the center of the machine, as the acceleration due to the Lorentz force in a strong magnetic field is directed toward the center of the circular orbit and normal to the velocity vector. The emission of photons due to this acceleration is a relativistic effect known as *bremsstrahlung* (braking radiation). Officially, this radiation arises from the 'braking' effect, or anti-breaking effect, of the accelerating force that is normal to the velocity of the electron. It is believed that the first treatment of an

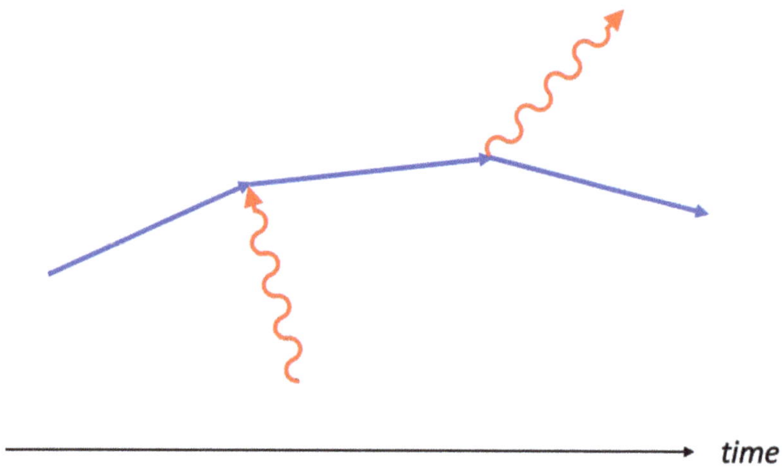

Figure 8.17. A diagrammatic representation of Compton scattering. The blue line represents the progression of the electron, while the curvy red lines represent the incoming and outgoing photons, distinguished by the directions of the arrows.

electron orbiting an ion while emitting a single photon was given by Sommerfeld [41]. For the case in which the acceleration is normal to the velocity, the radiated power produced by the electron is given by

$$P = \frac{e^2 a^2}{6\pi\epsilon_0 c^3}\left(1 - \frac{v^2}{c^2}\right)^{-2}, \tag{8.33}$$

where

$$a = \frac{dv}{dt} \tag{8.34}$$

is the acceleration of the electron. In the case of linear acceleration, where the acceleration is parallel to the velocity, such as that due to an electric field, the Lorentz factor in the parenthesis has the exponent changed from two to three. In figure 8.18, a diagrammatic illustration of *bremsstrahlung* radiation is shown for the perpendicular acceleration case. This radiation is extremely important to science, as it gives rise to the high-energy photons used for photoemission spectroscopy at storage ring sites.

8.6.2 Free-carrier absorption

In the above sections, the quantum interactions between the electron and the photon were considered. This can still be done for low-energy photons which interact with a free carrier. In a semiconductor, 'low energy' means waves of energy insufficient to produce an electron–hole pair. This is often microwave radiation, the absorption of which can only occur in a partially filled band, where there are free electrons or holes capable of gaining energy from the microwave field. However, simple absorption

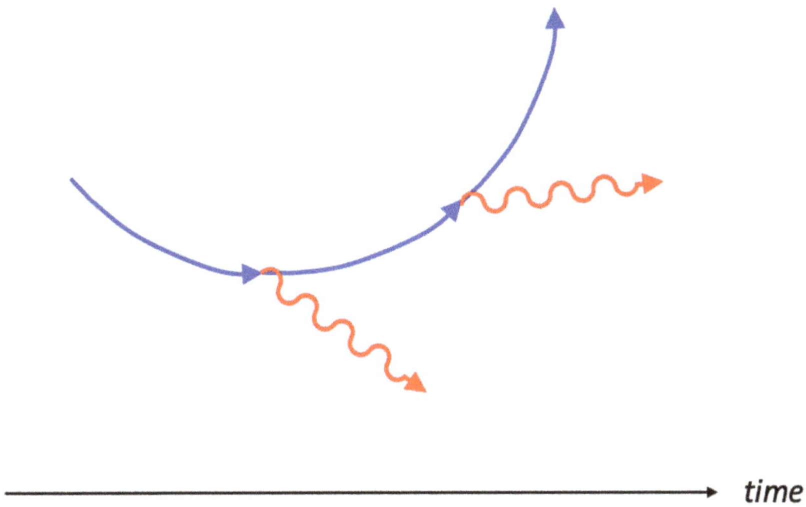

Figure 8.18. A diagrammatic representation of *Bremsstrahlung* radiation in which an accelerated electron emits a series of photons.

cannot conserve momentum in the semiconductor energy band, so that this absorption also requires a scattering process in order to conserve both energy and momentum. This process leads to microwave conductivity, although this 'microwave' frequency may well extend into the infrared region. This absorption is more easily treated via a semiclassical approach in which the quantum mechanics is limited to the details of the scattering process.

To discuss free-carrier absorption, the time variation of the distribution function in the Boltzmann equation is usually required. This will be done in the relaxation-time approximation, although it is quite easy to extend this to a detailed consideration of any and all scattering processes and a full Monte Carlo simulation. The Boltzmann equation gives the basic details of the process. It becomes simply

$$\frac{\partial f}{\partial t} + \frac{1}{\hbar}\boldsymbol{F} \cdot \frac{\partial f}{\partial \boldsymbol{k}} = -\frac{f - f_0}{\tau_{\mathrm{m}}}, \tag{8.35}$$

where $\boldsymbol{F}$ is the electric field and f_0 is the equilibrium distribution function. This equation holds whether we are dealing with a Maxwellian distribution or a Fermi–Dirac distribution. The electric field is assumed to vary according to $\boldsymbol{F} = \boldsymbol{F}_0\exp(-i\omega t)$. Even with this sinusoidal excitation, a steady state is reached in which both the velocity and the distribution function exhibit this same temporal behavior. The deviation from equilibrium is then written as $f_1 = f - f_0$, so that (8.35) becomes

$$(1 - i\omega\tau_{\mathrm{m}})f_1 = e\tau_{\mathrm{m}}(\boldsymbol{v} \cdot \boldsymbol{F}_0)\frac{\partial f_0}{\partial E}. \tag{8.36}$$

The exponential temporal variation has been omitted because it is common to all terms. It is important to note that the frequency ω and the momentum relaxation time τ_m appear as a product here. This implies that the absorption process *requires* a scattering event in order to occur. In fact, this requirement arises from the need to conserve both energy and momentum in the absorption process, as mentioned above. A microwave photon has almost zero momentum on the scale of the carrier. For example, a 10 GHz photon has a wave number of $k \sim (2\pi/3)\ \text{cm}^{-1}$, while the carrier has a wave number $>10^6\ \text{cm}^{-1}$. Because the energy band is dispersive, absorption cannot occur unless a scattering event provides the momentum shift that is needed to reach the higher energy $E + \hbar\omega$. As a result, free-carrier absorption requires both scattering and the photon in order to occur.

Assuming the electric field is oriented along the x-axis, the microwave current is determined by the nonequilibrium distribution function below [42]:

$$J_x = e^2 F_0 \int_0^\infty \frac{v_x^2 \tau_m}{1 - i\omega\tau_m} n(E) \frac{\partial f_0}{\partial E} dE, \tag{8.37}$$

By straightforward integration, the high-frequency conductivity is found to be complex and is given by

$$\sigma(\omega) = \frac{ne^2}{m_c}\left(\left\langle \frac{\tau_m}{1 + \omega^2\tau_m^2} \right\rangle + i\omega\left\langle \frac{\tau_m^2}{1 + \omega^2\tau_m^2} \right\rangle\right). \tag{8.38}$$

The real part has the normal form and falls off rapidly with frequency. This decay is often called a Drude tail. The imaginary part, the second term in the large parentheses, contributes to the dielectric function of the material. This becomes evident if the total current, including the displacement current appearing in Maxwell's equations, is written as

$$\boldsymbol{J} = \sigma \boldsymbol{F} - i\omega\varepsilon(\omega)\boldsymbol{F}, \tag{8.39}$$

so that the dielectric function, using the imaginary part of (8.38), may be written as

$$\varepsilon(\omega) = \varepsilon_s - \text{Im}\left\{\frac{\sigma}{\omega}\right\} = \varepsilon_s - \frac{\sigma(0)}{\langle \tau_m \rangle}\left\langle \frac{\tau_m^2}{1 + \omega^2\tau_m^2} \right\rangle. \tag{8.40}$$

A remarkable result is that the dielectric 'constant' can, in fact, become negative at low frequencies due to the effect of the free carriers. This is a well-known microwave effect and leads to what is known as the 'skin depth,' that is, the depth to which a microwave will penetrate into a conductor, which is limited by the conductivity of the conductor. In an infinite conductance, such as a superconductor, the skin depth becomes zero.

The very high frequency limit of (8.40), where $\omega\tau \gg 1$, gives another known result. The large-bracket term can be written out using the definition of the conductivity to obtain

$$\varepsilon(\omega) \sim \varepsilon_s\left(1 - \frac{\omega_p^2}{\omega^2}\right), \tag{8.41}$$

where

$$\omega_{\mathrm{p}}^2 = \frac{ne^2}{m_{\mathrm{c}}\varepsilon_{\mathrm{s}}} \tag{8.42}$$

defines the square of the plasma frequency. The electron 'gas' has a collective mode at this frequency which is an oscillation of the entire set of electrons. However, (8.41) is an approximation to the complete dielectric function and neglects effects due to the momentum of the free carriers [40]. Thus, when we consider energy loss to the plasmons, the quantized version of the plasma waves (8.41) is inadequate for the task. For this task, much more of the dielectric function is required.

The absorption coefficient for microwave fields may now be calculated using the wave vector itself, which is determined from the dielectric response as follows:

$$k_0 = \frac{\omega}{c}\sqrt{\frac{\varepsilon_{\mathrm{s}}}{\varepsilon_0}}. \tag{8.43}$$

Our interest here is in the weak free-carrier absorption at optical and terahertz frequencies well above the plasma edge (a relatively small free-carrier population), so that the imaginary part of the conductivity may be ignored. Still, the dielectric function is complex, and the absorption coefficient is determined from the imaginary part of the wave vector (8.42) as follows:

$$\alpha = 2k_i = 2\mathrm{Im}\{k_0\} \sim \frac{\omega_{\mathrm{p}}^2}{c\omega^2}\sqrt{\frac{\varepsilon_{\mathrm{s}}}{\varepsilon_0}}\left\langle\frac{1}{\tau_{\mathrm{m}}}\right\rangle. \tag{8.44}$$

8.6.3 Electron–hole creation

The properties of photoexcited carriers (both electrons and holes) in semiconductors are important for all optical devices today, from the vast and growing field of solar cells to advanced lasers. The role of these carriers is crucial to the operation of such devices. The behavior of optically created carriers is quite different from that of natural carriers excited by an electric field. Photoexcited carriers still create a hot distribution, but this distribution normally remains symmetric in momentum space. However, this electron system is no longer in thermal equilibrium, especially with the lattice. As a result, it is usually difficult to define an effective temperature for the distribution, even though it is symmetric. In many cases, especially those in which the polar optical interaction is present, some asymmetry in the distribution may even arise from this anisotropic scattering process.

In the creation of an electron–hole pair, there arise a number of different timescales that must be considered. The first important timescale is the time it takes for a photon to generate an electron–hole pair. This is *not* an instantaneous event. The photon does not simply cause an electron in the valence band to 'jump' across the bandgap to the conduction band. This would be too simple and is not the quantum mechanical process. Even when we consider normal perturbation theory, the absorption process is a two-step process. And this is, in fact, driven by the electric field of the photon, although it is often characterized by the vector potential (8.27) rather than the field. They are, of course, related through (8.6), especially when the coulomb gauge is adopted.

The first step of this two-step process is to create a polarization quantity that characterizes the correlation between a hole wave function in the valence band and an electron wave function in the conduction band. This polarization is generated by a coherent interaction between the light's electric field and the two wave functions of the conduction band and the valence band. In a common approach, the perturbing term in the Hamiltonian defines the interaction potential arising from the electric field as $\boldsymbol{r} \cdot \boldsymbol{F}$, so that the time-dependent interaction may be simply written as

$$H_{\text{int}} = -\frac{1}{2}\sum_{\boldsymbol{k}} F(t)\left(d_{\boldsymbol{k}} a^{\dagger}_{\boldsymbol{k},\,\mathrm{c}} a_{\boldsymbol{k},\,\mathrm{v}} e^{-i\omega t} + \text{h. c.}\right), \tag{8.45}$$

where the dipole element of the polarization is

$$d_{\boldsymbol{k}} = e\int d^3 r u^{*}_{\boldsymbol{k},\,\mathrm{c}} \boldsymbol{e}_{\boldsymbol{k},\,\mathrm{s}} \cdot \boldsymbol{r} u_{\boldsymbol{k},\,\mathrm{v}}. \tag{8.46}$$

Here, the spatial variation of $\boldsymbol{F}$ is not considered, because the wavelength of the light is much larger than the unit cell. The creation operator generates an electron state in the conduction band and the annihilation operator actually creates a hole state in the valence band. The u's are the cell periodic parts of the Bloch functions for the two states, and $\boldsymbol{e}$ is the field polarization unit vector of (8.28). The form of the interaction term (8.45) is often termed the rotating wave approximation in optics.

In normal perturbation forms, the interaction Hamiltonian term usually appears as the magnitude squared in the square of the matrix element. Thus, the complex conjugate of (8.45) generates the second form that appears in this process. This final term formalizes the creation of the electron pair by either destroying the polarization (the photon goes happily on its way) or by breaking up the correlation to create a free electron and a hole. The quantum efficiency defines the fraction of the photons that actually create free carriers. Like the actual time for scattering a phonon, the time required for this two-step process is a few femtoseconds. One of the earliest studies in this regard was by Geltman, who studied the time required for a photon to ionize an atom [43]. This follows the same two-step process but with different wave functions, such as that of an electron state bound in the atom and a free electron state, or an unoccupied higher-orbital state. In the atomic case, this may result in Rabi oscillations between these two states [44]. This is reflected in the polarization as the correlation of (8.45), which may be considered to be the entanglement of the unoccupied quantum state in the conduction band with the ground state (electron) in the valence band. Although it is the electromagnetic field in (8.44) that creates the interaction, it is actually the energy of the absorbed and annihilated photon that is transferred in an energy-conserving process to the electron–hole pair. The normal time dependence of this two-step absorption is governed by the semiconductor Bloch equations [45–47]. This process is basically coherent, even when the excitation is incoherent (such as that caused by solar radiation). It is important to point out that the ephemeral coherence discussed here in the context of photon absorption, for example, in a photovoltaic cell, does not persist once the electron–hole pairs (or excitons) have been created by incoherent and polychromatic sunlight.

The Bloch equations in the quantum version involve nonequilibrium Green's functions to handle the coherence of the system during the absorption process. However, this can be coupled to an ensemble Monte Carlo process to treat the resulting electrons and holes via normal particle methods [47]. This is not considered in depth here, and a full treatment is available in [48].

Once the polarization is broken up and the free electron and hole are created, the remaining part of the photon absorption is handled by standard perturbation theory. This occurs on a timescale longer than a few femtoseconds. The quantity that measures the strength of the optical transition is the absorption coefficient, which describes how much of the optical electromagnetic wave is turned into electron–hole pairs. To find this, the transition probability for the interband transition must first be determined. The absorption coefficient is primarily measured in units of cm^{-1}, as the wave is assumed to be attenuated as it passes through the material, according to

$$P \sim e^{-\int \alpha dx}. \tag{8.47}$$

Here, α is the absorption coefficient. The optical transition from the valence band to the conduction band is vertical, as discussed at the start of section 8.6. This is because the momentum of the photon is orders of magnitude smaller than the momentum of the electron and hole. The actual transition is made away from $k = 0$, because the densities of states for both the valence band and the conduction band are zero at $k = 0$.

For $\hbar\omega > E_G$, the energy in excess of the gap energy splits in the form of kinetic energy between the valence band and the conduction band; the factors for each are determined by the effective masses of the two bands. That is, the photon energy splits according to

$$\hbar\omega = E_G + \frac{\hbar^2 k^2}{2}\left(\frac{1}{m_e} + \frac{1}{m_h}\right), \tag{8.48}$$

where m_e and m_h are the electron and hole masses, respectively. The fraction that goes into, for example, the conduction band is

$$\frac{m_h}{m_h + m_e}. \tag{8.49}$$

In the presence of an electromagnetic wave, the Hamiltonian for the electrons includes the effects of this wave through both of the potentials and may be written as

$$H = \frac{1}{2m_0}(-i\hbar\nabla - e\boldsymbol{A})^2 + V(\boldsymbol{r}). \tag{8.50}$$

The scalar potential $V(\boldsymbol{r})$ and vector potential $\boldsymbol{A}$ were defined at the beginning of this chapter. While a mass is indicated in this equation, one should be careful about the meaning of this term, as at least two bands, each of which has its own effective mass, are under consideration. The mass shown comes from the original Hamiltonian and is the free electron mass. In general, the scalar potential is relatively slow to vary over a unit cell of the lattice; thus, the leading term in (8.50) is the vector potential term, which is the more important term in optical absorption situations. Hence, only the leading terms of importance are kept, so that (8.36) can be rewritten as

$$H = -\frac{\hbar^2}{2m_0}\nabla^2 + \frac{i\hbar e}{m_0}\boldsymbol{A}\cdot\nabla. \tag{8.51}$$

The last term in this equation is taken to be the perturbing potential that will be used in the matrix element.

The wave functions for each of the two bands are standard Bloch functions, which may be written in the form

$$\psi_{\mathrm{c,\,v}}(\boldsymbol{r}) = \frac{1}{\sqrt{N}}u_{\mathrm{c,\,v}}(\boldsymbol{r})e^{i\boldsymbol{k}\cdot\boldsymbol{r}}, \tag{8.52}$$

where the $u_{\mathrm{c,\,v}}$ are the cell periodic parts used already in (8.46). Using the vector potential term from (8.50), the matrix element may now be written as

$$\begin{aligned} H_{12} &= \frac{i\hbar e A_0}{2m_0 N}\int d^3r u_{\mathrm{c}}^{\dagger}(\boldsymbol{r})e^{-i\boldsymbol{k}\cdot\boldsymbol{r}}e^{-i\boldsymbol{q}\cdot\boldsymbol{r}}\boldsymbol{e}\cdot\nabla u_{\mathrm{v}}(\boldsymbol{r})e^{i\boldsymbol{k}'\cdot\boldsymbol{r}} \\ &= \frac{i\hbar e A_0}{2m_0 N}\int d^3r u_{\mathrm{c}}^{\dagger}(\boldsymbol{r})e^{-i(\boldsymbol{k}+\boldsymbol{k}_0-\boldsymbol{k}')\cdot\boldsymbol{r}}[\boldsymbol{e}\cdot\nabla u_{\mathrm{v}}(\boldsymbol{r}) + i\boldsymbol{e}\cdot\boldsymbol{k}'u_{\mathrm{v}}(\boldsymbol{r})] \end{aligned}. \tag{8.53}$$

Since the central terms in the Bloch functions are periodic over a unit cell, the integration can be separated into an integral over a unit cell plus a summation over all the unit cells. The summation term provides the closure condition for the Fourier transforms, and within a reciprocal lattice vector **G** the momentum conservation requires

$$\boldsymbol{k} + \boldsymbol{q} - \boldsymbol{k}' = 0. \tag{8.54}$$

Since the optical wave vector is orders of magnitude smaller than either the valence band vector or the conduction band wave vector, it may be ignored, which leads to the two particle wave vectors being equal. This gives the vertical transition previously discussed above. Using this selection rule and summing over the unit cells in the lattice, (8.52) may be rewritten as

$$H_{12} = \frac{i\hbar e A_0}{2m_0}\int d^3r u_{\mathrm{c}}^{\dagger}(\boldsymbol{r})[\boldsymbol{a}\cdot\nabla u_{\mathrm{v}}(\boldsymbol{r}) + i\boldsymbol{a}\cdot\boldsymbol{k}u_{\mathrm{v}}(\boldsymbol{r})]. \tag{8.55}$$

Normally, the first term in the brackets dominates the matrix element and is called the ‘allowed,’ or dipole, transition, as it defines certain required properties of the wave functions within the crystal. The second term is termed the ‘forbidden’ transition, as it can occur regardless of the relationship between the wave functions; this is the case because it basically breaks the required symmetry, even though it is a weaker term. This second term becomes especially important when the dipole transition is forbidden. The dipole term is similar to the manner in which this transition occurs in atomic systems as well as semiconductors. The presence of the momentum term (the spatial gradient term) means that the initial and final states must have different angular momentums. In a semiconductor, this refers to the atomic basis states that lead to the band structure, as discussed in section 6.2. The energy bands in tetrahedrally coordinated semiconductors involve the s and p atomic levels, and these differ in their angular momentums. The four tetrahedral bonds are composed of sp^3 hybrids. Thus, for example, in direct gap III–V semiconductors, the top of the valence band originates from anion p-functions, while the bottom of the conduction band originates from cation s-functions. Thus, the transition of an electron from the valence band to the conduction involves a change of one unit of angular momentum, which is accomplished by the dipole term in (8.55). So, this is an allowed transition, and the dipole term in (8.55) dominates. It is this term that will be pursued in the following.

The allowed dipole interaction usually leads to a high absorption coefficient. When this dipole transition is allowed, the matrix element (8.55) may be reduced to

$$H_{12} = \frac{i\hbar eA_0}{2m_0}\int d^3r u_c^{\dagger}(\boldsymbol{r})\boldsymbol{a}\cdot\nabla u_v(\boldsymbol{r}) \equiv -\frac{eA_0}{2m_0}\boldsymbol{a}\cdot\boldsymbol{P}_{12}. \tag{8.56}$$

The 'dipole contribution' $\boldsymbol{P}_{12}$ represents the integral over the unit cell of the central parts of the Bloch functions. For a given semiconductor, the wave functions can be found exactly and the integral can easily be obtained, giving this dipole contribution. This interaction term can now be used to determine the transition rate via the Fermi golden rule (the long time limit is not taken at this point), which is

$$\Gamma_{12} = \frac{e^2A_0^2t}{m_0^2\hbar^2}|\boldsymbol{a}\cdot\boldsymbol{P}_{12}|^2\frac{\sin^2[(\omega_{21}-\omega)t/2]}{(\omega_{21}-\omega)^2t^2}, \tag{8.57}$$

where the difference between the kinetic energies of the electron and hole is given by

$$\hbar\omega_{21} = E_c(k) - E_v(k) = E_G + \frac{\hbar^2k^2}{2}\left(\frac{1}{m_e} - \frac{1}{m_h}\right) = E_G + \frac{\hbar^2k^2}{2m_R}. \tag{8.58}$$

The effective masses of the electrons and holes have now entered the picture, and these determine how much of the excess photon energy (beyond the bandgap) is taken by each of these two carriers via (8.49).

For monochromatic radiation, a sum must be taken over the available states k that are allowed in the process. (Even for broadband radiation, this sum must still be taken, but this much more complicated in form, as it involves the joint density of states [49].) In this integration, the angle-averaged dipole moment is introduced, as follows:

$$\int d\Omega\,|\,\boldsymbol{a}\cdot\boldsymbol{P}_{12}\,|^2 = 4\pi\langle P_{12}^2\rangle. \tag{8.59}$$

The integration over the allowed states may now be performed as shown below:

$$\begin{aligned}
P(\omega) &= \frac{e^2 A_0^2 t}{m_0^2 \hbar^2} \int d^3\boldsymbol{k} \frac{|\boldsymbol{a} \cdot \boldsymbol{P}_{12}|^2}{4\pi^3} \frac{\sin^2[(\omega_{21} - \omega)t/2]}{(\omega_{21} - \omega)^2 t^2} \\
&= \frac{e^2 A_0^2}{m_0^2 \hbar^3} \frac{2k\langle P_{12}^2 \rangle m_R}{\pi^2} \int \frac{\sin^2 x}{x^2} dx \\
&= \frac{2e^2 A_0^2 k \langle P_{12}^2 \rangle m_R}{\pi m_0^2 \hbar^3}.
\end{aligned} \tag{8.60}$$

At this point, the oscillator strength is defined to be the ratio of the kinetic energy in the transition to the photon energy, as follows:

$$f_{osc} = \frac{2\langle P_{12}^2 \rangle}{m_0 \hbar \omega}. \tag{8.61}$$

This oscillator strength is usually near unity, but in semiconductors it can be of the order of $(1 + m_0/m_h)$. Using this result and replacing the wave number k with the energy allows (8.60) to become

$$P(\omega) = \frac{e^2 A_0^2 (2m_R)^{3/2} \omega f_{osc}}{2\pi m_0 \hbar^3} (\hbar\omega - E_G)^{1/2}. \tag{8.62}$$

Finally, it is necessary to normalize (8.62) to the number of incident photons arriving per second in order to calculate the absorption coefficient. The number of photons absorbed within the semiconductor sample, for a thickness of d, is just $P(\omega)d$, which is related to the absorption coefficient by $\alpha F_0 d$, where F_0 is the flux of incoming photons. The latter quantity is readily calculated from the power incident on the sample, given by the Poynting vector S of the plane wave, as follows:

$$F_0 = \frac{\boldsymbol{S} \cdot \boldsymbol{a}_k}{\hbar\omega} = \frac{\omega A_0^2}{2\hbar} \sqrt{\frac{\varepsilon_\infty}{\mu_0}} = \frac{\omega A_0^2}{2\hbar} G_\omega, \tag{8.63}$$

where G_ω is the wave admittance. Finally, the absorption coefficient can be found to be

$$\alpha = \frac{e^2 (2m_R)^{3/2} f_{osc}}{\pi m_0 \hbar^2 G_\omega} (\hbar\omega - E_G)^{1/2}. \tag{8.64}$$

The square-root variation in (8.64) is just that expected from the density of final states of either the conduction band or the valence band. The numerator mass involved is the reduced mass, which accounts for both the valence state and the conduction state. The denominator mass, however, arose from the Hamiltonian, and is obtained prior to the effective mass approximation, as it still contains the full potential of the crystal and thus should be the free electron mass. Excellent tabulations of the properties of materials, including the optical absorption coefficient (as a function of energy), can be found at [50, 51]. In figure 8.19, the allowed absorption coefficients for InAs doped to 10^{16} cm^{-3} [52], semi-insulating GaAs [53],

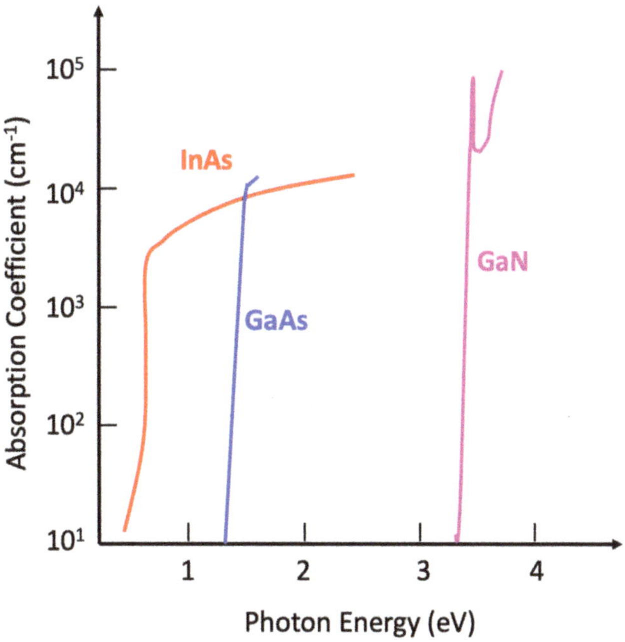

Figure 8.19. The absorption coefficients of some typical III–V materials. Reproduced from [42]. © IOP Publishing Ltd. All rights reserved.

and wurtzite GaN doped to 10^{17} cm^{-3} [54] are shown. For InAs and GaAs, the data refer to room temperature, while the GaN data were obtained at 77 K.

8.7 Electron–electron scattering

The coulomb interaction between a pair of electrons is an embodiment of Maxwell's equations. In this interaction, the transfer of momentum and energy is carried by a photon. Thus, one electron emits a photon while a second electron absorbs a photon, as is described diagrammatically in figure 8.20. The photon is the red wiggly line that transits the two sets of electron lines. The change of momentum of the first electron is mirrored in the second electron, as the photon momentum is negligible compared to that of the electrons. The resulting potential is then given by

$$V = \frac{e^2}{4\pi\epsilon_0 r}, \tag{8.65}$$

where r is the distance between the two electrons. The problem with this potential is that it can take effect at very long distances and is usually considered to be divergent upon integration. As a result, it is often treated by inserting a cutoff, such as a Debye or Fermi–Thomas screening length, which adds an exponential decay term to the above formula. However, this often results in improper use, as a great deal of the interaction is thrown away.

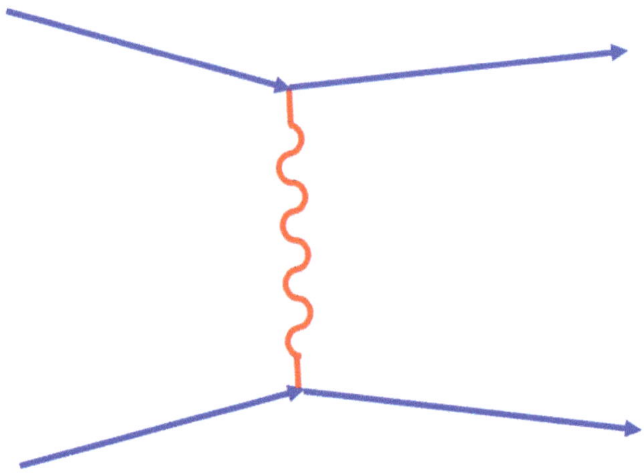

Figure 8.20. A diagrammatic representation of the single-particle electron–electron interaction. In this process, a 'photon' is exchanged between the two particles. The photon is the carrier of the coulomb interaction.

The recognition of the type of scattering follows from introducing the Fourier transform of (8.65) into a many-electron system, which leads to

$$V = \frac{e^2}{4\pi\epsilon_0}\sum_{i<j}\frac{1}{q^2}e^{i\mathbf{q}\cdot(\mathbf{r}_i-\mathbf{r}_j)}. \tag{8.66}$$

The limit on the sum is such that the interaction is not double counted. If a cutoff is introduced, such as the Debye screening factor, whose momentum is given by

$$q_D^2 = \frac{ne^2}{\epsilon_0 k_B T}, \tag{8.67}$$

where n is the electron density and k_BT is the thermal energy for a Maxwellian distribution, this results in the change

$$\frac{1}{q^2} \to \frac{1}{q^2 + q_D^2}, \tag{8.68}$$

that effectively cuts off the potential at the Debye screening momentum. But this does not mean that only the remaining component contains the entire interaction. Instead, the screened portion that remains after introducing this cutoff is only the single-particle scattering, which is a short-range part for which $q > q_D$. This short-range part is the proper coulomb interaction for the scattering of an electron by another electron, as in the single-particle interaction shown in figure 8.20.

However, the long-range interaction, for which $q < q_D$, has been ignored yet is still important. This long-range interaction is responsible for electron scattering due to the collective excitations of the total electron gas. These are the so-called plasmon modes, and they are also effective in scattering electrons. This plasmon mode is the fundamental mode of oscillation in the electron gas and arises from the interparticle coulomb interaction. These are boson modes, so scattering by plasmons can be treated in much the same fashion as scattering by phonons [55]. In the next two sections, we discuss the scattering of an electron by plasmons and by the single-particle interaction. Yet it is far better to include the electron–electron scattering in the real-space approach introduced later.

8.7.1 Electron–plasmon scattering

Long-range electron–plasmon scattering takes into account the scattering of an incident electron by the long-range collective excitations of the electron gas—the plasmons. Here, as in the single-particle case, the strength of the scattering is determined by the dielectric function. The zeroes of the dielectric function give the plasmon frequency (8.42). But this simple plasmon-pole approximation does not allow one to determine the scattering, as it gives a real dielectric function, and the scattering and dissipation arise from the imaginary part. Moreover, the plasmon frequency is a set quantity only in three dimensions. In fewer dimensions, the plasmon frequency depends upon the momentum in the scattering event. As a result, a more complete, momentum-dependent form for the dielectric function must be used. Indeed, the importance of the plasmons for transport in semiconductors has been known for quite some time [56–58]. The full dielectric function has zeroes at both zero frequency and at the plasmon frequency, which leads to the two parts of the interaction [59, 60]. Hence, in order to capture the details around the plasmon pole, especially in low-dimensional systems, the imaginary part of the inverse of the dielectric function must be properly incorporated. It is the poles of this latter function which give rise to the important scattering processes involving the plasmons. An abbreviated form of this dielectric function can be used, as follows [42]:

$$\varepsilon(\boldsymbol{q}) = \varepsilon(0)\left[1 - \frac{\omega_p^2}{(\boldsymbol{q} \cdot \boldsymbol{v} - i\alpha)^2}\right], \tag{8.69}$$

where $\boldsymbol{q}$ is the plasmon momentum wave number, $\boldsymbol{v}$ is the carrier velocity, and α is a damping factor that will drop out in the next step. The scattering strength is given by the imaginary part of the inverse of the dielectric function, as follows [42]:

$$\frac{1}{\pi}\mathrm{Im}\left(\frac{1}{\varepsilon(\boldsymbol{q})}\right) \cong \frac{1}{2\varepsilon(0)}\left[\delta(\boldsymbol{q} \cdot \boldsymbol{v} + \omega_\mathrm{p}) + \delta(\boldsymbol{q} \cdot \boldsymbol{v} - \omega_\mathrm{p})\right]. \tag{8.70}$$

This last expression arises from the relationship

$$\lim_{\alpha \to 0} \frac{1}{x - i\alpha} = P\frac{1}{x} + i\pi\delta(x). \tag{8.71}$$

Here, P denotes the principal part of the integral that will arise in summing over the set of final states, so that the damping factor disappears with this result. The two delta functions correspond to the absorption and the emission of a plasmon by the electron during the interaction. Restoring the full arguments of the delta functions, the scattering rate can be written as

$$
\begin{aligned}
\Gamma_{\mathrm{e-pl}}(k) = \frac{2\pi}{\hbar}\frac{1}{4\pi^2}\int q^2 dq \frac{\hbar\omega_{\mathrm{p}}}{2\varepsilon(0)q^2}\int_0^{\pi}\sin\vartheta d\vartheta \\
\times\left[N_{\mathrm{pl}}\delta\left(E(k+q) - E(k) + \hbar\omega_{\mathrm{p}}\right)\right. \\
\left. + \left(N_{\mathrm{pl}} + 1\right)\delta\left(E(k+q) - E(k) - \hbar\omega_{\mathrm{p}}\right)\right]
\end{aligned}
\quad . \tag{8.72}
$$

As usual, this is a coulomb interaction, so it is anisotropic. Thus, the arguments of the delta functions include the scattering angles and lead to limits on the remaining integrations and some numerical factors. In addition, the Bose–Einstein distribution applies to the plasmons themselves,

$$
N_{\mathrm{pl}} = \left[1 + \exp\left(\frac{\hbar\omega_{\mathrm{p}}}{k_{\mathrm{B}}T_{\mathrm{e}}}\right)\right]^{-1}. \tag{8.73}
$$

This distribution is for the plasmons, which are actually the electrons themselves; therefore, the distribution is a function of the *electron* temperature and not the lattice temperature. The plasmons are collective modes of the electron gas and should evolve to have the same electron temperature as that found in the single-particle scattering. With this in mind, the integrations follow an identical path to those of the polar optical phonons, yielding

$$
\begin{aligned}
\Gamma_{\mathrm{e-pl}}(k) &= \frac{m_{\mathrm{c}}e^2\omega_{\mathrm{p}}}{4\pi\hbar^2 k}\left[N_{\mathrm{pl}}\int_{q_-^a}^{q_{\mathrm{D}}}\frac{dq}{q} + \left(N_{\mathrm{pl}} + 1\right)\int_{q_-^e}^{q_{\mathrm{D}}}\frac{dq}{q}\right] \\
&= \frac{m_{\mathrm{c}}e^2\omega_{\mathrm{p}}}{4\pi\hbar^2 k}\left[N_{\mathrm{pl}}\ln\left(\frac{q_{\mathrm{D}}/k}{\sqrt{1 + \hbar\omega_{\mathrm{p}}/E} - 1}\right)\right. \\
&\quad \left. + \left(N_{\mathrm{pl}} + 1\right)\ln\left(\frac{q_{\mathrm{D}}/k}{1 - \sqrt{1 - \hbar\omega_{\mathrm{p}}/E}}\right)\right]
\end{aligned}
\quad . \tag{8.74}
$$

One important difference between this scattering and phonon scattering is that the Bose–Einstein distribution does not stay in equilibrium but evolves as the electron temperature evolves. Fortunately, studies have shown that the plasmon temperature comes into a quasi equilibrium with the free-carrier temperature at a given high electric field [57]. This provides a reasonable assurance that the approach is consistent with the single-particle scattering to be discussed next.

8.7.2 Single-particle scattering

We now consider the scattering of one carrier by the other carriers in the single-particle form. One must not forget that in the interaction between any pair of electrons, the total momentum and energy must be conserved. This does not imply that an individual electron cannot lose significant energy in the process, only that the process does not give rise to a net overall energy loss to the lattice. However, it does lead to a redistribution of the energy within the electron ensemble, such as that predicated in the decay to a Maxwellian distribution. As discussed above, the scattering matrix element is determined from the screened coulomb interaction

$$M(q) = \frac{e^2}{\epsilon_s\left(q^2 + q_D^2\right)}. \tag{8.75}$$

Here, ϵ_s is the static permittivity for the material of interest. Following Takanaka [61], the scattering rate can be written as

$$\begin{aligned}\Gamma_{ee}(k) = \frac{2\pi}{\hbar^2} \sum_{k_2,k',k_2'} \left[\frac{e^2}{\varepsilon_s\left(q^2 + q_D^2\right)}\right]^2 f(k_2) \\ \times \delta_{k+k_{12},\, k'+k_2'} \delta\left(E_{k'} + E_{k_2'} - E_k - E_{k_2}\right)\end{aligned} \tag{8.76}$$

In this expression, $\boldsymbol{k}$ and $\boldsymbol{k}_2$ are the momentum vectors of the initial particle and the particle with which it interacts, respectively. The other two wave vectors are the resulting final momenta of the two particles, as appropriate. The two delta functions ensure that momentum and energy are conserved in the scattering process. It may be assumed, without loss of generality, that the distribution function is nondegenerate; it is, of course, the distribution function for the energetic carriers and includes the electron temperature.

The summation over the final wave vector $\boldsymbol{k}_2'$ is carried out first using the momentum-conserving delta function. To accomplish this, two new wave vectors are introduced for convenience through the substitutions

$$\boldsymbol{g} = \boldsymbol{k}_2 - \boldsymbol{k}, \quad \boldsymbol{g}' = \boldsymbol{k}_2' - \boldsymbol{k}' = \boldsymbol{k} + \boldsymbol{k}_2 - 2\boldsymbol{k}'. \tag{8.77}$$

As a result of these substitutions, the arguments of the delta functions for energy and another for the momentum integration are changed to use the new variables, as follows:

$$\begin{aligned} E_{k'} + E_{k_2'} - E_k - E_{k_2} &= \frac{1}{2}\left(E_{g'} - E_g\right), \\ \boldsymbol{k}' &= \boldsymbol{k}_2 - \frac{1}{2}(\boldsymbol{g} + \boldsymbol{g}'). \end{aligned} \tag{8.78}$$

The summation over $\boldsymbol{k}'$ can now be changed into a summation over $\boldsymbol{g}'$ (there is a factor of eight in the new denominator arising from the various factors of ½ in each of the three directions). The scattering rate (8.75) then becomes

$$\Gamma_{ee}(k) = \frac{\pi}{4\hbar} \sum_{k_2,g'} \left[\frac{e^2}{\varepsilon_s\left(\frac{1}{4}\left|\boldsymbol{g} - \boldsymbol{g}'\right|^2 + q_D^2\right)}\right]^2 f(k_2)\delta\left(\frac{1}{2}\left(E_{g'} - E_g\right)\right). \tag{8.79}$$

Another change of variables is now made, namely

$$u = g' - g. \tag{8.80}$$

Following this change, the summation over g' can now be performed as shown below:

$$\begin{aligned}\sum_{g'} &\rightarrow \frac{1}{4\pi^2}\int u^2 du \int_0^{\pi} \sin\vartheta d\vartheta \frac{e^4}{\varepsilon_s^2 \left(u^2/4 + q_D^2\right)^2}\delta\left(\frac{\hbar^2}{4m_c}(u^2 + 2ug\cos\vartheta)\right) \\ &= \frac{e^4 m_c}{8\pi\hbar^3 q_D^2 \varepsilon_s^2}\sum_{k_2} f(k_2)\left[\frac{|k_2 - k|}{|k_2 - k|^2 + q_D^2}\right]\end{aligned}. \tag{8.81}$$

At this point, some assumptions have to be made about the distribution function that appears in this last expression. Here, it will be assumed to be a Maxwellian at an elevated electron temperature T_e, which is appropriate for nondegenerate statistics. This does not reduce the generality, as Fermi–Dirac statistics could easily be used at the cost of employing Fermi integrals for the final expression. Finally, the last change of variables is used to give the reduced quantities

$$\begin{aligned} y^2 &= \frac{\hbar^2 \xi^2}{2m_c k_B T_e}, \ \xi = k_2 - k, \\ x^2 &= \frac{\hbar^2 k^2}{2m_c k_B T_e}, \\ \lambda^2 &= \frac{\hbar^2 q_D^2}{2m_c k_B T_e}. \end{aligned} \tag{8.82}$$

The final integration can now be carried out, which yields the following expression for the single-particle scattering:

$$\begin{aligned}\Gamma_{ee}(k) &= \frac{e^4 m_c}{8\pi\hbar^3 q_D^2 \varepsilon_s^2}\int_0^{\infty}\frac{\xi^2 d\xi}{2\pi^2}\int_0^{\pi}\sin\vartheta d\vartheta f(\xi + k)\left(\frac{\xi}{\xi^2 + q_D^2}\right) \\ &= \frac{ne^4 m_c}{8\sqrt{\pi}\hbar^3 q_D^2 \varepsilon_s^2 k}e^{-x^2}\int_0^{\infty} e^{-y^2}\frac{y^2}{y^2 + \lambda^2}\sinh(2xy)dy\end{aligned}. \tag{8.83}$$

This result can be rewritten as a collection of error functions and exponential integrals, but such a result provides no new information, as it remains a set of tabulated functions and not a closed form that is easily evaluated. It is best just to numerically evaluate (8.82). In figure 8.21, the scattering rate (8.82) for Si is plotted as a function of the carrier energy using the Maxwellian distribution at an elevated temperature of 2500 K and a density of 10^{17} cm^{-3}. Curves are shown for the emission and absorption of energy by the initial carrier as well as for the total scattering rate. The fact that these two curves are almost identical strongly supports

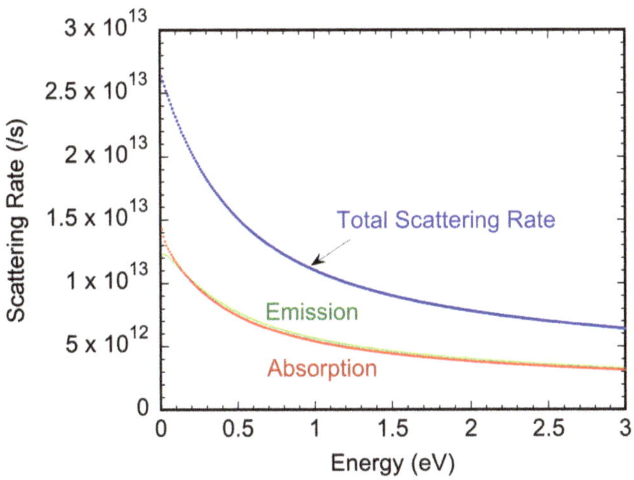

Figure 8.21. The emission, absorption, and total scattering rate for single-particle electron–electron scattering. This is shown for the case of Si doped to 10^{17} cm^{-3} at an electron temperature of 2500 K. Reproduced from [42].

the overall conservation of energy in this single-particle interaction. These values correspond to the situation in many metal–oxide–semiconductor field-effect transistors (MOSFETs) today.

8.7.3 Treating the interaction in real space

In the previous sections above, the electron–electron interaction is treated entirely in momentum and frequency space. Thus, scattering from the single-particle interaction as well as the plasmon absorption and emission were considered to be separate processes. The results are complicated. One important note is that these processes are largest when the particles are closest in momentum space. In fact, the coulomb interaction between two particles is largest when they are closest in real space. It is, in fact, possible to treat the complete electron–electron interaction in real space and time. In this treatment, approximations are not required, although other adaptations will be made. The real-space approach is simply molecular dynamics, in which the forces between the particles are computed directly and are updated at each time step in the simulation. That is, the force on each carrier due to the other carriers and the charged impurities is computed anew at each time step in order to maintain the temporal evolution of these forces. This force is then applied to the carrier during the next time step. It is important to note that this is not a perturbative approach; in addition, no separation into single-particle interactions and plasmons is required.

Consider an electron distribution (in an *n*-type semiconductor; the charged donors are ignored for the moment) in which normal transport and scattering is treated through an ensemble Monte Carlo process. The interelectronic coulomb interaction is retained as a real-space potential. The problem with this direct calculation in real space is that it can be compute intensive. Each particle scatters with $N - 1$ other particles, where N is the number of particles in the simulation. This leads to a downside, which is that only a finite number of particles can be included due to the large computational load of recomputing the interparticle forces at each time step. The potential between the electrons is given by the standard form in terms of the ith electron that has a position $\mathbf{r}_i$, as follows:

$$V = \frac{1}{2}\sum_{i,j}\frac{e^2}{4\pi\varepsilon_s\left|\mathbf{r}_i - \mathbf{r}_j\right|}, \tag{8.84}$$

although the case $i = j$ has to be omitted from the summation. The factor of two arises from the fact that summing completely over both indices counts each interaction twice. The force that arises from this potential is simply

$$F(\mathbf{r}_i) = -\nabla V = \sum_j \frac{e^2}{4\pi\varepsilon_s\left|\mathbf{r}_i - \mathbf{r}_j\right|^2}. \tag{8.85}$$

The most significant use of transport and simulation that is of interest here lies in modeling real semiconductor devices. In these simulations, the internal self-consistent potential that arises as charges (free carriers) move (or not when the impurities are included) is found from the solution of Poisson's equation. One may note that the potential (energy) is in fact a solution to the Lienard–Wiechert potential for the static wave equation for the scalar potential—and this is the normal solution to the Poisson equation. Thus, if the molecular dynamic force is computed and used for all the electrons and Poisson's equation is solved as well, then the interparticle forces have been computed twice! One cannot do both without some corrections. Generally, a grid is imposed on the device, as discussed in chapter 7, and the potential is determined by Poisson's equation at these grid points (the solution is often referred to as the mesh potential). Here, the grid is *not* part of a periodic structure, but must replicate the actual device structure. Imposing this grid naturally yields a cutoff in the accuracy of the Poisson equation solution at small distances. This is where the correction can be imposed.

The separation of the interparticle coulomb force from the Poisson forces has been discussed for quite some time. Such a split was used by Kohn for electronic structure calculations, where interelectronic forces had to be added to the potential solutions from Poisson's equation for the atomic potentials in the density-functional approach [62]. The split was also previously used in order to treat molecular dynamics problems in plasma physics, in which the short-range interaction also appears between ions and electrons [63]. In addition, this type of coulomb force split has been shown to be generally useful in many-body problems [64]. The general approach is similar to dividing a delta function, where one can write

$$\delta(r) \rightarrow \delta(r) + f(r) - f(r). \tag{8.86}$$

This suggests a split of coulomb forces according to the formula

$$\frac{1}{r} \rightarrow [1 - \mathrm{erf}(r)]\frac{1}{r} + \frac{1}{r}\mathrm{erf}(r). \tag{8.87}$$

The first term on the right-hand side is a short-range function that vanishes as the magnitude of r increases. On the other hand, the last term on the right-hand side is a long-range function that vanishes at short distances. This is, of course, the entire principle of the potential split in which the long-range function is found using Poisson's equation. Using a split of the form of (8.87) makes the use of Poisson's equation more difficult, as the actual charge density has the more cumbersome multiplicative function. On the other hand, the molecular dynamics part becomes much more efficient, as only the nearby particles interact with the single particle of concern. The range of the error function correction is adjusted or corresponds closely to the grid spacing that is chosen. One can use a slightly larger grid (the Poisson solution is usually the most time-consuming part of the simulation) to speed up the simulation; the short-range corrections come from the molecular dynamic terms. This makes both the Poisson solution and the molecular dynamics more efficient in terms of computer resources [65]. One can go further by also incorporating the degeneracy of the semiconductor into the molecular dynamics through the rejection for filled states [66]. For extremely high densities, one can also correct the interparticle force for the quantum mechanical exchange interaction energy [67].

8.8 The quantization of electrons and radiation

All the quantum developments that have been treated in the previous sections address either a quantum state for electrons or a quantum state for electromagnetic fields but not a quantum state for both. The interaction between them has been considered through external potentials, scattering phenomena or, at best, perturbative techniques. In this section, a unique quantum state for both electrons and radiation will be described. Such a compact treatment provides an understanding of the so-called strong matter–light coupling regime [68] that cannot be modeled through perturbative theories. The formal study of such strong coupling between light and matter was started by the quantum optics community a long time ago [69–71], but it is basically ignored by the electron device community, even today.

The starting point will be the development of a common Hamiltonian for electrons and electromagnetic fields. In fact, this Hamiltonian is just the one already mentioned in (8.49), to which the electromagnetic energy is added. To properly discuss the simultaneous quantization of radiation and matter, the Hamiltonian needs to be written in terms of canonical variables. The electron position and its momentum are already canonical variables, but the vector and scalar potentials appearing in (8.49) are not. Thus, most of the mathematical effort in this section will be devoted to rewriting the vector and scalar potentials in (8.49), and adding the electromagnetic energy, in terms of canonical variables. These canonical variables are basically the Fourier modes that were already presented in section 8.5.

Instead of working with the bra-ket notation as done in section 8.5, here the quantization of radiation and electrons will be addressed in terms of a wave function in the configuration space that includes both of the degrees of freedom for electrons and for electromagnetic fields. Indeed, at the end of this section, it will be shown that all these light–matter phenomena (even in the strong coupling regime) allow a description in terms of electrons as particles with well-defined positions and velocities and also as electromagnetic fields with well-defined amplitudes and phases [68].

8.8.1 Maxwell's equations in terms of transverse and longitudinal fields

The starting point is to distinguish the longitudinal and transverse components of the electromagnetic field in Maxwell's equations. The Helmholtz decomposition of vector fields into solenoidal and irrotational components [72] allows one to write the total electric field $\boldsymbol{E} = \boldsymbol{E}_{\perp} + \boldsymbol{E}_{\parallel}$ in (8.1) in terms of transverse $\boldsymbol{E}_{\perp}$ (zero divergence) and longitudinal $\boldsymbol{E}_{\parallel}$ (zero rotational) components. The transverse field $\boldsymbol{E}_{\perp}$ satisfies:

$$\nabla \cdot \boldsymbol{E}_{\perp} = \boldsymbol{0} \tag{8.88}$$

and

$$\nabla \times \boldsymbol{E}_{\perp} = \frac{\partial \boldsymbol{B}}{\partial t}, \tag{8.89}$$

where the last equation couples it to the magnetic field $\boldsymbol{B}$. The longitudinal electric field $\boldsymbol{E}_{\parallel}$ satisfies

$$\nabla \cdot \boldsymbol{E}_{\parallel} = \frac{\rho}{\epsilon}, \tag{8.90}$$

and the transverse field $\boldsymbol{E}_{\perp}$ satisfies

$$\nabla \times \boldsymbol{E}_{\perp} = 0. \tag{8.91}$$

The first equation (8.90) couples the longitudinal electric field to the charge density ρ, defined by

$$\rho \equiv \rho(\boldsymbol{r}, t) = -\sum_{j=1}^{N} e\delta(\boldsymbol{r} - \boldsymbol{r}_j(t)) \tag{8.92}$$

for an active device region with N electrons, each one with a charge of $-e$ located at $\boldsymbol{r}_j(t)$, where ϵ is the vacuum permeability. The magnetic field, on the other hand, only has a transverse component, that is $\boldsymbol{B}_{\perp} = \boldsymbol{B}$ and $\boldsymbol{B}_{\parallel} = 0$, and it satisfies

$$\nabla \cdot \mathbf{B} = 0 \tag{8.93}$$

and

$$\nabla \times \boldsymbol{B} = \mu\epsilon \frac{\partial \boldsymbol{E}}{\partial t} + \mu \boldsymbol{J}, \tag{8.94}$$

where μ is the vacuum magnetic permeability that defines the speed of light as $c = 1/\sqrt{\mu\epsilon}$. The (particle or conduction) current density in (8.94) is defined by

$$\boldsymbol{J} \equiv \boldsymbol{J}(\boldsymbol{r}, t) = -\sum_{j=1}^{N} ev_j(t)\delta(\boldsymbol{r} - \boldsymbol{r}_j(t)), \tag{8.95}$$

where $v_j(t)$ is the velocity of particle j. Notice that (8.92) and (8.95) are valid definitions of the charge and current densities not only when classical mechanics is used but also when Bohmian quantum mechanics [73, 74] is invoked, in which each particle has a well-defined position $\boldsymbol{r}_j(t)$ and velocity $v_j(t)$.

It was already mentioned in section 8.1 that Maxwell's equations imply the existence of vector $\boldsymbol{A}$ and scalar V potentials, which define the fields through

$$\boldsymbol{E} = -\frac{\partial \boldsymbol{A}}{\partial t} - \nabla V \tag{8.96}$$

and

$$\boldsymbol{B} = \nabla \times \boldsymbol{A}. \tag{8.97}$$

Since distinct potential pairs $[\boldsymbol{A}, V]$ may yield the same field pairs $[\boldsymbol{E}, \boldsymbol{B}]$, one needs to impose a *gauge*, as discussed in section 8.1, to uniquely define the electromagnetic field. The coulomb gauge, in which $\boldsymbol{A}$ is also a transverse vector satisfying $\nabla \cdot \boldsymbol{A} = 0$, will be used throughout this section. Thus, the electric field in (8.96) can be written as

$$\boldsymbol{E}_{\perp} = -\frac{\partial \boldsymbol{A}}{\partial t} \tag{8.98}$$

and

$$\boldsymbol{E}_{\parallel} = -\nabla V. \tag{8.99}$$

Gauss's law (8.90) can now be rewritten in terms of V as the well-known (and common in electronics) Poisson equation

$$\nabla^2 V = -\frac{\rho}{\epsilon}. \tag{8.100}$$

Notice that (8.100) is exactly (8.11) when the above-mentioned coulomb gauge $\nabla \cdot \boldsymbol{A} = 0$ is used. The explicit solution of (8.100) is given, in general, by the coulomb potential (8.84). The fact that the longitudinal electric field $\boldsymbol{E}_{\parallel}$ in (8.90) instantaneously follows the evolution of the charge distribution through (8.84) does not imply the existence of electrical phenomena propagating faster than light. Indeed, only the total electric field $\boldsymbol{E}$ in (8.96) can be measured. It can be shown that the transverse field $\boldsymbol{E}_{\perp}$ also has an instantaneous component which exactly balances that of $\boldsymbol{E}_{\parallel}$ in such a way that the total field $\boldsymbol{E}$ remains purely retarded (even in the coulomb gauge) [69].

8.8.2 A Hamiltonian for electrons interacting with the electromagnetic field

It is now necessary to develop a Hamiltonian that will provide an equation of motion not only for the electrons but also for the electromagnetic field (i.e. the Maxwell equations mentioned previously). The results will be valid for both classical and quantum scenarios. As indicated in (8.20), the inclusion of the interaction between electrons and radiation in the Hamiltonian needs the following substitution $\boldsymbol{p} \rightarrow \boldsymbol{p} - e\boldsymbol{A}$. Thus, the Hamiltonian for N electrons interacting with radiation is described by

$$H = \left(\sum_{j=1}^{N} \frac{1}{2m_0} \left(\boldsymbol{p}_j - e\boldsymbol{A} \right)^2 + U_e(\boldsymbol{r}_j, t) \right). \qquad (8.101)$$
$$+ V(\boldsymbol{r}_1, \boldsymbol{r}_2, \ldots, \boldsymbol{r}_N) + H_R$$

When comparing (8.101) to the previous equation (8.50) in section 8.6.3, one mainly notices the addition of the energy of the electromagnetic field, H_R, defined as

$$H_R = \frac{\epsilon_0}{2} \int d^3r (\boldsymbol{E}_\perp \cdot \boldsymbol{E}_\perp + c^2 \boldsymbol{B} \cdot \boldsymbol{B}). \qquad (8.102)$$

The electromagnetic energy only depends on the transverse component $\boldsymbol{E}_\perp$, because the energy assigned to the longitudinal field $\boldsymbol{E}_\parallel$ is given (in the coulomb gauge) by the scalar potential $V(\boldsymbol{r}_1, \boldsymbol{r}_2, \ldots, \boldsymbol{r}_N)$ solution of the Poisson equation (8.100). Thus, in the coulomb gauge, $\boldsymbol{E}_\parallel$ is not an independent variable but a variable totally fixed by the position $\boldsymbol{r}_1, \boldsymbol{r}_2, \ldots, \boldsymbol{r}_N$ of the electrons, as indicated in (8.84). Thus, only $\boldsymbol{E}_\perp$ and $\boldsymbol{B}$ have to be considered to be independent variables when defining the electromagnetic fields and the electrons. This issue will become relevant when discussing the canonical variables that describe the electromagnetic field.

The term $U_e(\boldsymbol{r}_j, t)$ in (8.101) accounts for the band structure 'seen' by the electron $\boldsymbol{r}_j$, which results from the interaction of the simulated electron $\boldsymbol{r}_j$ with the other non-simulated degrees of freedom (such as the atoms of the device within the Born–Oppenheimer approximation). Notice that $U_e(\boldsymbol{r}_j, t)$ is an external potential, while $V(\boldsymbol{r}_1, \boldsymbol{r}_2, \ldots, \boldsymbol{r}_N)$ is an internal potential.

The first simplification, to keep a didactic explanation, is to neglect the coulomb interaction between electrons. In the context of electronics, this approximation is typically invoked when ballistic transport is assumed. In our case, this approximation means that the Hamiltonian in (8.101) deals with only one ($N = 1$) electron. In this single-electron scenario, then, the term $V(\boldsymbol{r}_1, \boldsymbol{r}_2, \ldots, \boldsymbol{r}_N)$ is zero. Thus, from (8.99), one gets $\boldsymbol{E}_\parallel = \boldsymbol{0}$ and, from (8.98), $\boldsymbol{E} = \boldsymbol{E}_\perp = -\partial \boldsymbol{A}/\partial t$.

As already discussed in section 8.1, the Hamiltonian (8.100) provides identical results regardless of the choice of the gauge. Using the dipolar approximation $\boldsymbol{A} \equiv \boldsymbol{A}(\boldsymbol{r}, t) \approx \boldsymbol{A}(\boldsymbol{r}_0, t)$, the gauge $\boldsymbol{A}' = \boldsymbol{A} + \nabla\chi$, and $V = V' + \partial\chi/\partial x$ with $\chi = e\boldsymbol{A}(\boldsymbol{r}_0, t) \cdot \boldsymbol{r}/\hbar$, equation (8.103) can be rewritten as

$$H = \frac{1}{2m_0} \boldsymbol{p}^2 + U_e(\boldsymbol{r}, t) + \frac{\epsilon_0}{2} \int d^3r (\boldsymbol{E}_\perp \cdot \boldsymbol{E}_\perp (\boldsymbol{E}_\perp \cdot \boldsymbol{E}_\perp + c^2 \boldsymbol{B} \cdot \boldsymbol{B}) . \qquad (8.103)$$
$$+ c^2 \boldsymbol{B} \cdot \boldsymbol{B}) + e\boldsymbol{r} \cdot \boldsymbol{E}_\perp$$

The dipolar approximation used here just indicates that the matter–light interaction takes place in a spatial region smaller than the wavelength of the electromagnetic field [70]. As mentioned previously, this is a typical and reasonable assumption inside the active region of nanoscale devices.

8.8.3 Canonical variables for the description of one mode of the electromagnetic fields

The variables $\boldsymbol{r}$ and $\boldsymbol{p}$ are conjugate canonical variables for the electron, in the sense that they satisfy the Hamilton equations in expression (1.8) of chapter 1. The same is not true for $\boldsymbol{E}_{\perp}$ and $\boldsymbol{B}$ in (8.103) or the vector and scalar potentials in (8.101). To look for canonical degrees of freedom of the electromagnetic fields, it is convenient to work in the reciprocal space. When the electromagnetic field is confined within a cube of volume Ω, the electric and magnetic fields can be described by a discrete set of complex functions of time $\widetilde{\boldsymbol{E}}_{k,\,s}(t)$ and $\widetilde{\boldsymbol{B}}_{k,\,s}(t)$, i.e. the spatial Fourier components, defined as

$$\widetilde{\boldsymbol{E}}_{k,\,s}(t) = \frac{1}{V}\int d^3\boldsymbol{r}\boldsymbol{e}_s\boldsymbol{E}_{\perp}(\boldsymbol{r},\, t)e^{-i\boldsymbol{k}\cdot\boldsymbol{r}} \tag{8.104}$$

and

$$\widetilde{\boldsymbol{B}}_{k,\,s}(t) = \frac{1}{V}\int d^3\boldsymbol{r}\boldsymbol{e}_s'\boldsymbol{B}(\boldsymbol{r},\, t)e^{-i\boldsymbol{k}\cdot\boldsymbol{r}}. \tag{8.105}$$

The polarization of the magnetic field $\boldsymbol{e}_s'$ is different from that of the electric field $\boldsymbol{e}_s$. Finally, after relating $\widetilde{\boldsymbol{E}}_{k,\,s}(t)$ and $\widetilde{\boldsymbol{B}}_{k,\,s}(t)$ to the so-called normal variables of the normal modes, $g_{k,\,s}(t) = 1/(2\xi_{k,\,s})(-ic\widetilde{\boldsymbol{B}}_{k,\,s}(t) - i\widetilde{\boldsymbol{E}}_{k,\,s}(t))$, the transverse electrical field in real space can be written as

$$\boldsymbol{E}_{\perp}(\boldsymbol{r},\, t) = \sum_{\boldsymbol{k},s}\boldsymbol{e}_s\xi_{k,\,s}\left[ig_{k,\,s}(t)e^{i\boldsymbol{k}\cdot\boldsymbol{r}} - ig_{k,\,s}^*(t)e^{-i\boldsymbol{k}\cdot\boldsymbol{r}}\right]. \tag{8.106}$$

Here, $\boldsymbol{e}_s$ is just the polarization vector that has already been defined in (8.28), ω_k is the angular frequency given by the wave vector $\boldsymbol{k}$, and $\xi_{k,\,s}$ indicates the strength of each Fourier component of the field. In an identical manner, the vector potential can be written as

$$\boldsymbol{A}(\boldsymbol{r},\, t) = \sum_{\boldsymbol{k},s}\boldsymbol{e}_s\frac{\xi_{k,\,s}}{\omega_k}\left[g_{k,\,s}(t)e^{i\boldsymbol{k}\cdot\boldsymbol{r}} + g_{k,\,s}^*(t)e^{-i\boldsymbol{k}\cdot\boldsymbol{r}}\right]. \tag{8.107}$$

Expression (8.107) is very similar to expression (8.28), since the new variables $g_{k,\,s}(t)$ and $g_{k,\,s}^*(t)$ play a role similar to the operators $a_{k,\,s}(t)$ and $a_{k,\,s}^{\dagger}(t)$ in (8.28). The term $\xi_{k,\,s}/\omega_k$ in (8.107) is just an irrelevant constant [70]. Notice, however, that in (8.106) and (8.107), $g_{k,\,s}(t)$ and $g_{k,\,s}^*(t)$ are still just complex functions of time (i.e. Fourier components), because no quantization of the electromagnetic field has been invoked as yet. Such quantization will be done in section 8.8.6.

A similar expression to that above can be obtained for the magnetic field $\boldsymbol{B}(\boldsymbol{r},\, t)$ [70]. It can then be shown that the definitions of the fields in (8.106) and (8.107) satisfy Maxwell's equation (8.89) and the (transverse part of) (8.95) in the reciprocal space, as

$$\frac{\partial\widetilde{\boldsymbol{B}}_{\perp}(\boldsymbol{k},\, t)}{\partial t} = i\boldsymbol{k} \times \widetilde{\boldsymbol{E}}_{\perp}(\boldsymbol{k},\, t) \tag{8.108}$$

and

$$\frac{\partial \widetilde{\boldsymbol{E}}_{\perp}(\boldsymbol{k}, t)}{\partial \boldsymbol{t}} = \boldsymbol{i}\boldsymbol{c}^2\boldsymbol{k} \times \widetilde{\boldsymbol{B}}(\boldsymbol{k}, t) - \frac{\mathbf{1}}{\epsilon_0}\widetilde{\mathbf{J}}_{\perp}(\boldsymbol{k}, t). \tag{8.109}$$

The (normal mode) variable $g_{k,s}(t)$ satisfies

$$\frac{\partial g_{k,s}(t)}{\partial t} + \boldsymbol{i}\omega_k g_{k,s}(t) = \boldsymbol{i}\frac{\widetilde{\mathbf{J}}_{\perp,k,s}(\boldsymbol{k}, t)}{2\epsilon_0 \xi_{k,s}}. \tag{8.110}$$

The radiation Hamiltonian in (8.102) can now be written as

$$H_{\mathrm{R}} = \frac{\epsilon_0}{2}\int d^3r(\boldsymbol{E}_{\perp}\cdot\boldsymbol{E}_{\perp} + c^2\boldsymbol{B}\cdot\boldsymbol{B}) = 2\epsilon_0 V\sum_{k,s}|\xi_{k,s}|^2\left|g_{k,s}\right|^2. \tag{8.111}$$

The dependence of $g_{k,s}$ on time is eliminated when it represents a degree of freedom in the Hamiltonian (as did for $\boldsymbol{r}$). In principle, the number of normal modes is infinite [70]. Hereafter, for simplicity, it is assumed that only one mode (with one polarization) is relevant. Of course, such an assumption is made to obtain a manageable equation of motion for light and matter, but it mimics a real scenario in the laboratory in which the device's active region is embedded inside an optical cavity [68].

Finally, the energy of the radiation in (8.111) can be simplified (if only one mode is considered), as follows:

$$H_{\mathrm{R}} = \frac{\epsilon_0}{2}\int d^3r(\boldsymbol{E}_{\perp}\cdot\boldsymbol{E}_{\perp} + c^2\boldsymbol{B}\cdot\boldsymbol{B}) = 2\epsilon_0 V|\xi|^2|g|^2. \tag{8.112}$$

The next step is to replace the normal variables g and g^* (of the normal modes) by the new canonical variables q and s defined [70] as

$$q = \sqrt{\frac{\epsilon_0\ V}{\omega}}\xi\left(\boldsymbol{i}g^* - \boldsymbol{i}\ g\right) \tag{8.113}$$

and

$$s = \sqrt{\frac{\epsilon_0\ V}{\omega}}\xi\left(g + g^*\right). \tag{8.114}$$

It can be shown that these variables are canonical in the sense that they satisfy Hamilton's equations [70]. Then, with these new degrees of freedom, the previous Hamiltonian in (8.102) can be written as

$$H = \frac{1}{2m_0}p^2 + U_e(x, t) + \frac{\omega}{2}(q^2 + s^2) + \frac{\alpha}{\sqrt{\omega}}x\ q. \tag{8.115}$$

For simplicity, a one-dimensional (1D) electron in the x-direction is considered. The canonical electron degrees of freedom x and p are relevant in the description of the electron Hamiltonian $H_e = p^2/2m_0 + U_e(x, t)$, while the new canonical variables q and s take care of the description of the radiation Hamiltonian $H_R = \omega(q^2 + s^2)/2$, which has the shape of a harmonic oscillator. Finally, the term that incorporates the interaction between electrons and light is given by $H_I = -\alpha x\, q/\sqrt{\omega}$, which is the result of writing the electric field in (8.106) for a single mode as $\boldsymbol{E}_{\perp}(\boldsymbol{r}, t) = \boldsymbol{e}\xi[ig - ig^*]$ (where the dipolar approximation $e^{i\boldsymbol{k}\cdot\boldsymbol{r}} \approx 1$ has been used again), and the transformations (8.113) and (8.114) have been used. Thus, one gets $\boldsymbol{E}_{\perp}(\boldsymbol{r}, t) = \boldsymbol{e}\xi q$ by assuming the polarization $\boldsymbol{e}$ of a one-mode electromagnetic field lying in the x-direction. The parameter α controls the strength of the interaction through the expression

$$\alpha = e\omega\sqrt{\frac{1}{\epsilon_0 V}}. \tag{8.116}$$

The physical meanings of x and p are well-known, but what are the physical meanings of q and s? A simple understanding comes from the solution of (8.111) when $\tilde{\mathbf{J}}_{\perp,\,k,\,s}(\boldsymbol{k}, t) = 0$. One then gets $g(t) = |g|e^{i\theta}e^{-i\omega t}$ and $g^*(t) = |g|\, e^{-i\theta}e^{i\omega t}$, which can be directly linked to the values $s(t)$ and $q(t)$ through (8.113) and (8.114). When this solution is used in the one-mode expression (8.106), one obtains

$$\boldsymbol{E}_{\perp}(\boldsymbol{r}, t) = -\boldsymbol{e}\xi 2\,|\,g(t)|\,\sin(\boldsymbol{k}\cdot\boldsymbol{r} - \omega t + \theta(t)), \tag{8.117}$$

which is just a polarized monochromatic traveling plane wave [70]. Thus, q and s are the parameters that determine the amplitude and the phase of the electromagnetic field. The same conclusions apply when $\tilde{\mathbf{J}}_{\perp,\,k,\,s}(\boldsymbol{k}, t) \neq 0$, but with more complicated expressions.

Up to this point, the discussion that has led us to the Hamiltonian in (8.115) in terms of the canonical degrees of freedom x, p, q, and s is valid for either classical or quantum systems. No quantization has been invoked yet. Only a description of one mode of the electromagnetic field interacting with a 1D electron in terms of canonical variables has been developed.

In fact, several types of solutions for the light–matter interaction in (8.115) can be considered. First, a classical equation of motion for light and matter could be assumed. Second, a quantum equation of motion for electrons but a classical solution for electromagnetic fields could be considered, or vice versa. Third, an (orthodox) quantum treatment for both electrons and fields is possible. Finally, a Bohmian formulism can be added to the previous orthodox approach, which yields the ability to describe the most intricate light–matter phenomena using well-defined microscopic properties. These possibilities will be treated in the next several subsections.

8.8.4 The classical solution for matter–light interaction

In classical scenarios, the degrees of freedom x and p (for the position and momentum of the electron) and q and s (for the amplitude and phase of the transverse electric and magnetic field) lead to a coupled system of Hamilton equations. The practical solution just requires the initial conditions for all the variables $x(0)$, $p(0)$, $q(0)$, and $s(0)$ to be fixed, which gives the final trajectories $x(t)$, $p(t)$, $q(t)$, and $s(t)$ that specify the evolution of the light–matter system at all times.

The word 'trajectory' for $q(t)$ and $s(t)$ cannot literally be understood to mean the trajectory of the electromagnetic field in real (physical) space, but is rather a continuous definition of the amplitude and the phase of the electromagnetic field in q–s space as defined by (8.117). The unique Hamiltonian for light and matter implies that the energies of the electrons and the radiation can be interchanged: the electron's dynamics affects the light, and the light also affects the electrons. This type of classical solution offers enough physical insight in many scenarios, as, for example, in section 8.2, which provides an understanding of free-carrier absorption.

8.8.5 The semiclassical solution for matter–light interaction

The previous classical solution missed one very important ingredient for devices at the nanoscale: quantum wholeness. An intuitive explanation of this nonclassical phenomenon is that an electron corresponding to one experiment (called the a-experiment) subjected to the Hamiltonian (8.115) is affected by what happens to an electron in another experiment (the b-experiment) also governed by (8.115). In other words, the trajectory $x_a(t)$ is affected by the trajectory $x_b(t)$ (even if they are in different experiments). Similarly, the momentum $p_a(t)$ is affected by the momentum $p_b(t)$. The appropriate element that can capture such nonclassical behavior is a wave function whose extension (quantum coherence) includes $x_a(t)$ and $x_b(t)$ or equivalently $p_a(t)$ and $p_b(t)$. But how do we find such a wave function? It is well-known that such a wave function can be found by imposing a condition that the degrees of freedom of the previous Hamiltonian (8.115) are operators with a particular noncommutative property in order to ensure that it is not possible to simultaneously minimize the quantum coherence (wholeness) in the x and p dimensions. The most common example for the electron is to define the following noncommutative operators in the position representation:

$$\begin{aligned} x &\rightarrow x, \\ p &\rightarrow -i\hbar\frac{\partial}{\partial x}. \end{aligned} \tag{8.118}$$

Such operators satisfy the commutation relation $[x, p] = i\hbar\mathbb{1}$, where $\mathbb{1}$ is a unity operator. The light in (8.118) is treated as consisting of classical electromagnetic fields with well-defined values of $q(t)$ and $s(t)$. With such a substitution, the Hamiltonian in (8.116) is transformed into

$$H = \frac{\hbar^2}{2m_0}\frac{\partial^2}{\partial x^2} + U_e(x, t) + \frac{\alpha}{\sqrt{\omega}}q(t)x. \tag{8.119}$$

The electromagnetic energy $H_R(t) = \omega(q(t)^2 + s(t)^2)/2$ is not included because it is just a pure time-dependent potential (without spatial dependence) that has no practical effect on the electron dynamics. The transverse electromagnetic field affects the electron through the function $q(t)$, but the electron has no effect on the electromagnetic field. In other words, the electromagnetic field in (8.119) is an 'external' potential like $U_e(x, t)$. The state of the semiclassical system is given by the wave function $\Psi(x, t)$, which parametrically depends on $q(t)$.

It is important to note that the terms $\alpha q(t)x/\sqrt{\omega}$ in (8.118) have no relationship to the scalar potential typical of electronic devices but are related to the vector potential through the approximation leading to (8.103). This last term can be, for example, an external laser affecting the dynamics of electrons inside the electronic device.

8.8.6 The quantum solution for matter–light interaction

For a full quantum treatment of electrons and radiation, in addition to the noncommutative operators in (8.118) for the electron, the quantization of the electromagnetic field is also required. The typical way of quantizing the electromagnetic field in the literature is through the creation and annihilation operators $a^\dagger$ and a used previously in section 8.5. Here, to keep the knowledge of q and s, the degrees of freedom of light in (8.115) are replaced by the operators

$$q \rightarrow \sqrt{\omega} q$$
$$s \rightarrow -i\frac{\hbar}{\sqrt{\omega}}\frac{\partial}{\partial q}, \qquad (8.120)$$

where $[q, s] = i\hbar\mathbb{I}$ and $\mathbb{I}$ is the unity operator. Finally, we make the substitutions of (8.120) in (8.115) to give the new fully coherent Hamiltonian

$$H = -\frac{\hbar^2}{2m_0}\frac{\partial^2}{\partial x^2} + U_e(x, t) + \alpha x\, q - \frac{\hbar^2}{2}\frac{\partial^2}{\partial q^2} + \frac{\omega^2}{2}q^2. \qquad (8.121)$$

Note that the radiation Hamiltonian in (8.121) can be rewritten as $H_R = \hbar\omega(a^\dagger a + 1/2)/2$ when defining $a = (s + iq)/\sqrt{2\hbar}$ and $a^\dagger = (s - iq)/\sqrt{2\hbar}$ with the condition $[a, a^\dagger] = \mathbb{I}$, as was done in section 8.5. Of course, another possibility is to use a quantum treatment of the light and a classical treatment of the electrons, which will be done in the Monte Carlo method of chapter 9 that deals with collisions between classical electrons and quantized light.

In (8.121), the x and q information is kept as variables so that the state of the system is described by a time-dependent wave function $\Psi(x, q, t)$, whose evolution is given by the Schrödinger equation

$$i\hbar\frac{\partial\Psi(x, q, t)}{\partial t} = \left(-\frac{\hbar^2}{2m_0}\frac{\partial^2}{\partial x^2} + U_e(x, t) + \frac{\omega^2}{2}q^2 - \frac{\hbar^2}{2}\frac{\partial^2}{\partial q^2} + \alpha x q\right)\Psi(x, q, t). \qquad (8.122)$$

From a mathematical point of view, (8.122) is just a two-dimensional (2D) Schrödinger equation whose numerical solution can be worked out straightforwardly. In such a wave function, $\Psi(x,q,t)$ means that the position x of the electron no longer has a well-defined value but is given by a probability distribution. Similarly, the amplitude (and phase) of the electromagnetic field q no longer has a well-defined value but also satisfies a probability distribution. According to the orthodox view, a well-defined value of the position of the electron or the amplitude of the electromagnetic field is only recovered if it is measured explicitly. In the next section, 8.8.7, within a Bomian approach, well-defined properties of the electrons and fields will be recovered (without even measuring them).

Equation (8.122) gives us a deep understanding of the light–matter interaction. First, one can define the set of single-electron wave functions $\phi_n^{(e)}(x)$, for $n = 0, 1, \dots, N_e$, as the solutions of the electron Hamiltonian

$$H_e = \frac{\hbar^2}{2m_0}\frac{\partial^2}{\partial x^2} + U_e(x). \tag{8.123}$$

In the context of transport in an electronic device, $\phi_n^{(e)}(x)$ can be simply, for example, the (Landauer) scattering states. The radiation Hamiltonian

$$H_R = \frac{\omega^2}{2}q^2 - \frac{\hbar^2}{2}\frac{\partial^2}{\partial q^2} \tag{8.124}$$

is a harmonic oscillator with mass one, whose eigenstates $\phi_m^{(f)}(q)$ for $m = 0, 1, \dots, N_f$ are well-known. Each one has an energy of $\hbar\omega(m + 1/2)$. Of course, each eigenstate can be identified with the field state in (8.26).

Often, the eigenstate $\phi_m^{(f)}(q)$ is said to describe a system with m photons in the particular mode. In this sense, $m = 0$ corresponds to zero photons, $m = 1$ corresponds to one photon, etc. and the state is often called a Foch state. Using this convention, the eigenstate $\phi_0^{(f)}(q)$ describes a mode with zero photons, but it still has an energy of $\hbar\omega/2$, which is referred to as the electromagnetic quantum vacuum-state energy [70]. Finally, using $\phi_n^{(e)}(x)$ and $\phi_m^{(f)}(q)$, a general solution of (8.127) can be written as follows:

$$\Psi(x, q, t) = \sum_{n=0,m=0}^{N_e,N_f} c_{n,m}(t)\phi_n^{(e)}(x)\phi_m^{(f)}(q). \tag{8.125}$$

The Hamiltonian (8.122) ensures that the total energy and probability of the electron plus electromagnetic field are conserved. Such a conservation of the total energy and probability is fully compatible with the interchange of energy and probability between the electron and radiation subsystems.

In light of the above, one can understand a dynamic evolution of $\Psi(x, q, t)$ in (8.122) that results in the following transition

$$\begin{gathered}\Psi(x, q, 0) \rightarrow \Psi(x, q, t),\\ \phi_n^{(e)}(x)\phi_m^{(f)}(q) \rightarrow \phi_{n+1}^{(e)}(x)\phi_{m-1}^{(f)}(q).\end{gathered} \tag{8.126}$$

This corresponds to the absorption of a photon by an electron. Of course, we are assuming that the energy difference between $\phi_{n+1}^{(e)}(x)$ and $\phi_n^{(e)}(x)$ is the same as that between $\phi_m^{(f)}(q)$ and $\phi_{m-1}^{(f)}(q)$. Such absorption is not possible if the initial photon state corresponds to the electromagnetic vacuum $\phi_0^{(f)}(q)$.

Similarly, one can understand a dynamic evolution of $\Psi(x, q, t)$ that gives the reverse transition

$$\begin{aligned} \Psi(x, q, 0) &\rightarrow \Psi(x, q, t), \\ \phi_n^{(e)}(x)\phi_m^{(f)}(q) &\rightarrow \phi_{n-1}^{(e)}(x)\phi_{m+1}^{(f)}(q). \end{aligned} \tag{8.127}$$

This transition corresponds to the spontaneous emission of a photon by an electron. The case of spontaneous emission from an initial photon state $\phi_0^{(f)}(q)$ is somehow 'surprising' because a photon is 'emitted' from a region of space that (before the emission) had no photon ($m = 0$). This phenomenon has a quite problematic explanation in terms of classical electromagnetics, but here the mystery disappears because the 'absence of electromagnetic radiation' is defined by the wave function $\phi_0^{(f)}(q)$ in (8.124) linked to an electromagnetic energy $\hbar\omega/2$.

These two examples, emission and absorption, correspond to the discussion given in the previous section, 8.6. One clearly understands that no instantaneous quantum jump occurs between states. Equation (8.122) shows that these types of processes are continuous in time. The probability does not instantaneously disappear from one point in the configuration space to appear at another point far from the first. In fact, there is a continuity equation hidden in (8.122), as will be seen in the next subsection.

As indicated in the introduction to this section, there is a very rich phenomenology hidden in (8.122) that cannot be accessed using perturbation techniques. The application of (8.122) to develop new applications for electronic devices at terahertz or optical frequencies remains mainly unexplored [68]. Of course, in a realistic situation, with many electrons and many modes (each mode with many photon states), the direct solution of the many-electron and many-mode Schrödinger equation in (8.122) with (8.106) and (8.107) becomes computationally very demanding.

8.8.7 The quantum solution with particles and fields

At this point, the reader may wonder whether we must really definitively abandon the attempt to understand all these intricate light–matter quantum phenomena with well-defined microscopic properties for electrons and light [75]. Continuing the discussion begun in chapter 2, the answer is again 'no.' It is possible to understand all these complicated light–matter interactions without abandoning the fact that the microscopic properties of electrons and light are both well-defined.

The unitary evolution of $\Psi(x, q, t)$ in (8.122) ensures that there is a global conservation of probability. In fact, a straightforward manipulation of (8.122) shows that the probability is locally conserved because it has to satisfy the following (local) continuity equation

$$\frac{\partial|\Psi|^2}{\partial t} + \frac{\partial J^{(e)}}{\partial x} + \frac{\partial J^{(f)}}{\partial q} = \frac{\partial|\Psi|^2}{\partial t} + \frac{\partial(|\Psi|^2 v^{(e)})}{\partial x} + \frac{\partial(|\Psi|^2 v^{(f)})}{\partial q} = 0. \tag{8.128}$$

Here, $J^{(e)} = J^{(e)}(x, q, t)$ is the particle current. On the right-hand side of (8.128), once a particle current and a probability density are defined, one can use the Bohmian definition of velocities [73, 74] to define the velocity of the electron as follows:

$$v^{(e)} \equiv v^{(e)}(x, q, t) = \frac{J^{(e)}(x, q, t)}{|\Psi(x, q, t)|^2} = \frac{1}{m_0}\frac{\partial S(x, q, t)}{\partial x}. \tag{8.129}$$

In (8.128), $J^{(f)} = J^{(f)}(x, q, t)$ is the electromagnetic mode 'current' in the q-space. Thus, following the approach already taken for the electron velocity, the Bohmian theory [68] allows us to define a 'velocity' of the q-variable as follows:

$$v^{(f)} \equiv v^{(f)}(x, q, t) = \frac{J^{(f)}(x, q, t)}{|\Psi(x, q, t)|^2} = \frac{\partial S(x, q, t)}{\partial q}. \tag{8.130}$$

In the above expressions, the complex wave function is rewritten in a polar form $\Psi = Re^{iS/\hbar}$, where R and S are real functions. From the above velocities, it is straightforward to define the coupled trajectories

$$x^j(t) = x^j(0) + \int_0^t v^{(e)}(x^j(t'), q^j(t'), t')dt' \tag{8.131}$$

and

$$q^j(t) = q^j(0) + \int_0^t v^{(f)}(x^j(t'), q^j(t'), t')dt'. \tag{8.132}$$

The superscript j specifies the initial variables $x^j(0)$ and $q^j(0)$ that are different in each experiment [73]. Since all experiments are described by the same identically prepared wave function $\Psi(x, q, 0)$, the distributions of the different values $x^j(0)$ and $q^j(0)$ have to be consistent with the initial probability distribution $|\Psi(x, q, 0)|^2$.

It is important to understand that $x^j(t)$ is really a trajectory in the physical space, while $q^j(t)$ is just a trajectory in the (abstract) space of the properties of the electromagnetic field. In particular, $q^j(t)$ yields the exact value the amplitude and phase of the electromagnetic field at each time, as described in (8.117).

Finally, to illustrate the use of the Bohmian trajectories developed in (8.131) and (8.132), an electron inside a quantum well that has a length of 16 nm and an effective mass of 0.041 times the free electron mass is considered. The electron Hamiltonian is the one indicated in (8.119). The initial energy of the electron system corresponds to the first excited state. Such a quantum well is inserted into an optical cavity in which only one electromagnetic mode of the frequency ω = 160Trad/s (25 THz) is relevant. The Hamiltonian of the radiation is given by (8.120). The initial energy of the optical cavity corresponds to zero. The wave function $\Psi(x, q, t)$ is found numerically from (8.127) with the initial state $\Psi(x,q,0) = \phi_1^{(e)}(x)\phi_0^{(f)}(q)$. Once the initial positions $x^j(0)$ and $q^j(0)$ are selected, the trajectories can be computed from (8.131) and (8.132). Figure 8.22 shows examples of two such trajectories plotted in the 2D (x, q) space. The evolution of these trajectories over time follows a circular or semi-circular path which exhibits the typical Rabi evolution [68]. The electron in the first excited state with zero photons in the optical cavity corresponds to the white area of

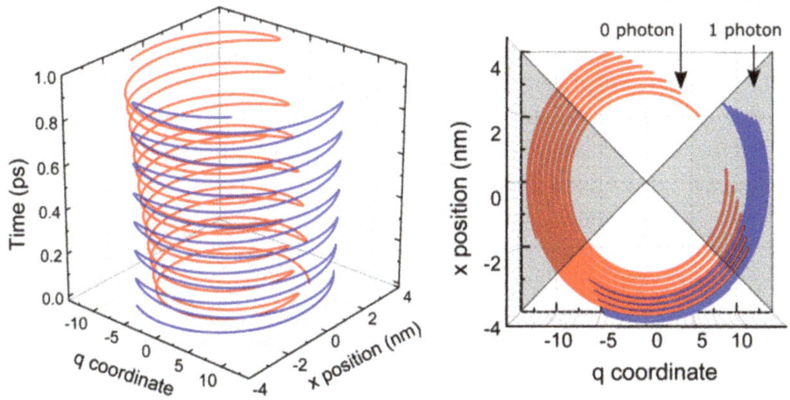

Figure 8.22. An electron inside a quantum well with two quantized states, namely the ground and excited states. The quantum well is embedded inside an optical cavity with one electromagnetic mode. (Left) Examples of two trajectories in the 2D (x, q) configuration space, guided by the analytical evolution of the light–matter wave function in (8.127). (Right) A schematic representation of the different quadrants of the (x, q) space inside the quantum well and optical cavity. The white quadrants correspond to the zero-photon state (and the electron in the excited state), while the gray ones correspond to the one-photon state (and the electron in the ground state).

figure 8.22(b). The electron in the ground state and one photon in the optical cavity correspond to the gray area of figure 8.22(b). It can be seen from the figure that the photon emitted by the electron remains inside the optical cavity, so that it is reabsorbed by the electron itself and is then re-emitted, repeating in an infinite sequence. This is the well-known Rabi oscillation. The limits of each oscillation (when the trajectory has null velocity) always alternate from one quadrant to the other in different ways depending on the initial conditions of the electron and electromagnetic field trajectories.

The physical explanation of the Rabi oscillations is that $\phi_1^{(e)}(x)\phi_0^{(f)}(q)$ is an eigenstate of the electrons and radiation in the absence of any interaction, but it is not an eigenstate of the entire system when light–matter interaction is considered. The Hamiltonian in (8.122) has its own energy eigenstates, which are different from the product of the electron eigenstates and the radiation eigenstates. Thus, the initial state $\phi_1^{(e)}(x)\phi_0^{(f)}(q)$ is not an eigenstate of the global system but a superposition of two global eigenstates. The oscillations seen in the figure are just natural oscillations between these two global eigenstates. At this point, it is clear that a strong coupling between light and matter cannot be understood using the simple perturbative techniques discussed in the previous sections because the light and the matter cannot be treated as separated elements, but only as a unique (polaritonic) wave function [68–70].

As indicated in the introduction to this chapter, most electronic device simulation tools only account for the interaction between electrons and electromagnetic fields by solving the Poisson equation. Indeed, such an electrostatic approximation is sufficient to successfully predict many device applications. There are other applications, mainly associated with working frequencies higher than hundreds of gigahertz, in which the previous electrostatic assumption is not sufficient to correctly model the interactions between electrons and electromagnetic fields. At such terahertz frequencies, for example, the conduction current does not adequately explain the behavior of electrons inside the device's active region, and the displacement current (given by the time derivative of the electric field) needs to be considered carefully [76]. As shown in section 8.4, some applications even require the electron dynamics to be coupled with the full solution of Maxwell equations, including the magnetic field.

In addition, at frequencies of a few terahertz, the energy of a photon is on the order of the energy of the electrons inside the device. Thus, the consideration of the quantization of the electromagnetic field discussed in section 8.5 is mandatory. In fact, optoelectronic devices are designed to fully utilize the interaction of electrons with the electromagnetic field, described in terms of the quanta of energy $\hbar\omega$. A scattering process involving the emission and absorption of photons of energy $\hbar\omega$ by an electron is somehow mandatory in most electron device simulators. Such a scattering process is indeed a perturbative treatment of the real interaction between electrons and electromagnetic fields.

The quantum optics community has developed more sophisticated formalisms with which to discuss new light–matter phenomenology (such as the Rabi oscillations discussed in figure 8.22) that can hardly be captured using perturbative techniques. In section 8.8, such a full quantum treatment of light and electrons was adapted to electron devices embedded in an optical cavity; this showed that there is plenty of room for engineers to develop new concepts of optoelectronic devices using quantized radiation and quantized electrons that share a common (polaritonic) wave function.

At the end of section 8.8, it was emphasized that a pure quantum treatment of light and matter is perfectly compatible with a treatment of quantized electromagnetic fields with well-defined microscopic properties (such as the amplitude and the phase), as conceptually discussed in the first sections of this chapter. Of course, the quantum radiation implies new phenomenology that cannot be understood using classical electromagnetism. However, the fact the electromagnetic fields are well-defined (even when they are not measured) does not need to be abandoned to understand phenomena such as spontaneous emission and electromagnetic vacuum fluctuations [68]. One can understand all these complicated light–matter phenomena without abandoning the idea that the elements of the microscopic world that are always well-defined [75]. More importantly, it is emphasized that all of this can be treated fully in terms of particles, whether they be electrons, holes, or photons.

References

[1] Chau R *et al* 2004 *IEEE Trans. Nanoelectron* **4** 153

[2] Akis R *et al* 2007 *IEEE Electron Device Lett.* **29** 306

[3] Qin G *et al* 2014 *J. Semicond.* **35** 125004

[4] Maroldt S *et al* 2013 *Jpn. J. Appl. Phys.* **52** 08JN13

[5] Aharonov Y and Bohm D 1959 *Phys. Rev.* **115** 485

[6] Ferry D K 2020 *Transport in Semiconductor Mesoscopic Devices* 2nd edn (Bristol: IOP Publishing)
[7] Kirchhoff G 1845 *Ann. Phys.* **64** 497
[8] Maxwell J C 1861 *Phil. Mag., Ser. 4* **23** 11
[9] Kirchhoff G 1857 *Phil. Mag., Ser. 4* **13** 393
[10] Graneau P and Assis A K T 1994 *Apeiron* **19** 19–25
[11] Read W T 1958 *Bell Syst. Tech. J.* **37** 401
[12] Gummel H K and Scharfetter A L 1966 *Bell Syst. Tech. J.* **45** 1797
[13] Saraniti M and Goodnick S M 2000 *IEEE Trans. Electron Devices* **47** 1909
[14] Kunc K and Nielson O H 1979 *Comput. Phys. Commun.* **17** 413
[15] Ayubi-Moak J S *et al IEEE Trans. Electron Devices* **54** 2327
[16] Marino F A *et al* 2010 *Phys. Status Solidi* C **7** 2445
[17] Akis R *et al* 2008 *J. Phys.: Condens. Matter* **20** 38201
[18] Akis R *et al* 2008 *Phys. Status Solidi* C **7** 2502
[19] Guerra D *et al IEEE Electron Device Lett.* **31** 1217
[20] Guerra D *et al* 2011 *Int. Conf. Simulation of Semiconductor Processes and Devices* (New York: IEEE Press) 87–90
[21] Palacios T *et al* 2006 *IEEE Electron Device Lett.* **27** 13
[22] Maas S A 2003 *Nonlinear Microwave and RF Circuits* (Norwood, MA: Artech House) https://us.artechhouse.com/Nonlinear-Microwave-and-RF-Circuits-Second-Edition-P1097.aspx
[23] Bertazzi F *et al* 2008 *Int. J. Comput. Math. Electr. Electron. Eng.* **27** 1319
[24] Guerra D *et al* 2012 *Int. Microwave Symp. Digest* (New York: IEEE Press)
[25] Taflove A and Hagness S C 2005 *Computational Electrodynamics: The Finite-Difference Time-Domain Method* 3rd edn (Norwood, MA: Artech House) https://us.artechhouse.com/Computational-Electrodynamics-The-Finite-Difference-Time-Domain-Method-Third-Edition-P2342.aspx
[26] Yee K S 1966 *IEEE Trans. Antennas Prop.* **14** 302
[27] Connolly K M *et al* 1990 *IEEE MTS-S Digest* (New York: IEEE Press) 295–8
[28] Goodnick S M *et al* 1995 *Int. J. Numer. Modeling* **8** 208
[29] Ayubi-Moak J S *et al* 2003 *J. Comput. Electron* **2** 183
[30] Ayubi-Moak J S, Goodnick S M and Saraniti M 2006 *J. Comput. Electron.* **5** 415
[31] Ayubi-Moak J S *et al* 2009 *13th Int. Workshop Computational Electronics* (New York: IEEE Press)
[32] Sule N *et al* 2013 *J. Comput. Electron.* **12** 563
[33] Jungemann C *et al* 2022 *Solid-State Electron.* **197** 108439
[34] Bahl M *et al* 2016 *Int. Conf. Numerical Simulations and Optoelectronic Devices* (New York: IEEE Press) 101–2
[35] Pace F *et al* 2019 *49th European Solid-State Devices Research Conf.* (New York: IEEE Press) 230–3
[36] Planck M 1901 *Ann. Phys.* **4** 553
[37] Einstein A 1905 *Ann. Phys.* **17** 132
[38] Bohr N 1913 *Phil. Mag., Ser. 6* **26** 1
[39] Dirac P A M 1927 *Proc. Roy. Soc. Lond.* A **114** 243
[40] Compton A H 1923 *Phys. Rev.* **21** 483
[41] Sommerfeld A 1931 *Ann. Phys.* **403** 257
[42] Ferry D K 2020 *Semiconductors: Bonds and Bands* 2nd edn (Bristol: IOP Publishing)

[43] Geltman S 1977 *J. Phys.* B **10** 831

[44] Haan S L and Geltman S 1982 *J. Phys.* B 15 1229

[45] Lindberg M and Koch S W 1988 *Phys. Rev.* B **38** 3342

[46] Haug H 1988 *Optical Nonlinearities and Instabilities in Semiconductors* ed H Haug (San Diego, CA: Academic) https://shop.elsevier.com/books/optical-nonlinearities-and-instabilities-in-semiconductors/haug/978-0-12-332915-8

[47] Kuhn T and Rossi F 1992 *Phys. Rev.* B **46** 7496

[48] Ferry D K 2017 *An Introduction to Quantum Transport in Semiconductors* (Singapore: Jenny Stanford Publishing)

[49] Kirk A P 2014 *Solar Photovoltaic Cells: Photons to Electricity* (London: Academic) https://shop.elsevier.com/books/solar-photovoltaic-cells/kirk/978-0-12-802329-7

[50] Madelung O 2004 *Semiconductors: Data Handbook* 3rd edn (Berlin: Springer)

[51] http://ioffe.ru/SVA/NSM/Semicond/

[52] Dixon J R and Ellis J M 1961 *Phys. Rev.* **123** 1560

[53] Sturge M D 1962 *Phys. Rev.* **127** 768

[54] Muth J F *et al* 1997 *Appl. Phys. Lett.* **71** 2572

[55] Ferry D K 1991 *Semiconductors* (New York: Macmillan)

[56] Varga B B 1965 *Phys. Rev.* **137** 1896

[57] Mooradian A and Wright G B 1966 *Phys. Rev. Lett.* **16** 999

[58] Kim M E, Das A and Senturia S D 1978 *Phys. Rev.* B **18** 6890

[59] Lugli P and Ferry D K 1985 *Appl. Phys. Lett.* **46** 594

[60] Lugli P and Ferry D K 1985 *IEEE Electron Device Lett.* **6** 25

[61] Takanaka N, Inoue M and Inuishi Y 1979 *J. Phys. Soc. Jpn.* **47** 861

[62] Kohn W, Meir Y and Makarov D E 1998 *Phys. Rev. Lett.* **80** 4153

[63] Kelbg G 1964 *Ann. Phys.* **468** 385

[64] Morawetz K 2002 *Phys. Rev.* E **66** 022103

[65] Gross W J, Vasileska D and Ferry D K 2000 *VLSI Des.* **10** 437

[66] Lugli P and Ferry D K 1986 *IEEE Trans. Electron Devices* **32** 2431

[67] Kriman A M *et al* 1990 *Phys. Rev. Lett.* **65** 1619

[68] Destefani C *et al* 2022 *Phys. Rev.* B **106** 205306

[69] Cohen-Tannoudji C *et al* 2004 *Photons and Atoms: Introduction to Quantum Electrodynamics* (New York: Wiley)

[70] Grynberg G *et al* 2011 *Introduction to Quantum Optics: From the Semi-Classical Approach to Quantized Light* (Cambridge: Cambridge University Press)

[71] Scully M O and Zubairy M S 2008 *Quantum Optics* (Cambridge: Cambridge University Press)

[72] Javid M and Brown P M 1963 *Fields Analysis and Electromagnetics* (New York: McGraw-Hill)

[73] Oriols X and Mompart J 2019 *Applied Bohmian Mechanics: From Nanoscale Systems to Cosmology* (Singapore: Jenny Stanford Publishing)

[74] Bohm D 1952 *Phys. Rev.* **85** 166

[75] Oriols X and Ferry D K 2021 *Proc. IEEE* **109** 955

[76] Eisenberg B, Oriols X and Ferry D 2017 *Mol. Based Math. Biol.* **5** 78

Part III

Simulation techniques

IOP Publishing

Quantum Transport in Semiconductor Devices
Simulation using particles
David K Ferry, Xavier Oriols and Josef Weinbub

Chapter 9

The Monte Carlo method

The use of the Boltzmann transport equation, in common with most other analytical approaches used as solutions for classical transport in semiconductor devices, is quite difficult to evaluate carefully in the real situation of a semiconductor with non-parabolic energy bands and complicated scattering processes. This becomes even more so with realistic full-band simulations. In no small part, this is because the equation is not only nonlinear, but is also noncausal due to phonon emission due to higher energies, so it cannot be solved by a simple energy integration working up from zero energy. However, since it is first order in the time derivative, temporal solutions are a valid approach to the problem [1]. When quantum transport becomes necessary, the task of solving the transport equation becomes even more difficult [2].

At least for classical solutions, an alternative approach is to use the computer to completely solve the transport problem via a stochastic methodology, and several methods have arisen to do this, two of which are an integral iteration technique [3, 4] and the Monte Carlo method [5]. Today, the most widely used approach is the latter. The ensemble Monte Carlo (EMC) technique, which is based on an ensemble of particles propagating in parallel, has now been used for more than five decades as a numerical method to simulate far-from-equilibrium transport in semiconductor materials and devices. It has been the subject of many reviews [6–9]. The methodology of the EMC approach is to spend time and effort on characterizing the correct physics, perhaps even taking this as far as using atomistic full-band energy behavior [10] and the scattering properties, and let the computational burden be handled by a modern parallel processing computer. Hence, no attempt is made to arrive at an *ab initio* analytical form. These approaches have become extremely sophisticated, and many factors such as degeneracy [11], discrete impurities in real space [12], and many-body effects [13] can be incorporated.

The immediate problem with adapting the EMC approach for quantum transport lies in the use of the particle paths in phase space in the classical approach. Generally, the uncertainty principle restricts our ability to simultaneously define

doi:10.1088/978-0-7503-5237-6ch9

both a position and momentum for a particle path (see sections 1.4 and 5.3). Nevertheless, this strict interpretation has been finessed by a variety of approaches, many of which clearly endorse some form of real path for a particle that exists with the wave. Not the least of these approaches is due to Feynman, who described the probability amplitude as arising from a complete particle path, rather than simply the position of a particle at a particular time [14]. Then, of course, there is the de Broglie view of quantum mechanics as a particle *and* a wave [15], as discussed in chapter 1. De Broglie described photons as 'atoms of light' that rode upon a nonmaterial light wave. This would be extended to particles with the development of the theory of atomic structure, and it was relatively easy for him to connect this momentum to a wavelength ($\lambda = c/f$) [16]

$$\lambda = \frac{h}{p}. \tag{9.1}$$

From this beginning, there has been a rich history for the use of particles in quantum mechanics.

The phase-space representation of the Wigner function automatically suggests a comparison between quantum dynamics and the corresponding classical motion [17, 18]. However, the Wigner function can have a non-unitary evolution, and off-diagonal terms in the density matrix (from which the Wigner function is found) lead to nonclassical propagation. Indeed, quantum coherence is reflected in the oscillatory behavior of the Wigner function and nonpositive definite regions in phase space, and this leads to complications in the use of the EMC approach. Nevertheless, particle methods and the EMC approach have been adapted to treat quantum transport via the Wigner function [19–21]. This has been quite successful, as illustrated by figure 5.1, where the quantum interference of waves passing around impurities is simulated using the particle approach.

In this chapter, the classical EMC method will be reviewed. The use of particles and their paths in connection with the Bohm quantum potential, or other effective potentials, in the EMC will be treated in chapter 10, while the quantum treatment via the Wigner function is left to chapter 11. In the following sections, the path integral equation for the distribution function is derived first, primarily due to the fact that the kinetic Monte Carlo being discussed is a solution of the resulting Fredholm integral equation. Following that, the various statistical approaches to determining stochastic evaluations of the two principal parts of the integral equation will be outlined, and then some applications will be described. It is important to understand that a full understanding of the classical EMC approach is required before we can begin to discuss the quantum methods.

9.1 The path integral

The Monte Carlo technique was first used to solve the Boltzmann equation in semiconductors by Kurosawa [5]. Shortly after this, Budd put the Boltzmann equation into a path integral form, which clarified the process steps used in the Monte Carlo approach [3]. In this approach, the streaming terms on the left-hand side, which are written as partial derivatives of the distribution, are converted to derivatives along a 'path' in a six-dimensional phase space. This is used to develop a closed-form Fredholm integral equation for the distribution function. To begin, the Boltzmann equation is written as

$$\begin{aligned}\left(\frac{\partial}{\partial t} + \boldsymbol{v} \cdot \nabla - \frac{e}{\hbar}\boldsymbol{E} \cdot \frac{\partial}{\partial \boldsymbol{k}}\right) f(\boldsymbol{k}, \boldsymbol{r}, t) &= -\Gamma_0 f(\boldsymbol{k}, \boldsymbol{r}, t) \\ &+ \int d^3\boldsymbol{k}' P(\boldsymbol{k}', \boldsymbol{k}) f(\boldsymbol{k}', \boldsymbol{r}, t)\end{aligned}, \tag{9.2}$$

where part of the scattering term has been simplified to

$$\Gamma_0 = \int d^3\boldsymbol{k} P(\boldsymbol{k}, \boldsymbol{k}'), \tag{9.3}$$

which represents the total *out-scattering* rate. This is the scattering by phonons or impurities that take carriers out of the momentum-space point $\boldsymbol{k}$ and move them to other points in phase space. The last term on the right-hand side is the *in-scattering* that brings carriers into this point in phase space, so that a summation over all other momentum-space points is required. Classically, it is assumed that the scattering is local in space, so that both in- and out-scattering occur at the same spatial point.

In order to transform (9.2) into a path-variable form, the motion of the distribution function along a trajectory in phase space must be considered (or, more normally, to think of the motion of the particles through phase space, we need path variables, i.e. variables along the particle's path). The distribution function itself may be described as the ensemble of particles that is utilized for the Monte Carlo simulation. This ensemble exists in phase space, and in the absence of scattering, the phase-space volume is conserved. Each particle thus has a unique path, and the coordinates of (9.2) are transformed to lie along this path that itself is defined by the position and momentum of the particle. In the distribution function, each path is considered to represent a fraction of the distribution. The coordinate along this trajectory is taken to be s, and the trajectory is rigorously defined by the semiclassical trajectory, which can be found by any of the techniques of classical mechanics (e.g. in Lagrangian mechanics, it corresponds to that path which is an extremum of the action). It is as easy to remember, however, that it follows Newton's laws, where the forces arise from all possible potentials—including any self-consistent ones that arise in device simulations. Each normal variable for the particle can be parameterized as a function of this path variable as follows:

$$\boldsymbol{r} \to \boldsymbol{x}^*(s), \quad \boldsymbol{k} \to \boldsymbol{k}^*(s), \quad t \to s, \tag{9.4}$$

and the partial derivatives along the path become

$$\frac{d\boldsymbol{x}^*}{ds} = \boldsymbol{v}, \quad \boldsymbol{x}^*(t) = \boldsymbol{r}, \quad \frac{d\boldsymbol{k}^*}{ds} = -\frac{e}{\hbar}\boldsymbol{E}, \quad \boldsymbol{k}^*(t) = \boldsymbol{k}. \tag{9.5}$$

With these changes, the chain of partial derivatives can be recombined into the total time derivative, and (9.2) can be rewritten as

$$\frac{\partial f}{\partial s} + \Gamma_0 f = \int d^3\boldsymbol{k}^{*\prime} P(\boldsymbol{k}^{*\prime}, \boldsymbol{k}^*) f(\boldsymbol{k}^{*\prime}, \boldsymbol{x}^*, s). \tag{9.6}$$

This last equation is now a relatively simple equation to solve using an integration factor. It should be recalled at this point that $P(\boldsymbol{k}^{*\prime}, \boldsymbol{k}^*)$ is the probability per unit time that a collision scatters a carrier from state $\boldsymbol{k}^{*\prime}$ to $\boldsymbol{k}^*$, and these variables are retarded due to the phase-space variations described above. The form (9.6) becomes

$$\frac{d}{ds}[f(\boldsymbol{k}^*)e^{\Gamma_0 s}] = \int d^3\boldsymbol{k}^{*\prime} P(\boldsymbol{k}^{*\prime}, \boldsymbol{k}^*) f(\boldsymbol{k}^{*\prime}, \boldsymbol{x}^*, s) e^{\Gamma_0 s}, \tag{9.7}$$

and, after restoring the time variables appropriate to the laboratory space, this last equation can be rewritten as

$$\begin{aligned} f(\boldsymbol{k}, t) &= f(\boldsymbol{k}, 0)e^{-\Gamma_0 t} \\ &+ \int_0^t dt' \int d^3\boldsymbol{k}' P\left(\boldsymbol{k}' + \frac{e}{\hbar}\boldsymbol{E}, \boldsymbol{k}\right) f\left(\boldsymbol{k}' + \frac{e}{\hbar}\boldsymbol{E}, \boldsymbol{r}, t'\right) e^{-\Gamma_0(t-t')}. \end{aligned} \tag{9.8}$$

This last form is often referred to as the Chambers–Rees path integral [22], as the development by Budd [3] and further development by Rees [22] both followed a much earlier path integral developed for motion in a magnetic field [23]. Notice that the distribution function within the integral is shifted from the momentum state $\boldsymbol{k}'$ due to the acceleration from the electric field. As a result, the Monte Carlo procedure can be seen to have two distinct steps. First, the distribution function, or rather the particles that represent the distribution function are accelerated by the electric field for a time that averages $t - t'$. That is, the exponential term

$$e^{-\Gamma_0 \tau} \tag{9.9}$$

represents the probability that a given particle has *not* scattered (this is an unnormalized probability but is easily simulated by a random number in the range from zero to one). This random time τ represents the duration of the current acceleration period. It is important to note that each particle has its own time τ. At the end of the acceleration period, the particle is scattered by one of the various scattering processes contained in the $P(\boldsymbol{k}' + e\boldsymbol{E}/\hbar, \boldsymbol{k})$ term. It is also important to note that this process represents *out-scattering* from the distribution function within the integral.

It is now clear that the Monte Carlo process is a two-step process: acceleration and scattering. Each of these two steps is repeated for as long as required to reach a steady-state solution to Boltzmann's equation (if it exists). However, this is a stochastic evaluation of the path integral that is only as good as the various probabilities representing the physical processes. Thus, attention must be paid to describing these physical processes in as exact a manner as possible before turning the computer loose to find the solution.

The first-term on the right-hand side of (9.8) is the initial condition, and memory of this initial state decays with the scattering processes. Thus, the distribution function represented by the ensemble of particles must first be initialized as a proper equilibrium distribution, weighted by the appropriate density of states. Typically, a nondegenerate Maxwellian is used, but there is no reason why the Fermi–Dirac distribution could not be used. In the initialization routine for the computer simulation, several column matrices of rank N are required in order to keep tabs on the particles as they evolve. Obviously, the column matrices $\overline{\overline{KX}}$, $\overline{\overline{KY}}$, $\overline{\overline{KZ}}$ are required to store the momentum values and $\overline{\overline{EN}}$ is required to store the energy values (here, the double overbar is used to indicate matrices to avoid confusion with simple variables). Each of these column matrices can also be understood as a vector is of order N, where N is the number of particles in the simulation and the column index gives the particle number. If we are interested in anything other than a homogeneous semiconductor, then the column matrices N $\overline{\overline{X}}$, $\overline{\overline{Y}}$, $\overline{\overline{Z}}$ give the positions of each of the particles. All of the above matrices have real values. Another important column matrix in multivalley bands is $\overline{\overline{val}}$, which carries integer values that denote the valley index; e.g. 1, 2, and 3 may stand for the Γ, L, and X valleys, respectively. It will be found to be useful, especially for the purpose of parallelizing any code, to declare another pair of matrices denoted by $\overline{\overline{delE}}$, $\overline{\overline{flag}}$, the first of which stores real numbers and the last of which stores integers. These represent the energy change in scattering events and a flag used to denote the type of scattering, respectively. They will be explained more fully in the following discussions. In the initialization stage, $\overline{\overline{val}}$ and $\overline{\overline{flag}}$ are set uniformly to one. The column matrix $\overline{\overline{delE}}$ is initialized to zero.

Another helpful consideration at the beginning of the procedure is to create a table of all scattering values. In the second part of the Monte Carlo procedure of (9.8), the scattering process becomes important. To compute all scattering probabilities at each time step is quite computer intensive. To ease this task, a scattering table can remove this computation requirement and is also helpful in picking the scattering process to be implemented. To accomplish this requires a grid for the energy. At room temperature, a 1 meV energy step is usually enough, but at low temperatures, a finer grid will likely be necessary. The average thermal energy of an electron at room temperature is $3k_BT/2 \sim 39$ meV. This is much larger than the 1 meV energy step, so the noise associated with this step is small. In figure 9.1, the scattering rates for the polar optical phonon emission and absorption scattering, impurity scattering, and combined acoustic and alloy scattering are plotted for $In_{0.53}Ga_{0.47}As$ at 300 K. Only the intervalley scattering for the $\Gamma \leftrightarrow L$ valleys is shown, which sets in at about 0.53 eV. The impurity density is 10^{16} cm^{-3}. The acoustic and alloy scattering are combined, as they are both elastic and have the same energy dependence [24]. These few scattering rates are sufficient to fully discuss the EMC approach.

9.2 Free-flight generation

As remarked above, the path integral (9.8) contains two separate processes. The first part of the integral addresses the dynamics of the particle motion, which is assumed to consist of free flights interrupted by instantaneous scattering events. Of course, the precise time that a carrier drifts before scattering is random and is actually

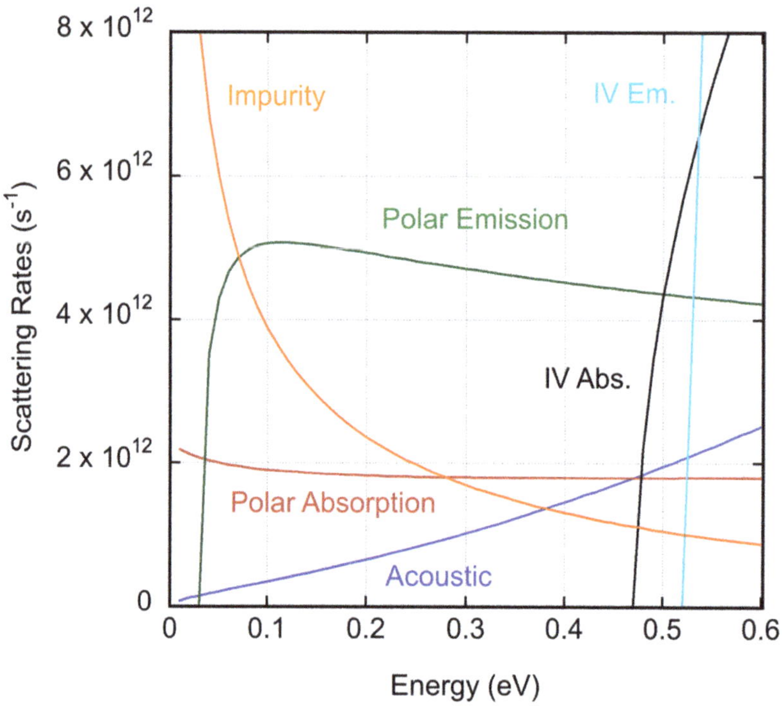

Figure 9.1. The scattering rates for electrons in $In_{0.53}Ga_{0.47}As$ at 300 K. The impurity density is 10^{16} cm^{-3}, and the L valleys lie some 0.5 eV above the central Γ valley.

determined by the scattering processes themselves. As remarked, the argument of the exponential in (9.9) is the total out-scattering rate, so that the drift time is inversely proportional to this rate. In fact, this integrand has been simplified, as Γ_0 is a function of the carrier energy. It may be seen in figure 9.1 that each scattering process has an energy dependence, so that (9.9) should properly be given as

$$\exp\left(-\int_0^{t'} \Gamma_0[E(t'')]dt''\right). \tag{9.10}$$

Hence, the scattering rate Γ_0 is a function of the energy E, and the energy E changes over time due to the field in (9.8) that produces acceleration. The total scattering rate for the processes shown for $In_{0.53}Ga_{0.47}As$ in figure 9.1 is plotted in figure 9.2. This does not vary as much with energy as the individual processes, but it still yields a complicated integral to evaluate.

Within the stochastic approximations, the actual transport is simulated by introducing a probability density $P(t)$, where $P(t)dt$ is the joint probability that a carrier both arrives at time t without scattering (after its last scattering event at $t = 0$) and then actually undergoes a scattering event at this time (i.e. within a time interval dt centered at t). This is the probability (9.10). The probability of actually scattering within this small time interval at time t may be written as $\Gamma[E(t)]dt$, where $\Gamma[E(t)]$ is

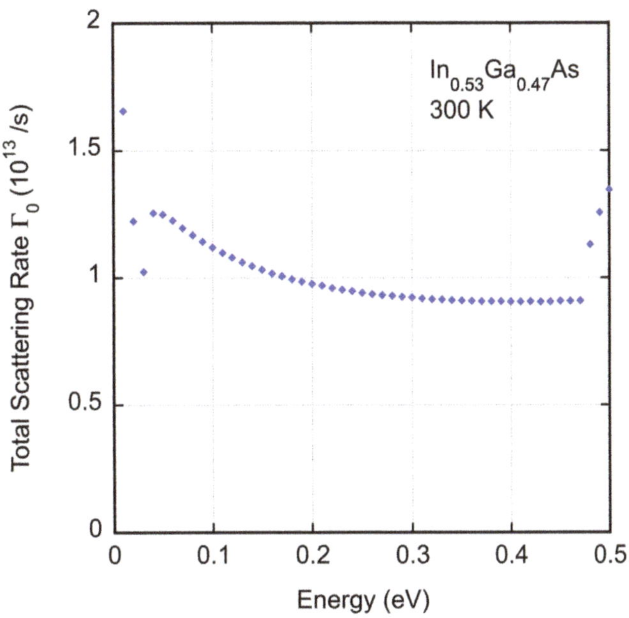

Figure 9.2. The total scattering rate Γ_0 for $In_{0.53}Ga_{0.47}As$, as determined from the individual scattering rates of figure 9.1.

the total scattering rate of a carrier of wave vector $\boldsymbol{k}(t)$ and energy $E(t)$. This scattering rate represents the sum of the contributions of each scattering process that can occur for a carrier of this wave vector (and energy). The explicit time dependence indicated is a result of the evolution of the wave vector (and energy) under any electric (and magnetic) fields that cause acceleration. In terms of this total scattering rate, the probability that a carrier has not suffered a collision after time t is given by (9.10). This function has not been normalized as a good probability would be, as the magnitude of (9.10) varies from zero to one, a proper range for a random number.

The random flight times τ for each particle are created by a random number generator, which gives a uniform distribution of real numbers between zero and one (including these two values). The normalization factor is extraneous, and the random flight time is sampled from $P(t)$ according to the random number r as follows:

$$\ln(r) = -\int_0^t \Gamma_0(E(t'))dt'. \tag{9.11}$$

This remains a difficult integral. While Γ_0 in figure 9.2 is relatively flat, it contains a number of rapid variations, especially at low energy and at the onset of intervalley scattering, which complicate the integral.

Because of the dependence of Γ_0 on energy and the difficult integration in (9.11), it is expedient to resort to a mathematical *trick*. Here, a fictitious scattering process that has no effect on the carrier is introduced. This process is called *self-scattering*, and the energy and momentum of the carrier are unchanged under this process [25]. An energy dependence is assigned to this process in just such a manner that the total scattering rate is a constant, as follows:

$$\Gamma_T - \Gamma_{self}(E) = \sum_i \Gamma_{0,i}(E), \tag{9.12}$$

where the sum runs over all the *real* scattering processes as a function of the energy E. The energy dependence of the self-scattering term is determined so that the quantity Γ_T is independent of energy and therefore independent of the acceleration time. Because the self-scattering process has no effect on the energy, momentum, or position of the particle, it has no effect upon the transport properties being evaluated by the Monte Carlo process. But the introduction of this imaginary scattering process simplifies the temporal integration in (9.11). Consider figure 9.3, in which the total scattering rate is again plotted, but now the values for Γ_T and $\Gamma_{self}(E)$ are shown for convenience. Using this constant Γ_T, the flight time τ may be determined as follows:

$$\tau = -\frac{1}{\Gamma_T}\ln(r). \tag{9.13}$$

In an ensemble Monte Carlo process, each individual particle in the ensemble carries its own flight time τ, and this time is selected at the start of each flight. This creates another problem in that the time of total transport is different for each particle.

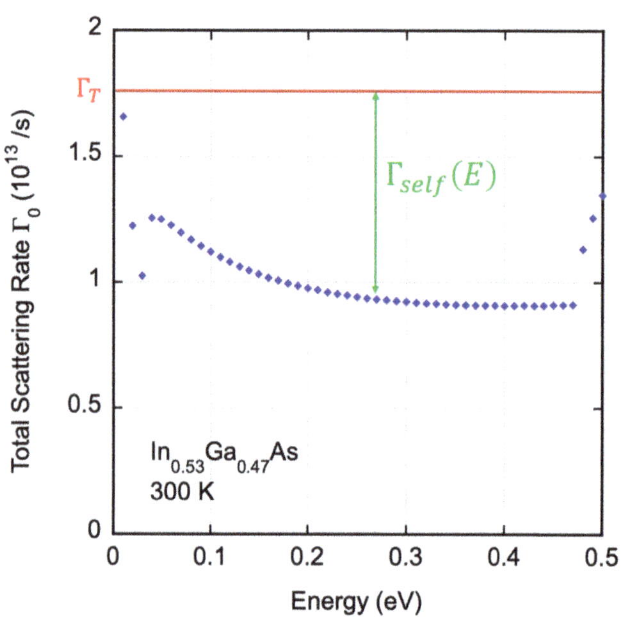

Figure 9.3. An illustration of the selection of the new total scattering rate Γ_T and the energy-dependent self-scattering rate $\Gamma_{self}(E)$, which are used in (9.12) to simplify the energy integration.

In order to obtain ensemble averages, it is necessary to stop the transport at uniform transport times. This will be discussed in subsequent sections.

Following the free flights, the momenta, energies, and positions are updated according to the general classical mechanics of Hamiltonian systems, as follows:

$$\begin{aligned} \boldsymbol{k}_i(t) &= \boldsymbol{k}_i(0) - \frac{e\boldsymbol{F}}{h}\tau_i \\ E_i(t) &= \frac{\hbar^2 k_i^2(t)}{2m^*} \\ \boldsymbol{r}_i(t) &= \boldsymbol{r}_i(0) + \frac{\hbar \boldsymbol{k}_i(t)}{m^*}\tau_i \end{aligned} \quad . \tag{9.14}$$

The electric field is written here as $\boldsymbol{F}$ to avoid confusion with the energy E. In these equations, parabolic bands have been assumed, but the hyperbolic non-parabolic bands could just as well be (and usually are) used. This is a detail that is easily incorporated into (9.14). Indeed, the basic definition of the effective mass is used in the last line to give the carrier velocity [24], and this mass has an energy dependence for non-parabolic bands. Equation (9.14) is the basic acceleration equation that is frequently used in EMC simulations.

9.3 Scattering

The second step of (9.8) is the scattering of the particles. This occupies the majority of the simulation code and the computation time in the EMC process (apart from the time taken to solve Poisson's equation in inhomogeneous situations, see chapter 7). It is represented by the factors

$$P(\boldsymbol{k}' + \frac{e}{\hbar}\boldsymbol{F}, \boldsymbol{k}). \tag{9.15}$$

Only the normal scattering processes need be described, and these are shown for $In_{0.53}Ga_{0.47}As$ in figure 9.1. To capitalize on parallel processing, all of the particles need to be simulated in parallel, which requires considering each step in the process at the same time. There is a step between the acceleration of the previous section and the scattering of this section that involves the use of a time grid to take averages. The intervening step involves determining whether or not a scattering event occurs within the current cell of the time grid. It may be observed from this temporal discussion that there are two parallel time assessments. One is the time that each particle experiences during its acceleration events. The second is the laboratory time, which has its own increments.

Ensemble averages for such quantities as the drift velocity, position, and average energy, etc. are computed on the various steps of the laboratory timescale. That is, a particle is accelerated only for the increment of the lab time and then halted to compute the various averages. A particular average, for example the velocity, is computed as

$$v_d(T) = \frac{1}{N}\sum_i v_i(T). \tag{9.16}$$

This is an ensemble average. In the far-from-equilibrium situation, and particularly when transients are involved, there is no ergodic theorem [26]. Therefore, only ensemble averages have statistically valid properties. To achieve such an average, all the particles need to be aligned at the same time T that appears in (9.16), which here runs from the beginning of the simulation. This is the rationale for the laboratory timescale, which is discretized into steps of $n\Delta T$. At integer multiples of this time step, all particles are stopped in their free flights, and ensemble averages are computed, as indicated above. Each particle has a flight which is composed of accelerations and scattering processes. The arrival at a synchronization time T usually occurs during one of the free flights. The imposition of the second laboratory timescale is illustrated in figure 9.4. This process is efficient, although at the expense of more bookkeeping in the algorithm.

If a scattering event is to occur for a particular particle within the time step ΔT, the flag $\left(\overline{\overline{flag}}\right)_i$ is set to 2 to indicate that a scattering event has to be carried out for particle i. Only particles with the flag set to 2 are considered in the scattering process. In the scattering event, the important process is selecting the individual scattering process that will be evaluated for the particular particle. This is done using another random number (an important point is that most modern compilers include the ability to select a large matrix of random numbers at the same time). The selection of the scattering event is demonstrated by the scattering processes shown in figure 9.1. While only a limited number of scattering processes are shown in this figure, the selection can easily be expanded to cover any arbitrary number of scattering events.

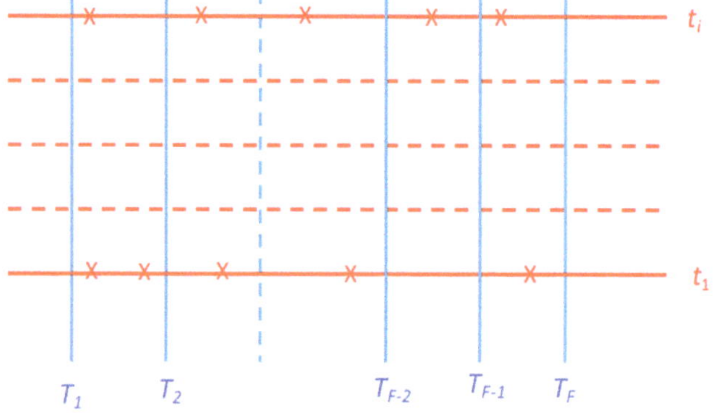

Figure 9.4. The alignment of particle timescales t with the global timescale T. The simulation and particle movement is stopped at each time step T_i and the ensemble averages are updated. Reproduced from [24].

To begin, the scattering rates of figure 9.1 are replotted in figure 9.5 as partial sums. That is, the lowest curve (blue) is solely impurity scattering. The next curve up (in red) adds acoustic (plus random alloy scattering) to this value. Each curve as one moves upward in the plot adds an additional scattering process. The selection of the actual process that scatters the particle in question is determined by a random number, which is multiplied by Γ_T. This is shown by the magenta arrow and label *r* in figure 9.5. For the particular particle in question, it is assumed that at the end of the acceleration stage, this particle has an energy of approximately 0.26 eV. The random number is chosen. In this case, the magenta arrow extends above the line corresponding to the sum of impurity and acoustic/alloy scattering and polar optical absorption, so these processes are *not* the source of the scattering. On the other hand, it lies below the sum of the previous three events plus polar phonon emission. Thus, polar optical phonon emission (POP_{em}) is chosen as the scattering process. If the arrow had been below the green line, then polar optical absorption would have been selected. If the arrow had extended above the black line (or the extended light blue line), then self-scattering would have been selected.

Since the selected scattering process is polar phonon emission, the energy change column matrix element is set to $\left(\overline{\overline{delE}}\right)_i = -\hbar\omega_0$. In addition, some recognition of whether or not the scattering is isotropic or anisotropic is required, and the flag is used for this; for this example, the flag element is set to $\left(\overline{\overline{flag}}\right)_i = 3$, which indicates that the scattering is anisotropic. Only piezoelectric scattering, impurity scattering, and the polar optical phonons have this property. If the process had been isotropic,

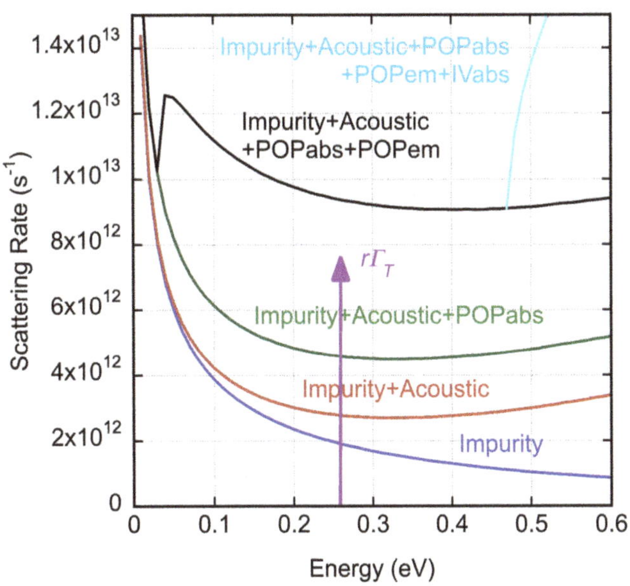

Figure 9.5. The selection of the scattering process is achieved by the following process. A random number produces the magenta arrow, whose length is $r\Gamma_T$. Since it lies above the blue curve but below the red curve, polar absorption is selected as the scattering process.

as in acoustic or intervalley scattering, the flag element would have been set to 4. If this flag had remained at 2, it would have indicated that self-scattering occurred for that particle, and the energy and momentum would remain unchanged. The need for the flag lies in the fact that determining the final-state energy and momentum involves quite different processes for isotropic and anisotropic scattering. If the scattering is also elastic, which is usually assumed for acoustic and impurity scattering, the energy change element remains at 0.

When the scattering is isotropic, all states on the final energy shell are available. The energy $E_i(t)$ is incremented by the energy change to give the new energy, and this is used to choose the new magnitude of the momentum as follows:

$$k_i = \sqrt{\frac{2m^*E}{\hbar^2}}. \tag{9.17}$$

This energy is projected onto the momentum coordinates using two additional random variables (spherical coordinates are used), as follows:

$$\begin{aligned} \vartheta &= r_3\pi \\ \varphi &= 2r_4\pi \end{aligned}. \tag{9.18}$$

These are now used to compute the momentum in the three directions.

With anisotropic scattering, determining the final angle ϑ is more difficult. The azimuthal angle (φ)remains as given in the second line of (9.18). However, the final angle ϑ depends upon the energy of the carrier involved in the scattering. Consider impurity scattering, which is anisotropic due to the coulombic nature of the scattering. The coulomb scattering potential in three dimensions varies in the matrix element according to

$$\frac{1}{q^2 + 4k^2\sin^2(\vartheta/2)}. \tag{9.19}$$

But the scattering rate depends upon the square of the matrix element, thus upon the square of the term in (9.19), as follows:

$$\left[\frac{1}{q^2 + 4k^2\sin^2(\vartheta/2)}\right]^2. \tag{9.20}$$

Hence, a scattering angle ϑ_{sc} can be selected using the integral of the probabilities as follows:

$$r = \frac{\int_0^{\vartheta_{sc}} [q^2 + 4k^2\sin^2(\vartheta/2)]^{-2} \sin\vartheta d\vartheta}{\int_0^{\pi} [q^2 + 4k^2\sin^2(\vartheta/2)]^{-2} \sin\vartheta d\vartheta}. \tag{9.21}$$

The integrals are tedious but not complicated, and it is possible to then determine that the angle is related to a random number r as follows:

$$\cos\vartheta_{sc} = \left[1 - \frac{2r\xi}{1+\xi-r}\right], \tag{9.22}$$

where

$$\xi = \frac{q_D^2}{4k^2}, \tag{9.23}$$

and q_D is the reciprocal of the Debye length. Almost 75% of the scattering is in the forward direction, which means that impurity scattering is not particularly efficient at relaxing the momentum. This has been well known for a long time, and this fact is handled in the classical Boltzmann equation by the introduction of the famous factor of $(1 - \cos\vartheta)$ in determining the momentum relaxation time [24].

With anisotropic scattering, there is another factor that comes into play. The scattering angle in (9.22), or the equivalent one for polar optical phonons is actually the angle with respect to the vector momentum prior to the collision! This vector momentum actually has its own angle with respect to the electric field. Thus, the polar coordinates in which ϑ_{sc} exists have to be rotated into the polar coordinates in which the electric field lies in the z-direction. Failure to do this leads to quite erroneous results for the transport simulation. This was not required for the isotropic scattering because of the fact that any point on the final energy surface was equally likely, so that the angles could be described as actually being in the laboratory coordinates. The amount of rotation is determined by the orientation of the vector momentum with respect to the electric field. The polar coordinate angles of this orientation define the coordinate angles determined above, and the coordinate transformation is straightforward and well known in classical mechanics [27].

To handle the anisotropic scattering of the polar optical phonons, several of the various column matrices come into play. Some of these are

$$\begin{gathered} E_i = \left(\overline{\overline{EN}}\right)_i \\ \left(\overline{\overline{delE}}\right)_i = \pm\hbar\omega_0 \\ \left(\overline{\overline{EN}}\right)_i = E_i + \left(\overline{\overline{delE}}\right)_i \end{gathered} . \tag{9.24}$$

In the example above, only the negative sign in the second line is used, as emission of the phonon was chosen. The magnitudes of the momentum values of both the initial and the final states are needed, due to the requirement for the coordinate rotation discussed in the previous paragraph. These are

$$\begin{gathered} k^2 = \frac{2m^*}{\hbar^2}E_i \\ k'^2 = \frac{2m^*}{\hbar^2}\left(\overline{\overline{EN}}\right)_i \end{gathered} . \tag{9.25}$$

Here, E_i is needed along with the prescattering momentum $\boldsymbol{k}$ in order to determine the angles for the subsequent coordinate rotation. Both of the momenta (and energies) are needed to determine the scattering angle ϑ_{sc} before the subsequent rotation. Like impurity scattering, the scattering rate is proportional to the square of the matrix element, so that in analogy with (9.20), this becomes

$$(k^2 + k'^2 - 2kk' \cos \vartheta)^{-2}, \tag{9.26}$$

and the actual angle is related to the random number r through the same type of integral, which is

$$r = \frac{\int_0^{\vartheta_{sc}} [k^2 + k'^2 - 2kk' \cos\vartheta]^{-2} \sin\vartheta d\vartheta}{\int_0^{\pi} [k^2 + k'^2 - 2kk' \cos\vartheta]^{-2} \sin\vartheta d\vartheta}. \tag{9.27}$$

As before, the integral is tedious, if not particularly difficult, and the scattering angle becomes

$$\cos\vartheta_{sc} = \frac{1 + \xi - 2r}{1 + \xi(1 - 2r)}, \tag{9.28}$$

where

$$\xi = \frac{2kk'}{k^2 + k'^2}. \tag{9.29}$$

It may be seen that this has the correct limits for the range of the random number. Once more, forward scattering is predominant in the process, and polar optical phonon scattering is not particularly efficient at relaxing the momentum, although it is more efficient than impurity scattering due to the energy loss in the predominant phonon emission.

Following determination of the states of the particles after scattering, the EMC procedure returns to the acceleration process of the previous section, and new final times are chosen for all the electrons. However, this process becomes slightly more complicated when the laboratory time grid is present, as the time synchronization discussed above is implemented within the overall temporal progression of the process.

9.4 Rejection techniques

While it was possible to actually integrate the angular probability functions for the anisotropic scattering, this may not always be the case for some unusual scattering processes, and one therefore has to resort to other statistical methods. Two examples of this are accounting for nonequilibrium phonons and degenerate statistics and the occupation of the final state. One method of treating these is the rejection technique. Suppose the probability density function for the process is quite nonlinear and not easily integrated to get the total probability. One can then use a pair of random numbers (r_1,r_2) to evaluate the angle. Consider figure 9.6, in which a complicated

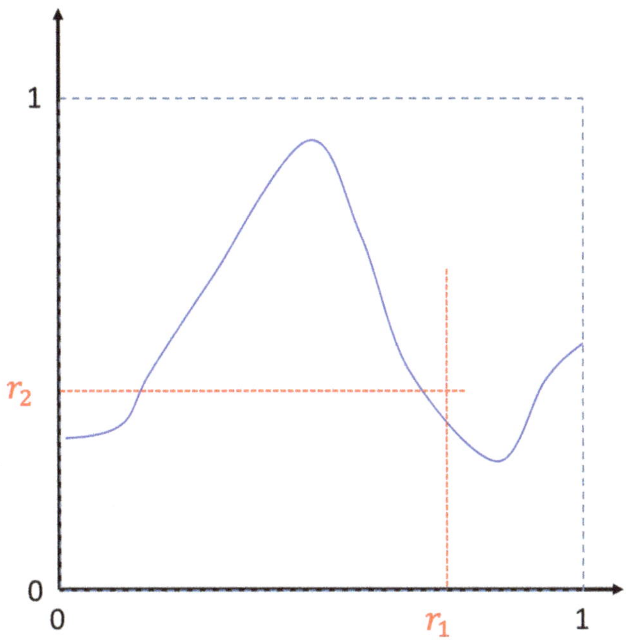

Figure 9.6. The rejection technique uses two random numbers. The first, r_1, selects a horizontal value at which to evaluate the nonlinear function. The second determines whether or not to use the selected value, depending upon whether or not this value is larger or smaller than the value of the function at the first number. In the case shown, the pair will be rejected. For acceptance, the value of the second random number must lie below the blue curve. Reproduced from [24].

probability density function is illustrated. The abscissa could be, for example, the polar angle θ, which has been normalized by π. In this manner, the maximum of the abscissa x is unity, so that the range of the function's argument is from zero to one. The maximum value of the ordinate (y axis) is also normalized to unity, but one can always renormalize this to the span of the random numbers if it is not zero to one. The first random number is taken to correspond to the abscissa of the plot (the x axis). This determines the argument of the function that is to be evaluated. For example, let us assume that this is $r_1 = 0.75$ (this is indicated by the vertical dashed line in the figure). We now use the second random number to determine whether $f(r_1) > r_2$. If this relationship holds, then the value r_1 is accepted for the argument of the function, and the scattering process proceeds with this value. On the other hand, if $f(r_1) < r_2$, then two new random numbers are chosen. One continues to repeat this process until a pair of random numbers is found that meets the criterion. Certainly, values of r_1 for which the function is large are more heavily weighted in this rejection process. This is considered in more detail for two processes: (1) state filling due to the degeneracy of the electron gas, and (2) nonequilibrium phonons. Both use the same basic process of creating a grid of values for the distribution, either the phonons or the various energy states for the Fermi–Dirac distribution.

9.4.1 State filling and degeneracy

Consider the case of state filling that occurs when the distribution function represents a degenerate carrier system [11]. The state filling appears in the selection of the final state after scattering, as the carrier is not allowed to scatter into a state that is already full. Degeneracy is introduced through a secondary self-scattering process [11, 28] based upon the rejection technique. It is termed secondary self-scattering, because if the condition $f(r_1) > r_2$ is not satisfied, the rejection proceeds in exactly the same way as a self-scattering process, which was introduced earlier. Each of the scattering processes must include a factor of $[1 - f(E)]$, where $f(E)$ is the time-dependent distribution function representing the probability that the final state after scattering is empty, and the energy E is the final-state energy after scattering. Since the nondegenerate scattering rate is always larger than the degenerate scattering rate due to the reduction by the $[1 - f(E)]$ factor, this secondary self-scattering is a useful and convenient method. In actual practice, a grid *in momentum space* is maintained and the number of particles in each state is tracked (each cell of this grid has its population divided by the total number of states in the cell, which depends on the cell size, to give the value of the distribution function in that cell). The scattering processes themselves are evaluated, but the acceptance of the process depends on the rejection technique. To make this statistically efficient, the number of particles within each cell must be fairly large.

The most delicate point of the degeneracy method involves the normalization of the distribution function $f(\mathbf{p})$. The extension of the secondary self-scattering method to the ensemble Monte Carlo algorithm involves the fact that there are N electrons in the simulation ensemble, which represent an electron density of n. The effective volume V of 'real space' being simulated is N/n. The density of allowed wave vectors of a single spin in **k**-space is just $V/(2\pi)^3$. In setting up the grid in the three-dimensional wave vector space, the elementary cell volume is given by $\Omega_k = \Delta k_x \Delta k_y \Delta k_z$. Every cell can accommodate at most N_c electrons, with $N_c = 2\Omega_k V/(2\pi)^3$, where the factor of two accounts for the electron spin. For example, if the density is taken to be 10^{17} cm^{-3}, $N = 10^4$, and $\Delta k_x \Delta k_y \Delta k_z = (2 \times 10^5$ cm$^{-1})^3$ ($k_F = 2.4 \times 10^6$ cm^{-1} at 77 K), then $V = 10^{-13}$ cm^3 and $N_c = 6.45$. N_c represents the maximum occupancy of a cell in the momentum-space grid. (Obviously, a more careful choice of parameters would result in an integer value of N_c, for convenience.) A distribution function is defined over the grid in momentum space by counting the number of electrons in each cell. The distribution function is normalized to unity by dividing the number in each cell by N_c for use in the rejection technique. As remarked above, N_c should be sufficiently large that rounding it to an integer (if the numbers do not work out properly, as in the case above) does not create a significant statistical error.

9.4.2 Nonequilibrium phonons

The second example is the consideration of nonequilibrium phonons, which can occur in high electric fields, especially in low-dimensional systems. The nonequilibrium phonon distributions can be incorporated into the EMC process through the use of a rejection technique [29]. Up to this point, the assumption has been made that the phonons are in equilibrium and characterized by N_q. When these phonons are driven out of equilibrium, the population at a given q becomes larger than N_q. This makes the evaluation a little more difficult. The most common implementation assumes a maximum value for this distribution, taken as $N_{q,\max}$, and this value is used to compute the phonon scattering rates. As with degeneracy, this distribution is projected onto a momentum-space grid, so that an individual cell in this discretized space has a volume of $\Delta q_x \Delta q_y \Delta q_z$. This small volume has a number of available states given by $V/(2\pi)^3$, where V is determined by the effective simulation volume N/n, as described previously. The difference between state filling for carrier degeneracy and phonon state filling is that there is no limit to the number of phonons that can exist within a given state, as they obey the Bose–Einstein distribution. A secondary self-scattering process is used to reject the scattering when the phonon population at the prescribed q and time is less than this maximum. This is implemented by rejecting the process if $r > N(q, t)/N_{q,\max}$. The number of phonons emitted or absorbed for wave vector **q** is carefully monitored with the distribution grid. At the synchronization times of the global timescale, the phonon population in each cell of momentum space is updated from the emission/absorption statistics that have been gathered during that time step. One must also include phonon decay (which takes place through a three-phonon process; see section 6.9.1) to other modes of the lattice vibrations, so that the update algorithm is

$$N(\boldsymbol{q}, t + \Delta t) = N(\boldsymbol{q}, t) + G_{\text{net}, \Delta t}(\boldsymbol{q}) - \left[\frac{N(\boldsymbol{q}, t) - N_0}{\tau_{\text{phonon}}}\right]\Delta t, \tag{9.30}$$

where N_0 is the equilibrium phonon distribution, $G_{\text{net},\Delta t}(\mathbf{q})$ is the net (emission minus absorption) generation of phonons by the carriers in the particular cell *during the global time step*, and τ_{phonon} is the phonon lifetime. In figure 9.7, the phonon distribution function is plotted at two values of the electric field for a GaN/AlGaN heterojunction device [30] at room temperature. Electrons in this structure show an experimental peak velocity near 3.2×10^7 cm s^{-1} at a field of 150 kV cm^{-1}, which is supported by detailed ensemble Monte Carlo simulations [30, 31]. The results shown in figure 9.7 indicate that the electrons only interact with a set of relatively low-wave-number phonons, but these can be driven quite far from equilibrium.

The generation of the nonequilibrium phonons can be estimated by going beyond (9.30). For example, in the approximation of an electron temperature, (9.30) may be expressed as [32]

$$\left.\frac{\partial N(q)}{\partial t}\right|_{\text{carriers}} = \frac{n}{\tau_{\text{e}}}\{N(q) - [N(q) + 1]e^{-x_{\text{e}}}\}, \tag{9.31}$$

where the power density that flows to the phonons from the carriers is given by

$$\left.\frac{\partial E}{\partial t}\right|_{\text{carriers}} = \frac{\hbar\omega_{\text{LO}}\Delta n}{\tau_{\text{e}}}\{N(q) - [N(q) + 1]e^{-x_{\text{e}}}\}, \tag{9.32}$$

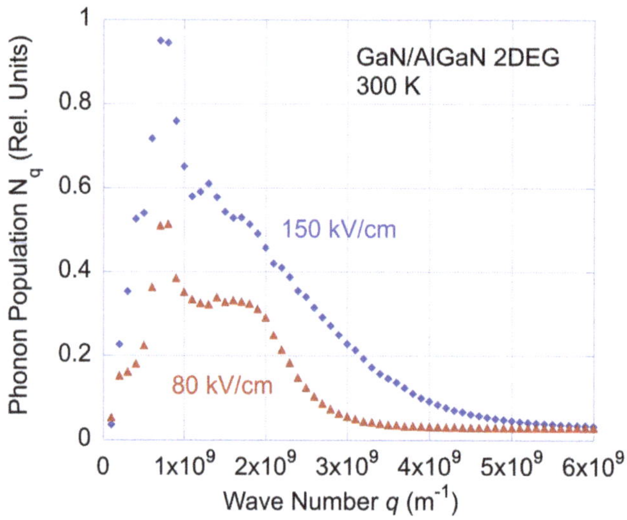

Figure 9.7. The nonequilibrium phonon distribution at two high values of the electric field as determined by Monte Carlo simulations for a GaN/AlGaN two-dimensional layer at the heterojunction interface.

where $x_e = \hbar\omega_{LO}/k_B T_e$, n is the local electron density, $\hbar\omega_{LO}$ is the polar optical phonon energy, and τ_e is the effective scattering time for the emission of an optical phonon by a carrier. The net emission of the carriers that generates phonons may then be combined with the net decay of the nonequilibrium optical phonons, given by

$$\left.\frac{\partial N(q)}{\partial t}\right|_{\text{decay}} = -N_M \frac{N(q) - N_{eq}}{\tau_{LO}}, \tag{9.33}$$

where N_{eq} is the equilibrium thermal Bose–Einstein form for the phonon distribution, τ_{LO} is the lifetime of the polar longitudinal optical (LO) modes, which decay via the abovementioned three-phonon process [33], and $N_M = 2N_{cell}/a_0^3$ is the number of phonon states in the LO branch. In this latter formula, N_{cell} is the number of atoms per unit cell, in this case the wurtzite cell, and a_0^3 is the volume of the wurtzite or face-centered cubic (FCC) cell. The wurtzite cell is a relaxation of the zinc-blende cell and has four atoms per primitive unit cell, so its volume is basically double that of the FCC cell.

In the case in which the various distributions are in a steady state, (9.31) and (9.33) can be set equal to one another, and the phonon distribution determined from this equality. Thus, one finds

$$N(q) = \frac{N_M N_{eq} + n\left(\frac{\tau_{LO}}{\tau_{eL}}\right)e^{-x_e}}{N_M + n\left(\frac{\tau_{LO}}{\tau_{eL}}\right)(1 - e^{-x_e})}. \tag{9.34}$$

If multiple sets of sub-bands are occupied, such as occurs in intervalley transfer situations, then the formula becomes more complicated [34]. Values for the various parameters can be determined from ensemble Monte Carlo simulations, leading to results such as those shown in figure 9.7. This means that a reasonably good estimate of the phonon population can be obtained from the above equations. The factor N_M for InGaAs can be evaluated to be $\sim 2 \times 10^{22}$ cm^{-3}, which is a very large number, but is of a similar magnitude to those of most semiconductors, differing only by the size of the edge of the FCC cell. If a heterojunction interface is used, such as that of GaN/AlGaN discussed above, which has an effective inversion layer thickness of ~5 nm, N_M drops to $\sim 10^{16}$ cm^{-2}. When the 2D density in the layer begins to be comparable to this, driving the phonons out of equilibrium becomes easier. These numbers are merely suggestive, and the proper dynamics needs to be followed to evaluate such nonequilibrium events. In the GaN/AlGaN interface, it is found that the phonon lifetime is unexpectedly dependent upon the 2D carrier density [35]. This means that the population of nonequilibrium phonons depends upon the carrier density and this, in turn, affects the effective velocity behavior of these carriers.

9.4.3 An example

To illustrate the data that can be determined from the EMC process, an example will be used to illustrate the behavior of transport in $In_{0.6}Ga_{0.4}N$ at room temperature. This material has a direct bandgap of approximately 1.9 eV and is a multivalley semiconductor in that there are two sets of upper valleys lying some 2.2 and 2.6 eV above the central valley. The material is used in a variety of electronics applications, including both high-electron-mobility transistors (HEMTs) and hot carrier solar cells. Here, the transport properties of the bulk material will be used as an example. The velocity as a function of the applied electric field is shown in figure 9.8. The most obvious features of the curve are a broad peak beginning at ~100 kV cm^{-1} and then a slow drop followed by a rapid drop. The latter is believed to signal the well-known onset of intervalley transfer, and the slow drop is due to the transfer of carriers to the second set of valleys, while the more rapid drop is likely to be due to the onset of population transfer to the third set of valleys. The mobility is significantly lower in the satellite valleys, and this is the reason the drift velocity drops rapidly. In figure 9.9, the distribution function for the carriers is shown for two different electric fields, one below the onset of transfer and one just above. The peak near zero energy is lower for the higher field, as the carriers have streamed to higher energies. This is typical for carrier heating in the applied electric field. To compute the distribution function, a grid is created in energy space and the particles are then projected onto this grid at the end of the simulation period. Note that this is not the true distribution function but is the population of carriers, which is the product of the distribution function and the density of states. The density of states is complicated in the hyperbolic bands, due to the energy dependence of the effective mass, but it can be found in a straightforward manner. For the hyperbolic bands of section 6.4, the standard parabolic band density of states is enhanced by a factor of

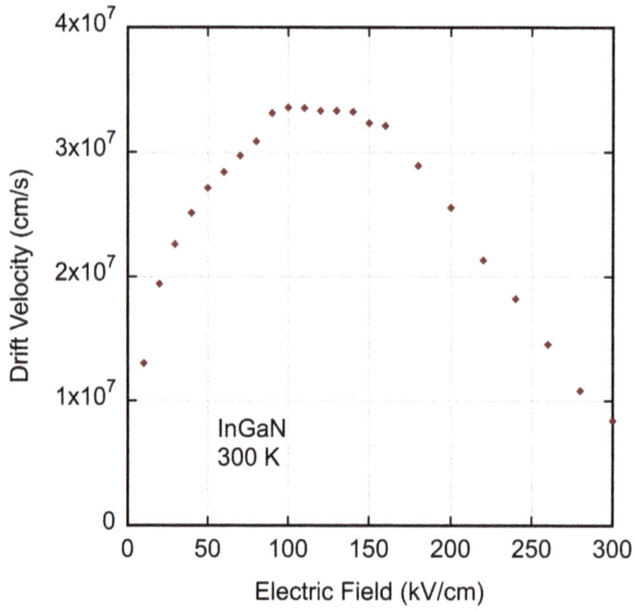

Figure 9.8. The drift velocity as a function of the electric field for In0.6Ga0.4N at room temperature. This material has the wurtzite crystal structure.

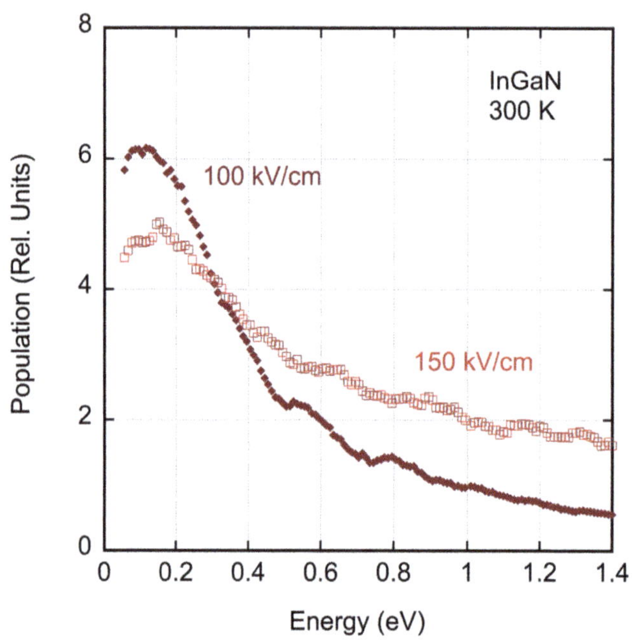

Figure 9.9. The relative populations of carriers as a function of the energy of these carriers.

$$\sqrt{1 + \frac{E}{E_G}}\left(1 + \frac{2E}{E_G}\right). \tag{9.35}$$

The population data can be divided by the density of states and renormalized to give the actual distributions.

9.5 Full-band approaches

In the above sections, the Monte Carlo approach used the effective mass approximation, usually with parabolic bands, but occasionally with the non-parabolic hyperbolic bands that arise in $\boldsymbol{k} \cdot \boldsymbol{p}$ theory (section 6.4). One can go further when it is desired to completely characterize electrons (or holes) that reach very high kinetic energies. The use of atomistic full-band energy behavior [10] allows the full band structure throughout the Brillouin zone to be fully determined and utilized. The first to do this were apparently Shichijo and Hess [10], who determined the band structure of GaAs using an empirical pseudopotential method [36], as discussed in chapter 6. This was a local pseudopotential that did not include the spin–orbit interaction [24]. The idea was to study impact ionization in GaAs, and the authors wanted a better description of the bands for carriers at the high energy required for this. The rationale is illustrated better in figure 6.7, where it can be seen that the actual band structure is quite far from an analytical band. To implement the full band structure, the grid used must fall within the Brillouin zone but, due to the set of symmetries of the Brillouin zone and the crystal structure, only a fraction (usually 1/48th) of the zone need be used. Nevertheless, scattering rates have to be computed and stored between all values of energy and momentum. Importantly, neither the velocity nor the effective mass is a simple function of momentum. The band in which the particle is located must be kept as a parameter and the velocity at any momentum determined from

$$v = \frac{1}{\hbar}\frac{\partial E}{\partial \boldsymbol{k}}. \tag{9.36}$$

Similarly, the effective mass must be determined at a given point in the Brillouin zone from the basic definition of effective mass, which is

$$\frac{1}{m^*} = \frac{v}{\hbar k} = \frac{1}{\hbar^2 k}\frac{\partial E}{\partial \boldsymbol{k}}. \tag{9.37}$$

One may note that this is not the usual textbook formula, as the latter is only valid near an extremum of the energy band. Instead, the fundamental mass relationship must equate the quasi-particle momentum (m^*v) with the quantum wave momentum ($\hbar k$). A vector/tensor form of (9.37) is easily derived [24]. As a result of these derivative requirements, the grid has to be fine enough to allow the velocity and mass to be determined with a minimum of noise. One major computational problem is the storage of the scattering tables. Another problem arises where the bands split (or cross in silicon, for example), such as at the X point in the valence band and near Γ in the second conduction band. The usual approach is to assert that the particle follows the band with approximately the same slope in momentum space; that is, the momentum vector in the Brillouin zone is conserved.

Perhaps the most well-known follow-on to the previous work is that of Fischetti and Laux, who developed the DAMOCLES code at IBM [37, 38]. This effort began with the local pseudopotential of Hess [10] without the spin–orbit correction, but was from the start aimed at the simulation of semiconductor devices, which means coupling to the Poisson equation. Here, they showed that storing the entire Brillouin zone at 41 000 points would provide a speed advantage by avoiding the symmetry operations required by using just a 1/48th wedge. However, the advantage of using self-scattering was lost, as the equations of motion could not be simply integrated and had to be numerically integrated. Nevertheless, this code allowed the pair to improve our knowledge of the band-structure parameters, masses, and phonon coupling constants in the principal valleys of the conduction band. Further work has allowed them to extend this development to a wider set of semiconductors and semiconductor devices.

Another advance in full-band calculations was made in the development of what is known as cellular Monte Carlo, in which the scattering rates are computed between cells in the Brillouin zone as a function of the initial momentum rather than tabulated with energy cells [39]. This code has been adapted with some quantum corrections and is capable of device simulations, as it also incorporates the Poisson equation [40, 41]. This method has been extensively used to study InGaAs strained-channel HEMTs [41, 42].

References

[1] Ferry D K 2021 *Hot Carriers in Semiconductors* (Bristol: IOP Publishing)

[2] Ferry D K 2018 *An Introduction to Quantum Transport in Semiconductors* (Singapore: Pan Stanford Publishing)

[3] Budd H F 1967 *Phys. Rev.* **158** 798

[4] Thornber K K and Feynman R F 1970 *Phys. Rev.* B **1** 4099

[5] Kurosawa T 1966 *J. Phys. Soc. Jpn. Suppl.* **21** 424

[6] Jacoboni C and Reggiani L 1983 *Rev. Mod. Phys.* **65** 645

[7] Jacoboni C and Lugli P 1989 *The Monte Carlo Method for Semiconductor Device Simulation* (Vienna: Springer)

[8] Hess K 1991 *Monte Carlo Device Simulation: Full Band and Beyond* (New York: Springer)

[9] Kosina H, Nedjalkov M and Selberherr S 2000 *IEEE Trans. Electron Devices* **47** 1898

[10] Shichijo H and Hess K 1981 *Phys. Rev.* B **23** 4197

[11] Lugli P and Ferry D K 1985 *IEEE Trans. Electron Devices* **32** 2431

[12] Joshi R P and Ferry D K 1991 *Phys. Rev.* B **43** 9734

[13] Jacoboni C 1976 *Proc. 13th Int. Conf. Physics of Semiconductors* (Rome: Tipo. Marves) p 1195

[14] Feynman R P 1948 *Rev. Mod. Phys.* **20** 367

[15] de Broglie L 1923 *Nature* **112** 540

[16] de Broglie L 1924 *Phil. Mag., Ser. 6* **47** 446

[17] Heller E J 1976 *J. Chem. Phys.* **65** 1544

[18] Dittrich T, Viviescas C and Sandoval L 2006 *Phys. Rev. Lett.* **96** 070403

[19] Vitanov P *et al* 1994 *Advances in Parallel Algorithms* ed B I Sendov and I Dimov (Amsterdam: IOS Press) pp 117–28
[20] Shifren L and Ferry D K 2002 *J. Comput. Electron* **1** 55
[21] Nedjalkov M *et al* 2002 *Proc. 7th Int. Conf. on Simulation of Semiconductor Processes* (New York: SISPAD) pp 187–90
[22] Rees H D 1972 *J. Phys.* C **5** 641
[23] Chambers R G 1950 *Proc. Roy. Soc.* A **202** 378
[24] Ferry D K 2020 *Semiconductors Bonds and Bands* 2nd edn (Bristol: IOP Publishing)
[25] Rees H D 1969 *J. Phys. Chem. Sol.* **30** 643
[26] Price P 1979 *Semiconductor and Semimetals* ed R K Willardson and A C Beer (New York: Academic) p 249 https://shop.elsevier.com/books/semiconductors-and-semimetals/willardson/978-0-12-752114-5
[27] Goldstein H 1950 *Classical Mechanics* (New York: Addison-Wesley)
[28] Bosi S and Jacoboni C 1976 *J. Phys.* C **9** 315
[29] Lugli P, Jacoboni C, Reggiani L and Kocevar P 1987 *Appl. Phys. Lett.* **50** 1251
[30] Barker J M *et al* 2004 *J. Vac. Sci. Technol.* **22** 2045
[31] Yu T-H and Brennan K F 2002 *J. Appl. Phys.* **91** 3730
[32] Goodnick S M and Honsberg C 2011 *Proc. Photo-Voltaics Specialists Conf. (New York)* (IEEE Press) p 2066
[33] Ferry D K 2021 *Appl. Phys. Rev.* **8** 921324
[34] Ferry D K, Whiteside V R and Sellers I R 2022 *J. Photonics Energy* **12** 02204
[35] Tsen K T *et al* 2006 *Appl. Phys. Lett.* **89** 112111
[36] Cohen M L and Bergstresser T K 1966 *Phys. Rev.* **141** 789
[37] Laux S E and Fischetti M V 1988 *IEEE Electron Device Lett.* **9** 467
[38] Fischetti M V and Laux S E 1988 *Phys. Rev.* B **14** 9721
[39] Rein A *et al* 1994 *Proc. Int. Electron Devices Meeting* (New York: IEEE Press) pp 351–4
[40] Wigner S J and Goodnick S M 2000 *Superlattices Microstruct.* **27** 417
[41] Yamakawa S *et al* 2004 *J. Comput. Electron.* **3** 299
[42] Akis R *et al* 2008 *J. Phys.: Condens. Matter* **20** 384201

IOP Publishing

Quantum Transport in Semiconductor Devices
Simulation using particles
David K Ferry, Xavier Oriols and Josef Weinbub

Chapter 10

Effective potentials and Bohmian trajectories

In many of the earlier chapters, the idea of an effective potential has been mentioned and discussed to some extent. It has been emphasized in previous chapters that quantum transport can, of course, be perfectly modeled using trajectories. The point is that the potentials that govern the dynamics of classical particles are different than the potentials that guide quantum particles. It is now time to fully clarify this concept and explain its usage, particularly in device modeling. It has been pointed out, especially in chapter 1, that approaches to quantum mechanics in which particles are utilized are actually quite common. Bohm [1], his predecessors (Madelung [2] and Kennard [3]), and Wigner [4] have all shown that such an effective potential arises naturally. But why is this so? At least in Wigner's case, along with later work on the topic, the idea was that the nonlocal behavior of the potential is much more important in quantum mechanics than in classical mechanics. Hence, expansions of the free energy lead to higher-order corrections to the free energy arising from spatial expansions of the potential. Although this makes a good argument and is likely true, other approaches to the effective potential are derived from the fact that the size of a quantum particle is dramatically larger than its size as given by classical mechanics. The importance of the effective potential possibly lies in the fact that nearly all of these arguments lead to the same form for the effective potential, at least within a numerical factor! This would seem to indicate that the effective potential is a crucial quantum correction that really does not depend much upon the manner in which it is derived. Rather, it provides the correcting quantum potential which Kennard has shown to be the crucial impact of quantum mechanics on dynamics [3].

The primary focus of this chapter will be these various forms of the effective potential and the use of such a potential in quantum transport, especially with particles. In the following sections, the various approaches to gaining an effective potential will be described, beginning with the argument about the particle size. This will be followed by the approaches of Bohm, Wigner, and Feynman [5]. The

doi:10.1088/978-0-7503-5237-6ch10

discussion will then turn to the actual use of the Bohm potential in quantum simulations before returning to the discussion of experiments which confirm the accuracy of such quantum potentials and trajectories.

10.1 On the role of particle size

In some ways, the idea of the size of a particle (or electron) is somewhat confusing. In classical physics, the electron is incredibly small. In fact, it is usually described as a 'point' particle, as if asking for its size is ridiculous. Quantum mechanically, this is unacceptable, because if one were to use the Heisenberg uncertainty relation for a free electron of this size, the uncertainty in velocity would be something like 2×10^7 m/s, or roughly 15% of the speed of light! It is highly unlikely that this is the case.

In quantum mechanics, the size of a particle can be understood, for example, as the distance at which particles start to 'feel' each other, or as the distance where a particle starts to 'feel' a potential barrier. Since the dynamics of quantum particles is mediated by the wave function, the somewhat confusing concept of the size of a particle in the classical regime can be translated to a less problematic concept connected to the size of the wave function. Of course, one can argue that the size of the wave function for an electron in a device is still controversial because we are dealing with an open system (as mentioned in section 3.4 of chapter 3). But such conceptual difficulties are not relevant here, as one can always understand the wave function as a type of 'conditional' wave function. In the atom, the wave packet size is comparable to the circumference of the electron orbit around the atom, which leads to de Broglie's quantum wavelength. Outside the atom, the packet may well be expected to be larger. In any case, this quantum wave packet is going to be larger than the classical size, if for no other reason than that it cannot be smaller in position and momentum space than dictated by the uncertainty relation. Therefore, it is quite reasonable to ask how large this wave packet might be. As in many things in life, size matters! If a potential barrier exists, a classical particle can sit next to the barrier and nothing of importance arises. On the other hand, placing a wave packet next to the barrier just means that one has to address the issue of size and then ask whether the wave function penetrates the barrier. This highlights the inability of the center of the wave packet to actually be next to the barrier. Just like a basketball next to a wall, the center of the particle's size cannot actually be at the barrier (or wall). The size of the quantum mechanical particle determines the distance. In modern nanoscale semiconductor devices, this size becomes quite important, as will be illustrated below.

10.1.1 Simply the size

The simple wave packet treatment discussed above considers an *isolated* packet for a single particle or electron. In a real semiconductor, the electron (or hole) density is composed of a range of plane waves, or momentum states, or wave packets, depending upon the description selected (and these contribute to the many-body wave function of all electrons). In order to describe the packets in real space, one must account for the contributions to the wave packet from all occupied plane wave states [6]. The states that exist in momentum space are the Fourier components of the real-space wave packets. For example, in a two-dimensional semiconductor system at low temperature, such as an inversion layer, all states up to the Fermi energy are occupied, and the Fermi wave vector is defined by the interface (surface) carrier density n_s as follows:

$$k_F = \sqrt{2\pi n_s}. \tag{10.1}$$

This means that, in the momentum representation, all states up to this value are occupied, and so plane wave states exist up to this value. From such a momentum space representation, a wave packet (centered at the origin) may be obtained by taking the two-dimensional Fourier transform (as the inversion layer is a quasi-two-dimensional system) of these Fermi states, which leads to

$$\psi(r) = \sqrt{\frac{2}{\pi}}\frac{1}{r}J_1(k_F r), \tag{10.2}$$

where J_1 is the Bessel function of the first kind and first order. It must be remembered that $\boldsymbol{k}$ and $\boldsymbol{r}$ are really two-dimensional vectors, and the scalar values shown are the magnitude of the radial component. The spatial extent of the real-space wave packet can be estimated to be the full width at half-maximum value, or twice the uncertainty in position, which leads to

$$\Delta r \sim \frac{2}{k_F} = \frac{\lambda_F}{\pi}. \tag{10.3}$$

This result is the central result for the degenerate two-dimensional electron gas and shows that the spatial extent of the wave packet is small and related to the de Broglie wavelength of the electron at the Fermi surface (where transport takes place). For an inversion layer of 10^{12} cm^{-2}, the Fermi wave number is approximately 2.5×10^6cm^{-1}, and the size is given by $\Delta r \sim 4$ nm. This is not small on the scale of the critical dimensions in modern semiconductor logic devices.

For nondegenerate semiconductors, the distribution of allowed momentum states follows the Maxwell–Boltzmann distribution. This could be, for example, an inversion layer at high temperature and relatively low density. This distribution brings the temperature into the problem. As the temperature is increased, higher momentum states become more occupied as the distribution spreads under the influence of the temperature. A wider momentum space distribution means a tighter distribution in real space for the electron wave packet. As previously discussed, we consider the momentum space distribution to be a description of the occupied plane wave states which contribute to the electron wave packet. The normalized momentum space distribution may then be defined to be

$$\varphi(k) = 2\left(\frac{\lambda_D}{2}\right)^{3/4} \exp\left(-\frac{\lambda_D^2 k^2}{4\pi}\right), \tag{10.4}$$

where

$$\lambda_D = \sqrt{\frac{2\pi\hbar^2}{m^* k_B T}} \tag{10.5}$$

is the thermal de Broglie wavelength [7]. This can now be Fourier transformed to obtain the real-space wave packet

$$\psi(r) = \frac{\pi}{\left(\sqrt{2}\lambda_D\right)^{3/2}} \exp\left(-\frac{\pi r^2}{\lambda_D^2}\right). \tag{10.6}$$

The expected size of this wave packet is given by

$$\Delta r \sim 0.6\lambda_D. \tag{10.7}$$

For electrons in Si at room temperature, $\lambda_D \sim 7.6$nm and $\Delta r \sim 4.6$nm. Thus, it appears that the thermal (classical) size is comparable to the quantum size, perhaps a not unexpected result. Of course, this latter size (10.7) is temperature dependent.

An interesting aspect of result (10.7) is that there is no density dependence in the thermal wavelength (10.5), and therefore the particle wave packet size is also independent of the density. While this may seem strange at first sight, it is quite natural, as the nondegenerate limit is one of a relatively low-density electron system. These particles can thus move around freely with a thermal motion without confining one another significantly. When the density becomes sufficiently large that this limit is no longer appropriate, then the distribution changes character to that of the Fermi–Dirac distribution. The particles are now sufficiently packed together that their motion is limited to a much larger extent; this motion is limited to the interparticle distances. In the latter situation, the density dependence reappears through the importance of the Fermi wavelength rather than the thermal wavelength, since the Fermi wavelength (and Fermi energy) depends explicitly upon the carrier density.

10.1.2 The effective potential

The next question is: how does the above 'size' of the particle give rise to an effective potential? The answer lies in the use of the appropriate many-body theory. In an inhomogeneous system, any potential enters the Hamiltonian that describes the total energy in density-functional theory as follows [8]:

$$H_V = e\int d^3\mathbf{r}\, U(\mathbf{r}) n(\mathbf{r}). \tag{10.8}$$

That is, the potential is weighted at each point in space by the local electronic density (hence the expression 'density function' theory). If the particles have a distribution of sizes, so that they are distributed around the values given above, this can be approximated by describing each particle as a Gaussian wave packet. The Gaussian shape is chosen as it is known to be a minimum-uncertainty wave packet [7]. Hence, in (10.8), the particles are gathered by number in a close region around r and then replaced by Gaussian weighted delta functions. This leads to

$$\begin{aligned} H_V &= e\int d^3 r U(\mathbf{r})\sum_i n_i(\mathbf{r}) \\ &= e\int d^3 r U(\mathbf{r})\sum_i \int d^3 r' \exp\left(\frac{|\mathbf{r}-\mathbf{r}'|^2}{(\Delta r)^2}\right)\delta(\mathbf{r}_i-\mathbf{r}') \\ &= e\int d^3 r' \exp\left(\frac{|\mathbf{r}-\mathbf{r}'|^2}{(\Delta r)^2}\right)U(\mathbf{r}')\sum_i \int d^3 r\,\delta(\mathbf{r}_i-\mathbf{r}) \end{aligned} \quad . \tag{10.9}$$

In the third line of this last equation, the two spatial variables have been interchanged, so that the last summation is now over the particles themselves, and each particle has its classical delta function form. The first integral only involves the potential and the Gaussian function, which now smooths the potential and thus produces the *effective potential.*

10.1.3 An example

As mentioned in our discussion above, we are interested in how this plays out in a modern semiconductor device. The traditional (or simplest) approach to quantization in the metal–oxide–semiconductor field-effect transistor (MOSFET) begins with the assumption of a triangular potential well [6]. The sharp barrier at $x = 0$ for electrons (holes have a different barrier) is between the Si conduction band edge and the SiO_2 conduction band edge—a barrier usually taken to be $V_0 = 3.2$ eV. This band discontinuity will be taken as an example of the effective potential. Within Si, the normal, classical conduction band rises linearly in response to an effective electric field of [9]

$$E_{\text{eff}} = \frac{e}{\epsilon_{\text{Si}}}\left[N_A W + \frac{1}{2}n_s\right], \tag{10.10}$$

where N_A and W are the background doping of the channel region (acceptors in an n-channel device) and the depletion depth arising from the gate voltage at inversion of the surface, respectively, and n_s is the inversion density. The factor of 1/2 appears because the electric field is split between the front and back of the inversion sheet density, while the field due to the depletion charge is entirely at the front (i.e. the gate). From a classical perspective, the inversion density is largest at the interface and decays into the semiconductor in an exponential manner.

The triangular barrier leads to an effective potential obtained by taking the integral over the real potential with Gaussian smoothing, as given by [10] using the formulation of the first integral in the last line of (10.9). This gives

$$\begin{aligned} U_{\text{eff}}(x) = \frac{V_0}{2}\left[1 - \text{erf}\left(\frac{x}{\sqrt{2}\lambda_D}\right)\right] + \frac{E_{\text{eff}}x}{2}\left[1 + \text{erf}\left(\frac{x}{\sqrt{2}\lambda_D}\right)\right] \\ + \frac{E_{\text{eff}}\lambda_D}{2}\exp\left(-\frac{x^2}{2\lambda_D^2}\right). \end{aligned} \tag{10.11}$$

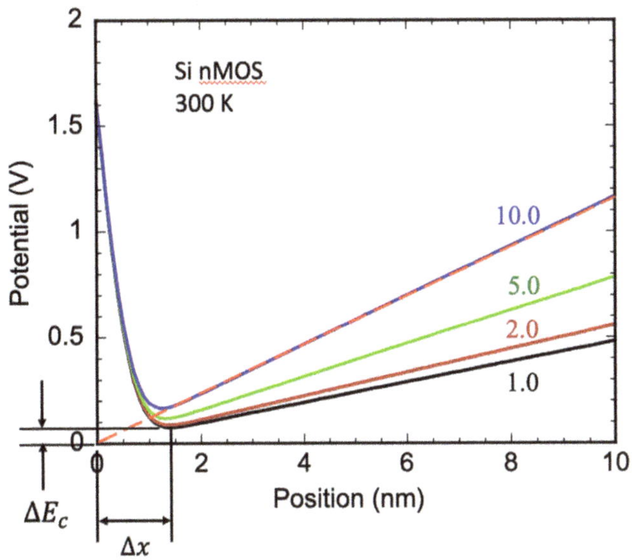

Figure 10.1. The effective potential for several different values of the inversion density (in units of 10^{12} cm^{-2}) (the parameter). The classical potential is the dashed magenta line for a density of 10^{13} cm^{-2} (at lower densities, it merges with the appropriate curve at large distances). The quantization leads to the charge setback Δx and the quantization energy ΔE_c. The oxide conduction band lies at 3.2 V.

Here, x is the direction normal to the interface (pointing into the semiconductor), with $x = 0$ taken at the oxide–semiconductor interface. The real and effective potentials are plotted in figure 10.1. The various curves are given for inversion densities of 10^{12} (black), 2×10^{12} (red), 5×10^{12} (green), and 10^{13} cm^{-2}. The dashed magenta curve visible in the lower-left corner represents the classical position of the conduction band edge for the highest density. The form (10.11) has two distinct advantages in that it provides the initial observables of quantization in the channel. These are the charge setback Δx and the quantization in the potential well ΔE_c. Here, this quantization in the potential well is about 75 meV at the lowest density and rises to about 125 meV at the highest density. The charge setback is about 1.3 nm at the lowest density and about 1.17 nm at the highest density. In this example, the depletion charge is approximately $3.4 \times 10^{12}$$cm^{-2}$.

10.2 The Bohm potential

In treating the quantum transport of carriers through small semiconductor devices, it is often desired to find an approach that gives the important aspects of the quantum effects without all the mathematical detail that is necessary for some of the more fundamental approaches. One approach to this preference that is closer to the ensemble Monte Carlo approach lies in the hydrodynamic approach outlined in section 1.2.3. Such an approach has been used for years in the simulation of semiconductor devices [11]. In this earlier section, the quantum potential (or, as it is usually known, the Bohm potential), may be described by

$$Q = -\frac{\hbar^2}{2mR}\nabla^2 R. \tag{10.12}$$

where $\psi = R\exp(iS/\hbar)$ and R is the amplitude of the wave function, while S is the action of the motion. The hydrodynamic connection was developed almost as early as the Schrödinger equation [2, 3], although it was later rediscovered by Bohm [1] and brought into modern discussions of quantum transport.

In devices, $| R |^2 = | \psi |^2$ is often taken to be the local particle density using a proper normalization of the wave function with the total number of particles in the system. In this case, the result (1.37) may be recognized as a continuity equation that conserves the probability within the system volume. This continuity equation is consistent with a connection between the momentum and the velocity given by

$$v = \frac{1}{m}\nabla S. \tag{10.13}$$

It is important to note that the velocity (and the momentum itself) is buried in the exponent of the wave function. As a result, it drops out of any formulation of the probability or the carrier density due to the magnitude requirements. Bohm referred to variables such as velocity or momentum as 'hidden' variables, as it was quite difficult to measure the phase of a wave function. This led to considerable acrimony from the proponents of Bohr's philosophy, which was particularly directed at those of the orthodox persuasion (see the discussion in chapter 1).

An important difference exists between the Bohm potential and the effective potential. The effective potential, discussed in section 10.1, arises from a smoothing of the real (classical) potential; it thus represents a smoothing that would be encountered by particles of realistic size. On the other hand, the Bohm potential arises from the wave function, and this is an important difference. If the classical potential were electrostatic, then it would be effective on charged particles, and a similar statement would be true for the smoothed effective potential. But this would have no effect upon bosons, such as phonons and photons. The Bohm potential is effective for these neutral particles, as they also have their own wave functions, which lead to (10.12). This will become clear below in section 10.3, where experiments on the classical two-slit system are probed with coherent photons.

10.2.1 Trajectories

Kennard brought quantum mechanics back to particles by showing that the hydrodynamic form was applicable to a system of particles [3]. In his view, the entire quantum theory of mechanical systems, and particularly those composed of particles, could be explained simply and without loss of generality by the use of Schrödinger's equation. In this view, the particles would follow their classical trajectories as given by Hamilton's equations in response to the classical potential plus the additional presence of a quantum potential (10.12). This quantum potential, which was derived from the wave function, served as the method by which the wave function *guided* the particles; i.e. the wave function was the pilot wave. The quantum potential is important to the particle trajectory approach, as one obtains the acceleration of the 'particle' trajectory from the total potential as follows:

$$\frac{dv}{dt} = -\frac{1}{m^*}\nabla[U(\mathbf{r}, t) + Q(\mathbf{r}, t)]. \tag{10.14}$$

This directly allows us to develop a time-resolved approach to electron transport using the quantum trajectories [12]. Once the Schrödinger equation is solved at each instant of time, the quantum potential can be computed along with the self-consistent potential energy (subject to any charge movement or rearrangement), and then the directional velocities of the various trajectories can be determined. These velocities now allow one to model the motion of the particles and the development of the trajectories during the next time increment. This is a consequence of the fact that the Bohm theory is both a wave *and* a particle approach. First, the wave mechanics is solved via the Schrödinger equation; then, the particle trajectories are determined. These resulting trajectories of the particles are real and may be observed in experiments, as will be discussed below.

This approach has been applied to a study of a resonant-tunneling diode (RTD) by Oriols *et al* [13]. The lower-left part of figure 10.2(a) shows the motion of a Gaussian wave packet, represented by an ensemble of particles, through a double-barrier, resonant-tunneling diode [13]. The barriers are each 2 nm thick and have an energy height of 0.3 eV. The enclosed quantum well is 18 nm thick. It is assumed that the material is GaAs and that the barriers are AlGaAs. The Gaussian wave packet is a coherent wave packet with a momentum corresponding to a central energy of 0.16 eV and a spatial dispersion of 10 nm. No bias is considered to be applied to the structure. Several trajectories are shown for particles that are either reflected or pass through the barriers. In figure 10.2(b), a phase-space distribution function for the full density of the tunneling device is shown for a bias of 0.39 V. Note the tunneling ridge in the lower right (designated by an arrow) that corresponds to the ballistically tunneling electrons. The increase in the momentum of this ballistic ridge arises from

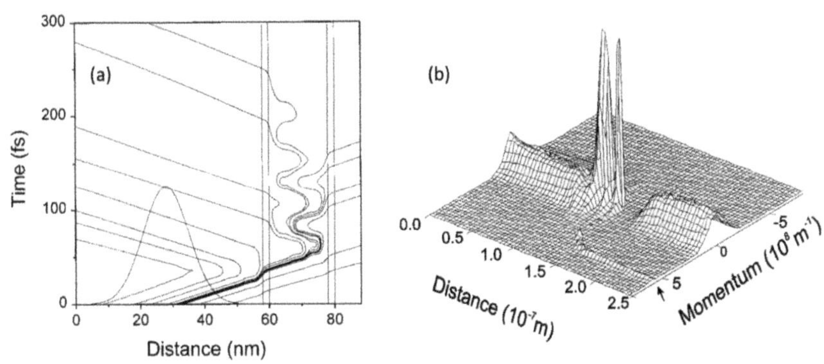

Figure 10.2. (a) Bohm trajectories for a Gaussian wave packet represented by an ensemble of particle trajectories tunneling through a double-barrier resonant-tunneling diode. Details are given in the text. (b) The phase-space distribution for such a diode with full density at a bias of 0.39 V. Reprinted from [13], with the permission of AIP Publishing.

the drop in electrostatic potential across the barriers. As the particles move ballistically, they travel at constant total energy and thus gain momentum as the potential falls away due to the applied voltage. It can be shown that the Bohm trajectory approach agrees very well with the direct use of the Schrödinger equation [13] and with the use of the Wigner phase-space distribution function [14].

The use of Bohm trajectories has also been applied to semiconductor nanostructures such as quantum wires and quantum point contacts. In figure 3.9, the results for a simple asymmetric quantum point contact are shown, and the trajectories are overlaid on the quantum potential itself [15]. In this study, the Schrödinger equation was solved by assuming hard walls for the boundaries; it was also solved both with and without consideration of a self-consistent rearrangement of the charge in response to the barrier potential. This was then used to find the trajectories and the quantum potential, both of which are shown in the top panel. Note that here, as well as in figure 10.2, none of the trajectories cross, which is necessary because the velocity field is uniquely determined; thus, only one trajectory can pass through any point.

10.2.2 The two-slit experiment

The Bohm machinery and the resulting particle trajectories are relevant to many interpretations of quantum mechanics. From the earliest days, consideration was given to the 'either/or' situation that relates to particles and waves. But with the guiding wave theory, this becomes an 'and' situation. Nowhere is this better depicted than in the two-slit experiment. The two-slit experiment has been one of the most confusing problems in quantum mechanics since its earliest days, when it was described for optical waves by Young [16]. Here, light passing through two small slits in a blocking panel is diffracted by the slits and interferes like waves, producing a observable interference pattern. Identical behavior results from electrons that diffract from multiple locations and interfere [17–19]. As mentioned in chapter 1, Taylor had investigated these diffraction effects with very weak light as early as 1909 [20] and found that even when the light was sufficiently weak that only a single photon at a time was in the volume of the experiment, the diffraction corresponded to the wave model and the arrival at the photographic plate corresponded to the particle model. Whether we consider photons or electrons, the interference pattern that is built up by the particles arriving at the detector clearly shows the effects of their wave nature as they pass through the diffraction region. Figure 10.3 shows the magnitude of the interference for Gaussian wave packets passing through two slits.

The causal interpretation that Bohm proposes describes the interference fairly well [1]. The wave function used is described in the usual manner but is regarded as a real field. From the theoretical perspective, it is a mathematical representation of a field that exerts a force upon the particle. This force is apparent in the experiments. The wave function generates this force in agreement with the Schrödinger equation through the quantum potential (10.13). It is the force from the quantum potential that leads to the motion of the particles which creates the interference fringes observed as the particles arrive at the detector. If it happens that there are places

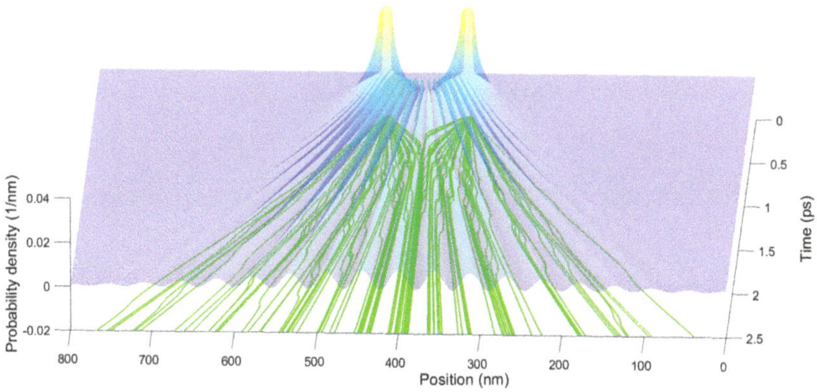

Figure 10.3. The intensity of the wave function passing through two slits. Individual Bohm trajectories are also shown by the green lines at the bottom. The wave function is initialized by two Gaussian wave packets at the slits. The spatial dispersion of the initial wave packets was 4 nm, and an effective mass of $0.4m_0$ was used.

where the wave function is zero, this leads to an infinitely large potential, which prevents the particles from approaching these places. Moreover, if one of the slits is closed, the diffraction of the wave function changes. This, in turn, dramatically changes the quantum potential, so that the particle is now allowed to reach places that it had no access to when both slits were open. The opening or closing of one of the slits affects the particles through the changes in the quantum potential caused by such an action. Thus, it is actually the real field of the wave function that results from the detailed state of the two slits which generates the forces affecting the particle motion. As a result, there is no strange mystery about the quantum behavior of the system.

The role of the quantum potential was investigated by Philippidis, Dewdney, and Hiley [21]. They calculated the actual wave function and quantum potential that resulted for the two-slit experiment. Their results for the quantum potential and trajectories can be found in many books and web sites. In short, there are three clear peaks in the quantum potential, and the major peak is located between the two slits. The slits themselves lie between these three peaks. The peaks are regions where the wave function is small and also regions that the particles cannot effectively reach. Clearly, there are troughs in the potential, and the idea is that the particles are focused into these troughs by the quantum potential. Others have repeated these simulations [22], and figure 10.4 shows the quantum potential for the two-slit experiment, assuming Gaussian wave packets. Similar behaviour for the Bohmian trajectories in another two-slit scenario is shown in figure 10.5 [22]. The corresponding particle trajectories run along the troughs but can jump from one trough to the next, which appears as 'zig-zags' in the trajectories. With particles, the trajectories correspond to different initial conditions within the Gaussian wave packets at the two slits. As a result of the potential troughs, the particles are focused into beams which create the interference fringes on the screen.

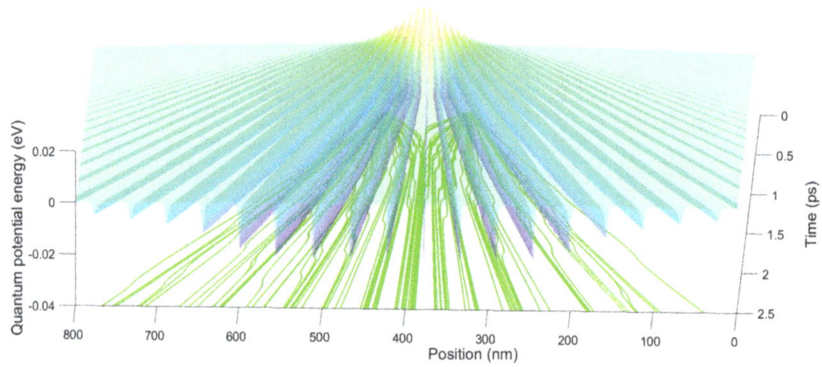

Figure 10.4. The Bohm potential for the two-slit experiment of figure 10.3 when the interference phenomena are present. As before, individual trajectories are shown in the green lines at the bottom.

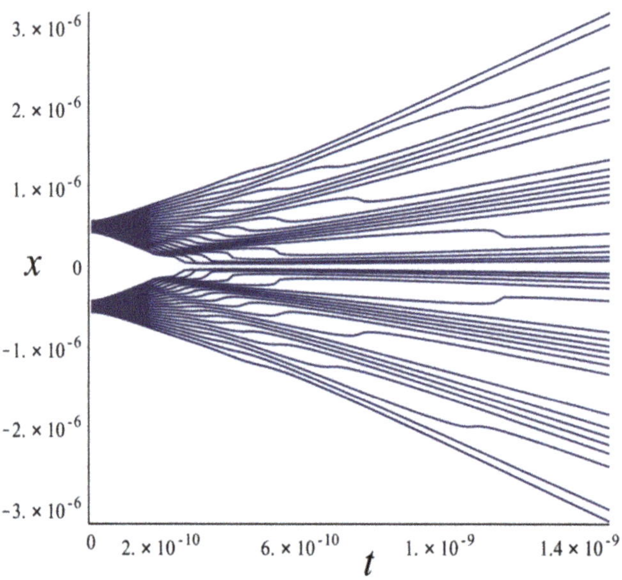

Figure 10.5. The trajectories in a two-slit interference experiment modeled by a one-dimensional Gaussian wave packet. Reproduced from [22].

10.2.3 Optical confirmation

It is important to understand that the ideas of Bohm were given short shrift by the dominant fathers of physics, who held dearly to the orthodox views. This was perhaps due more to the use of the words 'hidden variables' than to any misunderstanding of the ideas. Bohm clearly established that the particle *and* wave theory naturally fitted all known experiments on quantum effects. So the rejection of this approach could not be due to a failure to explain quantum mechanics. In addition, the idea of the trajectories through the two slits was considered to be wishful thinking, as it was held that no one could tell which slit any single photon passed through, and one certainly could not establish the photon trajectories, as that was thought to violate the uncertainty principle as well as causality. Thus, if the dominant fathers of physics were still alive, they would realize that the concept of a velocity for a particle (that they insisted was not a true element of the quantum world) can be measured in the laboratory through weak values. Of course, the dominant fathers of physics would still have an escape route that would allow them to avoid having to accept that they were wrong: the empirical value of the velocity is not obtained from a single experiment but after making an ensemble over many identical experiments. Thus, they could argue that such a 'velocity' is a type of ensemble velocity without ontological meaning for an individual quantum system. Of course, this escape route is intellectually valid, but it is also intellectually valid to argue that there is now a lot of new microscopic information about quantum systems that can be measured and thus leads to a better understanding on how quantum systems work. The relevant point is now that Bohmian velocities (and all other properties linked with them) are unquestionably measurable properties, not only post-processing tools to help in visualizing quantum phenomena. Consequently, these velocities are new empirical tools available in the laboratory to characterize quantum systems in general and quantum electron devices in particular. Appendix A, at the end of this chapter, provides an expanded discussion of the theoretical and experimental definition of weak values along these lines.

Perhaps the first relevant experiment was that carried out by Kocsis *et al* [23]. The experiment began with single photons that were produced by a semiconductor quantum-dot laser. These photons were then passed through a double-slit barrier. Of course, interference fringes were generated and appeared in subsequent detectors. Additional experiments were then carried out on the photons after they had passed through the slits. First, a set of weak measurements of the photon momentum was made, in addition to a set of strong measurements of the photon position as the photons passed a series of planes. Using a post-selection process on the coherent photons, they related a momentum to a particle's position over time. The details of the experiment explain how the correlated photons were obtained. The quantum dot used was an InGaAs dot embedded within a GaAs/AlGaAs heterostructure, which itself was patterned into a vertical optical cavity. The dot was excited by optical pumping, so that single photons at a wavelength of 943 nm were emitted. This single-photon nature was confirmed though correlation measurements made on the photons themselves.

In the experiments of Kocsis *et al* [23], the single photons were generated by the laser and then coupled to a single-mode optical fiber. The photons subsequently passed through a 50:50 beam splitter that directed them into one of two paths corresponding to the two slits. The fibers were intertwined in such a manner that the two photon paths exited the fibers and were then reflected into two parallel paths by mirrors. This produced two Gaussian beams, and these were adjusted so that the paths were separated by half a centimeter. The weak measurement started with a piece of birefringent calcite. As the beam was Gaussian, the incident angle of the photons on the crystal depended on their transverse momentum. This produced a phase shift proportional to the transverse momentum of the photon. As a result, the linear polarization was converted into an elliptical polarization, and this in turn served as a pointer to the photon momentum. The photons were then projected onto a quarter-wave plate, which effectively projected the elliptical polarization onto circular polarization, so that the signal was then an indicator of the transverse momentum. They were then passed through a beam displacer, which transmitted the right-hand circularly polarized component unchanged but displaced the left-hand circularly polarized component vertically by 2 mm. Thus, these two components could be separated in the charge-coupled device (CCD) detector array and the momentum determined. However, the ensemble of photons created interference fringes at the same time, which provided the strong measurement. Hence, the authors were able to obtain a measure of both the photon position and momentum. In the post-selection process, the researchers were able to reconstruct the photon trajectories. These trajectories were essentially identical to those calculated using Bohm trajectories. Such experiments clearly establish a strong argument for the viability of the Bohm approach. The simultaneous measurement of the position and the momentum (i.e. the velocity) is exactly what is discussed in appendix A. It is important to remark that such weak values also can be constructed from the orthodox machinery. In other words, orthodox quantum mechanics can also predict the results in figure 10.7. But then, is orthodox quantum mechanics also advocating for trajectories? Of course not! Bohmian mechanics and orthodox quantum mechanics predict the same numerical value for the weak value. However, for the Bohmian theory, such values coincide with the physical velocity of the particle, while in the orthodox theory the value is just a statistical average with no physical meaning that can be attached to an individual particle. In other words, orthodox and Bohmian mechanics agree with the empirical value but disagree with the ontological meaning of such an empirical value. Wanted or not, the fact that Bohmian velocities are now measurable elevates the relevance of the predictions obtained by the Bohmian theory.

These experiments have been repeated several times, which has confirmed the results. One such set of results has been obtained more recently [24] using a quite similar experimental setup, illustrated in figure 10.6. Measurements are made according to the photon positions in the normal Bohm theory as well as in a momentum-based version of the Bohm approach. The reconstructed Bohm trajectories in real space are shown in figure 10.7, and these may be compared with the theoretical versions shown in figure 10.5.

Further experiments have attempted to determine just which slit a particle actually passed through. One of these was performed by Menzel *et al* [25]. In this experiment, entangled photon pairs were generated in an optical parametric down-conversion process. They observed, in a different coincidence measurement, the idler photons produced by the parametric process, which, along with the signal photons, were used to create interference fringes. The interference fringes of the two separate processes allowed observation of the fringes and provided slit-identifying information. This surprising result violates much of the orthodoxy of the normal view of quantum mechanics. These authors have suggested this result is possible because the transverse mode structure produces a superposition of two intensity maxima for the signal photon. Hence, the resulting experiment may not exactly show a single photon but rather a superposition of photons.

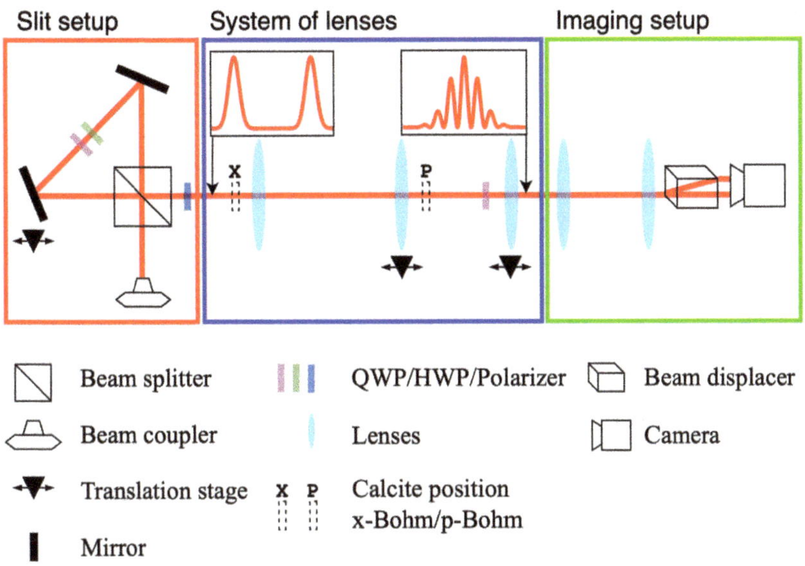

Figure 10.6. An illustration of the three-part experimental setup. The slit setup is used to initialize the double-slit experiment. A diagonally polarized beam is split into two co-propagating beams using a displaced Sagnac interferometer, in which the beam separation (set to 2 mm throughout the experiment) can be tuned by translating one of the mirrors. The system of lenses is used to simulate the propagation of the light along z over a large range. The two plots above the lens system indicate the intensity profiles of the beam before and after the lens transformation. A thin piece of calcite is used to weakly couple momentum to polarization (the weak measurement). The calcite is positioned in front of the lenses for the weak momentum measurement in the Bohm position experiment. The imaging setup, consisting of a polarizing beam displacer and a CCD camera, is used to obtain two interference patterns, one for each polarization. Reproduced from [24]. CC BY 4.0.

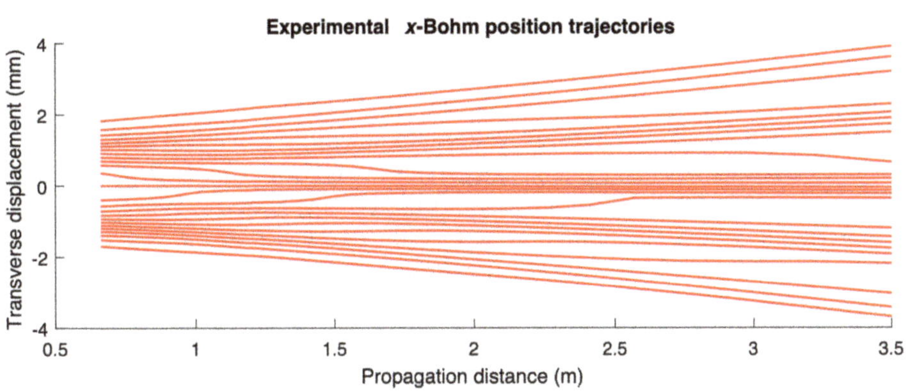

Figure 10.7. Experimentally reconstructed position trajectories based on the Bohm theory. Reproduced from [24]. CC BY 4.0.

10.2.4 Semiconductor devices

It should be recognized now that one of the better approaches to effective potential uses the Bohm potential and the resulting particle trajectories. However, there are other, more mathematical (analytical) approaches that can incorporate the Bohm or quantum potential [11]. The latter approach developed a set of hydrodynamic equations for the density, momentum, and energy. However, many simulations still use a drift-diffusion approach in which the Bohm potential is added to the driving terms [26], producing a result useful in studies of both the electronic and optical properties in quantum wells. Others have studied the mathematical properties of such quantum-corrected drift-diffusion models [27–30]. The quantum-corrected drift-diffusion equations have been used to model impact ionization avalanche transit time (IMPATT) diodes [31, 32] and carbon nanotube devices [33]. While these studies are interesting, they do not bring the full power of particle simulations to the detailed study of the physics of the device. A more powerful approach was introduced through adapting the Bohm potential to Monte Carlo particle simulations in the study of tunneling structures [13, 34, 35]. A somewhat different approach was followed by adding dissipative terms to create a non-Hermitian Hamiltonian and then using this to create particle trajectories with the added Bohm potential [36].

Using Bohm trajectories in addition to normal Monte Carlo simulation, it was possible to study quantum noise, which is particularly relevant to tunneling structures [37]. This was applied to gate oxide tunneling in a Si metal–oxide–semiconductor (MOS) structure. A later study developed a Monte Carlo approach to device simulation that incorporated a density gradient quantum correction [38] and applied it to double-gate quantum wire MOS transistors [39], in which the impurities were treated with real-space potentials. This led to the development of a particle-based simulation package for quantum devices using the Bohm potential [40]. This approach will be further elaborated in section 10.5.

10.3 The Wigner potential

As remarked previously, Wigner [4] also developed an approximation for quantum corrections to the potential via the equation of state. So far, only a Wigner function for a single particle state or for a single basis state has been discussed. If a number of basis states are occupied, the total Wigner function can be expanded in terms of a set of basis states, each of which is a Wigner function its own right. It is well known in quantum mechanics that the actual basis set used in any problem is not crucial, and any convenient set may be chosen. The transformations from one basis set to any other are easy due to the linearity of the Schrödinger equation [6].

The total Wigner function can be expressed in terms of the set of basis functions in the normal manner, as follows:

$$f_{\mathrm{W}}(\boldsymbol{r}, \boldsymbol{p}, t) = \sum_i c_i f_{\mathrm{W},\, i}(\boldsymbol{r}, \boldsymbol{p}, t). \tag{10.15}$$

This describes a mixture corresponding to the density matrix for such a mixture but differs from a normal wave function expansion. Here, c_i is the probability that the particular point in phase space is occupied.

If a classical system is in equilibrium at a temperature T, which is represented in the following equations by $\beta = 1/k_{\mathrm{B}}T$, where k_{B} is the Boltzmann constant, the relative probability of a particular state being occupied can be obtained from the energy E_i of that state through the function $\exp(-\beta E_i)$. That is, this exponential function is proportional to the coefficient c_i in (10.15) when it is in thermodynamic equilibrium. This exponential is related to the partition function which describes the occupations of the various states. This will become important in arriving at the Wigner corrections because they are derived as corrections to this partition function.

Each of the basis Wigner functions is related to the basis wave function for the particular state, and this connection appears through the application of the Wigner transform (1.54) of chapter 1. For the basis function $\varphi_i(\boldsymbol{r}, t)$, the Wigner function is

$$f_{\mathrm{w},\,i}(\boldsymbol{r}, \boldsymbol{p}, t) = \frac{1}{h^3} \int d^3\boldsymbol{r}' \varphi_i^\dagger\left(\boldsymbol{r} + \frac{\boldsymbol{r}'}{2}, t\right)\varphi_i\left(\boldsymbol{r} - \frac{\boldsymbol{r}'}{2}, t\right) e^{-i\boldsymbol{p}\cdot\boldsymbol{r}'/\hbar}. \tag{10.16}$$

When this basis Wigner function is weighted with the partition function, it gives the likelihood that a particular state is occupied. Because the energy is a function of momentum as well as position, the exponential function has to go inside the integral and between the two wave functions in the normal manner. Then (10.16) gives the matrix element of the exponential operator $\exp(-\beta H)$, and this latter is Fourier transformed into momentum by the first exponential. As an alternative approach, Weyl [41] suggests first transforming the Hamiltonian itself and then using the exponential of the transformed version. That is, the Hamiltonian would be transformed as follows:

$$\begin{aligned} H &= e^{i\boldsymbol{p}\cdot\boldsymbol{r}'/\hbar}\left(-\frac{\hbar^2}{2m}\nabla^2 + V(\boldsymbol{r}')\right)e^{-i\boldsymbol{p}\cdot\boldsymbol{r}'/\hbar} \\ &= E_c + \left(i\hbar\frac{\boldsymbol{p}\cdot\nabla}{m} - \frac{\hbar^2}{2m}\nabla^2\right) \end{aligned}, \tag{10.17}$$

where the 'classical' energy is given by

$$E_\mathrm{c} = \frac{\boldsymbol{p}^2}{2m} + V(\boldsymbol{r}'). \tag{10.18}$$

The integral is then taken over the partition function, using (10.16) and the second line of (10.17). The Barker–Campbell–Hausdorff expansion of the exponential operator is used to simplify the integration. Passing the various terms through the exponential operator gives rise to the several terms that will appear as correction terms to the partition function. To the lowest order, one reclaims the classical result for the probability, namely $\exp(-\beta E_c)$. Higher approximations are found to be

$$\frac{\hbar^2}{m}e^{-\beta E_c}\left(-\frac{\beta^2}{12}\nabla^2 E_\mathrm{c} + \frac{\beta^2}{24}(\nabla E_\mathrm{c})^2\right). \tag{10.19}$$

Thus, the new potential term involves the classical potential plus the quantum corrections and appears as

$$V(\boldsymbol{r}) = V_c(\boldsymbol{r}) - \frac{\beta^2\hbar^2}{12m}\nabla^2 E_c + \frac{\beta^2\hbar^2}{24m}(\nabla E_c)^2. \tag{10.20}$$

The second and third terms can be combined into a term that only differs from the Bohm potential (10.12) by a numerical factor. A slightly different combination can be obtained by noting that the spatially varying density is related to the partition function by $E_c \sim (1/\beta)\ln(n)$ (up to a normalization to average density). If this expression is used to replace the classical energy in (10.20), the correction term to the classical potential reduces to

$$Q_W = -\frac{\hbar^2}{8m}\nabla^2 \ln(n), \tag{10.21}$$

again up to a normalization term which appears only as a uniform shift in the potential. This latter form has been termed the *density gradient potential* and is another approach to the effective potential that has been applied to a wide variety of semiconductor devices [11, 42–44]. However, this form has many limitations in devices, particularly in heterostructure devices that contain potential discontinuities. At these points of discontinuity, gradient expansions are poor approximations, especially as there can also be strong discontinuities in the local carrier density.

10.4 Feynman's effective potentials

The effective potentials discussed in the above sections either use smoothing functions or depend upon additive functions obtained from a derivative expansion. These, of course, are integral and differential forms for correcting the classical potential. Each approach represents one way to obtain a quantum correction to the classical potential. Effective potentials have also been discussed by Feynman [5]. Feynman was concerned about the evaluation of path integrals in his approach to quantum mechanics and sought a way to impose quantization on the paths. He first introduced the mean value for the position along one of the paths, as follows:

$$\overline{x} = \frac{1}{\hbar\beta}\int_0^{\hbar\beta} x(\boldsymbol{r})d\boldsymbol{r}, \tag{10.22}$$

where $x(\boldsymbol{r})$ is a path variable representing the position along one of the paths in the integration. In essence, this definition represents a smoothing of the actual path position over a thermal spread in the action to get an effective value for this position. In the Feynman path integral, all paths need to have the same initial and final points, so that corrections to the classical potential are of the second order (in the spatial derivative) or higher. (The classical path would not be an extremum if a first derivative existed.) This then suggests that the correction to the free energy is of the order of

$$1 - \frac{(\hbar\beta)^2}{24m}\frac{\partial^2 V(\overline{x})}{\partial \overline{x}^2}. \tag{10.23}$$

This form has obvious similarities to the Wigner potential (10.20), as both approaches are actually trying to correct the partition function and energy. The result (10.23) may be transferred into a discussion of a corrected value for the Helmholtz free energy F'. These correction terms lead to a new value for the partition function

$$e^{-\beta F'} = \frac{1}{\lambda_D} \int e^{-\beta U(\bar{x})} d\bar{x}, \tag{10.24}$$

where the thermal de Broglie wavelength (10.5) has been introduced. Here, the effective potential is given by

$$U(\bar{x}) = \frac{\sqrt{12}}{\lambda_D} \int_{-\infty}^{\infty} V(\bar{x} + y) e^{-12\pi y^2/\lambda_D^2}. \tag{10.25}$$

An obvious connection to the effective potential of section 10.1 is now present through the Gaussian smoothing. But now there is also a smoothing over the path of the particle.

Feynman observes that an approximate value for the free energy F' can be achieved in an almost classical manner, and then a good estimate of the quantum corrections can be obtained using the smoothed classical potential (10.25). This approach basically corresponds to computing a local linear average of the actual potential when computing the mean position (10.22). This becomes inapplicable when the local potential may not have well-defined derivatives, such as in the case of a simple hard-wall potential, but this limitation can be avoided if (10.23) is avoided by not smoothing the local position in the path.

Many people have extended the Feynman approach to the case of bound particles in the presence of quantum confinement [45–47] and to the case of particles at interfaces [48, 49]. These approaches use the fact that the most likely trajectory in the path integral no longer follows the classical path when the electron is bound inside a potential well. The introduction of the effective potential is closely connected to the return to a phase-space description. It is clear that some modifications have to be made when non-parabolic energy bands or a magnetic field is present. However, the Gaussian approximation is a well-established way to incorporate the purely quantum fluctuations around the resulting path. The key ingredient for bound states (such as those in the potential well at the interface of a MOSFET) is the need to variationally determine the dominant path and hence the 'correct' value for the smoothing parameter α in (10.9). For the case in which the bound states are well defined in the potential, both Feynman and Kleinert [31] and Cuccoli *et al* [50] find that

$$\lambda^2 = \frac{\lambda_D^2}{8\pi}\left[\frac{\coth(f)}{f} - \frac{1}{f^2}\right], \quad f = \frac{1}{2}\hbar\beta\omega. \tag{10.26}$$

Here, $\hbar\omega$ is the spacing of the sub-bands. If the high-temperature limit is invoked, as appropriate for semiconductor devices at room temperature, the limit for small f is

$$\lambda^2 \sim \frac{\lambda_D^2}{24\pi}. \tag{10.27}$$

In Si, this gives a value of the smoothing parameter of 0.52 nm for the direction normal to the oxide interface. This is smaller than the values discussed above. For transport along the channel, a different mass appears, and this gives a value of 1.14 nm. Nevertheless, these values differ little from those discussed earlier in this chapter.

10.5 Determining effective potentials using Bohmian conditional wave functions

In section 10.2, the use of the quantum potential $Q(x, t)$ and classical potential $U(x, t)$ to compute Bohmian trajectories via (10.14) was presented. The shape of $U(x, t)$ is well defined from particle positions, while $Q(x, t)$ depends on the modulus of the wave function $R(x, t)$. Thus, the computation of $Q(x, t)$ requires previous knowledge of the wave function. One simple solution is to follow the arguments of 10.1 to obtain an approximate analytical description of the wave function and from that analytical shape develop an expression for $Q(x, t)$. This procedure has the great advantage of capturing quantum phenomena while only slightly increasing the computational burden. However, if one is interested in a more detailed description of the microscopic properties of individual electrons (and their many-body correlations), a more accurate description of the quantum potential $Q(x, t)$ is needed. Obtaining this more accurate evaluation is the final goal of this section.

One way of finding the quantum potential Q without the knowledge of the wave function is to use the so-called synthetic approach [51]. This is based on adopting a large ensemble of trajectories $j = 1, 2, \ldots, M$, each of which corresponds to a different experiment, where M is the number of such experiments (or particles from the above discussion). To simplify the discussion here, a one-dimensional (1D) system is considered. The many-body expression mentioned in (3.54) of chapter 3, which gives the equivariance of Bohmian trajectories [52, 53], can then be adapted as follows:

$$R(x, t)^2 = \frac{1}{M}\sum_{j=1}^{M}\delta(x - x^j(t)). \tag{10.28}$$

From the location of each of the trajectories $x^j(t)$ belonging to the group of experiments at time t, one can reconstruct the modulus of the wave function at different positions by counting trajectories inside the position interval $[x, x + dx]$ during the time interval $[t, t + dt]$, where dx and dt determine the spacings of the spatial and temporal grids. Once $R(x, t)$ is known from (10.28), the quantum potential $Q(x, t)$ can be evaluated from (10.12). From such a quantum potential (10.14), together with the classical $U(x, t)$ potential, all trajectories can be evolved during dt, so that $x^j(t + dt) = x^j(t) + v^j(t)dt$, $R^2(x, t + dt)$, and $Q(x, t + dt)$ can be reconstructed at the later time. Repeating this procedure for several dt, the entire evolution of the quantum system can be computed without previous knowledge of the wave function. This 'wave-function-free' algorithm is exact (i.e. it involves no approximation) but it has an important technical drawback. The quantum potential in (10.12) is inversely proportional to $R(x, t)$, so if there are no particles at $[x, x + dx]$ during the time interval $[t, t + dt]$, the computed value of $R(x, t) = 0$ gives $Q(x, t) \to \infty$ making the numerical algorithm very unstable. Of course, a better estimation of $R(x, t)$ can be obtained by increasing the number of experiments M in (10.28), but this comes at the cost of an increase in the computational burden [51]. For a proper description of tunneling, in which $R(x, t)$ varies by several orders of magnitude in different locations, the above technical difficulty makes the algorithm quite problematic. In addition, including several degrees of freedom in a many-body scenario does not reveal any advantage of the present algorithm over directly computing the wave function at all grid points.

Another method for obtaining sufficient knowledge of $Q(x, t)$ is to first compute the wave function by numerically solving the Schrödinger equation everywhere without introducing any approximations. Then, $Q(x, t)$ may be directly determined from the wave function using (10.12). However, this procedure for obtaining $Q(x, t)$ from the exact wave function has been criticized as being quite impractical: once the wave function has been defined everywhere at all times, the extra effort needed for the computation of $Q(x, t)$ and the Bohmian trajectories seems useless because the knowledge of the wave function is enough for most practical predictions. This argument is found in the literature as a justification that Bohmian trajectories are only useful as post-process visualization tools but not as predictive tools. However, this argument, which can be accepted for a 1D system, is invalid in general because it ignores the many-body problem introduced in section 3.1 of chapter 3. In other words, the exact many-body wave function solution of the Schrödinger equation is not accessible. Nevertheless, one can find interesting algorithms with which to compute approximate many-body trajectories without resorting to the 'impossible' many-body wave function [52]. Of course, the Bohmian conditional wave function described in chapter 3 seems to be an excellent starting point. In the next section, this last idea is developed.

10.5.1 Quantum transport based on Bohmian trajectories and conditional wave functions

In section 3.4.1 of chapter 3, an equation of motion for the Bohmian conditional wave function [54] of one electron $\varphi_1(x_1, t)$ interacting with the remaining ensemble of particles appears in (3.57). Such an equation of motion is exact (i.e. valid for Markovian and non-Markovian scenarios) when the potential terms $G_1(x_1, x_E(t), t)$ and $J_1(x_1, x_E(t), t)$ are known. These potentials are many-body potentials that include the entanglement of the particle x_1 with other degrees of freedom $x_E(t) = \{x_1(t), \ldots, x_N(t)\}$. For specific scenarios, one can make educated guesses for the many-body quantum potentials $G_1(x_1, x_E(t), t)$ and $J_1(x_1, x_E(t), t)$ without knowing the many-body wave function. One can assume, for example, that the dependence of these many-body quantum potentials on x_1 is negligible, so that $G_1(x_1, x_E(t), t) \approx G_1(t)$ and $J_1(x_1, x_E(t), t) \approx J_1(t)$ [54]. These potentials are then basically time-dependent functions that only affect a global phase of the many-body wave function and have no relevance to the computation of individual velocities of the trajectories. The equation of motion (3.57) for the conditional wave function $\varphi_1(x_1, t)$ can now be rewritten [52, 54] here as

$$i\hbar\frac{\partial\varphi_1(x_1, t)}{\partial t} = \left[-\frac{\hbar^2}{2m}\frac{\partial^2}{\partial x_1^2} + U_1(x_1, x_2(t), x_3(t), \ldots, x_N(t))\right]\varphi_1(x_1, t). \qquad (10.29)$$

Once $\varphi_1(x_1, t)$ is known, the trajectory $x_1(t)$ can be computed using the velocity given in (3.56). The conditional wave function in (10.29) still retains all the ingredients needed to discuss single-particle quantum phenomena, such as tunneling or energy quantization. In a sense, the approximation (10.29) includes all features of the quantum potential Q that apply to a single particle, but contains only some of the many-body quantum features. The term $U_1(x_1, x_2(t), x_3(t), \ldots, x_N(t)) = U_1(x_1, t)$ captures all coulomb interactions among electrons in an exact way, if the explicit dependence on the other positions of the electrons is known. Thus, to solve (10.29), one also needs the additional N Bohmian trajectories $x_2(t), x_3(t), \ldots, x_N(t)$. These other trajectories are computed, in turn, from the corresponding conditional wave functions $\varphi_2(x_2, t), \ldots, \varphi_M(x_M, t)$ found from the equations of motion

$$\begin{aligned} i\hbar\frac{\partial\varphi_2(x_2, t)}{\partial t} &= \left[-\frac{\hbar^2}{2m}\frac{\partial^2}{\partial x_2^2} + U_2(x_1, x_2(t), x_3(t), \ldots, x_N(t))\right]\varphi_2(x_1, t) \\ &\vdots \\ i\hbar\frac{\partial\varphi_N(x_2, t)}{\partial t} &= \left[-\frac{\hbar^2}{2m}\frac{\partial^2}{\partial x_N^2} + U_N(x_1, x_2(t), x_3(t), \ldots, x_N(t))\right]\varphi_N(x_1, t) \end{aligned} . \qquad (10.30)$$

The electrostatic potential energy $U_k(x_k, t)$ in each of the above equations is found by solving the corresponding Poisson equation given by

$$\nabla^2 U_k(x_k, t) = +\frac{e\rho_k(x_k, t)}{\varepsilon_0}, \qquad (10.31)$$

with the appropriate boundary conditions and the charge density $\rho_k(x_k, t)$ given by

$$\rho_k(x_k, t) = -\sum_{k'=1,\ k'\neq k}^{N} e\ \delta(x_k - x_{k'}(t)). \qquad (10.32)$$

For example, when there is no external potential (only a coulomb interaction among the M particles), the solution of $U_k(x_k, t)$ with the boundary conditions $U_k(x_k \to \pm\infty, t) = 0$ gives the typical form of Coulomb's law

$$U_k(x_k, t) = \sum_{k'=1, k'\neq k}^{N} \frac{1}{4\pi\epsilon_0}\frac{e^2}{|x_k - x_{k'}(t)|}. \qquad (10.33)$$

Thus, each particle has its own potential that includes the interactions with all other particles [52, 54, 55]. Note the condition $k' \neq k$ in (10.32) and (10.33). In the present development, the electrons are treated as quantum particles, while the electric field is a classical field. The quantum treatment (with Bohmian mechanics) of the electric and magnetic fields was explained in section 8.8.6.

From a computational point of view, the trajectories depend on the conditional wave functions for the computation of the velocity, as mentioned in (3.56). Equivalently, the conditional wave functions depend on the trajectories through the potential energy (10.31) present in the equation of motion (10.29). A self-consistent solution for both particles and waves is needed. It can be achieved by computing both conditional wave functions and trajectories at each temporal step of the simulation. This self-consistent loop clearly introduces many-body effects into the dynamics of each particle and confirms the non-Markovian nature of this approach [56].

At the end of the day, the microscopic information about the active region provided by this approach consists of the trajectories of each of the M electrons at all times $x_1(t), x_2(t), \ldots, x_N(t)$, plus the conditional wave functions $\varphi_1(x_1, t), \varphi_2(x_2, t), \ldots, \varphi_N(x_N, t)$, at all times. From this great amount of information, a number of features can be predicted. In order to predict expectation values over an ensemble of different experiments, different repetitions of the entire simulation would be needed. This is the motivation for the super-index of the index $\mathrm{j} = 1, \ldots, M$ in (10.28). However, when we test the performance of electronic devices in the laboratory, no repetition over different experiments is performed. Typically, one is interested in the time-averaged results for a single electronic device (a large simulation time for a single j device) and not the ensemble average over many identical electronic devices (simulations over an ensemble of different j devices). For this reason, the super-index j mentioned in (10.28) and in section 3.4 is omitted here.

A practical example of the use of this approach with Bohmian trajectories and conditional wave functions is the Bohmian Interacting Transport in non-equiLibrium eLEectronic Structures (BITLLES) simulator [57]. This is a simulator for nanoscale electronic devices operating at high frequencies (on the order of terahertz). The time resolutions of the electron dynamics and electrical currents are subpicosecond [54–56, 58]. Even electron–phonon decoherence effects can be included [59, 60] using a strategy very similar to that used in the Monte Carlo solution of the Boltzmann equation in chapter 9.

10.5.2 Application to resonant-tunneling diodes

To demonstrate the effectiveness of such an approach, numerical results obtained using BITLLES [54–60] will be shown for an RTD. At the computational level, the BITLLES simulator is basically a 'natural' generalization of the Monte Carlo simulators developed in chapter 9. Electrons enter the simulated active region according to some injection model and leave the active region according to their own equation of motion given in (10.30). Each electron in the Monte Carlo solution of the Boltzmann equation is described by a trajectory. Between collisions, the equation of motion that determines the trajectory originates from Newton's law and its response to the electrostatic potential (electric field). Here, each electron is described by its own trajectory plus the conditional wave function, as described in the previous section. The equation of motion between collisions includes the coulomb potential (10.33) that depends not only on the electrostatic potential but also on the quantum potential. The collisions change the properties of the conditional wave function and, indirectly, the evolution of the trajectories.

In section 2.7.1, a discussion showed how the total (conduction plus displacement) current was computed from the knowledge of a set of Bohmian trajectories. As a first example, a single electron crossing an RTD is presented. An RTD composed of two highly doped source–drain GaAs regions, two AlGaAs barriers with a length of 1.6 nm and a height of 0.5 eV, and a quantum well with a length of 2.4 nm is represented. A constant effective mass $0.067m_0$ is considered throughout the entire 3D structure, and room temperature in assumed in all the results of this section. Transport takes place in the x direction from source to drain. The initial state is a time-dependent Gaussian wave packet with a mean kinetic energy of 0.25 eV entering from the source contact [61]. Technically, this wave packet is a time-dependent conditional wave function as defined in section 3.4.1.

At each simulation time step, taken to be $\Delta t = 0.01$ ps, Poisson's equation is solved in the volume drawn in figure 10.8(a) using (10.31) with the appropriate boundary conditions to find $U_1(x_1)$ in (10.29). At the source and drain interfaces, the voltage is fixed to zero. At the other interfaces, an electric field of zero is assumed. This potential energy is used in (10.29) to compute the conditional wave function $\varphi_1(x_1, t)$ and the trajectory $x_1(t)$ seen in figure 10.9 (a spatial grid of $\Delta x = 2$ nm is used). The electric field (or potential energy U_1) that moves the particle $x_1(t)$ is not influenced by the particle $x_1(t)$ itself but by the electric field (or potential energy) computed from (10.31). For the computation of the displacement current, on the contrary, a new potential energy $U(\mathrm{x}, t)$ is required that solves Poisson's equation

$$\nabla^2 U(x, t) = \frac{e\rho(\mathrm{x}, t)}{\varepsilon_0}, \tag{10.34}$$

with a new charge distribution $\rho(x, t)$ computed, in general, as follows:

$$\rho(x, t) = -\sum_{k'=1}^{N} e\ \delta(x - x_{k'}(t)). \tag{10.35}$$

Note the absence of the condition $k' \neq k$ compared to (10.32). In the particular case of figure 10.8, with $N = 1$, one has $\rho(x, t) = -e\delta(x - x_1(t))$ and uses this in (10.35) with the same boundary conditions. From the knowledge of $U(\mathrm{x}, t)$, one obtains the electric field in a straightforward manner, and this allows the computation of the total current at all interfaces, as seen in figures 10.8(c)-10.8(h). (This uses 'method 1' for the direct computation of expression (2.33). Alternatively, the total current can be computed from 'method 2' using the Bohmian velocity through the Ramo–Shockley theorem [61–65] as explained in expression (2.34).) The Bohmian trajectory depicted in figure 10.8(b) is accelerated three times when traversing the barriers at high velocity. The increment in the velocity is shown in

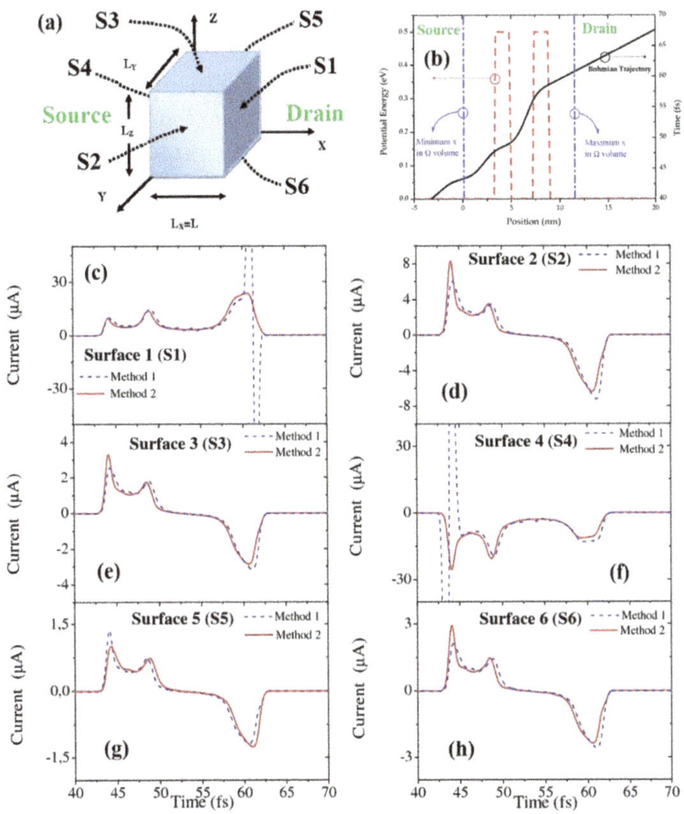

Figure 10.8. (a) A schematic representation of the closed volume and its six surfaces where the total (particle plus displacement) current is computed. (b) A Bohmian trajectory (solid black) crossing the volume in (a) with a potential energy profile of a double-barrier structure (dashed line). (c), (d), (e), (f), (g) and (h) Time-dependent total currents computed on the six surfaces of the volume of (a). The computation of the current using method 1 (dashed lines) creates spurious effects when the electrons enter (leave) the volume in (a) that are not present when the displacement and particle components are computed with method 2, i.e. using the Ramo–Shockley theorem. Reproduced from [61].

the peaks of the total current of 'method 2' since it is proportional to the velocity of the Ramo–Shockley [62–65] theorem in (2.34). The same results can be understood with 'method 1' as the result that the time derivative of the electric field is generated by the time-dependent charge $\rho(x, t) = -e\delta(x - x_1(t))$ that also depends on the time derivative (velocity) of $x_1(t)$. The total currents obtained using 'method 1' and 'method 2' represent basically identical results, except for the spurious peaks appearing at the time when the electron is entering (leaving) the volume through surface S1(S4). The sum of the six total currents in the six surfaces becomes zero [61], showing the accuracy of Bohmian trajectories in satisfying local current conservation. The sum of the particle plus the displacement currents is zero when integrated on any closed surface, showing the applicability of this approach to high-frequency scenarios [66].

In figure 10.9, the current–voltage behavior is shown for a continuous flux of electrons entering and leaving the active region for a typical RTD [67]. Here, a GaAs/AlGaAs RTD is considered with a 2 nm barrier length and 0.3 eV height. A 6 nm quantum well is considered. The DC current I_{DC} in figure 10.9(a) is computed by time averaging the total instantaneous current $I(t)$ computed using the algorithms mentioned in expressions (10.34) and (10.35) for N electrons. The noise in the current can be easily obtained by evaluating the autocorrelation function of such

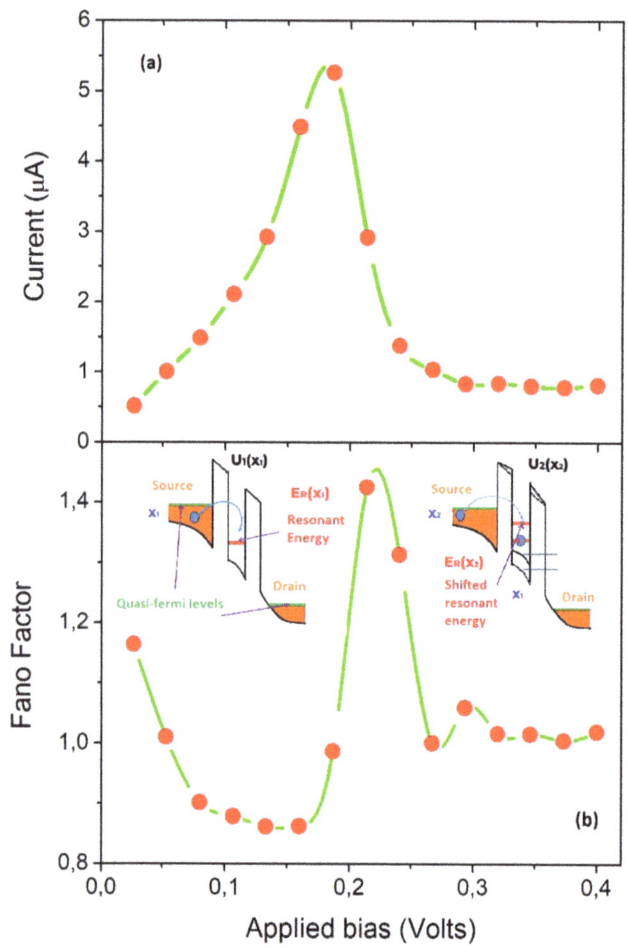

Figure 10.9. (a) Current–voltage characteristic for a RTD. A schematic representation of the volume considered as the simulation box. (b) Noise of the current quantified from the Fano factor explained in the text. The noise is generated by the coulomb correlation among electrons. Roughly speaking, as depicted in the left inset, just after a resonant bias, an electron trying to tunnel into the well 'sees' a potential energy $U_1(x_1)$ with a resonant energy $E_R(x_1)$. As depicted in the right inset, when the first electron is inside the well, as second electron 'sees' a potential energy $U_2(x_2)$ with a higher resonant energy $E_R(x_2) > E_R(x_1)$ implying that this second electron 'sees' a more transparent barriers due to the coulomb correlation with the first electron. This produces a 'bunching' of electrons which is seen in the current as an increase of the Fano factor. Reprinted from [67], with the permission of AIP Publishing.

current $I(t)$ and then taking its Fourier transform to obtain the (two-sided) power spectrum of the current fluctuations $S(\omega)$, as discussed in section 2.7 of chapter 2. The Fano factor is defined as the ratio $S(0)/S_{\text{shot}}$ of the noise at zero frequency $\omega \approx 0$ with $S_{\text{shot}} = 2qI_{\text{DC}}$. In the literature, this value $S_{\text{shot}} = 2qI_{\text{DC}}$ is the typical noise in many transport regimes (known as Poissonian noise).

The Fano factor for noise performance is plotted in the lower panel of figure 10.9. This simulation approach provides so much microscopic information as to what is happening inside the active region of the RTD that it explains clearly how the many-body coulomb interaction might affect the fluctuations of the current. Specifically, there is a correlation between the dynamics of one electron trapped in the resonant state during a dwell time and those remaining in the left reservoir. This correlation occurs essentially because the trapped electron x_1 perturbs the potential energy felt by other electrons in the reservoir. Without any electron in the well, the left-hand inset in the lower panel of figure 10.9 shows that the bottom of the well is determined by the solution of the Poisson equation $U_1(x_1)$. From such a potential energy, the resonant energy $E_R(x_1)$ of the double-barrier structure for electron x_1 is plotted in red in the inset. When an electron x_1 enters the well, the solution of the Poisson equation for a second electron $U_2(x_2)$ trying to enter the well is different. In this scenario, as plotted in right-hand inset of the lower panel of figure 10.9, the new resonant energy $E_R(x_2)$ is higher for this second electron than it was for the first one, $E_R(x_2) > E_R(x_1)$. In particular, if the resonant energy $E_R(x_1)$ is lower than the conduction band at the source (just after the resonant peak), the fact that an electron suddenly enters the well yields a barrier that is more transparent for subsequent electrons (because $E_R(x_2) > E_R(x_1)$). Therefore, many electrons in the source can tunnel into the well due to the first transmitted electron. Thus, the coulomb interaction tries to regroup the electrons (bunching) around the first transmitted electron, causing what is known as super-Poissonian noise ($S(0)/S_{\text{schot}} > 1$) as seen in the lower panel of figure 10.9. This type of behavior is found in experimental RTDs [68].

In figure 10.10, the ability of the BITLLES simulator to deal with dissipative phenomena is shown. Collisions can be implemented in the BITLLES simulator by a proper modeling of the potential $G_1(x_1, x_E^j(t), t)$ and $J_1(x_1, x_E^j(t), t)$, as mentioned in (3.57) of chapter 3. In particular, the main ingredient of the scattering with conditional wave functions is a modification of the central energy of the conditional wave function when the collision occurs. The modified conditional wave function then translates the effect of the collision into the trajectory [52, 59, 60].

In figure 10.10(a), the current in a RTD with scattering by impurities and acoustic and optical (absorption and emission) phonons is shown [59, 60]. The GaAs/AlGaAs double-barrier structure is formed by two barriers of 0.5 eV height and 1.6 nm width and the well width is 2.4 nm. The number and type of collisions are obtained from the Fermi golden rule used in GaAs materials [69]. The expected reduction of current at resonance seen in panel (a) arises because of the randomization of the momentum due to acoustic phonons and to the energy dissipation due to the emission of optical phonons. These collisions degrade the coherence and reduce the transmission coefficient of electrons. As can be seen in panel (b), the number of collisions at resonance is three times larger than the number out of

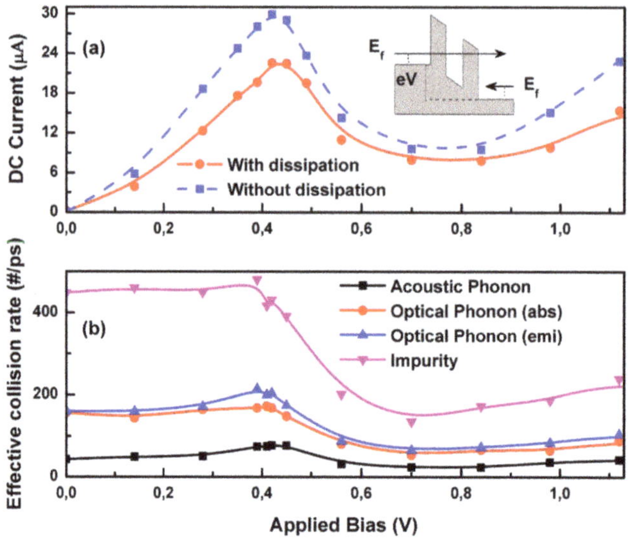

Figure 10.10. (a) The current–voltage characteristic for an RTD with (solid red curve) and without (blue dashed curve) dissipation due to acoustic and optical phonons and impurities. (b) The effective collision rate as a function of the applied bias. Optical phonons lead to an inelastic change of the electron energy of ± 0.036 eV. Reprinted with permission from [59]. Copyright (2017) by the American Physical Society.

resonance, showing that the ballisticity of tunneling devices also depends on the electron transit time, which varies from one voltage to another [59].

10.5.3 Final remarks

It is well known that the Monte Carlo solution of the Boltzmann equation presented in chapter 9 provides extensive microscopic details of what is happening inside an electron device [69]. Because of the large amount of detailed information it produces, the Monte Carlo solution of the Boltzmann equation is considered to extract a wide variety of performance data for semiclassical electron devices. In fact, in the literature, the Monte Carlo solution of the Boltzmann equation is so powerful that it is known as a 'numerical experiment' in the sense that some simpler semiclassical transport models are directly tested using Monte Carlo results, instead of real experiments. It has been shown in this section that the same richness of microscopic details offered by the Monte Carlo semiclassical techniques can be generalized to the quantum regime using conditional wave functions and Bohmian trajectories (as a type of quantum Monte Carlo solution of the Schrödinger equation). In chapter 9, the semiclassical trajectories were guided by a velocity field computed from a classical potential (the solution of the Poisson equation), while here the Bohmian trajectories are guided by a velocity field determined from a classical plus a quantum (many-body) potential.

Finally, it is important to remark that the large amount of microscopic information provided by this Bohmian approach is now accessible in the laboratory through the weak value measurements mentioned in section 10.2. For example, these weak values provide empirical access to the velocities of individual particles, as discussed in section 10.5. A more technical discussion of the nature of these weak values is provided in appendix A. Of course, an orthodox follower is free to ignore the rich amount of empirical microscopic information provided by weak values, but it is not clear to the authors that any fundamental advantage is obtained by ignoring such information when trying to understand how electronic devices work.

Appendix A Empirical and theoretical definitions of weak values

Appendix A of chapter 2 included a detailed discussion of the operations needed in the laboratory to perform strong and weak measurements. These concepts will be used here to define a weak value from an empirical and operational perspective. This appendix will focus only on the weak value of the momentum post-selected in position that provides an empirical value in the laboratory that coincides with the Bohmian velocity discussed in chapter 10 [70].

To begin, the resolution of the identity $\int | x\rangle\langle x \mid dx$ is introduced into the definition of the expectation value of the momentum. This gives

$$\begin{aligned}\langle p\rangle(t) &= \langle\psi(t)|p|\psi(t)\rangle \\ &= \int \langle\psi(t)|x\rangle\langle x|p|\psi(t)\rangle \; dx \\ &= \int | \psi(x, t)|^2 \frac{\langle x|p|\psi(t)\rangle}{\langle x|\psi(t)\rangle} \; dx\end{aligned} \quad (A.1)$$

An expression appears on the right-hand side of (A.1) that gives us the theoretical definition of a weak value:

$$p_W(x) = \mathrm{Re}\left(\frac{\langle x|p|\psi(t)\rangle}{\langle x|\psi(t)\rangle}\right). \quad (A.2)$$

The real part of the expression that is introduced in (A.2) is used to simplify the discussion in this appendix (but a physical meaning for the imaginary part $\mathrm{Im}(\langle x|p|\psi(t)\rangle/\langle x|\psi(t)\rangle)$ can also be discussed). By construction, the expression $\langle p\rangle(t) = \int| \psi(x, t)|^2 \, p_W(x)dx$ strongly suggests that we should interpret $p_W(x)$ as a type of spatial distribution of the momentum at the different positions [71, 72]. In other words, if we divide it by the mass, $p_W(x)/m$ can be interpreted as the velocity of the quantum particle. Up until this point, the definition of the weak value was just a mathematical curiosity. But one can ask: can $p_W(x)$, as defined in (A.2), really be measured in the laboratory? As anticipated by the figures of section 10.2, the answer is yes.

The operational protocol used to define a weak value $p_W(x)$ in the laboratory is defined through the expression

$$p_W(x) = \frac{\int p\mathcal{P}(p, x, t) \; dp}{\int \mathcal{P}(p, x, t) \; dp}. \quad (A.3)$$

Here, 'operational' means that there are some operations done in the laboratory which are not linked to any quantum theory. The four steps of the operational protocol used to compute the probability distribution $\mathcal{P}(p, x, t)$ are the following [72]:

(i) Many identical initial states $\psi(x, t)$ are prepared at time t,
(ii) For each initial state, a weak measurement of the momentum is done, yielding the value p at time t,
(iii) Subsequently, a strong measurement of the position is done, yielding the value x at time $t + dt$, and
(iv) Finally, the probability distribution $\mathcal{P}(p, x, t)$ is constructed by counting how many p and x occur when repeating (ii) and (iii) on ensemble (i).

The probability $\mathcal{P}(x, t)$ of measuring a value in all experiments can be defined as $\mathcal{P}(x, t) = \int \mathcal{P}(p, x, t)dp$. The conditional probability then becomes

$$\mathcal{P}(p \mid x, t) = \frac{\mathcal{P}(p, x, t)}{\mathcal{P}(x, t)} = \frac{\mathcal{P}(p, x, t)}{\int \mathcal{P}(p, x, t)\ dp}, \tag{A.4}$$

which gives the value p, conditioned to the fact of achieving the value x afterwards. Thus, the weak value in (A.4) can alternatively be written as

$$p_{\mathrm{W}}(x) = \frac{\int p\mathcal{P}(p, x, t)\ dp}{\int \mathcal{P}(p, x, t)\ dp} = \int p\mathcal{P}(p \mid x, t)\ dp. \tag{A.5}$$

Thus, $p_{\mathrm{W}}(x)$ can be understood to be an average of the measured momentum p of the particle conditioned to the fact that the particle is later found at the location x. Note that the protocol mentioned above is not linked to any interpretation of quantum mechanics. It is just a set of operations to be used in the laboratory by somebody interested in empirical information on the momentum at one position. Now, by using the common explanation of the previous protocol to find $\mathcal{P}(p, x, t)$ using wave functions and operators, it will be shown that the theoretical expression (A.2) coincides with the empirical expression (A.3).

After the initial state $\Psi(x', x_{\mathrm{E}}, x_M, t)$ is prepared in step (i) of the aforementioned operational protocol, the second step (ii) is a weak measurement of the momentum p. In appendix A of chapter 2, a weak measurement protocol was developed. Following the weak measurement of the momentum p, the wave function $\Psi_p(x', x_{\mathrm{E}}, x_M, t)$ can be rewritten using expression (A.8) in the appendix of chapter 2 in terms of the momentum instead of electrical current, as follows:

$$\Psi(x', x_{\mathrm{E}}, x_M, t) \to \Psi_p(x', x_{\mathrm{E}}, x_M, t)$$

$$\sum_{i,j} c_{p_i}\phi_{p_i}(x')b_{(p_j - p_i)}\phi_{p_j}(x_{\mathrm{E}})\ \phi_M\left(x_M - p_j'\right) \to \sum_i c_{p_i}\phi_{p_i}(x')b_{(p-p_i)}\phi_p(x_{\mathrm{E}})\phi_M(x_M - p). \tag{A.6}$$

The third step (iii) in the aforementioned operational protocol is a subsequent strong measurement of the position x. To simplify the discussion, it is assumed that the unitary evolution of the system between the first and second measurement takes a sufficiently small time that it may assumed that $dt \to 0$. Therefore, the second strong measurement of the position x basically forces the variable x' in $\phi_{p_i}(x')$ to be fixed to the value x in (A.6). This leads to

$$\begin{gathered}\Psi_p(x', x_{\mathrm{E}}, x_M, t) \to \Psi_{p,x}(x', x_{\mathrm{E}}, x_M, t)\\ \sum_i c_{p_i}\phi_{p_i}(x')b_{(p-p_i)}\phi_p(x_{\mathrm{E}})\phi_M(x_M - p) \to \sum_i c_{p_i}\phi_{p_i}(x')b_{(p-p_i)}\phi_p(x_{\mathrm{E}})\phi_M(x_M - p)\delta(x' - x)\cdot\end{gathered} \quad (A.7)$$

According to Born's law, the probability $\mathcal{P}(p, x, t)$ computed in step (iv) is just the integral of the modulus of the final wave function $\Psi_{p,x}(x', x_{\mathrm{E}}, x_M, t)$ in (A.7), which after all operations, may be expressed as

$$\begin{aligned}\mathcal{P}(p, x, t) &= \iint |\Psi_{p,x}(x', x_{\mathrm{E}}, x_M, t)|^2 \; dx'dx_{\mathrm{E}}\; dx_M\\ &= \sum_{i,i'} c^*_{p_i}c_{p_{i'}}\phi^*_{p_i}(x)\phi_{p_{i'}}(x)b_{(p-p_{i'})}b_{(p-p_i)}\end{aligned} \quad . \quad (A.8)$$

We now use $b_{(p-p_i)} = N_{\mathrm{b}}\cdot\exp\left(-(p - p_i)^2/\sigma_{\mathrm{E}}^2\right)$ as discussed in appendix A of chapter 2, and the numerator of (A.3) becomes

$$\begin{aligned}\int p\mathcal{P}(p, x, t)dp &= \sum_{i,i'} c^*_{p_i}c_{p_{i'}}\phi^*_{p_i}(x)\phi_{p_{i'}}(x)N_{\mathrm{b}}^2\\ &\times\int p\;\exp\left(-\frac{(p-p_i)^2}{\sigma_{\mathrm{E}}^2}\right)\exp\left(-\frac{(p-p_{i'})^2}{\sigma_{\mathrm{E}}^2}\right)\;dp\\ &= \sum_{i,i'} c^*_{p_i}c_{p_{i'}}\phi^*_{p_i}(x)\phi_{p_{i'}}(x)\frac{\left(p_i + p_i^{'}\right)}{2}\\ &=\mathrm{Re}\left(\sum_{i,i'}\langle\psi(t)|p_i\rangle\langle p_{i'}|\psi(t)\rangle p_i\langle x|p_{i'}\rangle\langle p_i|x\rangle\right)\\ &=\mathrm{Re}(\langle\psi(t)|x\rangle\langle x|p|\psi(t)\rangle)\end{aligned} \quad , \quad (A.9)$$

where $\sum_i |p_i\rangle p_i\langle p_i| = p$ has been used as the definition of the operator p and $\sum_i |p_i\rangle\langle p_i| = 1$ is a resolution of the identity. In the same way, we can compute the denominator of (A.3) as follows:

$$\int p\mathcal{P}(p, x, t)dp = \sum_{i,i'} c^*_{p_i} c_{p_{i'}} \phi^*_{p_i}(x)\phi_{p_{i'}}(x) N_b^2$$
$$\times \int \exp\left(-\frac{(p - p_i)^2}{\sigma_E^2}\right)\exp\left(-\frac{(p - p_{i'})^2}{\sigma_E^2}\right) dp$$
$$= \sum_{i,i'} c^*_{p_i} c_{p_{i'}} \phi^*_{p_i}(x)\phi_{p_{i'}}(x) \qquad . \tag{A.10}$$
$$= \mathrm{Re}\left(\sum_{i,i'} \langle \psi(t)|p_i\rangle\langle p_{i'}|\psi(t)\rangle\langle x|p_{i'}\rangle\langle p_i|x\rangle\right)$$
$$= \langle\psi(t)|x\rangle\langle x|\psi(t)\rangle$$

Finally, we use (A.9) and (A.10) in (A.3), and the operational protocol gives the expression

$$\frac{p_W(x)}{m} = \frac{1}{m}\frac{\int p\mathcal{P}(p, x, t)\ dp}{\int \mathcal{P}(p, x, t)\ dp}$$
$$= \frac{1}{m}\frac{\mathrm{Re}(\langle\psi(t)|x\rangle\langle x|p|\psi(t)\rangle)}{\langle\psi(t)|x\rangle\langle x|\psi(t)\rangle} = \frac{J(x, t)}{|\ \psi(x, t)\ |^2}, \tag{A.11}$$

which coincides with the theoretical expression (A.2). Here, $p_W(x)$ is divided by the mass of the particle m to get a local velocity (instead of a local momentum). In the last expression of (A.11), the definition $\langle x|p|\psi(t)\rangle = -i\hbar\partial\psi(x, t)/\partial x$ is used to give the particle current density, as follows:

$$J(x, t) = \frac{1}{m}\mathrm{Re}(\ \langle\psi(t)|x\rangle\langle x|p|\psi(t)\rangle)$$
$$= \frac{1}{m}\mathrm{Re}\left(\ \psi^*(x, t)(-i\hbar)\frac{\partial\psi(x, t)}{\partial x}\right) = \ \frac{1}{m}R^2(x, t)\frac{\partial S(x, t)}{\partial x}. \tag{A.12}$$

where, on the right-hand side of (A.12), a polar form of the wave function $\psi(x, t) = R(x, t)\exp[iS(x, t)/\hbar]$ is used. Finally, substituting (A.12) into (A.11) leads to the final result:

$$\frac{p_W(x)}{m} = \frac{J(x, t)}{|\ \psi(x, t)\ |^2} = \frac{1}{m}\frac{\partial S(x, t)}{\partial x} = v(x). \tag{A.13}$$

The weak value of the momentum post-selected by the position is just the Bohmian velocity discussed in (A.13).

At this point, the relevant question is: what is the actual ontological meaning of the empirical value in (A.3)? There are several possible answers that depend upon the ontology of the quantum theory used to explain the results. Those who believe in the orthodox postulates, which explicitly forbid the notion of velocity, can still argue that the value (A.3) is obtained in the laboratory as an average over many experiments. As such, they conclude that (A.13) is a type of measure of an ensemble velocity but not an ontological velocity. Of course, those who believe in Bohmian mechanics, assume that the whole protocol giving (A.3) provides a measure of the ontological Bohmian velocity in (A.13). However, independently of the ontological interpretation given to this velocity, there is a connection between the theoretical values (A.2) and the empirical value (A.3). Such a connection is a mathematical connection that is valid regardless of any ontology. Thus, one is allowed to make predictions about the microscopic properties of the system using (A.2), knowing that such predictions can be tested later using (A.3). This connection gives a new perspective on the use of trajectories in modeling quantum transport. Nevertheless, an orthodox follower is free to ignore the mathematical (not ontological) connection between (A.2) and (A.3), but it is not clear to the authors that there is any advantage in ignoring the amount of information that this mathematical (not ontological) connection between (A.2) and (A.3) offers. Examples of how these weak values can be used to produce new theoretical and empirical characterizations of quantum electron devices are shown in references [71, 72].

References

[1] Bohm D 1952 *Phys. Rev.* **85** 166

[2] Madelung E 1926 *Z. Phys.* **40** 322

[3] Kennard E H 1928 *Phys. Rev.* **31** 876

[4] Wigner E 1932 *Phys. Rev.* **40** 749

[5] Feynman R P and Hibbs A R 1965 *Quantum Mechanics and Path Integrals* (New York: McGraw-Hill) Section 10.3

[6] Ferry D K 2021 *Quantum Mechanics* 3rd edn (Boca Raton, FL: CRC Press)

[7] Fetter A L and Walecka J D 1971 *Quantum Theory of Many-Particle Systems* (New York: McGraw-Hill) pp 277

[8] Kadanoff L P and Baym G 1962 *Quantum Statistical Mechanics* (Reading, MA: Benjamin) Ch 6

[9] Ando T, Fowler A B and Stern F 1982 *Rev. Mod. Phys.* **54** 437

[10] Ferry D K 2000 *Superlattices Microstruct.* **28** 419

[11] Zhou J R and Ferry D K 1992 *IEEE Trans. Electron Devices* **39** 473

[12] Albareda G *et al* 2013 *J. Comput. Electron.* **12** 405

[13] Oriols X *et al* 1998 *Appl. Phys. Lett.* **72** 806

[14] Colomés E, Zhan Z and Oriols X 2015 *J. Comput. Electron.* **14** 894

[15] Shifren L, Akis R and Ferry D K 2000 *Phys. Lett.* A **274** 75

[16] Young T 1807 *Course of Lectures on Natural Philosophy and Mechanical Arts* 1 (London: Johnson) Lecture XXXIX

[17] Davisson C and Kunsman C H 1921 *Science* **54** 522

[18] Elsasser W 1925 *Naturwissenschaften* **13** 711

[19] Thomson G P and Reid A 1927 *Nature* **119** 890

[20] Taylor G I 1909 *Proc. Camb. Phil. Soc.* **15** 114

[21] Philippidis C, Dewdney C and Hiley B J 1979 *Il Nuovo Cim.* **52B** 15

[22] Kaloyerou P N and Ilunga A M 2019 *Phys. Scr.* **94** 105003

[23] Kocsis S *et al* 2011 *Science* **332** 1170

[24] Pang A O T *et al* 2020 *Quantum* **4** 365

[25] Menzel R *et al* 2012 *Proc. Natl Acad. Sci.* **109** 9314
[26] Wiedenhaus M *et al* 2001 *IEEE J. Quantum Electron.* **37** 684
[27] Jüngel A and Li H 2004 *Quart. Appl. Math.* **42** 569
[28] Li H and Marcati P 2004 *Commun. Math. Phys.* **245** 215
[29] Huang F, Li H-L and Matsumura A 2006 *J. Diff. Equat* **225** 1
[30] di Michele F *et al* 2016 *Int. J. Num. Anal. Model.* **13** 898–925 https://www.global-sci.org/intro/article_detail/ijnam/471.html#
[31] Bannerjee S *et al* 2015 *Proc. Third Int. Conf. Comp. Comm. Cont. Info. Tech.* (New York: IEEE Press) 7060141
[32] Archaryya A *et al* 2015 *J. Comput. Electron* **14** 309
[33] Buryk I P *et al* 2022 *J. Nano Electron Phys.* **14** 01024
[34] Suñe J *et al* 1996 *Appl. Surf. Sci.* **102** 255
[35] Suñe J *et al* 1997 *Microelectron. Eng.* **36** 125
[36] Barker J R and Watling J R 2000 *Superlattices Microstruct.* **27** 347
[37] Oriols X 2003 *IEEE Trans. Electron Devices* **50** 1830
[38] Albareda G, Sune J and Oriols X 2009 *Phys. Rev.* B **79** 075315
[39] Albareda G *et al* 2010 *J. Appl. Phys.* **108** 043706
[40] Traversa F L *et al* 2011 *IEEE Trans. Electron Devices* **58** 2104
[41] Weyl H 1927 *Z. Phys.* **46** 1
[42] Grubin H L *et al* 1993 *Solid-State Electron.* **36** 1697
[43] Grubin H L *et al* 1994 *Semicond. Sci. Technol.* **9** 855
[44] Kosina H, Langer E and Selberherr S 1995 *Microelectron. J.* **26** 217
[45] Giachetti R and Tognetti V 1986 *Phys. Rev.* B **33** 7647
[46] Feynman R P and Kleinert H 1986 *Phys. Rev.* A **34** 5080
[47] Kleinert H 1986 *Phys. Lett.* A **118** 267
[48] Kriman A *et al* 1989 *Solid-State Electron.* **32** 1603
[49] Kriman A M, Zhou J R and Ferry D K 1989 *Phys. Lett.* A **138** 8
[50] Cuccoli A *et al* 1992 *Phys. Rev.* B **45** 2088
[51] Wyatt R E 2005 *Quantum Dynamics with Trajectories: Introduction to Quantum Hydrodynamics* (New York: Springer)
[52] Oriols X and Mompart J 2019 *Applied Bohmian Mechanics: From Nanoscale Systems to Cosmology* (Singapore: Pan Stanford))
[53] Dürr D and Teufel S 2009 *Bohmian Mechanics: The Physics and Mathematics of Quantum Theory* (Berlin: Spinger)
[54] Oriol X 2007 *Phys. Rev. Lett.* **98** 066803
[55] Albareda G, Suñé J and Oriols X 2009 *Phys. Rev.* B **79** 075315
[56] Pandey D *et al* 2019 *Entropy* **21** 1148
[57] https://europe.uab.es/bitlles/
[58] Albareda G *et al* 2013 *J. Comput. Electron* **12** 405
[59] Colomes E *et al* 2017 *Phys. Rev.* B **96** 075135
[60] Villani M *et al* 2021 *Entropy* **23** 408
[61] Alarcón A and Oriols X 2009 *J. Stat. Mech. Theor. Exp.* **2009** P01051
[62] Ramo S 1939 *Proc. IRE* **27** 584
[63] Shockley W 1938 *J. Appl. Phys.* **9** 635
[64] Pellegrini B 1986 *Phys. Rev.* B **34** 5921
[65] Marian D, Zanghì N and Oriols X 2016 *Phys. Rev. Lett.* **116** 110404

[66] Villani M *et al* 2021 *IEEE Electron Device Lett.* **42** 224
[67] Oriols X, Trois A and Blouin G 2004 *Appl. Phys. Lett.* **85** 3596
[68] Iannaccone G *et al* 1998 *Phys. Rev. Lett.* **80** 1054
[69] Jacoboni C and Lugli P 1989 *The Monte Carlo Method for Semiconductor Device Simulation* (Vienna: Springer)
[70] Wiseman H W 2007 *New J. Phys.* **9** 165
[71] Pandey D *et al* 2021 *Phys. Rev.* A **103** 052219
[72] Destefani C F and Oriols X 2023 *Phys. Rev.* A **107** 012213

IOP Publishing

Quantum Transport in Semiconductor Devices

Simulation using particles

David K Ferry, Xavier Oriols and Josef Weinbub

Chapter 11

Wigner functions

Quantum mechanics began, and has evolved, primarily as a physical process that is represented in position space. This is in contradiction to classical mechanics, in which the dynamics is presented in phase space, and the evolution of both the position and the momentum is described by Hamilton's equations of motion. However, as early as 1932, Wigner had suggested that there was no reason why one could not adapt a phase-space description of quantum mechanics [1]. Indeed, he suggested that such a description would be far more effective in spanning the transition from the classical world to the quantum world. While the Wigner approach received relatively little acceptance, it has been studied for more than nine decades. Historically, it was developed from the Schrödinger wave equation, although it could have been developed in its own right [2, 3]. The Wigner function approach is generally felt to offer a number of advantages for use in modeling the behavior of physical processes in many fields of science [4, 5] (see also a recent special issue on Wigner functions in computational electronics and optics [6]). First, it is a phase-space formulation similar to classical formulations based upon the Boltzmann equation. By this, we mean that the Wigner function involves both real-space and momentum-space variables, in which respect it is distinctly different from the Schrödinger equation. Because of the phase-space nature of the distribution, it is conceptually possible to identify where quantum corrections enter a problem by comparison with the classical version, an approach that has been used to provide an effective quantum potential (chapter 10) that can be used as a correction term to classical simulations [7].

The Wigner function has already been introduced in chapters 1 and 5. In this earlier discussion, the density matrix was described as

$$\rho(\boldsymbol{r}, \boldsymbol{r}', t) = \psi^{\dagger}(\boldsymbol{r}', t)\psi(\boldsymbol{r}, t), \tag{11.1}$$

where the total wave functions have been used. To proceed, the center-of-mass coordinates (1.53) are introduced as follows:

doi:10.1088/978-0-7503-5237-6ch11

$$\begin{aligned} \boldsymbol{R} &= \frac{1}{2}(\boldsymbol{r} + \boldsymbol{r}'). \\ \boldsymbol{s} &= \boldsymbol{r} - \boldsymbol{r}' \end{aligned} \tag{11.2}$$

With this change, Wigner defined a phase-space distribution that was analogous to the Boltzmann distribution function [1], using these center-of-mass coordinates and a subsequent Fourier transform of the difference variable, as follows:

$$f_{\mathrm{W}}(\boldsymbol{R}, \boldsymbol{p}, t) = \frac{1}{h^3} \int ds \rho\left(\boldsymbol{R} + \frac{\boldsymbol{s}}{2}, \boldsymbol{R} - \frac{\boldsymbol{s}}{2}, t\right) e^{i\boldsymbol{p}\cdot\boldsymbol{s}/\hbar}. \tag{11.3}$$

This yields a distribution function that is defined in position, momentum, and time. That is, the distribution is defined in a six plus one-dimensional coordinate system (three for momentum, three for position, and one for time). Once we introduce the wave functions, the Wigner function becomes

$$f_{\mathrm{W}}(\boldsymbol{R}, \boldsymbol{p}, t) = \frac{1}{(2\pi\hbar)^3} \int ds \psi^{\dagger}\left(\boldsymbol{R} + \frac{\boldsymbol{s}}{2}\right) \psi\left(\boldsymbol{R} - \frac{\boldsymbol{s}}{2}\right) e^{i\boldsymbol{p}\cdot\boldsymbol{s}/\hbar}. \tag{11.4}$$

In this chapter, the properties of the Wigner function in simulations in physics will be discussed. First, some general properties of the Wigner function and its equation of motion will be discussed. Then, other phase space functions, which are variants of the Wigner function, will be introduced and discussed. Finally, the use of particles for simulations of the Wigner function in devices will be presented.

11.1 Some properties of the Wigner function

In general, the Wigner function is not a positive definite function. This is a consequence of the uncertainty relationship and the development of entanglement away from equilibrium. The off-diagonal terms of the density matrix carry the entanglement, and it is exactly these off-diagonal terms that give rise to the components of the Fourier transform momentum $\boldsymbol{p}$. It is also important to understand the normalization of the Wigner function. To demonstrate that the function is properly normalized, the one-dimensional form will be utilized. This one-dimensional form may be expressed as

$$f_{\mathrm{W}}(R, p, t) = \frac{1}{2\pi\hbar} \int ds \psi^{\dagger}\left(R + \frac{s}{2}\right) \psi\left(R - \frac{s}{2}\right) e^{ip\cdot s/\hbar}. \tag{11.5}$$

If this is now integrated over the momentum, the result is

$$\begin{aligned} \int dp f_{\mathrm{W}}(R, p, t) &= \frac{1}{2\pi\hbar} \int dp \int ds \psi^{\dagger}\left(R + \frac{s}{2}\right) \psi\left(R - \frac{s}{2}\right) e^{ip\cdot s/\hbar} \\ &= \int ds \psi^{\dagger}\left(R + \frac{s}{2}\right) \psi\left(R - \frac{s}{2}\right) \delta(s) \\ &= \psi^{\dagger}(R) \psi(R) = P(R). \end{aligned} \tag{11.6}$$

The last line is, of course, the normal probability arising from the Schrödinger equation in normal quantum mechanics.

To develop the normalization in momentum space requires somewhat more detail. For this, the first step is to introduce the Fourier transform of the wave function. The Fourier transform pair is

$$\begin{aligned}\varphi(k) &= \frac{1}{\sqrt{2\pi}}\int dr\psi(r)e^{ikr} \\ \psi(r) &= \frac{1}{\sqrt{2\pi}}\int dk\varphi(k)e^{-ikr}\end{aligned}. \tag{11.7}$$

The Wigner function (11.5) now becomes

$$\begin{aligned}f_{\mathrm{W}}(R, p, t) = \frac{1}{2\pi\hbar}\int ds\frac{1}{2\pi}\int dk\varphi^{\dagger}(k)e^{-ik(R+s/2)} \\ \times\int dk'\varphi(k')e^{ik'(R-s/2)}e^{ip\cdot s/\hbar}\end{aligned}. \tag{11.8}$$

The integration over s produces a delta function which sets

$$\frac{p}{\hbar} = \frac{k+k'}{2} \equiv K. \tag{11.9}$$

This is obviously the average wave number and suggests letting

$$\kappa = k - k'. \tag{11.10}$$

Integration over this delta function can be done with ease once the connect $dkdk' \to dKd\kappa$ is made. The integration over the delta function is done with the K integration, which connects the remaining integration to one over κ. With this definition, one obtains

$$f_{\mathrm{W}}(R, p, t) = \frac{1}{2\pi\hbar}\int d(\hbar\kappa)\varphi^{\dagger}\left(p + \frac{\hbar\kappa}{2}\right)\varphi\left(p - \frac{\hbar\kappa}{2}\right)e^{-i\kappa R}. \tag{11.11}$$

Integrating over R now yields a delta function on κ which produces a similar result to (11.7), as follows:

$$\int dRf_{\mathrm{W}}(R, p, t) = \varphi^{\dagger}(p)\varphi(p) = P(p). \tag{11.12}$$

Let us examine a simple wave function that has a propagation velocity associated with it. The form of this is a modified Gaussian so that there is an imaginary part which contains the velocity. Such a wave packet is often called a coherent wave packet, and it can be written as

$$\psi(r) = \left(\frac{1}{2\pi\sigma^2}\right)^{1/4} e^{-r^2/4\sigma^2 - ik_0 r}, \tag{11.13}$$

where σ is the standard deviation determined by the uncertainty in position, and $\hbar k_0$ is the momentum of the wave packet. The velocity is connected to this momentum, just as discussed in previous chapters, and is given by $v = \hbar k_0/m^*$, where m^* is the effective mass. To proceed, the Fourier transform version of (11.4), where the wave functions are in momentum space, will be used. This is expressed in (11.11). Using this approach, we find that the Wigner function is now given by

$$f_W(r, p) = \frac{2}{\hbar}e^{-2\sigma^2(k-k_0)^2-r^2/2\sigma^2}. \tag{11.14}$$

This result is a positive definite function, which is appropriate for what is basically a ground state function.

Let us now explore how two particles would appear in the Wigner formulation. To describe these two particles in a statistical way in the physical space (not in the too complicated configuration space), a single wave function is used, as it is assumed that these two particles have either interacted in the past or were created together via another process. In any case, it may be assumed that they are entangled. Each of the two particles is a coherent wave packet, such as (11.13), so that the total wave packet is

$$\psi(r) = \frac{1}{\sqrt{2}}[\psi_1(r - a, k_0) + \psi_1(r + a, -k_0)]. \tag{11.15}$$

Thus, one particle is assumed to be at a and moving in the positive r-direction, while the other particle is at $-a$ and moving in the negative r-direction. Of course, we really don't know which of the two particles is which, and this is preserved by the inversion symmetry of the combined wave function. It is well known that when one generates the Wigner function from such a composite wave function, extra terms arise [8]. If this last equation is Fourier transformed, the composite Wigner function becomes

$$f_W(r, p) = \frac{1}{h}[e^{-(r-a)^2/2\sigma^2-2\sigma^2(k-k_0)^2} + e^{-(r+a)^2/2\sigma^2-2\sigma^2(k+k_0)^2} + 2e^{-r^2/2\sigma^2-2\sigma^2k^2}\cos(ak)] \tag{11.16}$$

The first two exponentials are clearly identified with the Wigner functions of the individual wave packets in (11.15). The third term describes a Gaussian peak, centered around $x = 0$ and $k = 0$, which has an oscillatory behavior in momentum and the initial offset of the Gaussians (see figure 11.1 [4]). The fact that this extra term has both positive and negative values is clear evidence that it is not related to any probability function. Moreover, as the value a increases, the oscillation becomes faster, but its amplitude is not affected. Hence, in this case it becomes hard to argue that the Wigner function approaches a classical phase space distribution. The motion of the particles can be recovered (or introduced) by letting $a = vt$, so that they interacted (or were created) at the origin at $t = 0$. Just what does the extra term in (11.16) represent? If this term were ignored, there would be just two independent Gaussian coherent packets going on their way. But the wave function (11.15) deals

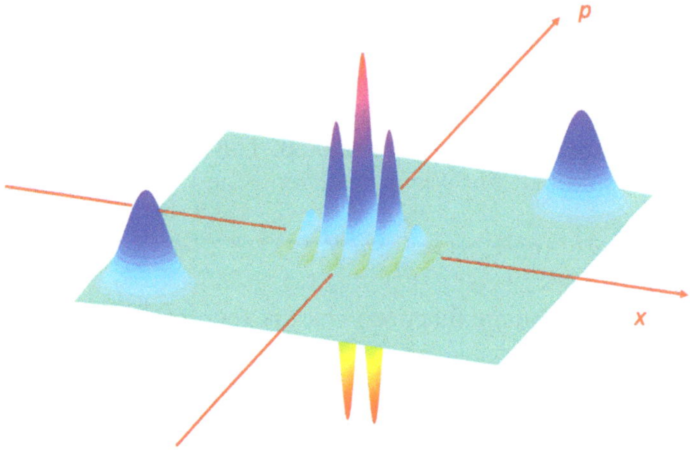

Figure 11.1. The Wigner transform of the composite wave function (11.15). The rapidly oscillating part centered around the origin is thought to the be representation of the entanglement between the two 'particles.' Reproduced from [4]. © IOP Publishing Ltd. All rights reserved.

with the entangled particles, and this extra term must correspond to the memory that these two packets are entangled. If an average is made over the momentum, this term vanishes and makes no contribution to the average momentum (or to the average position). Hence, it is not a measurable contribution to the probability density. One must conclude that this extra term represents an embodiment of the entanglement between the two particles.

11.1.1 The Moyal star product and the equation of motion

It was remarked in appendix A (of chapter 5) that the extension of the classical Poisson bracket to quantum mechanics actually resulted in the Moyal bracket, which is an expression of this star product [3, 9]. The Star product is itself an extension of the Poisson bracket. Usually, a double curly bracket is used to denote this expression, as follows:

$$\begin{aligned}\{\{f, g\}\} &= \frac{1}{i\hbar}(f\star g - g\star f) = \{f, g\} + O(\hbar^2)\\ &= \frac{2}{\hbar}f(r, p)\sin\left(\frac{\hbar}{2}\frac{\overleftarrow{\partial}}{\partial r}\frac{\overrightarrow{\partial}}{\partial p} - \frac{\hbar}{2}\frac{\overleftarrow{\partial}}{\partial p}\frac{\overrightarrow{\partial}}{\partial r}\right)g(r, p)\end{aligned} \quad . \tag{11.17}$$

The arrows over the top indicate that the first partial derivative operates *to the left* and the second partial derivative operates *to the right*. Hence, the lowest-order term of the sine function gives the classical Poisson bracket. The higher-order terms in the sine expansion give quantum corrections. In connecting with the standard equation for the Wigner function, one has to understand that both $\boldsymbol{p}$ and $\boldsymbol{r}$ are operators, the former giving the derivative in position and the latter a derivative in momentum, as in (11.17).

This star product can be used to obtain the equation of motion for the Wigner function. From (A.5) in chapter 1, the use of the star product replaces the Poisson bracket and, in the absence of collisions, one has

$$\frac{\partial f_{\mathrm{w}}(r, p, t)}{\partial t} + \left\{\left\{f_{\mathrm{w}}, H\right\}\right\} = 0. \tag{11.18}$$

As is normal, the Hamiltonian is taken to be a standard form

$$H(r, p) = \frac{p^2}{2m^*} + V(r). \tag{11.19}$$

The sine function can now be expanded in a power series as follows:

$$\sin(x) = x - \frac{x^3}{3!} + \frac{x^5}{5!} - \frac{x^7}{7!} + \cdots. \tag{11.20}$$

The first term in the argument of the sine function in (11.17) occurs only in the first term of this series, because there is no power higher than two in the momentum of the Hamiltonian. Thus, the first set of derivatives in the sine function gives the term, in vector form, as

$$\frac{\partial H}{\partial \boldsymbol{p}} \cdot \nabla f_{\mathrm{w}} = \boldsymbol{v} \cdot \nabla f_{\mathrm{w}}. \tag{11.21}$$

This term is unchanged from the classical term in the Boltzmann equation. Obviously, the quantum corrections will arise from the potential term, as discussed in chapter 10. Taking this latter term out of the sine function of (11.18) thus leaves the equation of motion as

$$\frac{\partial f_{\mathrm{w}}(r, p, t)}{\partial t} + \boldsymbol{v} \cdot \nabla f_{\mathrm{w}} - \frac{2}{\hbar} f_{\mathrm{W}}(r, p) \sin\left(\frac{\hbar}{2} \frac{\overleftarrow{\partial}}{\partial p} \frac{\overrightarrow{\partial}}{\partial r}\right) H(r, p) = 0. \tag{11.22}$$

There are now two approaches that can be taken. One is to use the differential form directly, and the second is to manipulate the equation into an integral form. Consider the differential form first, which directly arises from the expansion (11.20). In this case, (11.22) becomes

$$\frac{\partial f_{\mathrm{w}}(\boldsymbol{r}, p, t)}{\partial t} + \boldsymbol{v} \cdot \nabla f_{\mathrm{w}} = \frac{2}{\hbar} \sum_{n=0}^{\infty} (-1)^n \frac{(\hbar/2)^{2n+1}}{(2n+1)!} \times \frac{\partial^{2n+1} V(\boldsymbol{r})}{\partial \boldsymbol{r}^{2n+1}} \frac{\partial^{2n+1} f_{\mathrm{W}}(\boldsymbol{r}, \boldsymbol{p})}{\partial \boldsymbol{p}^{2n+1}}. \tag{11.23}$$

This form will be useful in developing a set of hydrodynamic equations below. But, it is clear that this form also leads to the effective potentials discussed in chapter 10 as corrections to the classical potential that arises solely from the first term in the series.

As pointed out, developing the integral form requires some manipulation. The first step is to introduce an integral representation of unity. This will be done in one dimension for clarity. First, the last term of (11.22) is written as

$$\begin{aligned}\theta \cdot f_{\mathrm{W}} &= \frac{2}{\hbar} f_{\mathrm{W}}(r, p)\sin\left(\frac{\hbar}{2}\frac{\overleftarrow{\partial}}{\partial p}\frac{\overrightarrow{\partial}}{\partial r}\right)H(r, p). \\ &= \frac{2}{\hbar}\int dp' f_{\mathrm{W}}(r, p')\delta(p - p')\sin\left(\frac{\hbar}{2}\frac{\overleftarrow{\partial}}{\partial p}\frac{\overrightarrow{\partial}}{\partial r}\right)H(r, p)\end{aligned} \quad . \tag{11.24}$$

The delta function is now replaced by its Fourier transform as follows:

$$\begin{aligned}\theta \cdot f_{\mathrm{W}} &= \frac{1}{\pi\hbar}\int dp' f_{\mathrm{W}}(r, p')\int ds e^{i(p-p')s/\hbar} \\ &\times \sin\left(\frac{\hbar}{2}\frac{\overleftarrow{\partial}}{\partial p}\frac{\overrightarrow{\partial}}{\partial r}\right)V(r)\end{aligned} \quad . \tag{11.25}$$

Clearly, all of the effort is around the sine and potential terms. Here, a connection to quantum mechanics is made by the operator definitions [10]

$$\begin{aligned}p &= -i\hbar\frac{\partial}{\partial r} \\ s &= i\hbar\frac{\partial}{\partial p}\end{aligned} \quad . \tag{11.26}$$

Thus, the sine function can be rewritten as

$$\sin\left(\frac{s \cdot p}{2\hbar}\right). \tag{11.27}$$

A fundamental aspect of quantum mechanics is the displacement operator [10]

$$D_x \mid r\rangle = e^{ipx/\hbar} \mid r\rangle = \mid r + x\rangle. \tag{11.28}$$

This means that the sine function is actually composed of a pair of displacement operators, such that

$$\sin\left(\frac{s \cdot p}{2\hbar}\right)V(r) = \frac{1}{2i}\left[V\left(r + \frac{s}{2}\right) - V\left(r - \frac{s}{2}\right)\right]. \tag{11.29}$$

The equation of motion (11.22) can now be written as

$$\frac{\partial f_{\mathrm{W}}(\mathbf{r}, p, t)}{\partial t} + v \cdot \nabla f_{\mathrm{W}} - \frac{1}{h}\int dp' f_{\mathrm{W}}(r, p')W(r, p - p') = 0, \tag{11.30}$$

where

$$W(r, p') = \frac{1}{i}\int ds e^{isp'/\hbar}\left[V\left(r + \frac{s}{2}\right) - V\left(r - \frac{s}{2}\right)\right]. \tag{11.31}$$

Equation (11.30) can be rewritten with a change of variable as

$$\frac{\partial f_{\mathrm{W}}(\boldsymbol{r}, p, t)}{\partial t} + v \cdot \nabla f_{\mathrm{W}} - \frac{1}{h} \int dp f_{\mathrm{W}}(r, p + p') W(r, p') = 0. \tag{11.32}$$

This is now in the form that appears in chapters 1 and 5. While these equations determine the time variation of the Wigner function, they require an initial condition. This latter function must be determined separately from the evolution equations developed here.

11.1.2 Initial conditions

One method of finding the initial condition is to develop the adjoint equation to (11.32) [11]. This is an interesting but somewhat difficult prospect. An alternative method is to utilize well-known solutions of the Schrödinger equation for an arbitrary potential to generate the density matrix adjacent to a contact at the end of any device. When the potential is constant, it is known that the solutions are plane waves. These plane waves are utilized in any standard introductory discussion of tunneling [10]. Another standard and basic tenet of quantum mechanics is that any convenient basis set may be used to solve the Schrödinger equation, as the latter is linear, even in the presence of potentials. Thus, it is certainly valid to utilize the plane waves used near the contacts to propagate completely through the regions where the potential varies. The continued orthonormality of these plane waves is guaranteed by the Lipmann–Schwinger equation [12, 13]. Thus, waves entering the device from the left-hand side can be represented by

$$\psi(x) = A\sqrt{\frac{1}{2\pi}} \begin{cases} e^{ikx} + r(k)e^{-ikx}, & x \ll 0, \\ t(k)e^{ikx}, & x \gg 0. \end{cases} \tag{11.33}$$

Here, it is assumed that the region where the potential is varying is located in the vicinity of $x \sim 0$. The coefficients $r(k)$ and $t(k)$ are the reflection and transmission coefficients, both of which build up during the transit of the potential region. Similarly, waves entering from the right-hand side of the device can be represented by

$$\psi(x) = B\sqrt{\frac{1}{2\pi}} \begin{cases} e^{-ikx} + r(k)e^{ikx}, & x \gg 0, \\ t(k)e^{-ikx}, & x \ll 0. \end{cases} \tag{11.34}$$

The two coefficients A and B are actually the amplitude of the Fermi–Dirac distribution at the appropriate energy (wave vector k). Thus, the density matrix may be constructed as follows:

$$\rho(x, x') = \frac{1}{Z} \sum_n \psi_n^\dagger(x) \psi_n(x') f_{\mathrm{FD}}[E(k_n)]. \tag{11.35}$$

In this expression, Z is the partition function, which may be expressed as [12]

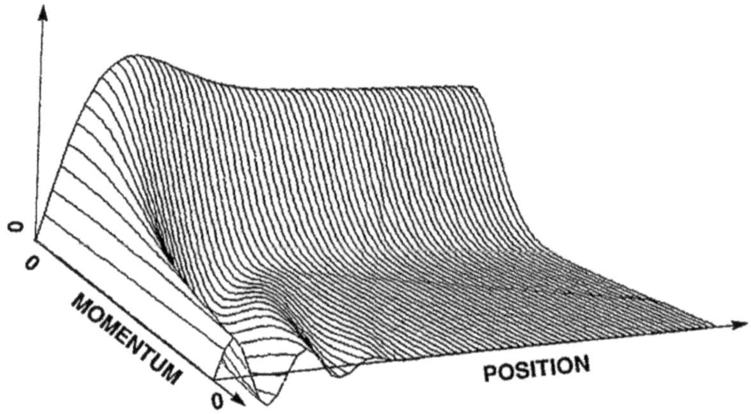

Figure 11.2. The Wigner function for an infinite barrier in the region $x \leqslant 0$. The amplitude is in relative units. The perspective is chosen to illustrate the quantum repulsion near the barrier. Reprinted with permission from [12], Copyright (1987) by the American Physical Society.

$$Z^{-1} = \lambda_D \lim_{x \to \pm\infty} [\rho(x, x)e^{\beta V(x)}], \tag{11.36}$$

where λ_D is the thermal de Broglie wavelength (10.5) that arises in discussions of the effective potential.

Once the density matrix is found, it can be transformed into the Wigner function throughout the device. In figure 11.2, the Wigner function is shown for an infinite potential that is assumed to exist for $x \leqslant 0$, so that the Wigner function exists only for positive x. It is known from introductory quantum mechanics that the first peak in the wave function occurs at a distance from the potential that depends upon the wave number and energy. This is observed in the Wigner function near the potential barrier, where sharp oscillations are observed and the Wigner function's amplitude extends toward larger momenta. In figure 11.3, the Wigner function for a resonant-tunneling diode potential is shown [14]. In both cases, the density and potential are solved self-consistently with one another; the variation in density affects the local potential through the appropriate Poisson equation, and peaks may be seen along $k \sim 0$ in the amplitude of the Wigner function.

11.1.3 Wigner–Weyl methods

Just as in classical mechanics, it is important in quantum mechanics to understand how to reach the expectation values for various operators. In the case of an arbitrary operator O, the expectation value is found from the wave functions according to

$$\langle O \rangle = \int d^3\mathbf{r}\psi^\dagger(\mathbf{r})O(\mathbf{r})\psi(\mathbf{r}). \tag{11.37}$$

In the Wigner formulation, the Wigner function involves the same wave functions but uses the center-of-mass coordinates (10.2). In quantum mechanics, Weyl pointed

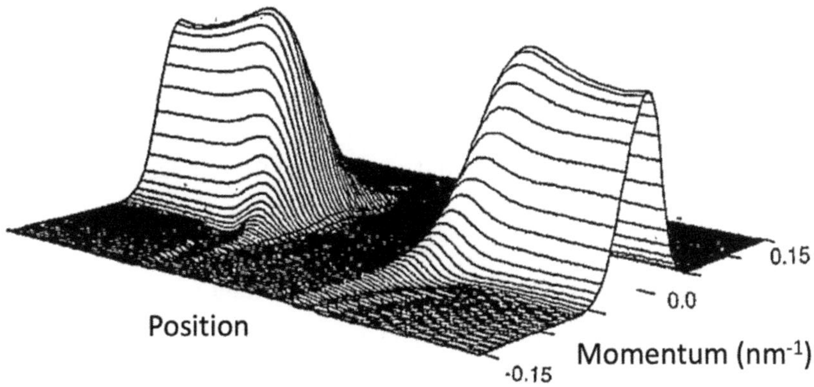

Figure 11.3. The Wigner function for a resonant-tunneling diode in the case of zero bias and for a self-consistent density and potential. The depletion near the barrier (in the center of the image) is due to quantum repulsion from the barrier and this is compensated by a local peak in density represented by the peaks along zero momentum in the figure. Reprinted from Kluksdahl *et al* [14], Copyright (1989) by the American Physical Society.

out that one first has to transform the operator O into the classical phase-space equivalent via

$$O(\mathbf{r}, \mathbf{p}) = \int d^3\mathbf{r}'\psi^\dagger\left(\mathbf{r} + \frac{\mathbf{r}'}{2}\right)O(\mathbf{r})\psi\left(\mathbf{r} - \frac{\mathbf{r}'}{2}\right)e^{-i\mathbf{p}\cdot\mathbf{r}'/\hbar}. \tag{11.38}$$

The expectation value for this operator then becomes

$$\langle O\rangle = \int d^3\mathbf{r}\int d^3\mathbf{p}O(\mathbf{r}, \mathbf{p})f_W(\mathbf{r}, \mathbf{p}). \tag{11.39}$$

To demonstrate that this gives the desired results, (11.4) is introduced:

$$\langle O\rangle = \frac{1}{h^3}\int d^3\mathbf{r}\int d^3\mathbf{p}O(\mathbf{r}, \mathbf{p})\int d^3\mathbf{s}\psi^\dagger\left(\mathbf{r} + \frac{\mathbf{s}}{2}\right)\psi\left(\mathbf{r} - \frac{\mathbf{s}}{2}\right)e^{i\mathbf{p}\cdot\mathbf{s}/\hbar}. \tag{11.40}$$

In general, the operator is at best a slowly varying function of momentum, so that the integration over the momentum is dominated by the exponential producing a delta function on $\mathbf{s}$. Thus, we have

$$\begin{aligned}\langle O\rangle &= \int d^3\mathbf{r}\int d^3\mathbf{s}\psi^\dagger\left(\mathbf{r} + \frac{\mathbf{s}}{2}\right)O(\mathbf{r})\psi\left(\mathbf{r} - \frac{\mathbf{s}}{2}\right)\delta(\mathbf{s}) \\ &= \int d^3\mathbf{r}\psi^\dagger(\mathbf{r})O(\mathbf{r})\psi(\mathbf{r})\end{aligned}. \tag{11.41}$$

When we have two operators, the expectation of their product can be found from an extension of the formula (10.39) as follows:

$$\langle O_1O_2\rangle = \int d^3\mathbf{r}\int d^3\mathbf{p}O_1(\mathbf{r}, \mathbf{p})O_2(\mathbf{r}, \mathbf{p}). \tag{11.42}$$

11.2 Generalizing the Wigner function

It is possible to gain further insight into the nature of the Wigner function by using a generalization of the formulation. It turns out that there are several phase space functions that have arisen since Wigner's original paper [1]. To ease the mathematics, this will be treated in one spatial and momentum direction by generalizing the Weyl operator approach.

11.2.1 The general Weyl operator and the Wigner function

To begin, a general characteristic function $C(\tau, \vartheta)$ is defined by utilizing the Weyl transformation (11.38). This characteristic function is thus given by [15]

$$C(\tau, \vartheta) = \int_{-\infty}^{\infty} dx\psi^{\dagger}(x)e^{i(\tau p+\vartheta x)}\psi(x), \tag{11.43}$$

where p and x in the exponential are the normal quantum mechanical operators. One of these becomes a differential operator when either the position or momentum representation is chosen for the Schrödinger equation. On the other hand, τ and ϑ may be thought of as classical transform variables for these two operators and are properly normalized in that any factors of $\hbar$ are incorporated into these transform variables. The Wigner function can now be obtained from this characteristic function via the double integration

$$f_{\mathrm{W}}(x, p) = \frac{1}{4\pi^2}\int_{-\infty}^{\infty} d\tau \int_{-\infty}^{\infty} d\vartheta C(\tau, \vartheta)e^{-i(\tau p+\vartheta x)}, \tag{11.44}$$

where x and p are now the semiclassical c-numbers associated with the Wigner function. In order to obtain the normal Wigner function from the integrals on the right-hand side of (11.44), the exponential can be rewritten using the Baker–Campbell–Hausdorff formula [16] as

$$e^{i(\tau p+\vartheta x)} = e^{i\tau p/2}e^{i\vartheta x}e^{i\tau p/2}. \tag{11.45}$$

This expansion is now inserted into (11.42) to give the characteristic function

$$\begin{aligned} C(\tau, \vartheta) &= \int_{-\infty}^{\infty} dx e^{-i\tau p/2}\psi^{\dagger}(x)e^{i\vartheta x}e^{i\tau p/2}\psi(x) \\ &= \int_{-\infty}^{\infty} dx\psi^{\dagger}\left(x - \frac{\tau}{2}\right)e^{i\vartheta x}\psi\left(x + \frac{\tau}{2}\right) \end{aligned} \tag{11.46}$$

This result is now introduced into (11.44) to give

$$\begin{aligned} f_W(x, p) &= \frac{1}{4\pi^2}\int_{-\infty}^{\infty} d\tau \int_{-\infty}^{\infty} d\vartheta \int_{-\infty}^{\infty} dx' \\ &\times \psi^\dagger\left(x' - \frac{\hbar\tau}{2}\right)e^{i\vartheta x'}\psi\left(x' + \frac{\hbar\tau}{2}\right)e^{-i(\tau p+\vartheta x)} \\ &= \frac{1}{2\pi}\int_{-\infty}^{\infty} d\tau \int_{-\infty}^{\infty} dx'\psi^\dagger\left(x' - \frac{\hbar\tau}{2}\right)\psi\left(x' + \frac{\hbar\tau}{2}\right)e^{-\tau p}\delta(x - x') \\ &= \frac{1}{h}\int_{-\infty}^{\infty} ds\psi^\dagger\left(x' - \frac{s}{2}\right)\psi\left(x' + \frac{s}{2}\right)e^{-sp/\hbar} \end{aligned} \quad . \quad (11.47)$$

In the last line, the replacement $s = \hbar\tau$ has been made to recover the normal variables seen in the Wigner function.

11.2.2 General uses of the characteristic function

The characteristic function can be used in a variety of situations, including arriving at some alternate formulations of phase space functions. But it can also be extended to discussions of more applications in quantum mechanics. It can be related to an observable, but only through a specific wave function. If an observable with the operator A is considered, it can be used to generate a characteristic function with a specific wave function through the relation

$$C_A(\xi) = \langle\psi|e^{i\xi A}|\psi\rangle, \quad (11.48)$$

where ξ is a real parameter, which could be position, for example. If the operator A has an eigenvalue equation, such as

$$A \mid k\rangle = A_k \mid k\rangle, \quad (11.49)$$

then the characteristic function can be determined as follows:

$$\begin{aligned} C_A(\xi) &= \sum_k \sum_{k'} \langle\psi \mid k\rangle\langle k|e^{i\xi A}|k'\rangle\langle k' \mid \psi\rangle \\ &= \sum_k \sum_{k'} \langle\psi \mid k\rangle e^{i\xi A_k}\langle k' \mid \psi\rangle = \sum_k e^{i\xi A_k} \mid \psi \mid^2 \end{aligned} . \quad (11.50)$$

Of course, there is an integral version of this relation for continuous eigenvalue equations. What is important is that the characteristic equation for an operator is the Fourier transform of the probability distribution of eigenvalues for that operator.

This approach suggests an approach to generating different phase-space functions from operators that appear in the Weyl approach above. But this becomes complicated because of the noncommutative properties of most operators. As an illustration, consider the case for two arbitrary operators A and B, for which a characteristic function can be defined from (11.47) as follows:

$$C_{AB}(\xi_1, \xi_2) = \langle\psi|e^{i(\xi_1 A+\xi_2 B)}|\psi\rangle. \tag{11.51}$$

If the assumption is made that each of these two operators has an eigenvalue equation for a set of basis functions which is different for each operator, according to

$$\begin{aligned} A \mid k\rangle &= A_k \mid k\rangle \\ B \mid k'\rangle &= B_{k'} \mid k'\rangle' \end{aligned} \tag{11.52}$$

then the noncommutative nature of these two operators requires some special assumptions. It is helpful if these two operators satisfy the requirement $[A, [A, B]] = [B, [A, B]] = 0$. The Baker–Campbell–Hausdorff identity can then be used to write the exponential in (11.51) as [16]

$$e^{i(\xi_1 A+\xi_2 B)} = e^{i\xi_1 A}e^{i\xi_2 B}e^{-\xi_1\xi_2[A,\, B]/2}. \tag{11.53}$$

This can now be used to find the characteristic function as follows:

$$\begin{aligned} C_{AB}(\xi_1, \xi_2) &= e^{-\xi_1\xi_2[A,\, B]/2}\sum_k\sum_{k'}\langle\psi \mid k\rangle\langle k|e^{i\xi_1 A}e^{i\xi_2 B}|k'\rangle\langle k'\mid \psi\rangle \\ &= e^{-\xi_1\xi_2[A,\, B]/2}\sum_k\sum_{k'}\langle\psi \mid k\rangle\langle k|e^{i\xi_1 A_k}e^{i\xi_2 B_k}\langle k|k'\rangle\langle k'\mid \psi\rangle \end{aligned}. \tag{11.54}$$

Since the two operators have different eigenvalue equations, there is no delta function in the central function. Nevertheless, this characteristic function can be used to define a corresponding phase-space function

$$\begin{aligned} f(A_k, B_k) &= \langle\psi \mid k\rangle\langle k \mid k'\rangle\langle k'\mid \psi\rangle \\ &= \frac{1}{4\pi^2}\int d\xi_1\int d\xi_2 e^{\xi_1\xi_2[A,\, B]/2}C_{AB}(\xi_1, \xi_2)e^{-i\xi_1 A_k}e^{-i\xi_2 B_k} \end{aligned}. \tag{11.55}$$

The actual form the function takes depends upon the manner in which the last exponentials are expanded. This function differs from the Wigner function by the presence of the first exponential involving the commutator.

As a different phase-space function, consider the form (11.55) and take the two operators as $A \to x$ and $B \to p$. Since the c-number commutator exponential cancels that arising from the use of the Baker–Campbell–Hausdorff relation in the last exponentials, (11.55) can be shown to become

$$\begin{aligned} f(x, p) &= \frac{1}{h}\int ds\psi^\dagger(x)e^{ips/\hbar}\psi(x-s) \\ &= \frac{1}{\sqrt{h}}\psi^\dagger(x)e^{ipx/\hbar}\varphi(p) \end{aligned}. \tag{11.56}$$

In this last expression, the momentum-space wave function has appeared as the last term. Hence, this phase-space function involves both a real-space function and a momentum-space function, along with a phase factor. This will be common in other forms of the phase-space function.

As may be seen from (11.56), the Wigner function is not the only phase-space function that can be imagined. In fact, there remain other possibilities. Some of these can be obtained from the Weyl exponential form used in the above paragraphs. A starting place is the symmetrical form

$$e^{i(\xi_1 x+\xi_2 p)} = \frac{1}{2}(e^{i\xi_1 x}e^{i\xi_2 p} + e^{i\xi_2 p}e^{i\xi_1 x}e^{i\xi_2 p}). \tag{11.57}$$

The detailed ordering of operators is well known to be important in quantum mechanics. It has already been shown to be important in the discussion of Green's functions in chapter 5, but it is just as important in the expansion of the above exponential operators. This is the case even for the linear expansion terms, which give *px* and *xp* products, as these must obey a commutator relationship. Higher-order terms become increasingly more complicated. In the absence of any accepted ordering for these exponential products, several have appeared in the literature. A few are discussed here.

In reviewing the use of the Wigner function, Mehta has suggested a 'standard ordering' that retains only the first term on the right-hand side of (11.57) [17]. In this 'standard' definition, the expansion of the two exponentials has all the powers of x preceding all the powers of p. This does not mean that this is the only choice or even the best choice, only that it is *a* choice. Using this form of (11.57), the characteristic function becomes (the factor of 1/2 is discarded in this ordering)

$$\begin{aligned} C_S(\xi_1, \xi_2) &= \int_{-\infty}^{\infty} dx\psi^\dagger(x)e^{i\xi_1 x}e^{i\xi_2 p}\psi(x) \\ &= \int_{-\infty}^{\infty} dx\psi^\dagger(x)e^{i\xi_1 x}\psi(x+\hbar\xi_2) \end{aligned}. \tag{11.58}$$

This may now be used to determine a phase-space function using

$$\begin{aligned} f_S(x, p) &= \frac{1}{4\pi^2}\iint d\xi_1 d\xi_2 C_S(\xi_1, \xi_2)e^{-i(\xi_1 x+\xi_2 p)} \\ &= \frac{1}{4\pi^2}\iint d\xi_1 d\xi_2 \int dx'\psi^\dagger(x')e^{i\xi_1 x'}\psi(x'+\hbar\xi_2)e^{-i(\xi_1 x+\xi_2 p)} \\ &= \frac{1}{2\pi}\iint d\xi_2 dx'\psi^\dagger(x')e^{-i\xi_2 p}\psi(x'+\hbar\xi_2)\delta(x-x') \\ &= \frac{1}{2\pi}\int d\xi_2\psi^\dagger(x)e^{-i\xi_2 p}\psi(x+\hbar\xi_2) \end{aligned}. \tag{11.59}$$

This latter form differs from (11.56) by a factor of $\hbar$, but this can be recovered by the change of variable $s = \hbar\xi_2$. Nevertheless, this standard ordering produces the same result as (11.56). There is some concern that neither of these two results actually gives the proper normalization [18], but this is certainly not clear here.

If keeping the first term of (11.57) corresponds to the 'standard' ordering, then keeping only the second term, as suggested by Kirkwood [19], must be the anti-standard ordering. In this case, all powers of p precede all powers of x when the exponentials are expanded. With this term retained, the characteristic function becomes

$$\begin{aligned} C_{AS}(\xi_1, \xi_2) &= \int_{-\infty}^{\infty} dx\psi^\dagger(x)e^{i\xi_2 p}e^{i\xi_1 x}\psi(x) \\ &= \int_{-\infty}^{\infty} dx\psi^\dagger(x - \hbar\xi_2)e^{i\xi_1 x}\psi(x) \end{aligned} \quad . \tag{11.60}$$

Following the prescription used above, the phase-space function now becomes

$$\begin{aligned} f_{AS}(x, p) &= \frac{1}{4\pi^2}\iint d\xi_1 d\xi_2 C_{AS}(\xi_1, \xi_2)e^{-i(\xi_1 x + \xi_2 p)} \\ &= \frac{1}{4\pi^2}\iint d\xi_1 d\xi_2 \int dx'\psi^\dagger(x' - \hbar\xi_2)e^{i\xi_1 x'}\psi(x')e^{-i(\xi_1 x + \xi_2 p)} \\ &= \frac{1}{2\pi}\iint d\xi_2 dx'\psi^\dagger(x' - \hbar\xi_2)e^{-i\xi_2 p}\psi(x')\delta(x - x') \\ &= \frac{1}{2\pi}\int d\xi_2\psi^\dagger(x - \hbar\xi_2)e^{-i\xi_2 p}\psi(x) \end{aligned} \quad . \tag{11.61}$$

While this result is also similar to (11.56), it is the complex conjugate of the momentum wave function that is found here. (One introduces the displacement operator to remove the x from the complex wave function and then recognizes the Fourier transform of $\hbar\xi_2$ to obtain the momentum wave function.) Of course, a variety of results can be obtained by keeping different forms of (11.57).

There is still another phase-space function that was suggested by Husimi [20], which arises from the coherent wave packets (11.13). Husimi worried about the fact that these wave packets do not make an orthonormal set suitable for being position basis functions. This is because one cannot make one Gaussian packet orthogonal to another because they are both positive-definite functions. However, despite the fact that the use of orthonormal sets is convenient, there is no requirement in quantum mechanics to use orthonormal sets. Although the mathematics is more difficult with non-orthogonal sets, it is not impossible. Non-orthogonal sets were introduced into linear algebra set theory as 'fuzzy sets' [21]. They have been used by many for quantum mechanics, which already has its own fuzziness in uncertainty. Hence, there is no reason not to use a set of Gaussians centered at different spatial points as a basis set with which to solve problems [22]. The use of these Gaussians is normally separated in space and momentum according to

$$\begin{aligned} \Delta x &= \sqrt{\pi\sigma} \\ \Delta k &= \frac{\sqrt{\pi}}{\sigma} \end{aligned} \quad . \tag{11.62}$$

Here, σ is the standard deviation of the Gaussian (11.13). Husimi then used an additional Gaussian to smooth the density matrix according to

$$\overline{\rho}(x) = \frac{1}{\sqrt{2\pi}\sigma} \int dx' \psi^{\dagger}\left(x' - \frac{x}{2}\right)\psi\left(x' + \frac{x}{2}\right)e^{-(x-x')^2/2\sigma^2}. \tag{11.63}$$

From this density function, the Husimi function is obtained as follows:

$$f_{\mathrm{H}}(x, k) = \int dx' \overline{\rho}(x') e^{-(x-x')^2/8\sigma^2 - ikx'}. \tag{11.64}$$

In essence, the Husimi function is actually a smoothed version of the Wigner function, so that this smoothing removes the negative excursions of the Wigner function. This has the result that when the momentum is integrated out of the Husimi function, the resulting probability function in space is broader than the probability function that would have been obtained from the same integration over the Wigner function.

11.3 The use of particles with the Wigner function

One advantage of the phase-space representation known as the Wigner function is that it facilitates a comparison between complex quantum dynamics and the corresponding classical motion [23, 24]. Because the density matrix can have a non-unitary evolution and off-diagonal terms, it can readily lead to nonclassical propagation in the Wigner function obtained from it. Indeed, quantum coherence is reflected in oscillatory behavior of the Wigner function and nonpositive definite regions in phase space [25]. The classical δ-function representation of a particle evolves in phase space just like a 'spot.' A trajectory generated by the movement of this 'spot' has its own evolutionary behavior and exhibits oscillatory Airy function interference fringes [26]. If a pair of trajectories is considered [27, 28], the oscillatory behavior can be thought of as arising from the interference between this pair. The inclusion of these oscillatory, nonpositive parts of the Wigner function are a challenge to the use of normal particles and must be faced. However, this occurs naturally with the Wigner function, as shown in figure 5.1. The task with particles, however, is to create the nonpositive parts of the Wigner function that illustrate the coherence. There are several methods of achieving this.

11.3.1 Weighted Monte Carlo

One of the earliest usages of particles to simulate transport via the Wigner function occurred in 1992 [29]. In this work, both the 'amplitude' of the particle and its trajectory were weighted. The paths considered for this approach were based upon a series expansion of the path integral for the Wigner function. The problem they addressed was the regions in phase space in which the distribution function has little weight. This means that the particles in the Monte Carlo simulation have low probabilities of reaching these areas and the consequent solutions are very noisy in these phase space areas. In some sense, this approach arises from older ideas of variations in importance sampling to reduce variances in Monte Carlo [30]. This approach also has some similarity to particle splitting [31–34], in which a particle arriving at a low-probability region is split into a number of particles, each of which has reduced weight governed by the number of particles into which the original is split. The weighted Monte Carlo procedure has been shown to be especially useful in the backward Monte Carlo, where one uses the series expansion to guide a propagation path from the final state back to the initial state [29].

The weighted Monte Carlo can also be used for the Wigner equation of motion. The path integral is expanded in a Neumann series [35–37]. The use of this series corresponds to a set of paths that contain different numbers of scattering events. The summation over these paths is similar to a Feynman path integral with the same notoriously slow convergence. This is overcome by the weighted Monte Carlo method—a large number of paths is chosen, each weighted by a suitable factor. Since the paths involve a number of scattering events, each scattering event introduces a phase proportional to

$$2\cos[q(x - x') - \omega_0(t - t')], \tag{11.65}$$

where q and ω_0 are the wave number and radian frequency of the phonon that is emitted or absorbed, respectively, and the spatial and temporal differences correspond to the nonlocal position shift and time duration of the scattering discussed in section 3.1.3.

The problem with this approach arises because a Wigner function may have negative excursions even without the presence of scattering. This is known to occur merely due to the passage of a localized Wigner packet through a tunneling barrier but also when the Wigner function is used to describe two entangled particles [25]. It is hard to fathom how the weighted Monte Carlo can provide the negative values to these problems by using purely phenomenological considerations.

11.3.2 The affinity parameter

In some sense, the *affinity* method might be considered to be an extension of the weighted Monte Carlo method. In this approach, each particle carries an affinity, which may itself be negative. This immediately solves the problem discussed in the last paragraph, since a negative part of the Wigner function is clearly represented by particles whose affinity is also negative. The Monte Carlo approach is set up so that two systems are solved simultaneously. The first system is the particle system, which resembles a standard classical ensemble Monte Carlo (EMC) approach. The second system is the wave properties of the particles—the affinity. That is, all particles in the system are treated classically as particles, but they accompany a wave, just as in the de Broglie view. They are scattered using normal EMC scattering techniques and are drifted and accelerated using the standard field term that derives from the solution of the Poisson equation in the presence of the real potential. Discontinuities in the potential are handled as boundary conditions on the solutions of the Poisson equation. However, these potential discontinuities also affect the particle motion through the Wigner potential term. As in the classical case, the Wigner distribution function is calculated from the particle's position and affinity according to [38, 39]

$$f_W(\boldsymbol{r}, \boldsymbol{p}) \sim \sum_i \delta(\boldsymbol{r} - \boldsymbol{r}_i)\delta(\boldsymbol{p} - \boldsymbol{p}_i)A_i, \tag{11.66}$$

where $\boldsymbol{r}_i$, $\boldsymbol{p}_i$, and A_i are the position, momentum, and affinity of the particular particle, respectively. The sum runs over the ensemble of particles, as is usual.

In the use of the above representation in the Wigner equation of motion, as time evolves, one needs to update both the classical properties (position, momentum, etc.) and the quantum properties (affinity). The first of these updates is done by the normal Monte Carlo technique discussed in chapter 9. The second is the affinity update. As a result of constructing the distribution function, we are then able to utilize the Wigner potential (11.31) to calculate the updated Wigner potential term for the next time step. In a sense, this is a time-splitting approach in which the particle properties are updated, then the quantum properties are updated by the updated potential, and the solution to Poisson's equation is utilized as discussed above. *This new potential now determines the change of affinity that each particle experiences due to the quantum structure in the system at that time.* That is, it is the Wigner potential term of the Wigner equation of motion that updates the wave-like properties of the particles through the update of their affinity. This can be summarized as follows. All particles in the system are drifted, accelerated, and scattered, regardless of their affinities. The particle affinities are changed by the Wigner potential at that point in time. It is clear that all the quantum mechanics is incorporated into the method via the nonlocal potential and the variation of the particle affinity. An alternate interpretation of the method is to recognize that the particles themselves do not see the quantum barriers, only their affinities 'see' the barriers. However, it is important to note that the use of this potential that changes over time does not differ much from the use of the Bohm potential, in which the time changing wave function is used to find the Bohm potential at that point in time.

The affinities are initially assigned to the particles from the 'normalized' Wigner distribution according to

$$A_i = \frac{f_W(\boldsymbol{r}_i, \boldsymbol{p}_i)}{N(\boldsymbol{r}_i, \boldsymbol{p}_i)}, \tag{11.67}$$

where the 'normalization factor' is

$$N(\boldsymbol{r}, \boldsymbol{p}) = \sum_i \delta(\boldsymbol{r} - \boldsymbol{r}_i)\delta(\boldsymbol{p} - \boldsymbol{p}_i). \tag{11.68}$$

That is, the normalization is the number of particles actually in the grid cell at the location of the particle for which the affinity is being determined. This was previously termed the maximal envelope. The affinity is then updated (in time) by the action

$$A_i(t + \Delta t) = A_i(t) + \frac{\theta(\boldsymbol{r}_i, \boldsymbol{p}_i)}{N(\boldsymbol{r}_i, \boldsymbol{p}_i)}, \tag{11.69}$$

where $\theta(\boldsymbol{r}_i, \boldsymbol{p}_i)$ represents the last term in (11.30); e.g. the value of the convolution integral over the Wigner distribution and the Wigner potential in the grid cell of the particle.

While the semiclassical Monte Carlo method appears in this approach, it is important to remember that this is a quantum ensemble Monte Carlo method, in that the full particle nature of the EMC technique is coupled to the full quantum behavior of the Wigner function. Full ensemble statistics are obtained by noting that any ensemble average takes the form

$$\langle O \rangle = \frac{\sum_i A_i O_i}{\sum_i A_i}. \tag{11.70}$$

where O_i value of the dynamic variable O for particle i. In this approach, care must be taken to establish an extended set of boundary conditions, so that not only are the Monte Carlo and Wigner function boundary conditions satisfied but also the normalization (11.68) must have its own set of boundary conditions that are consistent with the former set [40]. This is achieved by randomly distributing particles in the contact region during injection to satisfy the Monte Carlo condition, then randomly distributing these particles in momentum according to the normalization requirement, and finally assigning the particle affinities based on a thermal distribution function in the contact so that charge neutrality at the contacts is met. This last step incorporates not only the Wigner function boundary condition but also the need for charge neutrality. Current is calculated from the probability current, which is determined in the normal manner but with the velocity of each particle weighted by its affinity.

In figure 11.4, a simulation of a Gaussian Wigner function tunneling through a single barrier structure is shown. The structure is composed of GaAs active regions and an AlGaAs barrier 0.3 eV high and 5 nm thick. Several parts of this figure are

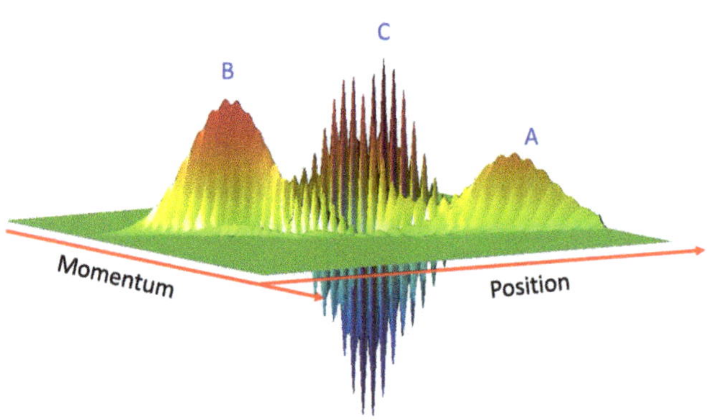

Figure 11.4. A simulation of a single AlGaAs barrier embedded in GaAs at a time just after the main Gaussian has passed the barrier. It is easy to identify the transmitted pulse (A), the reflected pulse (B), and the correlation/entanglement between the transmitted and reflected pulses (C).

interesting. The total Wigner function is shown for a time just after the incident pulse has completed its tunneling process. The transmitted pulse with a positive momentum is denoted by (A). The reflected packet is (B). A rapid oscillatory part (C) remains after the transmitted and reflected packets emerge. In the absence of scattering, both the reflected and transmitted packets are formed from the incident packet. In this sense, the rapid oscillations reflect the entanglement of the transmitted and reflected packets. It is this entanglement that allows the two packets to reform into a single packet if time is reversed [39]. It is quite clear that this Monte Carlo simulation gives rise to extensive negative excursions of the Wigner function which are associated with this entanglement [25].

Recently, the affinity method was used to study challenges related to boundary conditions which might result in nonphysical solutions of the Wigner transport equation [41]. The authors highlighted a limitation in the use of inflow boundary conditions which, particularly in cases of abrupt barriers, require that the contact length has to dictate the minimal wave number uncertainty.

11.3.3 Signed particles

Both the weight and the affinity used in the preceding paragraphs are really artificial numerical quantities whose purpose is to simulate the quantum phase interference that occurs in the Wigner function during particle propagation. A wide range of possibilities can be conceived to solve this problem. An important alternative approach is the 'signed particle' approach [42, 43]. In this approach, the Wigner potential is considered to be a 'scattering' potential, much like phonon scattering. Hence, one introduces the separation of the Wigner potential (11.31) into three subsidiary quantities:

$$\begin{aligned} W^{+}(\boldsymbol{r}, \boldsymbol{p}) &= \max(0, W(\boldsymbol{r}, \boldsymbol{p})) \\ W^{-}(\boldsymbol{r}, \boldsymbol{p}) &= W^{+}(\boldsymbol{r}, -\boldsymbol{p}). \\ \gamma_0 &= \int d^3\boldsymbol{p} W^{+}(\boldsymbol{r}, \boldsymbol{p}) \end{aligned} \quad (11.71)$$

In this separation, the quantity γ_0 plays a role exactly like self-scattering in a normal Monte Carlo process. The Wigner equation of motion (11.30) can then be rewritten as

$$\begin{aligned} \frac{\partial f_{\mathrm{W}}(\boldsymbol{r}, \boldsymbol{p}, t)}{\partial t} + \boldsymbol{v} \cdot \nabla f_{\mathrm{W}} = -\gamma_0 f_{\mathrm{W}} + \frac{1}{h} \int d^3\boldsymbol{p}' f_{\mathrm{W}}(\boldsymbol{r}, \boldsymbol{p}')[W^{+}(\boldsymbol{r}, \boldsymbol{p} - \boldsymbol{p}') \\ - W^{-}(\boldsymbol{r}, \boldsymbol{p} - \boldsymbol{p}') + \gamma_0 \delta(\boldsymbol{p} - \boldsymbol{p}')] \end{aligned} \quad (11.72)$$

Equation (11.72) formally resembles the normal Boltzmann equation with the use of a self-scattering term and can thus be processed in a similar way. Indeed, the integral path equation now contains the exponential term

$$\gamma_0 e^{-\gamma_0(t-t')}, \quad (11.73)$$

which provides a similar accelerative term for the Monte Carlo process. The square brackets of (11.72) contain the three terms that constitute the scattering interpretation. If integrated over the momentum, these terms yield a value of unity. However, the minus sign of the middle term precludes a strict interpretation of this quantity as a probability for the selection of the final state after 'scattering.' The goal is to interpret the signs of the terms as a weight and associate it with the trajectory. Nevertheless, the terms on the right can be given a probabilistic interpretation: with a probability of 1/3 (because of the three terms) the type of interaction with the Wigner potential is chosen to be one of the three terms and then the final state is selected after 'scattering' according to the term chosen. Since the replacement of the original term in (11.30) by that of (11.72) modifies the kernel, this needs to be compensated by a corresponding weight. If the term W^- is chosen, the weight becomes -1. An evolving trajectory then accumulates weight, which after each scattering event is multiplied by three to compensate for the 1/3 in the probability.

In this approach, the weights are accompanied by signs that are associated with them. This process leads to high variance in the results. To correct this, it is necessary to adopt a weight decomposition approach, which limits the value of a weight by storing part of it on a phase-space grid. This significantly improves the variance during simulations performed using this method. This variance problem finally leads to the adoption of a pure sign convention [44, 45]. Since each of the three terms in the bracket of (11.72) corresponds to a form of a probability distribution, they generate three different states after scattering. This corresponds to a branching of the trajectory, which can be conveniently interpreted as the generation of two novel particles. For the δ function term, the final state coincides with the initial state, which means that the interacting particle survives unchanged. As the weight of each particle is now given by its sign, this sign is taken into account in the physical averaging process. The classical sum over the particles now involves a sign prefactor, just like the affinity that appears in (11.66). This gives rise to a very important property of the signed particles. Two particles of opposite sign annihilate each other if they arrive at the same phase space point, and this primarily arises due to their common probabilistic future. The annihilation greatly reduces the computational burden. This creates a model in which the Wigner function is considered to include a Boltzmann-like scattering term, but which now includes a generation term. The quantum information is carried by the sign of the particles. This approach is most useful when treating the interaction with the nonlocal potential as a scattering event. When scattering occurs, even scattering due to the potential, two new particles are created, one with its momentum increased by q and one whose momentum is reduced by q, where q is randomly determined from the probability distribution of the potential's spatial Fourier transform, which is similar to a spectral decomposition method. The sign on one of these new particles is taken to be the same as that of the incident particle, while the sign on the other is the opposite. These signs are taken into account in each averaging process that is used to find average values. Equivalent particles with opposite signs annihilate one another when they meet in phase space. In accordance with the properties of the Wigner equation, the signed-particle approach maintains the antisymmetry of the solution during the evolution. This signed-particle approach was used to compute the Wigner function in figure 5.1.

The signed-particle theory was recently proposed for the treatment of general electromagnetic fields; it was used to devise a gauge-invariant formulation of the transport equation [46, 47], which was then also used in some early results for coherent electron control [48]. These theoretical advances provide a path for an intuitive use of electromagnetic field terms in the transport equation, which is particularly attractive when considering, for instance, a coupling with Maxwell solvers (see section 8.4). A computational approach for investigating coulomb interactions using Wigner–Poisson coupling was also recently developed, which represents a first step towards modeling the transport dynamics of entangled electrons in one to three dimensions with practicable and acceptable simulation runtimes [49].

11.3.4 Spectral methods

Spectral methods highlight the fact that the nonlocal potential creates significant Fourier coefficients. They also address the problem that the potential has values in regions where the Wigner function does not exist (such as outside the device). This creates a numerical problem, as one must ask just how significant these values actually are. While this suggests a cutoff problem, this will not be discussed here. Here, we will discuss the spectral method, following a very recent approach due to Van de Put *et al* [50]. It can be recognized from (11.31) that the potential energy $V(\boldsymbol{r})$ creates a nonlocal potential, and this affects the momentum change of the Wigner distribution. It is important to note from (11.31) that it is not the potential itself but the difference between two versions of the shifted potential at the points $\boldsymbol{r} \pm \boldsymbol{r}'/2$ that is the quantity of interest. From this difference in the two potentials, one can then *define* a version of the classical potential as follows:

$$V\left(\boldsymbol{r} + \frac{\boldsymbol{r}'}{2}\right) - V\left(\boldsymbol{r} - \frac{\boldsymbol{r}'}{2}\right) = -\int_C d^3\boldsymbol{s} \cdot \boldsymbol{F}(\boldsymbol{s}), \tag{11.74}$$

where C is an arbitrary contour from the first position to the second. That is, the potential on the left-hand side arises from the integration of the field over a contour that extends from the point $\boldsymbol{r} - \boldsymbol{r}'/2$ to the point $\boldsymbol{r} + \boldsymbol{r}'/2$. These are explicit points in space and are centered on the central point $\boldsymbol{r}$.

As is known from electromagnetics, the potential is a conservative quantity, so that the choice of the path between the above two points is not critical and can be any continuous path between the two end points. The basis of the spectral method is to introduce the Fourier transform of the force $\boldsymbol{F}(\boldsymbol{s})$, as follows (in one dimension now for clarity):

$$V\left(x + \frac{x'}{2}\right) - V\left(x - \frac{x'}{2}\right) = -\int_C ds \cdot \frac{1}{\sqrt{2\pi}} \int dkF(k)e^{iks}. \tag{11.75}$$

Since the path can be any continuous path between the two end points, it is convenient to take the linear path $s = x + \alpha x'$, in which α runs over the interval $-1/2 \leqslant \alpha \leqslant 1/2$, which is consistent with the two end points of the integral. If this value of s is used in (11.75), one obtains

$$\begin{aligned} V\left(x + \frac{x'}{2}\right) - V\left(x - \frac{x'}{2}\right) &= -\frac{1}{\sqrt{2\pi}} \int dk F(k) e^{ikx} x' \int d\alpha e^{i\alpha k x'} \\ &= -\frac{i}{\sqrt{2\pi}} \int dk F(k) e^{ikx} \frac{\partial}{\partial k} \int d\alpha e^{i\alpha k x'} \end{aligned} . \tag{11.76}$$

In the last line, the substitution of the position operator

$$x' \to i\hbar \frac{\partial}{\partial p} = i \frac{\partial}{\partial k} \tag{11.77}$$

has been used.

With this formulation of the Wigner potential, the potential itself can be rewritten from (11.31) as follows:

$$\begin{aligned} W(x, p) &= -\frac{1}{\sqrt{2\pi}} \int ds e^{isp/\hbar} \int dk F(k) e^{ikx} \frac{\partial}{\partial k} \int d\alpha e^{i\alpha k s} \\ &= \int dk F(k) \frac{\partial}{\partial k} \int d\alpha \delta(p - \alpha\hbar k) \end{aligned} . \tag{11.78}$$

The integral over the delta function could, of course, be taken immediately. However, there is a reason to hold it for future use, as will be seen. This now gives a result that is fully equivalent to (11.31), except that the Wigner potential is now written in spectral form. While the force term in (11.78) appears similar to the linear electric field, this is not the case, and this force contains all the derivatives of the potential as described in (11.74). The next step is to connect this Wigner potential with a generation–recombination process, similar to that of signed particles. To show this, (11.78) is used in (11.30) for the Wigner equation of motion, which gives

$$\begin{aligned} \frac{\partial f_W(\mathbf{r}, p, t)}{\partial t} + \mathbf{v} \cdot \nabla f_W &= \frac{1}{h} \int dp' W(r, p') f_W(r, p + p') \\ &= \int dp' \int dk F(k) \frac{\partial}{\partial p'} \\ &\quad \times \int d\alpha \delta(p - \alpha\hbar k)\, f_W(r, p + p') \\ &= \int dk F(k) \frac{\partial}{\partial p} \int_{-1/2}^{1/2} d\alpha f_W(r, p - \alpha\hbar k, t). \end{aligned} \tag{11.79}$$

Note that the integration over the primed momentum also converts the derivative with respect to momentum that is located outside the last integral. On the other hand, the integral over α can also be carried out to yield

$$\frac{\partial f_{\mathrm{W}}(\mathbf{r}, p, t)}{\partial t} + \mathbf{v} \cdot \nabla f_{\mathrm{W}} = \int dk \frac{F(k)}{\hbar k} \int dp' \frac{\partial f_{\mathrm{W}}(r, p - p', t)}{\partial p}$$

$$= -\int dk \frac{F(k)}{\hbar k} \left[f_{\mathrm{W}}(r, p + \frac{\hbar k}{2}, t) \right. \quad . \tag{11.80}$$

$$\left. -f_{\mathrm{W}}(r, p - \frac{\hbar k}{2}, t) \right]$$

In the limit of vanishing k, the term in the integral on the right-hand side of (11.80) is just the average value of the force field and relates to the classical version of this force field. For other values of k, an equivalent generation (and annihilation) process that transfers weight from one momentum state to another is created within the square brackets, where the generation and annihilation levels are separated by $\hbar k$. For this purpose, the generation process may be defined as [50]

$$G = \frac{F(k)}{\hbar k}. \tag{11.81}$$

This is also the annihilation process weight for the second term in the square brackets of (11.80). A positive value for the force field's Fourier component creates Wigner strength at the higher momentum value and annihilates strength at the lower momentum value, as shown in figure 11.5. A problem arises in that it is thought that a single such process does not conserve probability density. In fact, Van de Put *et al* [50]

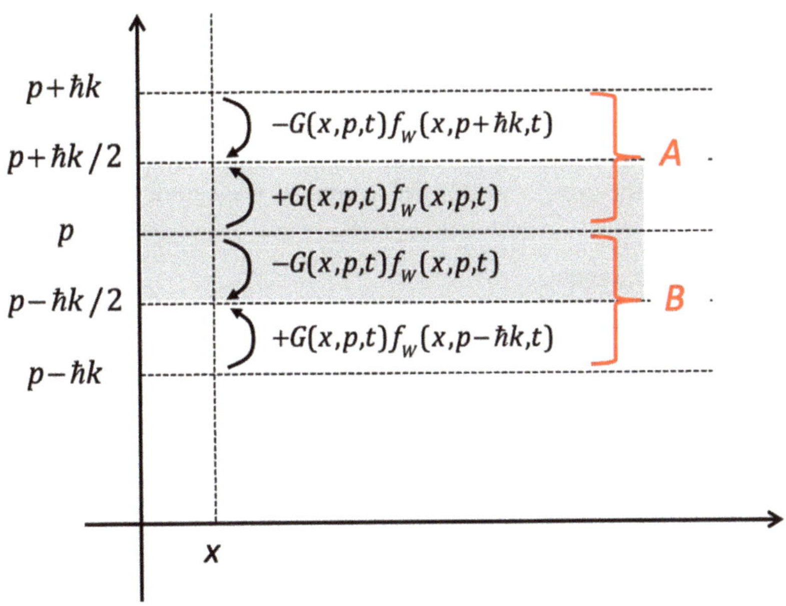

Figure 11.5. The various creation and annihilation processes inherent in (11.80). *A* and *B* represent two symmetric processes that ensure conservation of probability density. The gray area corresponds to the original single process in (11.79). Reproduced from [4]. © IOP Publishing Ltd. All rights reserved.

suggest that the use of two processes, A and B, can achieve this conservation of probability density if they are set up as shown in the figure. With this setup, a symmetric process launches transitions to higher and lower momentums from a central momentum and conserves probability density by construction. The central two processes in the figure represent the original transition, while the outer two assure that conservation is maintained. As noted above, this generation process is quite similar to the concept of signed particles used in the Monte Carlo process, although it is somewhat more detailed. The above process is clear as long as $p > \hbar k$. If the reverse is true, then the creation and annihilation processes move weight from one momentum direction to the opposite momentum direction, i.e. across the zero momentum line. It is necessary to ensure that this does not upset the balance and conservation of the probability density as well as the conservation of current density in the device. Recent advances in spectral methods have included a pseudo spectral scheme [51] and a higher-order accurate operator-splitting method [52].

11.4 Particles in Wigner optics

The use of Wigner functions in optics is old [53]. Particles occur naturally in optical physics: they are called photons. But most treatments of optics have utilized Maxwell's equations, the vector potential in particular. Particle approaches are not naturally used even in quantum optics, although they originated with the work of Planck. On the other hand, in quantum optics there is a recognition that non-Gaussian states can be represented by Fock states, which correlate to the number of photons in a given state [54, 55]. But these photons are not quite the same as the Monte Carlo particles used to simulate an actual photon wave function. In this section, the latter view of particles is developed for use with the photon Wigner function. A particular problem here is the lack of the normal classical potential term which is generally thought of as an electric potential. As photons have no charge, they do not respond to an electric potential. What, then, is their confining potential? In quantum optics, the answer to this lies in optical waveguides.

11.4.1 The optical waveguide

Optical waveguides come in many forms. Here, it is the dielectric waveguide that is of interest. The simplest mathematical model is the slab waveguide, which is also utilized in many applications such as waveguides coupled to ring resonators, both of which are fabricated using silicon technology. In figure 11.6, such a waveguide is illustrated. The waveguide is the rectangular material sitting atop the substrate. In this case, both the substrate and guide are assumed to be Si, with an index of refraction of n_1. The region outside the guide is air with an index of refraction $n_0 \sim 1 < n_1$. In this figure, the x-axis is vertical, the y-axis is horizontal, and the z-axis points into the page along the guide. The lowest waveguide mode is a transverse electric mode (TE_{10}) in which the wave electric field is oriented in the x-direction and varies in the y-direction according to the reduction of the Helmholtz equation to

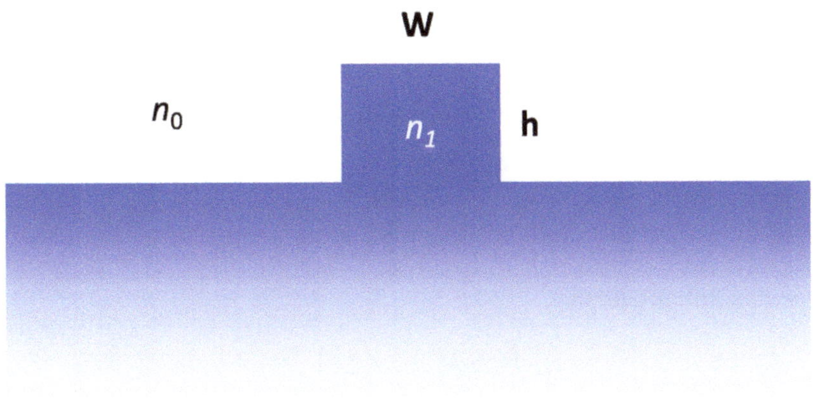

Figure 11.6. A dielectric waveguide (blue), here assumed to be fabricated in Si. The area outside the guide is assumed to be air. The guide has width W and height h. The indexes of refraction n_1 and $n_0 \sim 1$ are those for Si and air, respectively. The width is assumed to be 0.5 μm in the numerical simulations.

$$\frac{d^2E(y)}{dy^2} - n^2E(y) = 0. \tag{11.82}$$

In the waveguide of figure 11.6, the index of refraction may be written as

$$n^2 = 1 + (n_1^2 - 1)\left[\theta\left(y + \frac{W}{2}\right) - \theta\left(y - \frac{W}{2}\right)\right], \tag{11.83}$$

where W is the waveguide width indicated in the figure. The nature of the guide is that the wavelength is such that the wave is confined within the guide. The wave function in this situation is basically the electric field, which is determined from (11.82) to be

$$\psi(y) = E(y) = e^{ik_z z}\begin{cases} Be^{\alpha(y+W/2)}, & y < -W/2, \\ A\cos(k_y y), & -W/2 \leqslant y \leqslant W/2, \\ Be^{-\alpha(y-W/2)}, & y > W/2. \end{cases} \tag{11.84}$$

The remaining question concerns the values of k_y, k_z, and α. This requires some algebra. In free space, the frequency and wave number are related by

$$\omega^2 = c^2k_0^2, \tag{11.85}$$

where $c^2 = \epsilon_0\mu_0$, in metre–kilogram–second (MKS) units. When a dielectric guide is used, this equation is modified to

$$\omega^2\left(\frac{\epsilon_1}{\epsilon_0} - 1\right) = c^2k_0^2(n_1^2 - 1), \tag{11.86}$$

and is coupled to the cutoff condition $\lambda/2 = W$, or

$$\frac{k_0 W}{2}\sqrt{n_1^2 - 1} = \frac{\pi}{2}. \tag{11.87}$$

At cutoff, $k_z = 0$, so there is no propagation down the guide. Two of the eigenvalues required satisfy

$$\alpha = k_y \tan\left(\frac{k_y W}{2}\right). \tag{11.88}$$

The general solutions require

$$\begin{aligned} c^2\left(k_y^2 + k_z^2\right) &= \omega^2\left(\frac{\epsilon_1}{\epsilon_0}\right), \\ c^2(-\alpha^2 + k_z^2) &= \omega^2. \end{aligned} \tag{11.89}$$

After we subtract the second equation from the first and use (11.88), the frequency and lateral wave number are related by the transcendental equation

$$\frac{\omega}{c} = k_y \sqrt{\frac{\epsilon_0}{\epsilon_1 - \epsilon_0}} \sqrt{1 + \tan^2\left(\frac{k_y W}{2}\right)}. \tag{11.90}$$

The solution for this is shown in figure 11.7 for a Si waveguide in which $W = 0.5$ μm. Once this is determined, then the wave number along the guide, k_z, can be determined from the first line of (11.89). This is shown in figure 11.8, again plotted versus the radian frequency, with the latter on the vertical axis.

The Wigner function for the waveguide is directly determined from the wave function (11.84) using the Wigner transform. This gives

$$f_W\left(y, k_y\right) = \begin{cases} B^2 e^{\alpha(2y+W)}, & y < -W/2, \\ \frac{A^2}{4\pi}\left\{\sin\left[k_y(W + 2y)\right] + \sin\left[k_y(W - 2y)\right]\right\}, & -W/2 \leqslant y \leqslant W/2, \\ B^2 e^{-\alpha(2y-W)}, & y > W/2. \end{cases} \tag{11.91}$$

The forms in (11.91) neglect some smaller complex parts, so that this equation only contains the dominant parts of the Wigner function. One may note that the Wigner function becomes completely enclosed within the waveguide at cutoff.

11.4.2 The ring resonator

In many quantum optical applications, both a linear waveguide and a circular waveguide (the ring resonator) are used. Such a device depends upon the evanescent coupling between the guide and the resonator, or, to use more familiar language, tunneling between the two guides. As seen above in (11.84), the electric field has an exponential component outside the guide, and a similar behavior is found in the circular guide used for the ring resonator. The overlap between these two evanescent

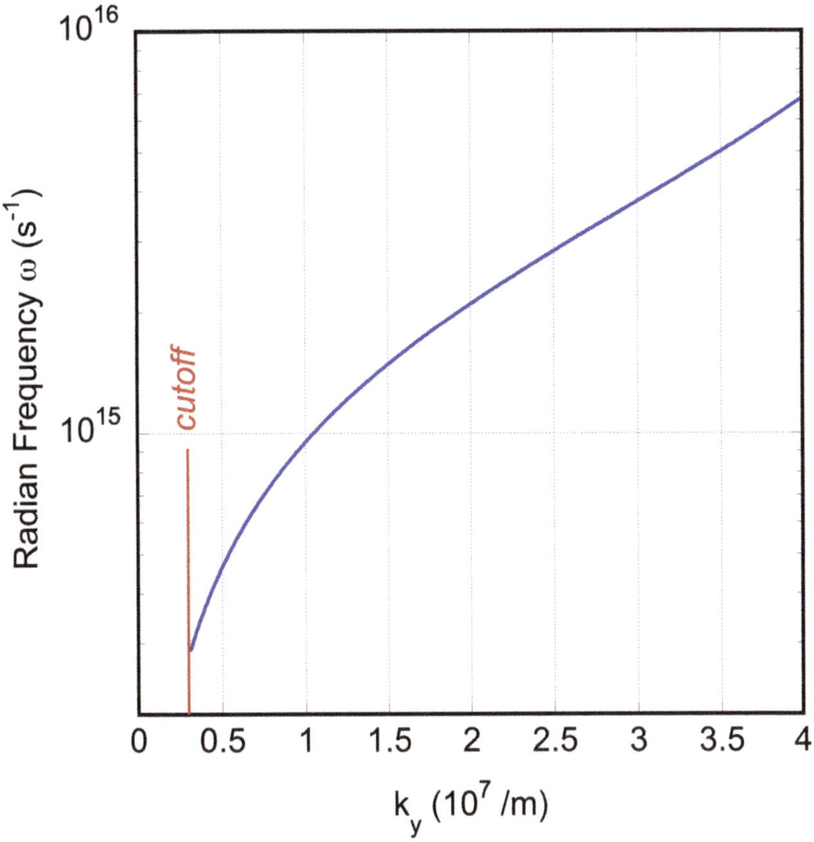

Figure 11.7. The frequency of the guided wave is plotted versus the transverse wave number k_y. The cutoff wave number is indicated by the vertical red line; no guided waves can propagate for wave numbers below this point.

parts of the field provides the tunneling effect. However, the ring is a bounded system, and therefore has a discrete eigen-spectrum. The Sturm–Liouville problem may be solved and has been analytically evaluated, but the eigen-spectrum can be inferred from the fact that any bound state of the ring must have a stationary phase around the ring. That is, just as for electrons in orbits around an atom, the wave momentum around the ring must be limited to integer values of its wavelength. Its eigenvalues are then given by

$$k_n = \frac{2n\pi R}{\lambda_g}, \tag{11.92}$$

where λ_g is the central wavelength of the guide and is determined by the radius R of the ring. Due to the evenly spaced eigen-spectrum of the ring, the spectral density is often referred to as an optical frequency *comb*. This is an important feature for nonlinear optics and requires the dispersion of the waveguide (see figure 11.8) to be in the linear regime and to have a nearly constant group velocity. This allows for

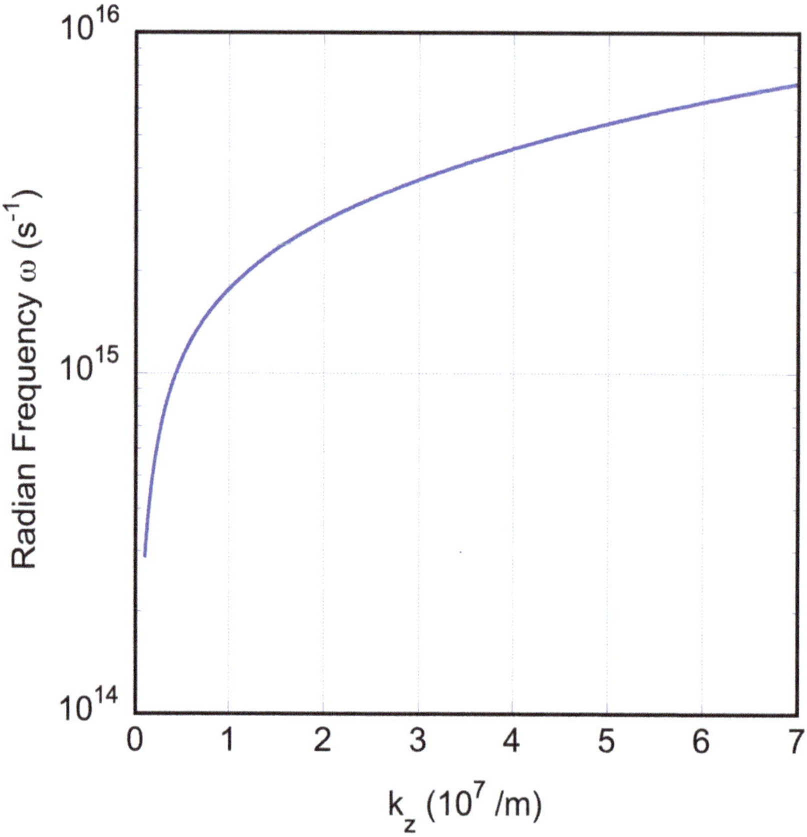

Figure 11.8. The longitudinal wave number along the guide is plotted here for given frequencies. The slope of this curve is the group velocity in the guide.

matching velocities, as needed in four-wave mixing, in which two incoming photons are parametrically converted to two photons at the sum and difference frequencies. This can be used, for example, to create two entangled photons on widely separated sidebands. Such a ring is illustrated in figure 11.9 [56]. The two guides shown are both 0.5 μm wide and are spaced at 0.25 μm. This ring-resonator combination is assumed to be for a free-space wavelength of 1.5 μm. Here, the Si guides are assumed to be coated with SiO_2 to avoid degradation due to surface oxidation.

Nonlinear optics depends upon the guide having an electric-field-dependent dielectric function. Generally, the polarization of a dielectric material is proportional via a constant to the electric field, as follows:

$$\boldsymbol{P} = \epsilon_0 \chi^{(1)} \boldsymbol{E}, \tag{11.93}$$

where $\chi^{(1)}$ is the linear susceptibility. This leads to the normal idea of a dielectric constant below:

$$\epsilon = \epsilon_0 (1 + \chi^{(1)}) = \epsilon_0 \epsilon_r. \tag{11.94}$$

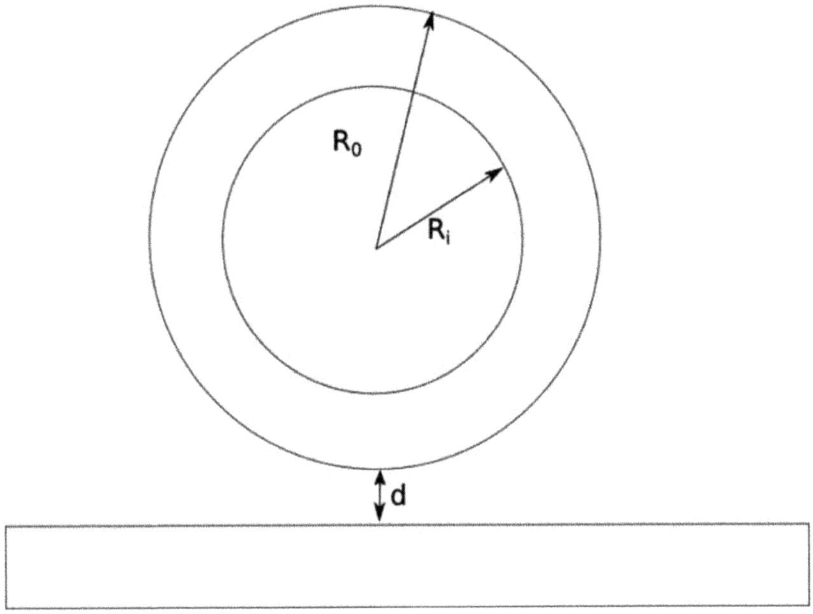

Figure 11.9. A typical ring-resonator combination. The inner ring diameter is 4.75 μm and the outer diameter is 5.25 μm. The spacing d is 0.25 μm. Reprinted from [56], with the permission of AIP Publishing.

This, of course, fails in the nonlinear regime in which quantum optics is mostly applied. The total susceptibility may be expanded in the electric field, leading to a very different polarization, as follows:

$$\boldsymbol{P} = \epsilon_0[\chi^{(1)}\boldsymbol{E} \ + \chi^{(2)}\boldsymbol{E}\boldsymbol{E} + \chi^{(3)}(\boldsymbol{E}\cdot\boldsymbol{E})\boldsymbol{E} + \cdots], \tag{11.95}$$

where it has been assumed that the material is isotropic. This is often not the case, so that $\chi^{(2)}$ is actually a dyadic (second-rank) tensor, $\chi^{(3)}$ is a third-rank tensor, and so on. The ring resonator will be utilized in the simulations below.

11.4.3 Monte Carlo methods

The Monte Carlo approach is based upon the use of particles that follow quasi-classical trajectories similar to those of the normal electron Monte Carlo. Technically speaking, however, these particles are taken to be small Gaussians corresponding to phase-space wave functions. The simulation begins with the initial wave function, or initial Wigner function. This distribution of particles is allowed to drift a short distance and the initial quantization is enhanced through the pair correlation function, which is supposed to fit the form

$$\sum_{i,j}\cos(\boldsymbol{p}_i\cdot\boldsymbol{r}_j - \boldsymbol{p}_j\cdot\boldsymbol{r}_i)f_W(z-\langle z\rangle, \boldsymbol{p}-\langle\boldsymbol{p}\rangle), \tag{11.96}$$

where the Wigner function is the Gaussian of (11.14). If N particles are used, there are $\sim N^2$ pairs. It is found that using less than 10^5 particles leads to significant noise.

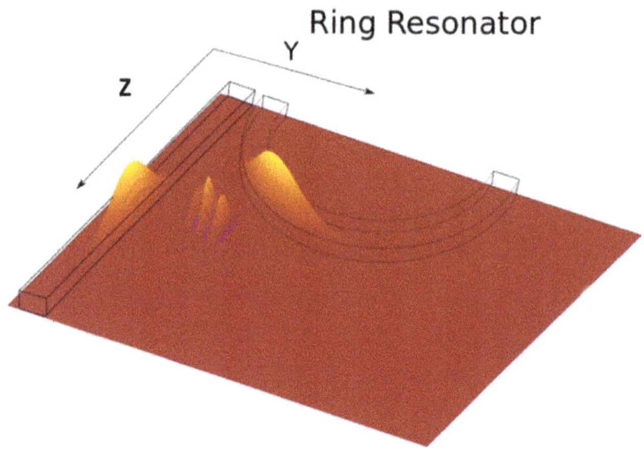

Figure 11.10. Early propagation of the waveguide pulse and tunneling into the ring. The z-direction is taken to be along the waveguide, as described in the text, while the y-direction is normal to this propagation direction. The plot covers a 13 × 13 μm region. Reprinted from [56], with the permission of AIP Publishing.

The use of larger numbers means that the above correlation function needs to be under-sampled for numerical efficiency. The results here use 2×10^6 particles, and these are assumed to be generated in the waveguide by a 1.15 fs semiconductor laser pulse.

The key to the optical simulations is the Wigner potential. Here, the potential is assumed to arise from the dielectric constants that describe the waveguides and their cladding (if any). However, the confinement is low where n is large and higher where n is low. While the shape is determined by the spatial dependence of the dielectric function, the potential in the Wigner sense must be governed by the inverse of the dielectric function. That is, the Wigner potential (11.31) becomes [57]

$$W(y, p) = \frac{V_0}{i} \int ds e^{isp/\hbar} \left[\frac{1}{n^2\left(y + \frac{s}{2}\right)} - \frac{1}{n^2\left(y - \frac{s}{2}\right)} \right], \tag{11.97}$$

and there is a z-variation as the two guides become further apart away from the point of closest approach. Tunneling from the waveguide to the ring is assumed to incur a time delay given by d/v_y, where d is the separation of the ring from the guide and v_y is the group velocity in the cladding material (or air). There is an additional complication for the particles that couple to the ring in that they must now follow curved trajectories around the ring. In figure 11.10, the total Wigner function is shown shortly after the incident pulse in the guide has passed the tunneling region. The portions in the ring and in the correlation/entanglement are clearly observable.

11.5 Scattering with Wigner functions

The generally accepted formulation used to add scattering terms to the Wigner equation of motion derives from a form developed by Levinson [58]. He generally worked out the situation in the presence of both an electric field and a magnetic field, although he ignored the diffusion term as a simple addition. The magnetic field will not be considered here, although it can easily be incorporated via the Peierls' substitution [59]. Levinson observes that any matrix elements obtained by parallel transport through the variables $\boldsymbol{r}$ and $\boldsymbol{r}'$ contain no new information about the system, *if the system is homogeneous*, which he assumes. This becomes apparent when it appears as the Wigner function

$$\begin{aligned} f_{\mathrm{W}}(\boldsymbol{p}, t) &= \frac{1}{h^3} \int d^3 s e^{i\boldsymbol{p}\cdot\boldsymbol{s}/\hbar} \left\langle \boldsymbol{r} + \frac{\boldsymbol{s}}{2} \mid \rho(t) \mid \boldsymbol{r} - \frac{\boldsymbol{s}}{2} \right\rangle \\ &= \frac{1}{h^3} \int d^3 s e^{i\boldsymbol{p}\cdot\boldsymbol{s}/\hbar} \left\langle \frac{\boldsymbol{s}}{2} \mid \rho(t) \mid -\frac{\boldsymbol{s}}{2} \right\rangle \end{aligned}, \tag{11.98}$$

where $\rho(t)$ is the density matrix appearing in (11.3). This leads to an important observation: in the homogeneous system, the Wigner function does not depend upon $\boldsymbol{r}$ and $\boldsymbol{p}$ separately but only depends upon them through the kinetic momentum that includes the magnetic field

$$\hbar \boldsymbol{k} = \boldsymbol{p} - e\boldsymbol{A}(\boldsymbol{r}). \tag{11.99}$$

Even this becomes moot if we do not consider the presence of any magnetic field (which is the usual case in semiconductor devices).

Here, only the true scattering processes are of interest, and these are assumed to be representable by a function S. This collision function is very different from the classical form, and the scattering term in the Wigner equation of motion can be written in greater detail as

$$S\left[f_{\mathrm{W}}(\boldsymbol{k}, t)\right] = \int_0^t dt' \int d^3 k' f_{\mathrm{W}}(\boldsymbol{k}', t) I(\boldsymbol{k}, \boldsymbol{k}'; t, t'). \tag{11.100}$$

The kernel term I within the integral involves four separate terms: there are terms for the emission and absorption of a phonon leaving state **k** and terms for emission and absorption from another state into state **k**. These, of course, are the *out*-scattering and *in*-scattering terms we discussed earlier. In addition, the scattering kernel must be summed over the types of phonons present in the system. This much is normal. What is new is the finite collision time and the nonlocality of the scattering in quantum systems. The out-scattering terms for emission and absorption may be written as [58]

$$\begin{aligned} I(\boldsymbol{k}, \boldsymbol{k}'; t, t') &\to A^{(\pm)}(\boldsymbol{k}, \boldsymbol{k}'; t, t') \\ &= \frac{2\pi}{\hbar} \int d^3 q \mid M(\boldsymbol{k}, \boldsymbol{q}) \mid^2 \Delta\left(\boldsymbol{k} \mp \frac{\boldsymbol{q}}{2}, \boldsymbol{k}' \pm \frac{\boldsymbol{q}}{2}; t, t'\right) \end{aligned} \tag{11.101}$$

where $\mid M(\boldsymbol{k}, \boldsymbol{q}) \mid^2$ is the normal matrix element for a scattering process and the joint spectral density is given by

$$\Delta(\boldsymbol{k}, \boldsymbol{k}'; t, t') = \frac{1}{h}\int d^3\boldsymbol{x} \int d^3\boldsymbol{x}' e^{i[\omega_0(t-t') - \boldsymbol{q}\cdot(\boldsymbol{x}-\boldsymbol{x}')]} \\ \times\ \delta(\hbar(\boldsymbol{k} - \boldsymbol{k}') - e\boldsymbol{F}(t - t'))e^{i[(\boldsymbol{k}+\boldsymbol{k}')\cdot(\boldsymbol{x}-\boldsymbol{x}')/2]} . \tag{11.102}$$

The in-scattering terms $B^{(\pm)}(\boldsymbol{k}, \boldsymbol{k}'; t, t')$ are the same as (11.101) except that the $\mp$ and $\pm$ are reversed in their positions within the various momenta and, of course, there is a difference in the forms of the Bose–Einstein term corresponding to emission and absorption (in the matrix element). Obviously the time difference accounts for the duration of the collision, and the spatial difference accounts for the nonlocality of the collision. Both of these effects were discussed in section 3.1.3. This latter section described numerical techniques which could be used with particles, each of which accounts for either the spatial or temporal needs of the joint spectral density function (11.102). These needs arise as scattering occurs between states that are evolving in the applied field $\boldsymbol{F}$ and not between plane waves of a free electron. This scattering between evolving states has often been called the intracollisional field effect [60, 61]. This effect reflects the evolution of the states during the actual collision process over a time termed the collision duration. Other approaches than the analytical approach described here have been used for particles, the most important of which were discussed in chapter 3. In [62], surface roughness models in quantum wires were investigated by means of a signed particle approach. It was shown that a quantum picture is essential in order to correctly describe the impact of surface roughness on the transport dynamics.

11.6 Device simulation with Wigner particles

Simulating electronic devices with Wigner functions using particle methods has a long tradition [4, 5, 63]. Quite often, the prototypical quantum device is the resonant tunneling diode (RTD), and this device has been treated by quantum methods for more than half a century. The concept of this device was formulated by Esaki and his colleagues [64]. It remained an interesting physics problem for many years, but more interest arose after it was shown that the device would exhibit oscillations in the terahertz regime [65]. In this device, two tunneling barriers (typically GaAlAs) surround a quantum well (typically GaAs). At least one bound state is found in the quantum well. From introductory quantum mechanics, it is known that the transmission coefficient is unity for energies at the bound state (the resonant state) and for a symmetrical tunneling barrier (under bias). Layers of GaAs are grown on either side of the tunneling structure to provide connections to the actual contacts. Since the actual form of the barrier changes with the voltage applied across the device, simulations have to include solutions of Poisson's equation in order to maintain the self-consistency of the potential and the charge. Hence, the Wigner potential changes as the bias changes, and this affects the transport equation [15, 66]. Several particle simulations have used the Wigner function for the RTD. One of the earliest approaches used particles only in the cladding area and employed quantum transmission probabilities to decide whether or not a particle passed the barrier [67].

Hence, the tunneling was actually treated as a scattering process, which was suggested earlier by Barker [68].

A full quantum simulation for the RTD, incorporating the affinity approach to handle the particles within the quantum barriers, appeared in the late 1980s [41, 69]. In this study, the tunneling structure was composed of a 5 nm undoped GaAs quantum well and 3 nm AlGaAs barriers 0.3 eV high. The tunneling structure was clad with 5 nm undoped spacer layers on either side to focus the applied bias onto the quantum region. The normal doping of 10^{18} cm^{-3} was maintained outside this portion of the device. A simulation length of 80 nm was used. In the early studies of this device, scattering was not incorporated, so that the basic quantum properties could be studied. The current–voltage characteristic that resulted from this structure is shown in figure 11.11(a). Negative differential conductance is clearly evident in the form of the current 'valley' in the curve at 0.2–0.3 V. The ratio of the peak current (at ~0.16 V) to the valley minimum (at ~0.2 V) is known as the peak-to-valley ratio (used as a characterization of the device), and is ~2 for this device. The Wigner distribution is shown in figure 11.11(b) for a voltage bias near the peak of the curve. There is a significant negative excursion of the Wigner function to the left of the barrier (labelled 'A'). This is the input side (electrons are incident from the left), and this minimum represents the true quantum correlation. One can also see a build-up of charge in the well region (labelled 'B').

Scattering was introduced into the device within both the quantum region and the cladding regions. The detailed emission and absorption of polar-optical phonons in the GaAs and GaAlAs regions were included, as well as other scattering processes [40]. Boundary conditions remain critical, especially to ensure that the current and electron number are both conserved. Because of the scattering, more time is required to reach a steady state at each bias point, but this steady state is necessary to ensure

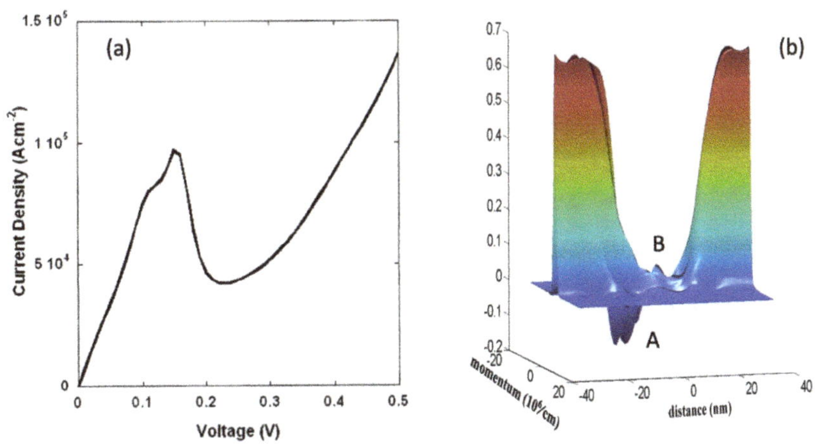

Figure 11.11. (a) The current–voltage characteristic for the RTD described in the text, taken at 300 K. Reprinted from [69], Copyright (2002), with permission from Elsevier. (b) The Wigner function for a bias near the peak (~1.6 V) of the current in panel (a). This clearly shows a negative excursion that represents correlation (A) and a peak of charge in the well (B).

the self-consistency that is required of the potential and the charge. The bias voltage is swept upward and then reduced stepwise back to zero to check for any hysteresis, especially in the region of negative differential conductance. While hysteresis may be expected because of the difference in the charge in the well between the up-sweep and the down-sweep, none was found. A longer device of 150 nm is considered in this case. In figure 11.12(a), the current–voltage characteristic is shown for both the up-sweep and the down-sweep of the potential. The peak-to-valley ratio is smaller with the presence of scattering, since phonon-assisted tunneling can take place, and the nonlocal character of the scattering actually helps particles to surmount the barrier. This peak-to-valley ratio is now about 1.5, which is closer to that observed in experimental devices. In figure 11.12(b), the distribution of charge and potential is shown for a bias of 0.225 V, which is approximately at the current peak shown in panel (a). Clearly, there is a depletion on the input (left) side of the device due to the lower doping near the barrier and the tendency for the actual potential to create a small well in this region. The actual Wigner function at this peak bias is shown in figure 11.13. In this figure, the color scale from red to blue represents high to low values of the relative amplitude of the Wigner function. The ragged top of the potential near zero momentum is actually due to noise in the simulation. While this is an instructive figure, variations near and within the quantum structure are much more difficult to pick out in such a plot.

Wigner functions, coupled with the particle Monte Carlo technique, have been used to simulate a number of semiconductor devices. One of the earliest studies continued to simulate the prototypical RTD [70]. The RTD was also studied by Koba *et al* [71]. The former (French) group also studied a nanoscale double-gate metal-oxide semiconductor field-effect transistor (MOSFET) [72, 73]. The source and drain regions were heavily doped to degenerate levels, while the channel region was undoped. The gate length was 6 nm, and the Si channel layer was 3 nm thick. Results were obtained that compared quite favorably to other approaches and device saturation was found even when approaches such as nonequilibrium Green's functions (NEGFs) did not show this effect. The sub-threshold behavior was also

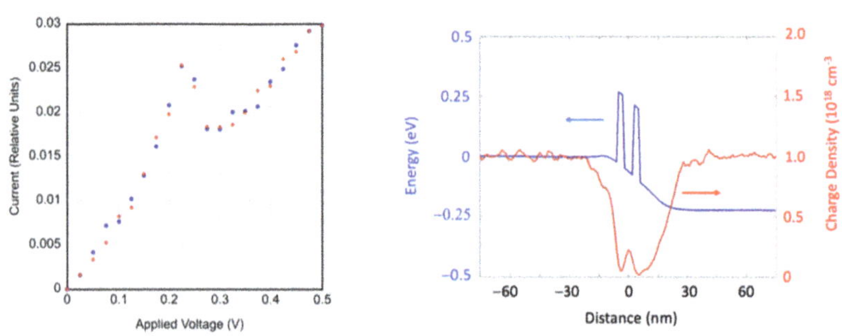

Figure 11.12. (a) (Left) The current–voltage characteristic of the RTD at 300 K with scattering included. The blue dots represent the up-sweep and the red diamonds represent the down-sweep. The small differences are due to the current noise present in the simulation. (b) (Right) The spatial distribution of the potential (energy) and the charge itself for an applied bias of 0.225 V, which is approximately at the peak shown in the left panel. The small bump at zero is the charge in the well. Depletion is apparent to the left of the first barrier.

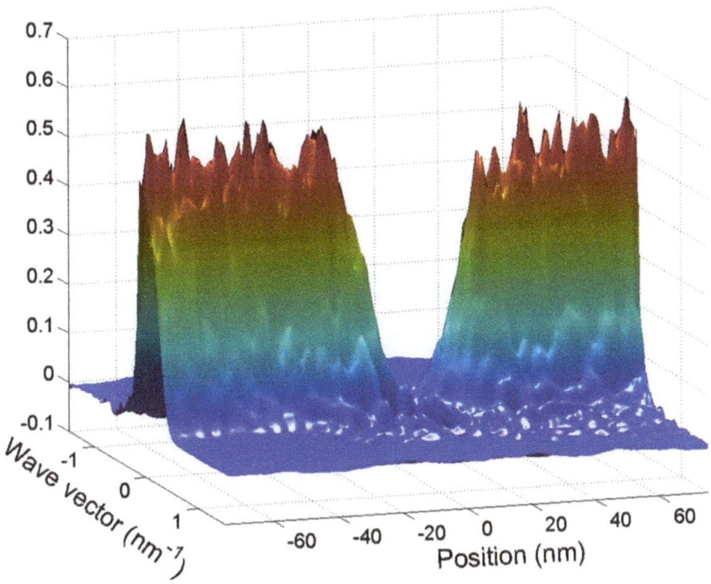

Figure 11.13. The Wigner distribution for an applied bias of 0.225 V, near to the peak in the *I–V* curve for the device. The ragged top is a representation of current noise due to the simulation, and no attempt to smooth the plot has been made.

fairly well predicted (this is often difficult due to the small number of particles required for the sub-threshold current). It was found that the lowest sub-band dominated the current through the device. Both ballistic transport and scattering due to impurities and acoustic phonons were important in the transport. An important point was that the current-carrying carriers did *not relax* even well into the drain region, a common effect in small devices [74, 75].

RTDs are studied to this day to advance Wigner function-based solution approaches, e.g. [76–80]. In addition to RTDs, Wigner function approaches have also been applied to study field-effect transistors [75, 81–84] (for a recent review of quantum transport in field-effect transistors and the importance of particle Wigner approaches, see [85]), transport in quantum cascade lasers [86], and thermoelectric properties in periodic quantum structures [87]. Wigner transport approaches also lend themselves to material studies, such as those of graphene [88–91] and topological insulators [92].

In recent work, a signed-particle Wigner approach was used to propose a gate-controlled electron quantum interference logic structure [93]. The fundamental principle is based on electrostatically controlling the single-electron charge interference pattern of coherently injected electrons by top-mounted gates (see figure 11.14). It was shown that logic gates can be realized by guiding the interference maxima/minima to specific output gates. The main quantum transport process of charge interference can be naturally modeled by the chosen particle Wigner approach, which also allows for practical simulation runtimes of the necessary coherent two-dimensional (2D) transport dynamics simulations.

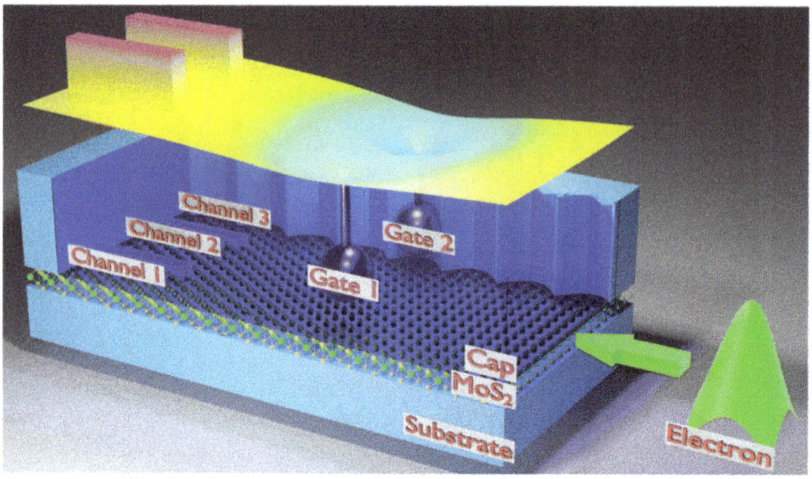

Figure 11.14. (Left) A schematic representation of the electron-interference logic structure. An exemplary potential energy distribution of the waveguide derived from biased gates and the structure is shown on top. The blue cap layer is made partially transparent to show the inside. (Right) The distribution of quantum electron density (in arbitrary units) for two asymmetric gate configurations. Green and yellow circles indicate the positively and negatively charged gates, respectively. Reproduced from [93]. CC BY 3.0.

In recent work in the field of *electron quantum optics* (for reviews on this topic, see [94, 95]), Wigner signed-particles were used to study the transport dynamics of a confined electron in a tunable barrier (including a comparison with the Schrödinger solution) and its particle ejection behavior [96]. It was shown that the ejection timing has a direct impact on the ejected electron's state shape.

References

[1] Wigner E P 1932 *Phys. Rev.* **40** 749
[2] Weyl H 1931 *Group Theory and Quantum Mechanics* (Boston, MA: E. P. Dutton)
[3] Moyal J E 1949 *Proc. Camb. Phil. Soc.* **45** 99
[4] Ferry D K and Nedjalkov M 2018 *The Wigner Function in Science and Technology* (Bristol: IOP Publishing)
[5] Weinbub J and Ferry D K 2018 *Appl. Phys. Rev.* **5** 041104
[6] Weinbub J, Everitt M and Ferry D K 2021 *J. Comput. Electron* **20** 2019
[7] Kennard E H 1928 *Phys. Rev.* **31** 876
[8] Ballentine L E 1998 *Quantum Mechanics: A Modern Development* (Singapore: World Scientific)
[9] Groenewold H J 1946 *Physica* **12** 405
[10] Ferry D K 2021 *Quantum Mechanics* 3rd edn (Boca Raton, FL: CRC Press)
[11] Carruthers P and Zachariasen F 1983 *Rev. Mod. Phys.* **55** 245
[12] Kriman A M, Kluksdahl N C and Ferry D K 1987 *Phys. Rev.* B **36** 5953
[13] Merzbacher E 1970 *Quantum Mechanics* 2nd (New York: Wiley)
[14] Kluksdahl N C *et al* 1989 *Phys. Rev.* B **39** 7720
[15] Iafrate G J, Grubin H L and Ferry D K 1982 *Phys. Lett.* **87A** 145
[16] Messiah A 1961 *Quantum Mechanics* 1st edn (New York: Wiley) pp 442

[17] Mehta H 1964 *J. Math. Phys.* **5** 677
[18] Lee H-W 1995 *Phys. Rep.* **259** 147
[19] Kirkwood J G 1933 *Phys. Rev.* **44** 31
[20] Husimi K 1940 *Proc. Math. Soc. Jpn.* **22** 246
[21] Zadeh L A 1967 *Inf. Control* **8** 338
[22] von Neumann J 1955 *Mathematical Foundations of Quantum Mechanics* Beyer R T transl. (Princeton, NJ: Princeton University Press) https://press.princeton.edu/books/ebook/9781400889921/mathematical-foundations-of-quantum-mechanics
[23] Heller E J 1976 *J. Chem. Phys.* **65** 1544
[24] Dittrich T, Viviescas C and Sandoval L 2006 *Phys. Rev. Lett.* **96** 070403
[25] Zurek W H 2001 *Nature* **412** 712–7
[26] Berry M V and Balazs N L 1979 *J. Phys.* A **12** 625
[27] de M Rios P P and Ozorio de Almeida A M 2002 *J. Phys.* A **35** 2609
[28] Ozorio de Almeida A M and Brodier O 2006 *Ann. Phys.* **321** 1790
[29] Rossi F, Poli P and Jacoboni C 1992 *Semicond. Sci. Technol.* **7** 1017
[30] Kalos M H and Whitlock P A 1986 *Monte Carlo Methods Volume 1: Basics* (New York: Wiley)
[31] MacDonald J L and Cashwell E D 1976 *Trans. Am. Nucl. Soc* **24** 198
[32] Prettyman T H, Gardner R P and Verghese K 1990 *Nucl. Instrum. Methods Phys. Res* **A299** 516
[33] Painter S 1993 *Comput. Phys. Commun* **77** 142
[34] Hirayama H 1995 *J. Nucl. Sci. Technol* **32** 1201
[35] Vitanov P *et al* 1994 *Advances in Parallel Algorithms* ed I Dimov and O Tonev (Amsterdam: IOS Press) pp 117–28 https://www.iospress.com/catalog/books/advances-in-parallel-algorithms
[36] Dimov I T 1996 *Monte Carlo Methods for Applied Scientists* (Singapore: World Scientific)
[37] Jacoboni C and Bordone P 2004 *Rep. Prog. Phys.* **67** 1033
[38] Shifren L and Ferry D K 2001 *Phys. Lett.* A **285** 217
[39] Shifren L and Ferry D K 2002 *J. Comput. Electron* **1** 55
[40] Shifren L, Ringhofer C and Ferry D K 2003 *IEEE Trans. Electron Devices* **50** 769
[41] Eryilmaz M K *et al* 2021 *J. Comput. Electron* **20** 2039
[42] Nedjalkov M *et al* 2002 *Proc. 7th Int. Conf. on Simulation of Semiconductor Processes* (New York: SISPAD) pp 187–90
[43] Nedjalkov M, Kosina H and Selberherr S 2004 *Large Scale Scientific Computing* er Lirkov I *et al* (Heidelberg: Springer) pp 178–84
[44] Nedjalkov M *et al* 2004 *Semicond. Sci. Technol.* **19** S226
[45] Nedjalkov M *et al* 2004 *Phys. Rev.* B **70** 115319
[46] Nedjalkov M *et al* 2019 *Phys. Rev.* B **99** 014423
[47] Nedjalkov M *et al* 2022 *Phys. Rev.* A **106** 052213
[48] Ballicchia M, Nedjalkov M and Weinbub J 2020 *IEEE 20th Int. Conf. Nanotechnology* (New York: IEEE Press) 73–6
[49] Benam M *et al* 2021 *J. Comput. Electron.* **20** 775
[50] Van de Put M L, Sorée B and Magnus W 2017 *J. Comput. Phys.* **350** 314
[51] Thomann A and Borzi A 2016 *Numer. Methods Part. Differ. Equ.* **11** 62–87
[52] Chen Z, Jiang H and Shao S 2022 *J. Comput. Electron* **21** 756
[53] Mandel L and Wolf E 1965 *Rev. Mod. Phys.* **37** 231
[54] Straka I *et al* 2018 *npj Quantum Inf.* **4** 4
[55] Souquet J R and Clerk A A 2016 *Phys. Rev.* A **93** 060301
[56] Welland I and Ferry D K 2020 *J. Appl. Phys.* **128** 153102

[57] Welland I and Ferry D K 2021 *J. Comput. Electron* **20** 267
[58] Levinson I B 1970 *Zh. Exp. Teor. Fys.* **57** 660
transl. *Sov. Phys. JETP* **30** 362 http://jetp.ras.ru/cgi-bin/e/index/e/30/2/p362?a=list
[59] Peierls R 1933 *Z. Physik* **80** 763–91
[60] Thornber K K and Feyman R P 1970 *Phys. Rev.* B **1** 4099
[61] Barker J R 1973 *J. Phys.* C **6** 2663
[62] Nedjalkov M *et al* 2018 *Comput. Phys. Commun* **228** 30
[63] Qucrlioz D and Dollfus P 2013 *The Wigner Monte Carlo Method for Nanoelectronic Devices: A Particle Description of Quantum Transport and Decoherence* (London: Wiley-ISTE)
[64] Chang L L, Esaki L and Tsu R 1974 *Appl. Phys. Lett.* **24** 593
[65] Sollner T C L G *et al* 1983 *Appl. Phys. Lett.* **43** 588
[66] Ravaioli U *et al* 1985 *Physica B+C* **134B** 36
[67] Jensen K L and Buot F A 1991 *IEEE Trans. Electron Devices* **38** 2337
[68] Barker J R 1985 *Physica* 134**B** 22
[69] Shifren L and Ferry D K 2002 *Physica* B **314** 72
[70] Querlioz D *et al* 2008 *Phys. Rev.* B **78** 165306
[71] Koba S, Aoyagi R and Tsuchiya H 2008 *J. Appl. Phys.* **108** 064504
[72] Querlioz D *et al* 2006 *IEEE Trans. Nanotechnol.* **5** 737
[73] Querlioz D *et al* 2008 *Phys. Status Solidi* C **5** 150
[74] Croitoru M D *et al* 2003 *J. Appl. Phys.* **93** 1230
[75] Croitoru M D *et al* 2008 *Solid State Commun* **147** 31
[76] Moscato O 2021 *J. Comput. Electron* **20** 2062
[77] Lee J-H and Shin M 2017 *IEEE Trans. Nanotechnol* **16** 1028
[78] Schulz L and Schulz D 2016 *IEEE Trans. Nanotechnol* **15** 801
[79] Woloszyn M and Spizak B J 2017 *Phys. Rev.* B **96** 075440
[80] Schulz L and Schulz D 2020 *J. Comput. Electron* **19** 1399
[81] Kosina H 2006 *Int. J. Comput. Sci. Eng.* **2** 100
[82] Queerlioz D *et al* 2009 *J. Comput. Electron* **8** 324
[83] Jiang H and Cai W 2010 *J. Comput. Phys.* **229** 4461
[84] Lee J-H *et al* 2018 *Solid-State Electron* **139** 101
[85] Ferry D K *et al* 2022 *Semicond. Sci. Technol.* **37** 043001
[86] Jonasson O and Knezevic I 2015 *J. Comput. Electron* **14** 879
[87] Kominni A and Aksamija Z 2017 *J. Phys.: Condens. Matter* **30** 044004
[88] Morandi O and Schürrer F 2011 *J. Phys. A: Math. Theor.* **44** 265301
[89] Camiola V D, Mascali G and Romano V 2021 *J. Comput. Electron* **20** 2135
[90] Zamponi M and Barletti L 2011 *Math. Methods Appl. Sci.* **34** 807
[91] Diaz-Bautista E and Betancour-Ocampo Y 2020 *Phys. Rev.* B **101** 125402
[92] Cepellotti A and Kozinsky B 2021 *Math. Phys. Today* **19** 100412
[93] Weinbub J, Ballicchia M and Nedjalkov M 2022 *Nanoscale* **14** 13520
[94] Bocquillon E *et al* 2014 *Ann. Phys.* **526** 1
[95] Bäuerle C *et al* 2018 *Rep. Prog. Phys.* **81** 056503
[96] Ballicchia M, Nedjalkov M and Weinbub J 2022 *IEEE 22nd Int. Conf. Nanotechnology* (New York: IEEE Press) p 565

Chapter 12

Why not Green's functions?

Over time, it is usual to become familiar with the Boltzmann equation that is used for classical transport. In the case of low electric fields, the scenario is generally near equilibrium, resulting in a Maxwellian or Fermi–Dirac distribution function given in both cases by a temperature that may be slightly above that of the lattice. In devices, this transport equation is always connected to the assumption that the contacts are in near-equilibrium, although this is not always true. It is even less likely to be true in the quantum case. But in most devices, the carrier distribution is never near equilibrium but rather well out of equilibrium. This implies that it is not really known. Throughout this book, many methods for finding this distribution have appeared. However, in the case of quantum transport, other approaches are necessary, and several of these have been discussed. The discussion now turns to one of the more popular approaches used to treat quantum transport and to find the quantum distributions, namely the nonequilibrium Green's functions (NEGFs) approach.

Green's functions were already introduced in chapter 5, at least for the equilibrium situation. Importantly, that chapter clearly stated that the use of these equilibrium Green's functions assumes time reversibility. This has important consequences, as no dissipative mechanisms can be treated with this requirement. Dissipation assures the breaking of time-reversal symmetry. It is also important to note that the equilibrium Green's functions assume that the temperature is at absolute zero. Hence, temperature-dependent processes cannot be treated by these functions. The restrictions on the use of equilibrium Green's functions are many and stringent. Although temperature can be introduced through the use of the Matsubara Green's functions, or temperature Green's functions [1, 2], their structures, which are based on the Fermi–Dirac distribution, are difficult to use and almost certainly inconducive to inhomogeneous systems. The formulations that have proved to be somewhat useful, as they include both temperature and dissipation, are NEGFs, first introduced by Schwinger [3, 4]. With NEGFs, new functions will be needed and

doi:10.1088/978-0-7503-5237-6ch12

must be evaluated. One of these will yield the nonequilibrium distribution. As was noted earlier, these Green's functions are described by the wave functions ψ and $\psi^\dagger$, but these go beyond the simple forms and become fermion creation and annihilation operators. These wave functions are now referred to as field operators [2].

As pointed out in chapter 5, the Green's functions begin with the retarded and advanced functions

$$\begin{aligned} G_r(\boldsymbol{x}, \boldsymbol{x}'; t, t') &= -i\Theta(t - t')\left\langle \left\{ \hat{\psi}(\boldsymbol{x}, t), \hat{\psi}^\dagger(\boldsymbol{x}', t') \right\} \right\rangle \\ G_a(\boldsymbol{x}, \boldsymbol{x}'; t, t') &= i\Theta(t_0 - t)\left\langle \left\{ \hat{\psi}^\dagger(\boldsymbol{x}', t'), \hat{\psi}(\boldsymbol{x}, t) \right\} \right\rangle \end{aligned}. \tag{12.1}$$

In chapter 5, the angular brackets were explained as denoting averaging over the ground-state wave functions. *This is no longer the case here*, as the averaging has to be done over the full nonequilibrium distribution function. This is a major change in the properties of these two functions. While they are not dependent on the temperature, they are dependent upon this distribution function. As a result, they cannot be fully evaluated in the absence of the full set of NEGFs. Yet, they still determine the spectral density

$$A(k, \omega) = -\mathrm{Im}\{ G_r(k, \omega) - G_a(k, \omega)\}. \tag{12.2}$$

While this is written in the Fourier-transformed mode, it must be remembered that it can be a function of position and time, especially in semiconductor devices. This is because it depends upon interactions, and the strength of these interactions can vary with position and time as the carrier densities vary in position and time, as can any temperature inferred in the distribution function.

With NEGFs, one must deal with two new functions, which are correlation functions containing single wave function products. These are defined by

$$\begin{aligned} G^<(\boldsymbol{x}, \boldsymbol{x}'; t, t') &= i\langle \hat{\psi}^\dagger(\boldsymbol{x}', t')\hat{\psi}(\boldsymbol{x}, t)\rangle \\ G^>(\boldsymbol{x}, \boldsymbol{x}'; t, t') &= -i\langle \hat{\psi}(\boldsymbol{x}, t)\hat{\psi}^\dagger(\boldsymbol{x}', t')\rangle \end{aligned}. \tag{12.3}$$

As before, the angular brackets denote averaging over the full nonequilibrium distribution function. The connection between these two functions and the non-equilibrium distribution function is not really known for the far-from-equilibrium situation but has been hypothesized to maintain the same form as that found near equilibrium [4]. Thus, an assumption must be made, and this is known as an ansatz. The 'less-than' function (the first of the two equations (12.3)) is presumed to connect to the quantum distribution through the Kadanoff–Baym ansatz (KBA):

$$G^<(k, \omega) = f(\omega)A(k, \omega). \tag{12.4}$$

Similarly, the 'greater-than' function (the second of the two equations) is assumed to take the form

$$G^>(k, \omega) = [f(\omega) + 1]A(k, \omega). \tag{12.5}$$

In (12.4) and (12.5), the distribution function is defined as $f(\omega)$. There is an obvious limitation here, as no spatial variation is assumed, which implies that there is no inhomogeneity in any temperature or spatial dependence of the system, and the Fourier form certainly assumes no variation in the average time in the system. This time is determined from the spatial and temporal center-of-mass transformations, taken from (1.53) as

$$\begin{aligned} \boldsymbol{R} &= \frac{1}{2}(\boldsymbol{x} + \boldsymbol{x}'), \; \boldsymbol{s} = \boldsymbol{x} - \boldsymbol{x}', \\ T &= \frac{1}{2}(t + t'), \quad \tau = t - t', \end{aligned} \tag{12.6}$$

with the Fourier transforms being taken for **s** and τ. Neither of these assumptions is suitable for semiconductor devices, because effects occur at specific positions in the device, and the central positions are not known to convey this important fact. What is worse, however, is that the ansatz (12.4) is not thought to preserve causality, so that a modified version has been suggested [5] (in the untransformed domain, which remains more physical and contains all the position and time variables so that it also applies to inhomogeneous systems):

$$G^{<}(\boldsymbol{x}, \boldsymbol{x}'; t, t') = -G_{\mathrm{r}}(\boldsymbol{x}, \boldsymbol{x}'; t, t')\rho(\boldsymbol{x}, \boldsymbol{x}', t') + \rho(\boldsymbol{x}, \boldsymbol{x}', t)G_{\mathrm{a}}(\boldsymbol{x}, \boldsymbol{x}'; t, t'). \tag{12.7}$$

This is termed the generalized KBA (or GKBA). As discussed several times already in this book, causality is a basic assumption in electronics and in general life. Time has a single arrow from the past to the future and nobody wants the present state of a semiconductor device to be subject to what might happen in the future. Further, however, (12.7) does not give us the distribution function. Still further processing is needed to extract it from the density matrix.

So, why the question in the title of this chapter? The principal reason is that the authors are not aware that particles ever have been used to simulate NEGFs (however, a beginning to particle simulation is given in section 12.5). This then becomes a significant limitation on the use of NEGFs in particle simulations of semiconductor devices. All of these Green's functions in (12.1) and (12.3) have differential equations whose solutions provide the particular function. The basic equations for the retarded and advanced functions were given in (5.46) for the case in which no dissipative interactions were present. These have to be modified in order to take account of dissipation. Differential equations can often be transformed into path variable integro-differential equations, such as is done with the Boltzmann equation to produce ensemble Monte Carlo techniques. This means that there probably is no rational reason to say that particles cannot be used with NEGFs. However, there is the issue that the Green's functions above have two time and two space variables. Any particle trajectory has a beginning (and end) in space and time, so it can be described in exactly the same form. So the problem lies in arriving at the necessary path variable formulation that is useful for devising a particle basis for simulation. Indeed, this view was explicitly remarked upon by Feynman with the assumption that a path is connected to the entire motion of a particle from beginning to end [6].

In the remainder of this chapter, several limitations on the use of NEGFs are described, in spite of the fact that these have often been ignored in practice (this naturally raises questions about the accuracy of solutions based on these approaches). Some approaches that have arrived at possible points from which particles can be used are then described in section 12.4.

12.1 Equations of motion

Here, the equations of motion will be developed for the simplest case, in which any two-particle NEGF will be ignored and a small electric field will be assumed. The goal is to produce the quantum version of the Boltzmann equation, i.e. the *quantum Boltzmann equation.* This is achieved through a truly large set of approximations. Some of these approximations will be discussed along the way. Phonon scattering will be incorporated into this approach. There are two differential equations for the retarded Green's function due to the fact that there are two times involved in the function. These two equations can be written down as follows:

$$\begin{aligned}
&\left[i\hbar\frac{\partial}{\partial t} - H(\mathbf{x})\right]G_{\mathrm{r}}(z, z') = \delta(z - z') \\
&\quad + i\int d^4z'' G_{\mathrm{r}}(z, z'')D_{\mathrm{r}}(z, z'')G_{\mathrm{r}}(z'', z') \\
&\left[-i\hbar\frac{\partial}{\partial t'} - H(\mathbf{x}')\right]G_{\mathrm{r}}(z, z') = \delta(z - z') \\
&\quad + i\int d^4z'' G_{\mathrm{r}}(z, z'')G_{\mathrm{r}}(z'', z')D_{\mathrm{r}}(z'', z')
\end{aligned} \tag{12.8}$$

In the above, a shorthand notation $z = (\mathbf{x}, t)$ is used and the quantity D_{r} is the retarded phonon Green's function. The integrations over space and time on the right-hand sides of the two equations go from the initial position and time to the final position and time, as defined in the functions on the left-hand sides of the equations, that is, from z' to z. The first approximation has been made already, as the two Green's functions in the integral on the right-hand side of each equation should actually be a two-particle Green's function. What is missing here is the correlation between the wave functions before the scattering and that following the scattering. It is probably sufficiently accurate to ignore this correlation for the nonpolar phonon scattering which is isotropic in momentum space. But it is certainly inappropriate to ignore the two-particle form of this Green's function for coulomb scattering, which includes polar optical phonon scattering. If the correlation may be ignored, then the retarded self-energy can be defined as

$$\Sigma_{\mathrm{r}}(z, z') = iG_{\mathrm{r}}(z, z')D_{\mathrm{r}}(z, z'). \tag{12.9}$$

These equations then become

$$\begin{gathered}\left[i\hbar\frac{\partial}{\partial t}-H(\boldsymbol{x})\right]G_{\mathrm{r}}(z,z')=\delta(z-z')+\int d^4z''\Sigma_{\mathrm{r}}(z,z'')G_{\mathrm{r}}(z'',z')\\ \left[-i\hbar\frac{\partial}{\partial t'}-H(\boldsymbol{x}')\right]G_{\mathrm{r}}(z,z')=\delta(z-z')+\int d^4z''G_{\mathrm{r}}(z,z'')\Sigma_{\mathrm{r}}(z'',z')\end{gathered}. \tag{12.10}$$

The differential equations for the 'less-than' functions follow in a similar manner. Assuming the same caveat about two-particle Green's functions as that used for (12.10), these equations become

$$\begin{aligned}\left[i\hbar\frac{\partial}{\partial t}-H(\boldsymbol{x})\right]G^{<}(z,z')&=\int d^4z''\Sigma_{\mathrm{r}}(z,z'')G^{<}(z'',z')\\ &+\int d^4z''\Sigma^{<}(z,z'')G_{\mathrm{a}}(z'',z')\end{aligned}. \tag{12.11}$$

$$\begin{aligned}\left[-i\hbar\frac{\partial}{\partial t'}-H(\boldsymbol{x}')\right]G^{<}(z,z')&=\int d^4z''G_{\mathrm{r}}(z,z'')\Sigma^{<}(z'',z')\\ &+\int d^4z''G^{<}(z,z'')\Sigma_{\mathrm{a}}(z'',z')\end{aligned}. \tag{12.12}$$

One should note that the collision integrals on the right-hand side of the equations are more complicated, both in number and composition, as they involve both retarded (and advanced) and 'less-than' functions. The advanced functions will be shown to be just the complex conjugates of the retarded functions, as in chapter 5. For an electric field that is not too large, the Hamiltonians for the above equations can simply be expressed as

$$H(\boldsymbol{x})=-\frac{\hbar^2}{2m^*}\frac{\partial^2}{\partial \boldsymbol{x}^2}+e\boldsymbol{E}\cdot\boldsymbol{x}, \tag{12.13}$$

where $\boldsymbol{E}$ is the applied electric field.

12.1.1 The retarded Green's function

At this point, it is useful to introduce the center-of-mass coordinate transformation given in (12.6). The sum and difference of the two equations in (12.10) are then taken, using the relationships

$$\begin{aligned}&\frac{\partial}{\partial t}+\frac{\partial}{\partial t'}=\frac{\partial}{\partial T},\ \frac{\partial}{\partial t}-\frac{\partial}{\partial t'}=2\frac{\partial}{\partial\tau},\\ &H(\boldsymbol{x})+H(\boldsymbol{x}')=-\frac{\hbar^2}{2m^*}\frac{\partial^2}{\partial s^2}-\frac{\hbar^2}{4m^*}\frac{\partial^2}{\partial \boldsymbol{R}^2}-2e\boldsymbol{E}\cdot\boldsymbol{R},\\ &H(\boldsymbol{x})-H(\boldsymbol{x}')=-\frac{\hbar^2}{m^*}\frac{\partial}{\partial s}\cdot\frac{\partial}{\partial \boldsymbol{R}}-e\boldsymbol{E}\cdot s.\end{aligned} \tag{12.14}$$

With these transformations and the above relationships, the two equations (12.10) become

$$\begin{gathered}\left(i\hbar\frac{\partial}{\partial\tau}+\frac{\hbar^2}{2m^*}\frac{\partial^2}{\partial \boldsymbol{s}^2}+\frac{\hbar^2}{8m^*}\frac{\partial^2}{\partial \boldsymbol{R}^2}+e\boldsymbol{E}\cdot\boldsymbol{R}\right)G_r(\boldsymbol{s},\boldsymbol{R},T,\tau)\\=\delta(\tau)\delta(s)+\frac{1}{2}\int d^4z''(\Sigma_r G_r+G_r\Sigma_r)\end{gathered}\tag{12.15}$$

$$\begin{gathered}\left(i\hbar\frac{\partial}{\partial T}+\frac{\hbar^2}{m^*}\frac{\partial}{\partial \boldsymbol{s}}\cdot\frac{\partial}{\partial \boldsymbol{R}}+e\boldsymbol{E}\cdot\boldsymbol{s}\right)G_r(\boldsymbol{s},\boldsymbol{R},T,\tau)\\=\int d^4z''(\Sigma_r G_r-G_r\Sigma_r)\end{gathered}\tag{12.16}$$

The integrals have not yet been transformed. This latter is a complicated process, and will be dealt with further below. At present, we still need to work on the left-hand sides of the equations.

Normally, with Green's functions, such as those dealt with in chapter 5, only the relative coordinates are of interest, as the system is assumed to be homogeneous. This is not the case in devices, so that all of the variables must be retained. It is now useful to Fourier transform the relative coordinates, taking $(\boldsymbol{s},\tau)\to(\boldsymbol{k},\Omega)$. Equation (12.16) then becomes

$$\begin{gathered}\left(\hbar\Omega-E_k+\frac{\hbar^2}{8m^*}\frac{\partial^2}{\partial \boldsymbol{R}^2}+e\boldsymbol{E}\cdot\boldsymbol{R}\right)G_r(\boldsymbol{k},\boldsymbol{R},T,\Omega)\\=1+\frac{1}{2}\int d^4z''(\Sigma_r G_r+G_r\Sigma_r)\\i\hbar\left(\frac{\partial}{\partial T}+\frac{\hbar\boldsymbol{k}}{m^*}\cdot\frac{\partial}{\partial \boldsymbol{R}}+\frac{e\boldsymbol{E}}{\hbar}\cdot\frac{\partial}{\partial \boldsymbol{k}}\right)G_r(\boldsymbol{k},\boldsymbol{R},T,\Omega)\\=\int d^4z''(\Sigma_r G_r-G_r\Sigma_r)\end{gathered}\tag{12.17}$$

Here, the energy has been introduced as follows:

$$E_k=\frac{\hbar^2k^2}{2m^*}.\tag{12.18}$$

The second equation of (12.17) has the same streaming terms on the left-hand side as the Boltzmann equation. However, the first equation has some problems to be addressed, notably with the field streaming term. This term introduces what can be described as a size dependence to the equations, which is unphysical. In a sense, this is a result of the gauge chosen for the field representation. To remove this effect, a gauge transformation will be introduced [7]. In this gauge, a ballistic electron travels horizontally, effectively remaining at constant energy, while the conduction band edge falls due to the electric field. The transformations that are required for this are

$$\Omega+\frac{e\boldsymbol{E}\cdot\boldsymbol{R}}{\hbar}\to\omega,\ \frac{\partial}{\partial \boldsymbol{R}}\to\frac{\partial}{\partial \boldsymbol{R}}+\frac{e\boldsymbol{E}}{\hbar}\frac{\partial}{\partial\omega}.\tag{12.19}$$

This gauge transformation changes equation (12.17) significantly. At this point, the Fourier transforms are introduced to the terms on the right-hand side of the equations, although the evaluation of these transformations will be dealt with below. The new form of the equations is as follows:

$$\begin{gathered}\left[\hbar\omega - E_k + \frac{\hbar^2}{8m^*}\left(\frac{\partial}{\partial \boldsymbol{R}} + \frac{e\boldsymbol{E}}{\hbar}\frac{\partial}{\partial\omega}\right)^2\right]G_r(\boldsymbol{k}, \boldsymbol{R}, T, \omega) \\ = 1 + \frac{1}{2}\int d^3 s e^{i\boldsymbol{k}\cdot\boldsymbol{s}}\int d\tau e^{-i(\omega - e\boldsymbol{E}\cdot\boldsymbol{R}/\hbar)\tau}\int d^4 z''(\Sigma_r G_r + G_r\Sigma_r) \\ i\hbar\left[\frac{\partial}{\partial T} + \frac{\hbar \boldsymbol{k}}{m^*}\cdot\frac{\partial}{\partial \boldsymbol{R}} + \frac{e\boldsymbol{E}}{\hbar}\cdot\left(\frac{\partial}{\partial \boldsymbol{k}} + \frac{\hbar\boldsymbol{k}}{m^*}\frac{\partial}{\partial\omega}\right)\right]G_r(\boldsymbol{k}, \boldsymbol{R}, T, \Omega) \\ = -i\int d^3 s e^{i\boldsymbol{k}\cdot\boldsymbol{s}}\int d\tau e^{-i(\omega - e\boldsymbol{E}\cdot\boldsymbol{R}/\hbar)\tau}\int d^4 z''(\Sigma_r G_r - G_r\Sigma_r)\end{gathered} . \tag{12.20}$$

The positional variation is now introduced on the right-hand sides as a phase variable, as may be seen in the exponential terms in the integrals. Hence, phase interference during scattering is now a real possibility, especially within the nonzero time duration of an individual scattering event. Hence, it is necessary to ensure that this nonlocality is retained in the evaluation of the two scattering integrals (which differ solely by the sign change in their integrands).

As in classical transport, the field term in the second equation of (12.20) is linear in the electric field. Strong field effects, such as hot carriers and intervalley scattering, originate not from high orders in the electric field but from the significant modifications of the carrier distribution function that arise from the high electric field. Since the interest here is in relatively small values of this field, it is possible to expand the collision terms using an approach known as gradient expansion [4]. That is, the self-energy and Green's functions are expanded in a Taylor series around the center-of-mass position and the time variables $\boldsymbol{R}$ and T. In some sense, this limits the results to systems in which the potentials weakly vary in space and time, and this is a critical limitation for small devices, which can clearly violate this assumption. Therefore, the reader should be aware that this is a major limitation, but the development is meant to be illustrative of the general approach that must be taken with NEGFs.

In order to evaluate the collision integrals on the right-hand side of (12.20), a series of coordinate transformations will be made [8], and it is these terms that will be expanded in the gradient expansion. To begin, the first term in each of the two equations of (12.20) may be expressed as

$$I = \int d^3 s e^{i\boldsymbol{k}\cdot\boldsymbol{s}}\int d\tau e^{-i\Omega\tau}\int d^4 z''\Sigma_r(z, z'')G_r(z'', z'). \tag{12.21}$$

Note that the center-of-mass transformations and the gauge transformation have been backed out of the integrals. The effect of the first transformation is to set the four variables $\boldsymbol{u} = z - z''$, so that the integrand is $\Sigma_r(z, z - \boldsymbol{u})G_r(z - \boldsymbol{u}, z')$ and the integration is now over u. The transformations (12.6) are then made using the four vector $\boldsymbol{w} = (\boldsymbol{s}, \tau)$, which gives the product $\Sigma_r(z + \boldsymbol{w}/2, z + \boldsymbol{w}/2 - \boldsymbol{u})$ $G_r(z + \boldsymbol{w}/2 - \boldsymbol{u}, z - \boldsymbol{w}/2)$. The final transformation is to let $y = z + \boldsymbol{w}/2 - \boldsymbol{u}$, $y = (\boldsymbol{r}, \tau')$, so that the integral of (12.21) becomes

$$I = \int d^3s d^3r e^{i\boldsymbol{k}\cdot(\boldsymbol{w}+y)} \int d\tau' d\tau e^{-i\Omega(\tau+\tau')} \Sigma_r(\boldsymbol{R} + \boldsymbol{s}/2, \boldsymbol{r}, T + \tau/2, \tau') \\ \times G_r(\boldsymbol{r}, \boldsymbol{R} - \boldsymbol{s}/2, T - \tau/2, \tau') . \tag{12.22}$$

It may be seen that the center-of-mass variables now all appear with offsets (the reduced variables in each case), and it is around these variables that the expansion is taken. The Fourier-transformed version of the collision integral then can be written as

$$I = \Sigma_r(\boldsymbol{R}, \boldsymbol{k}, T, \Omega) G_r(\boldsymbol{R}, \boldsymbol{k}, T, \Omega) \\ + \frac{i}{2}\left[\frac{\partial \Sigma_r}{\partial \boldsymbol{R}} \cdot \frac{\partial G_r}{\partial \boldsymbol{k}} - \frac{\partial \Sigma_r}{\partial \boldsymbol{k}} \cdot \frac{\partial G_r}{\partial \boldsymbol{R}} + \frac{\partial \Sigma_r}{\partial \Omega} \cdot \frac{\partial G_r}{\partial T} - \frac{\partial \Sigma_r}{\partial T} \cdot \frac{\partial G_r}{\partial \Omega}\right] + \ldots. \tag{12.23}$$

One may recognize that the second term, in the brackets, has the natural form of the Poisson brackets that appear in classical mechanics. This second term in parentheses will be denoted by $\{\Sigma_r, G_r\}$ in shorthand notation, and one should note that the reverse of the two functions gives the same results as the first term on the right-hand side of (12.23).

At this point, the gauge transformation has still not been made. When this transformation is made, it introduces extra terms into the electric field, and the term in brackets becomes

$$\{\Sigma_r, G_r\} \to \{\Sigma_r, G_r\} + \frac{e\boldsymbol{E}}{\hbar} \cdot \left(\frac{\partial \Sigma_r}{\partial \omega}\frac{\partial G_r}{\partial \boldsymbol{k}} - \frac{\partial \Sigma_r}{\partial \boldsymbol{k}}\frac{\partial G_r}{\partial \omega}\right). \tag{12.24}$$

Hence, the first term in the gradient expansion already provides a correction term to the drift term of the left-hand side of (12.20). With these expressions, the two equations (12.20) can be written as

$$\left[\hbar\omega - E(\boldsymbol{k}) + \frac{\hbar^2}{8m^*}\left(\frac{\partial}{\partial \boldsymbol{R}} + \frac{e\boldsymbol{E}}{\hbar}\frac{\partial}{\partial \omega}\right)^2 - \Sigma_r\right] G_r(\boldsymbol{R}, \boldsymbol{k}, T, \omega) = 1, \\ \hbar\left\{\frac{\partial}{\partial T} + \frac{\hbar \boldsymbol{k}}{m^*} \cdot \frac{\partial}{\partial \boldsymbol{R}} + \frac{e\boldsymbol{E}}{\hbar} \cdot \left[\left(1 - \frac{1}{\hbar}\frac{\partial \Sigma_r}{\partial \omega}\right)\frac{\partial}{\partial \boldsymbol{k}} + \right.\right. \\ \left.\left. + \left(\frac{\hbar \boldsymbol{k}}{m^*} + \frac{1}{\hbar}\frac{\partial \Sigma_r}{\partial \boldsymbol{k}}\right)\frac{\partial}{\partial \omega}\right]\right\} G_r(\boldsymbol{R}, \boldsymbol{k}, T, \omega) = -i\{\Sigma_r, G_r\}. \tag{12.25}$$

These equations satisfy the general case for weak inhomogeneity and slow time variation. If it is assumed that the system is, in fact, homogeneous and in a steady-state (non-oscillatory) condition, they become even simpler, as the derivatives with respect to $\boldsymbol{R}$ and T vanish. In this case, the retarded Green's function is found from the first of equations (12.25) to be

$$G_r(\boldsymbol{R}, \boldsymbol{k}, T, \omega) = \frac{1}{\hbar\omega - E(\boldsymbol{k}) - \Sigma_r(\boldsymbol{R}, \boldsymbol{k}, T, \omega)}, \tag{12.26}$$

which is the same *form* as that in equilibrium. But there is a major difference from the latter case, which is that the system is not in equilibrium, which in turn means that the phonon scattering does not disappear because there is no detailed balance in the nonequilibrium situation. Thus, while the form is the same, the details are far from the same. The self-energy now encompasses all the scattering processes, and this leads to a very different form for the spectral density.

12.1.2 The 'less-than' function

The distribution function that is needed for normalization of the current is obtained from the 'less-than' function that appears in (12.11). This is transformed using exactly the same steps as those used for the retarded function of the previous section. The difference that arises lies in the different collision integral terms of (12.11) that arise for the 'less-than' function. Nevertheless, following the same steps as above, the equations of motion for this function become

$$\begin{aligned}
&\left[\hbar\omega - E(\boldsymbol{k}) + \frac{\hbar^2}{8m^*}\left(\frac{\partial}{\partial \boldsymbol{R}} + \frac{e\boldsymbol{E}}{\hbar}\frac{\partial}{\partial \omega}\right)^2\right]G^< = \frac{1}{2}\int d^3 s e^{i\boldsymbol{k}\cdot\boldsymbol{s}} \int d\tau e^{-i(\omega - e\boldsymbol{E}\cdot\boldsymbol{R}/\hbar)\tau} \\
&\qquad \times \int d^4 z''(\Sigma_r G^< + \Sigma^< G_a + G_r\Sigma^< + G^<\Sigma_a), \\
&i\hbar\left[\frac{\partial}{\partial T} + \frac{\hbar \boldsymbol{k}}{m^*}\cdot\frac{\partial}{\partial \boldsymbol{R}} + \frac{e\boldsymbol{E}}{\hbar}\cdot\left(\frac{\partial}{\partial \boldsymbol{k}} + \frac{\hbar \boldsymbol{k}}{m^*}\frac{\partial}{\partial \omega}\right)\right]G^< = \int d^3 s e^{i\boldsymbol{k}\cdot\boldsymbol{s}} \int d\tau e^{-i(\omega - e\boldsymbol{E}\cdot\boldsymbol{R}/\hbar)\tau} \\
&\qquad \times \int d^4 z''(\Sigma_r G^< + \Sigma^< G_a - G_r\Sigma^< - G^<\Sigma_a).
\end{aligned} \tag{12.27}$$

Some people like to use the ballistic situation in which there is no scattering, thus ignoring the right-hand side of each of these equations. But this is not appropriate for any real semiconductor device. Except for the frequency derivative in the square brackets of the second of these equations, the bracket is the same as that for the Boltzmann equation.

The distribution function may be obtained from the 'less-than' function by first defining the Wigner distribution from the density matrix, as done in chapter 11. This definition follows from the density matrix as follows:

$$f_W(\boldsymbol{R}, \boldsymbol{k}, T) = \frac{1}{(2\pi\hbar)^3}\int d^3 s e^{i\boldsymbol{k}\cdot\boldsymbol{s}} \rho(\boldsymbol{R} + s/2, \boldsymbol{R} - s/2, T). \tag{12.28}$$

When the density matrix is written in terms of the wave functions and then compared to the Green's function, one can see that the Wigner function may be written as

$$\begin{aligned} f_W(\boldsymbol{R}, \boldsymbol{k}, T) &= -\frac{i}{(2\pi\hbar)^3}\lim_{\tau\to 0} G^<(\boldsymbol{R}, \boldsymbol{k}, T, \tau) \\ &= -\frac{i}{(2\pi\hbar)^3}\frac{1}{2\pi}\int_{-\infty}^{\infty} d\omega G^<(\boldsymbol{R}, \boldsymbol{k}, T, \omega) \end{aligned} . \tag{12.29}$$

The last equation requires that the 'less-than' function be integrable and thus not diverge at the limits. Moreover, the function is usually assumed to be symmetric in frequency [4], so that the integral of the frequency derivative vanishes in the second equation of (12.27), and the bracketed term becomes the same as that of the Boltzmann equation. Thus, the spatially varying density may then also be found to be

$$\begin{aligned} n(\boldsymbol{R}, T) &= \int d^3\boldsymbol{k} f_W(\boldsymbol{R}, \boldsymbol{k}, T) \\ &= \frac{-i}{(2\pi)^4}\int d^3\boldsymbol{k}\int_{-\infty}^{\infty} d\omega G^<(\boldsymbol{R}, \boldsymbol{k}, T, \omega) \end{aligned} . \tag{12.30}$$

The collision integrals can be evaluated by following exactly the same steps as those used for the retarded function. In order to more clearly follow the development, only the homogeneous, steady-state situation will be treated, so that the derivatives with respect to $\boldsymbol{R}$ and T vanish. The inhomogeneous and time-varying solutions can easily be extrapolated from those below. In addition, the real and imaginary parts of the retarded (and advanced) functions will be separated using

$$\begin{aligned} G_r &= G_a^\dagger = W - i\frac{A}{2}, \\ \Sigma_r &= \Sigma_a^\dagger = \Sigma - i\Gamma, \end{aligned} \tag{12.31}$$

where Σ (without any subscript) is the self-energy correction arising from the interactions, and A is the spectral density. This should not cause any confusion. The resulting equations of motion for the 'less-than' function may be written as

$$\begin{aligned} &\left[\hbar\omega - E(\boldsymbol{k}) - \Sigma - \frac{e\boldsymbol{E}}{2\hbar}\cdot\left(\frac{\partial\Gamma}{\partial\boldsymbol{k}} - \frac{\partial\Gamma}{\partial\omega}\frac{\partial}{\partial\boldsymbol{k}}\right)\right]G^< = W\Sigma^< \\ &\qquad - \frac{e\boldsymbol{E}}{4\hbar}\cdot\left(\frac{\partial\Sigma^<}{\partial\omega}\frac{\partial A}{\partial\boldsymbol{k}} - \frac{\partial A}{\partial\omega}\frac{\partial\Sigma^<}{\partial\boldsymbol{k}}\right), \\ &e\boldsymbol{E}\cdot\left[\left(1 - \frac{1}{\hbar}\frac{\partial\Sigma}{\partial\omega}\right)\frac{\partial}{\partial\boldsymbol{k}} + \left(\frac{\hbar\boldsymbol{k}}{m^*} + \frac{1}{\hbar}\frac{\partial\Sigma}{\partial\boldsymbol{k}}\right)\frac{\partial}{\partial\omega}\right]G^< + 2\Gamma G^< = A\Sigma^< \\ &\qquad + \frac{e\boldsymbol{E}}{\hbar}\cdot\left(\frac{\partial\Sigma^<}{\partial\omega}\frac{\partial W}{\partial\boldsymbol{k}} - \frac{\partial W}{\partial\omega}\frac{\partial\Sigma^<}{\partial\boldsymbol{k}}\right). \end{aligned} \tag{12.32}$$

It is now possible to make another assumption for the case near equilibrium. In this situation, higher powers of the electric field can be ignored. Hence, in the first term on the left-hand side of the second equation of (12.32), it is possible to replace the 'less-than' function with the near-equilibrium form [4]

$$G_0^< = iF_{\rm FD}(\omega)A(\boldsymbol{k},\,\omega), \tag{12.33}$$

in which the first term on the right is the Fermi–Dirac distribution. Now, for each term in which the electric field appears, either the distribution or its derivative appears. In this near-equilibrium situation, the important terms are those in which the derivative of the distribution function appears, just as in the classical case of the Boltzmann equation. So, only the terms in the second equation of (12.32) are retained, and this equation becomes [9]

$$\begin{aligned} ie\boldsymbol{E}\cdot\left[\frac{1}{\hbar}(\hbar\omega - E(\boldsymbol{k}) - \Sigma)\frac{\partial\Gamma}{\partial\boldsymbol{k}} + \left(\frac{\hbar\boldsymbol{k}}{m^*} + \frac{1}{\hbar}\frac{\partial\Sigma}{\partial\boldsymbol{k}}\right)\Gamma\right]A^2\frac{\partial F_{\rm FD}(\omega)}{\partial\omega} \\ = A\Sigma^< - 2\Gamma G^<. \end{aligned} \tag{12.34}$$

In arriving at this last equation, it has also been assumed that

$$\frac{\partial\Sigma^<}{\partial\omega} = 2\Gamma\frac{\partial F_{\rm FD}(\omega)}{\partial\omega}, \tag{12.35}$$

which can be shown to be exact for a number of scattering mechanisms. Equation (12.34) is the steady-state, homogeneous version of the quantum Boltzmann equation (QBE). This equation is, within the approximations assumed, almost exact and represents another form of the Kubo formula used in near-equilibrium transport.

12.1.3 The solution of the quantum Boltzmann equation

In the square bracket of (12.34), one may observe that the gradient of the Γ term (the first term) is small at the maximum value of the spectral density A. This is also true for the term in the gradient of Σ because scattering terms tend to be proportional to the density of final states of the scattering event. As a result, these two scattering terms reach a maximum where the spectral density peaks, and thus the derivatives are quite near zero at this point. As a result, (12.34) can be simplified to the easier equation

$$e\boldsymbol{E}\cdot\frac{\hbar\boldsymbol{k}}{m^*}A^2\frac{\partial F_{\rm FD}(\omega)}{\partial\omega} = -i[A\Sigma^< - 2\Gamma G^<]. \tag{12.36}$$

The terms on the right-hand side are assumed to vanish in equilibrium, so it is the electric field that drives the system out of equilibrium. In this condition, it is reasonable to assume that the form of the solution is similar to that of the Boltzmann equation for fields that are not too big. Hence, it is postulated that the form of the 'less-than' function will be something like

$$G^<(\boldsymbol{k},\,\omega) = iA(\boldsymbol{k},\,\omega)\left[F_{\rm FD}(\omega) - e\boldsymbol{E}\cdot\frac{\hbar\boldsymbol{k}}{m^*}\Lambda(\boldsymbol{k},\,\omega)\frac{\partial F_{\rm FD}(\omega)}{\partial\omega}\right], \tag{12.37}$$

where $\Lambda(\boldsymbol{k}, \omega)$ is a function to be determined. This result is predicated on the form of (12.26), in which the *form* of the retarded Green's function is not changed from that of equilibrium, and therefore the *form* of the spectral density is not expected to be changed. Of course, the energy dependence of the scattering events is likely to actually modify the exact shape, but it is the form that is taken to be important. This means that it is the distribution function whose form is changed by the field and provides the important result (12.37). In addition, it is expected that $\Sigma^<$ will have the same form as (12.9), which involves the 'less-than' Green's function for the electrons and the phonons. The phonon term includes the matrix element and a propagator. This allows the use of (12.37) in (12.36) to produce an integral equation for the unknown function, as follows:

$$\Lambda(\boldsymbol{k}, \omega) = \frac{1}{2}A(\boldsymbol{k}, \omega) + \frac{i}{2\Gamma(\hbar \boldsymbol{k}/m^*)}\sum_{\pm}\int \frac{d^3\boldsymbol{q}}{(2\pi)^3} \mid M(\boldsymbol{k}, \boldsymbol{q}) \mid^2 \\ \times \frac{\hbar(\boldsymbol{k} \pm \boldsymbol{q})}{m^*} A(\boldsymbol{k} \pm \boldsymbol{q}, \omega \pm \omega_q)\Lambda(\boldsymbol{k} \pm \boldsymbol{q}, \omega \pm \omega_q). \tag{12.38}$$

This last equation provides the quantum replacement for the simple scattering time that appears in the Boltzmann equation. In some sense, this is that scattering time corrected by a vertex correction that introduces the quantum effects. As a result, the solutions to this equation can be connected by analogy to the momentum relaxation time that is commonly used with the Boltzmann equation to arrive at a mobility. However, a great many approximations have been made to reach this point. Not all of these approximations are expected to be useful, but the result does establish a connection with the Boltzmann equation of classical physics.

12.2 A high-field solution

As may be inferred from the previous paragraphs, it has proved to be difficult to generate a tractable quantum transport equation based on NEGFs that: (i) incorporates both the collisional broadening and the intracollisional field effect and (ii) is adaptable to the small space and time behaviors of modern semiconductor devices. There have been many so-called formulations of this problem that begin with the integration path sketched in figure 5.4. One well documented and pursued approach breaks the time-reversal symmetry of this latter figure by breaking the path at the latest time and folding it back to $t \to -\infty$ [10]. Thus, the new path begins and ends in the infinite past. Others have added extra folds to treat the products of NEGFs [11]. The problem with these paths, like the original of figure 5.4, is that they assume that the interactions and the fields are turned on adiabatically. Unfortunately, semiconductor devices do not behave in such a manner. Hence, such path modifications cannot replace the need to perform the integrations over the exact times involved in the NEGFs.

When the center-of-mass transformations (12.6) are introduced and homogeneity constraints in space and time are used, this corresponds, in some sense, to coarse graining of the relevant NEGF. This makes such formulations difficult to apply to small structures, such as today's digital devices, in which the critical lengths are comparable to, or smaller than, the characteristic transport lengths, such as the inelastic mean free path. Further, the application of voltages or the resulting electric fields breaks time-reversal symmetry, and the resulting currents signal a phase transition in the device. In addition, treating the phonon interactions in a strong field leads to the mixture of space and time coordinates, and many simplifying assumptions are just not appropriate to this situation. This was seen in the last sections where the gauge transformation to $\omega = \Omega + e\boldsymbol{E} \cdot \boldsymbol{R}/\hbar$ was introduced.

This suggests a different approach to the high-field situation, one which builds upon the gauge transformation previously used. In the case of a constant electric field, one can utilize the Airy transformation, using the scalar potential gauge for the field, to change the Hamiltonian into one more appropriate to the transport problem. This was done in the early treatments of the Franz–Keldysh effect, in which a large applied electric field modulates the bandgap in semiconductors [12]. The introduction of this transform leads to the proper symmetries at the beginning of the problem and ultimately makes the Dyson equation easier to separate. This approach will be discussed in the following parts of this section. Moreover, it will be the basis for the development of a particle-based approach to evaluating NEGFs.

12.2.1 The Airy transform

The first goal is the diagonalization of the equation of motion within the Hilbert space defined by the normalized eigenfunctions of the unperturbed Hamiltonian for the simple system, given by (12.13). The electric field will be oriented in the z-direction for convenience. The one-dimensional eigenfunctions in this direction are found from the one-dimensional Airy equation:

$$\left(-\frac{\hbar^2}{2m^*}\frac{\partial^2}{\partial z^2} + eFz\right)\varphi(z) = E\varphi(z), \tag{12.39}$$

where the field is now denoted by the symbol F in order to avoid confusion with the energy E. The wave functions in the directions normal to this primary axis (x,y) are normal plane waves, and the variation of this transverse momentum with energy gives the transverse band structure. For this latter, the normal $\boldsymbol{k}$ and $\boldsymbol{r}$ are used to denote the wave vector and the position in the transverse plane of (x,y)-coordinates, respectively. The solutions to (12.39) are fairly well known to be the Airy functions, which may be represented in reduced coordinates as

$$\varphi(\boldsymbol{r}, s) = \frac{1}{2\pi L}e^{i\boldsymbol{k}\cdot\boldsymbol{r}}Ai\left(\frac{z-s}{L}\right), \tag{12.40}$$

and the corresponding energy levels are given by (for parabolic bands)

$$E_{k,s} = \frac{\hbar^2k^2}{2m^*} + eFs. \tag{12.41}$$

Here, it will be seen that the second term on the right-hand side of (12.41) plays the gauge role of the field in the transformation used in section 12.1 on the frequency (energy). In this sense, the approach used here follows exactly that introduced in the low-field case. The factor L is a length which appears when normalizing the function (12.40) and is given by

$$L = \left(\frac{\hbar^2}{2m^* eF}\right)^{1/3}. \tag{12.42}$$

(The function $Ai(\cdot)$ is the Airy function of the first kind, and s is a spatial variable related to the zero crossings of this function [13].)

In most textbooks, one encounters the Airy functions when solving the Schrödinger equation for triangular potential wells (such as those encountered in a metal–oxide–semiconductor field-effect transistor (MOSFET)) [14]. This leads to a set of discrete bound-state eigenvalues and eigenfunctions in which the energy levels are determined by the discrete zeroes of the Airy function. This process generates a set of basis functions for use in further calculations. *This is not the approach to be used here.* Rather, a continuous *Airy transform* will be introduced. The difference between the Airy functions and the Airy transform is the same as the difference between a Fourier series and a Fourier transform. The series of functions in both case can be an infinite (or not in some cases) set of discrete functions. The latter (transform) case is a single continuous function. For the Airy transform, the continuous transform variable s appears and is not connected to the discrete set of zeroes. This transform is defined in the same manner as a Fourier transform. Here, the Airy transform of the function $f(\mathbf{r}, z)$ is defined as

$$U(\mathbf{k}, s) = \int d^2\mathbf{r} \int \frac{dz}{2\pi L} e^{i\mathbf{k}\cdot\mathbf{r}} Ai\left(\frac{z-s}{L}\right) f(\mathbf{r}, z). \tag{12.43}$$

In the Green's functions of (12.1) and (12.3), there are two spatial variables, and therefore one needs a double Airy transform. For simplicity, homogeneity in the transverse dimensions is assumed. This transform may then be expressed as

$$U(\mathbf{k}, s, s') = \int d^2\mathbf{r} \int \frac{dz}{2\pi L} \int \frac{dz'}{2\pi L} e^{i\mathbf{k}\cdot\mathbf{r}} Ai\left(\frac{z-s}{L}\right) Ai\left(\frac{z'-s'}{L}\right) f(\mathbf{r}, z, z'). \tag{12.44}$$

It may be noticed that the Airy function part of the Green's function is only being developed for the single direction along the electric field. Normal Fourier transformation is used for the transverse directions. A function that is diagonal in both $\mathbf{k}$ is translationally invariant, and this is the case in the transverse plane. This is not necessarily true for the Airy transform variable s in the longitudinal direction along the electric field.

12.2.2 The retarded Green's function

As the Green's function is translationally invariant in the transverse direction, it is a only a function of $\mathbf{r} - \mathbf{r}'$. In (12.44), the integration over the intermediate variables in (12.8) is a convolution integral. It is known in Fourier transform theory that the convolution becomes a simple product after transformation, so that the retarded Green's function satisfies the inhomogeneous integral equation

$$\begin{aligned} &G_r(E_{k,\,s},\, z,\, z',\, t,\, t') = G_0^{\mathrm{F}}(E_{k,\,s},\, z,\, z',\, t,\, t') \\ &+ \int du_1 \int du_2 G_0^{\mathrm{F}}(E_{k,\,s},\, u,\, u_1)\Sigma_r(E_{k,\,s},\, u_1,\, u_2)G_r(E_{k,\,s},\, u_2,\, u') \end{aligned}. \tag{12.45}$$

As before, the shorthand notation $u = (z, t)$ is reintroduced for the various z and t variables. The equilibrium (unperturbed) Green's function, i.e. the first term on the right-hand side of (12.45), includes the electric field and may be defined as [15]

$$\begin{aligned} G_0^{\mathrm{F}}(E_{k,\,s},\, s,\, s',\, t,\, t') &= G_{\mathrm{r},\,0}^{\mathrm{F}}(E_{k,\,s},\, s,\, s',\, t,\, t') \\ &= -\frac{i}{\hbar}\theta_0(t-t')e^{-iE_{k,\,s}(t-t')/\hbar}\delta(s-s') \end{aligned}. \tag{12.46}$$

The energy term here is given by (12.41). The advantage is that the Fourier transform over time is easily taken, which yields

$$G_0^{\mathrm{F}}(E_{k,\,s},\, s,\, s',\, t,\, t') = \frac{\delta(s-s')}{\hbar\omega - E_{k,\,s} + i\eta}, \tag{12.47}$$

where the limit $\eta \to 0$ is usually taken. In (12.46), the function $\theta(\xi)$ is the normal Heavyside step function, which is one for $\xi \geqslant 0$, and zero otherwise. The convergence factor η is often confusing. The use of this factor implies that the electric field was turned on in the infinite past. But this is not the case, especially for devices. As may be seen in (12.46), the field is turned on at time t'. While this convergence factor is useful, one must be aware that it is not the correct interpretation. This point will be clarified when the self-energy is introduced. The full transformation with the Airy functions transform for the retarded Green's function can now be written as

$$\begin{aligned} &G_{\mathrm{r}}(E_{k,\,s},\, s,\, s',\, \omega) = G_0^{\mathrm{F}}(E_{k,\,s},\, s,\, \omega)\delta(s-s') \\ &+ G_0^{\mathrm{F}}(E_{k,\,s''},\, s,\, \omega)\int ds''\Sigma_{\mathrm{r}}(E_{k,\,s},\, s,\, s'',\, \omega)G_{\mathrm{r}}(E_{k,\,s'},\, s'',\, s',\, \omega) \end{aligned}. \tag{12.48}$$

It is now necessary to consider the self-energy term.

The retarded self-energy in (12.9) is the product of the retarded Green's function and the retarded phonon Green's function. In most semiconductors, the scattering is weak, and the self-energy can be calculated to the lowest order using perturbation theory. This means that the Green's function propagator in (12.9) is approximated by the free propagator (in the presence of the field) (12.47). While this may give the impression that the physical effects of the scattering are being minimized, this is incorrect. The retarded Green's function (12.48) is still being solved in a self-consistent manner, and the energy corrections arising from the self-energy will certainly appear in Dyson's equation. In the present treatment, a nondegenerate electron gas is considered. This is equivalent to neglecting any screening of the optical phonons. Moreover, the phonons will be assumed to be in equilibrium and described by the normal Bose–Einstein distribution at the lattice temperature. The retarded phonon propagator is then

$$D_{0,\,\mathrm{r}}(\boldsymbol{q},\,\omega') = -i\pi\sum_{\nu=\pm1}|\,M_q\,|^2\left(N_q+\frac{1+\nu}{2}\right)\delta(\omega+\nu\omega_0-\omega'), \tag{12.49}$$

where the summation is for the emission and absorption of a phonon by the electron. The phonon energy is considered (in the Einstein approximation) to be constant. Here, the quantity M_q is the electron–phonon matrix element, and N_q is the Bose–Einstein distribution. This now leads to the following self-energy term:

$$\begin{aligned}\Sigma_{\mathrm{r}}(\boldsymbol{k},\,z,\,z',\,\omega) = \int\frac{d^3\boldsymbol{q}}{8\pi^2}\sum_{\nu=\pm1}|\,M_q\,|^2\left(N_q+\frac{1+\nu}{2}\right)\\ \times G_{\mathrm{r}}\left(E_{\boldsymbol{k}+k_z\boldsymbol{a}_z-\nu\boldsymbol{q},\,s'},\,\omega-\nu\omega_0\right).\end{aligned} \tag{12.50}$$

This form remains relatively simple, but it is still necessary to apply the Airy transform, which will provide adequate complication. The Airy transform (12.43) is applied to both longitudinal variables, and the self-energy may now be expressed as

$$\begin{aligned}\Sigma_r(\boldsymbol{k},\,s,\,s',\,\omega) = \int\frac{d^3\boldsymbol{q}}{8\pi^2}\sum_{\nu=\pm1}|\,M_q\,|^2\left(N_q+\frac{1+\nu}{2}\right)\\ \times\int ds_1\int ds_2 G_r(\boldsymbol{k}-\nu\boldsymbol{q},\,s,\,s',\,\omega-\nu\omega_0)\\ \times\int dz\int dz' As(z-s)As(z-s_1)As(z'-s')As(z'-s_2),\end{aligned} \tag{12.51}$$

where

$$As(y) = \frac{1}{L}Ai(y),\ y = \frac{E_{k,\,s}-\hbar(\omega+\nu\omega_0)}{\Theta},\ \Theta = \left[\frac{3(e\hbar F)^2}{2m^*}\right]^{1/3} = eFL. \tag{12.52}$$

Because G_0^{F} is diagonal in s, the resultant self-energy is also diagonal in s even after the transform has been taken. This just reflects the fact that in a homogeneous semiconductor, the uniform electric field enters the energy (12.41) through the term $eF(z-z')$, which should then transform into $eF(s-s')$, which is then essentially diagonal. However, both must be retained, for purposes that will become clear. The integrations can be performed using the known/published properties of the Airy function integrals [12]. The self-energy is then found to be [16]

$$\Sigma_{\mathrm{r}}(\boldsymbol{k},\,s,\,\omega) = \frac{(m^*)^{3/2}\,|\,M_q\,|^2\sqrt{\Theta}}{\sqrt{2}\hbar^2}\sum_{\nu=\pm1}\left(N_q+\frac{1+\nu}{2}\right)F(s,\,\omega), \tag{12.53}$$

with

$$\begin{aligned}
\mathrm{Im}\{F(s, \omega)\} &= Ai'(\zeta)Bi'(\zeta) + \zeta Ai(\zeta)Bi(\zeta), \\
\mathrm{Re}\{F(s, \omega)\} &= Ai'^2(\zeta) + \zeta Ai^2(\zeta), \\
Bs(\zeta) &= \frac{1}{L}Bi(\zeta) \quad . \\
\zeta &= [E_{k,\,s} - \hbar(\omega + \eta\omega_0)]/\Theta
\end{aligned} \tag{12.54}$$

Here, the primes denote derivatives with respect to the arguments, and *Bi* is the Airy function of the second kind.

The real part of the self-energy is plotted in figure 12.1 for an applied field of 100 kV cm^{-1} and for the parameters appropriate to a single type of equivalent intervalley phonon scattering in Si at room temperature (remember that the charge is negative). The dominant phonon is assumed to be the so-called *f*-phonon, which scatters between valleys on different {100} axes; e.g., the (001) and (010) axes. The oscillatory nature leads to the conclusion that the triangular potential is trying to create sub-bands in the energy structure, just like those that occur in Si MOSFETs. The positive parts are trying to lower the energy, while the negative parts are trying to raise the energy. These sub-bands are normal in quasi-two-dimensional materials and in the inversion layer of the transistor. This same tendency to form sub-bands is apparent in the imaginary part of the self-energy plotted in figure 12.2. This is evident from the tendency toward a step-like structure as it deviates from the normal square-root behavior of energy (dashed line). This arises from the differences between the density of final states in two dimensions versus three dimensions. The normal scattering rate is proportional to the three-dimensional density of states.

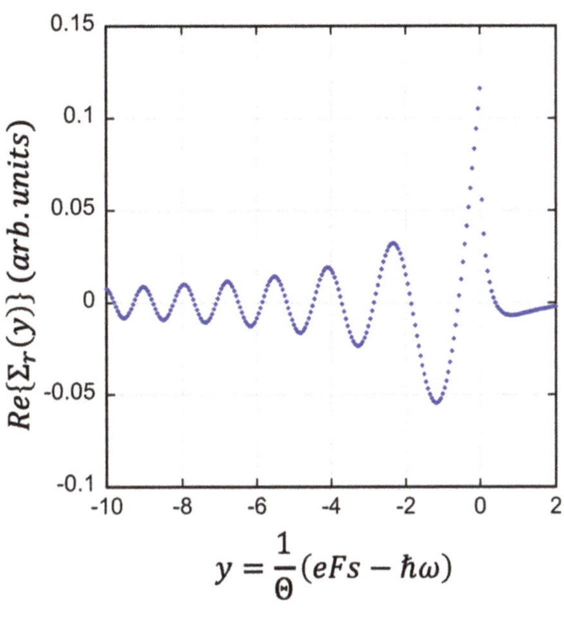

Figure 12.1. The real part of the self-energy for equivalent intervalley phonon scattering in Si at 300 K. Adapted from [16].

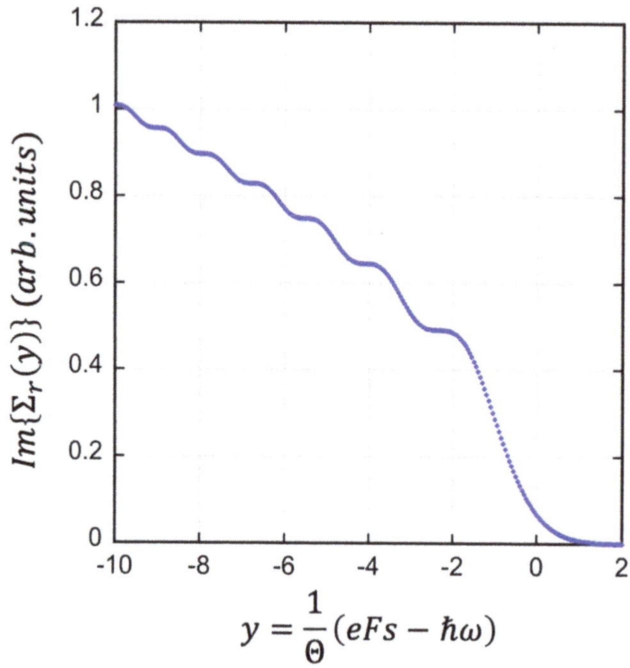

Figure 12.2. The imaginary part of the self-energy for Si at 300 K under an applied field of 100 kV cm^{-1}.

Since the self-energy is assumed to be diagonal in the Airy variable s, as discussed above, the integral (12.48) can be taken by assuming that a delta function in the two Airy variables is implied in the integrand. The result is then Dyson's equation, from which the retarded Green's function can be obtained as follows [16]:

$$G_r(\boldsymbol{k}, s, \omega) = \frac{1}{\hbar\omega - E_{k,s} - \Sigma_r(\boldsymbol{k}, s, \omega)}. \tag{12.55}$$

It may be noted immediately that this has the same *form* as the equilibrium retarded function, although the details are very different due to the presence of the inelastic phonon scattering. In equilibrium, there are no inelastic scattering events, but they are dominant in the nonequilibrium situation. The spectral density may immediately be found from the retarded Green's function using (12.2) (the spectral density is twice the imaginary part of the last equation). This spectral density is plotted in figure 12.3 for three different electric fields applied to the same Si sample as used in the previous two figures. These fields are 20 kV cm^{-1} (blue curve), 50 kV cm^{-1} (red curve), and 100 kV cm^{-1} (green curve). Each curve is shifted away from the zero point due to the intracollisional field effect, which provides additional acceleration during the collision event. Surprisingly, this shift is larger for the lower electric fields, which is probably due to the normalization to Θ, which is directly proportional to the field. Part of the spectral density is due to the deceleration of some carriers during the collision, and this appears as a shoulder at a lower energy than the major

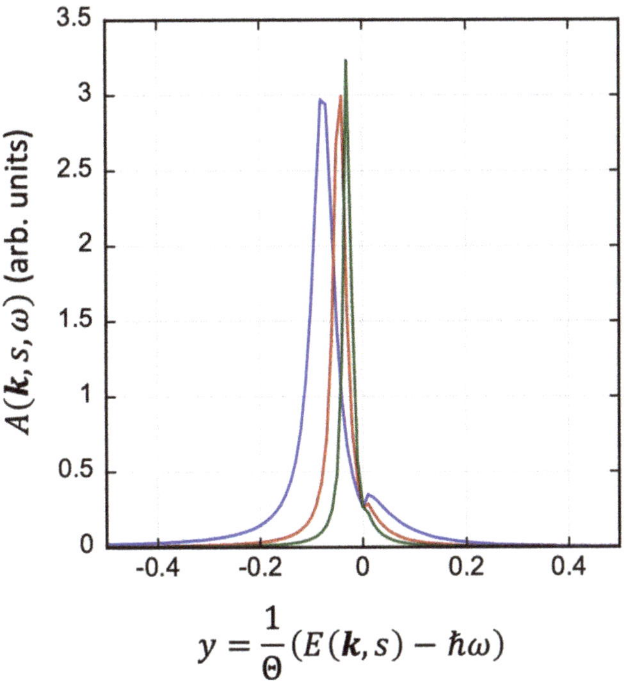

Figure 12.3. The spectral density obtained from the retarded Green's function (12.55) for the Si case. Three different values of the electric field are plotted: 20 kV cm^{-1} (blue curve), 50 kV cm^{-1} (red curve), and 100 kV cm^{-1} (green curve).

part of the curve, as seen in the 20 kV cm^{-1} curve. In equilibrium, the spectral density would simply be a Lorentzian line centered at zero energy.

12.2.3 The 'less-than' function

The 'less-than' Green's function originates from the above retarded functions, but it satisfies its own differential equation (12.3). This function gives a distribution function through the KBA (12.4), but this distribution is not the Wigner function used in chapter 11. Following the above treatment, the differential equations are Fourier transformed in the transverse variables and Airy transformed in the longitudinal (along the electric field). They are now written as

$$\begin{aligned}[\hbar\omega - E(\boldsymbol{k}, s)]G^{<}(\boldsymbol{k}, s, s', \omega) &= \int ds''[\Sigma_{\mathrm{r}}(\boldsymbol{k}, s, s'', \omega)G^{<}(\boldsymbol{k}, s'', s', \omega) \\ &\quad + \Sigma^{<}(\boldsymbol{k}, s, s'', \omega)G_{\mathrm{a}}(\boldsymbol{k}, s'', s', \omega)], \\ [\hbar\omega - E(\boldsymbol{k}, s')]G^{<}(\boldsymbol{k}, s, s', \omega) &= \int ds''[G_{\mathrm{r}}(\boldsymbol{k}, s, s'', \omega)\Sigma^{<}(\boldsymbol{k}, s'', s', \omega) \\ &\quad + G^{<}(\boldsymbol{k}, s, s'', \omega)\Sigma_{\mathrm{a}}(\boldsymbol{k}, s'', s', \omega)].\end{aligned} \tag{12.56}$$

It has already been established in the previous section that the retarded and advanced functions depend upon only a single Airy variable, and this eliminates the integration (the convolution integrals) on the right-hand side of the above equation. Thus, the above equation can be rewritten as

$$\begin{aligned}[\hbar\omega - E(\boldsymbol{k}, s)]G^{<}(\boldsymbol{k}, s, s', \omega) &= [\Sigma_{\mathrm{r}}(\boldsymbol{k}, s, \omega)G^{<}(\boldsymbol{k}, s', \omega) \\ &\quad + \Sigma^{<}(\boldsymbol{k}, s, \omega)G_{\mathrm{a}}(\boldsymbol{k}, s', \omega)], \\ [\hbar\omega - E(\boldsymbol{k}, s')]G^{<}(\boldsymbol{k}, s, s', \omega) &= [G_{\mathrm{r}}(\boldsymbol{k}, s, \omega)\Sigma^{<}(\boldsymbol{k}, s', \omega) \\ &\quad + G^{<}(\boldsymbol{k}, s, \omega)\Sigma_{\mathrm{a}}(\boldsymbol{k}, s', \omega)].\end{aligned} \tag{12.57}$$

At this point, it is normal to compute the sum and difference of these two equations. In the present case, this leads to the same equation, and this may then be directly solved for the 'less-than' function. Taking the sum of the two equations (12.57), one finds, after a little algebra, that the 'less-than' function reduces to [17]

$$G^{<}(\boldsymbol{k}, s, s', \omega) = G_{\mathrm{r}}(\boldsymbol{k}, s, \omega)G_{\mathrm{a}}(\boldsymbol{k}, s', \omega)\Sigma^{<}(\boldsymbol{k}, s, s', \omega). \tag{12.58}$$

It is easy to demonstrate that the difference of the two equations (12.57) leads to the same result (again, after a little algebra). From figure 12.3, it can be seen that the peak of the spectral density is sufficiently narrow as to confirm that the two Airy variables must be the same, hence it is not a severe approximation to assume that $s = s'$. One can also show that $G_{\mathrm{r}} - G_{\mathrm{a}} = 2\mathrm{Im}\{\Sigma_{\mathrm{r}}(\boldsymbol{k}, s, \omega)\}G_{\mathrm{r}}G_{\mathrm{a}}$. These results allow one to rewrite the 'less-than' function in the generalized Kadanoff–Baym manner as

$$\begin{aligned}G^{<}(\boldsymbol{k}, s, \omega) &= A(\boldsymbol{k}, s, \omega)f(s, \omega), \\ f(s, \omega) &= \frac{\Sigma^{<}(,s, \omega)}{2\mathrm{Im}\{\Sigma_{\mathrm{r}}(s, \omega)\}}.\end{aligned} \tag{12.59}$$

In this last line of the above equation, a quantum distribution function is defined that is consistent with the equilibrium form, although the present distribution function is clearly out of equilibrium. While the form of this distribution function appears to be simple enough, one must take care, as the details of the 'less-than' self-energy certainly make this equation far more difficult. Normally, (12.59) appears as the KBA [4] and only the energy ω is present in the argument of the distribution function. Here, however, the parameter s appears for convenience, where it will be important for the particle approach to be described in section 12.5.

The 'less-than' self-energy is found in the same manner as that used for the retarded self-energy. This will be developed for the case of the nonpolar intervalley phonons in Si. For the case in which the function is diagonal in the Airy variables, as assumed above, the self-energy can be written as [17]

$$\begin{aligned}\Sigma^{<}(\boldsymbol{k}, s, \omega) &= \frac{|M_q|^2}{3^{1/6}} \sum_{\nu=\pm 1}\left(N_q + \frac{\nu + 1}{2}\right)\int d^2\boldsymbol{q}_t \\ &\quad \times \int ds' As^2\left(\frac{s - s'}{3^{1/3}}\right)G^{<}\left(\boldsymbol{k} - \nu\boldsymbol{q}_t, s', \omega + \nu\omega_0\right)\end{aligned}. \tag{12.60}$$

This latter equation depends upon the 'less-than' Green's function. Hence, equation (12.59) leads to an integral equation for the distribution function (which is part of the 'less-than' Green's function). The integration over the phonon momentum $\boldsymbol{q}_t$ (which may equally be written as an integration over the transverse momentum) removes the dependence on the transverse momentum. As a result, the integral equation for the distribution function can be written as [17]

$$f(s, \omega) = \sum_{\nu=\pm 1}\left(N_q + \frac{\nu + 1}{2}\right)\int ds' K(s, s', \omega - \nu\omega_0) f(s', \omega - \nu\omega_0). \tag{12.61}$$

The kernel function is defined as

$$\begin{aligned} K(s, s', \omega) = &\frac{\sqrt{3}\,|\,M_q\,|^2\, m^*}{\hbar \mathrm{Im}\{\Sigma_r(s, , \omega)\}} A s^2\left(\frac{s - s'}{3^{1/3}}\right) \\ &\times\left[\frac{\pi}{2} + a\tanh\left(\frac{\hbar\omega - E_{k,s'} - \mathrm{Re}\{\Sigma_r(,s', \omega)\}}{\mathrm{Im}\{\Sigma_r(,s', \omega)\}}\right)\right]. \end{aligned} \tag{12.62}$$

A representation of the 'less-than' Green's function is shown in figure 12.4 [18]. It is clear from this latter figure that the tendency to form quasi-two-dimensional sub-bands is present in this function as well, as indicated by the quasi plateaus that appear at higher energy.

The Wigner function cannot be determined from this distribution function because the Wigner function exists in phase space (position and momentum). Thus, the full form of the 'less-than' Green's function (12.59) must be used, as it

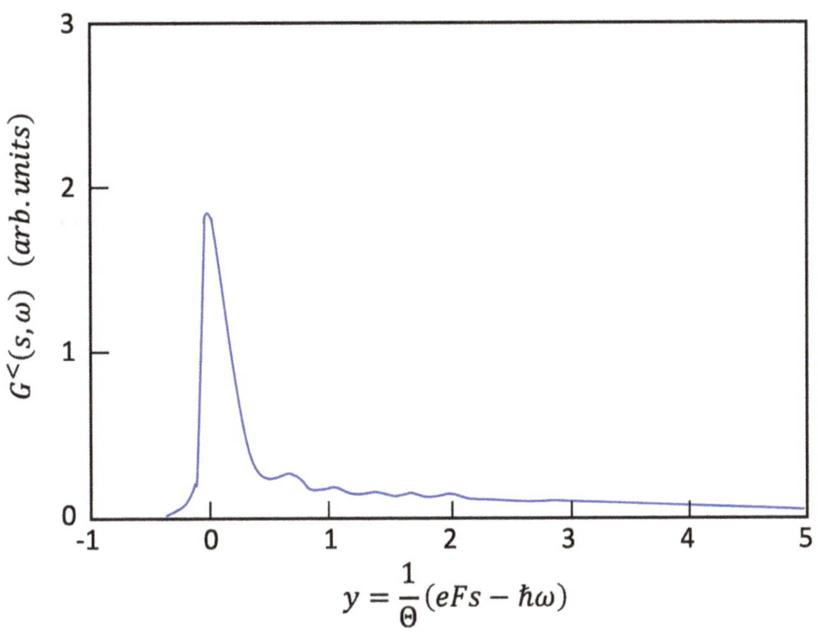

Figure 12.4. The 'less-than' Green's function in the Airy representation for a high electric field of 500 kV cm^{-1}. Adapted from [18].

is the spectral density that transforms the energy ($\hbar\omega$) into the momentum, which is necessary to gain the full phase space. The 'less-than' function must then be retransformed from the Airy coordinates to the spatial coordinates.

While an integral equation has resulted from the above development, this at first sight is not the path variable type of integral equation that is usual for a Monte Carlo particle simulation of NEGFs. The normal particle Monte Carlo has alternating processes of drift (transport) and scattering. Obviously, a significant part of $K(s, s', \omega)$ relates to the scattering processing, but it is not clear that it also contains a transport part. The prefactor and the term in the square brackets are parts of the scattering process (the square brackets represent part of the final state density). So, the only part that could relate to transport is the $As(\cdot)$ term. The argument of this function varies according to $F^{2/3}$ (and is squared as well), while the amplitude varies according to $F^{1/3}$. And it is an oscillatory function, so that it might be recognizable as a 'probability' of transport through a distance $s - s'$. This connection will be explored in section 12.5. This possibility is enhanced because this is the first form of NEGF that has been achieved which has even a hint of possible use in a particle simulation approach.

12.3 The limitations of NEGFs

Many groups today use NEGFs in the simulation and modeling of semiconductor devices. This approach is usually based on the Keldysh [10] formulation and integration path. It has already been seen that the scattering 'less-than' self-energies Σ depend on the correlation functions $G^<$ and $G^>$ and on the Green's functions of the phonons. This accounts for the occupancy of the phonon states and also depends upon the phonon energies. Keldysh himself used the S-matrix from equilibrium theory, but this approach *only works when the system is close to equilibrium and the Hamiltonian remains Hermitian*. It cannot be expected to be correct in the far-from-equilibrium case. Keldysh also assumed that the interaction representation would lead to a unitary time-translation operator. Again, this assumption can be expected to fail when the Hamiltonian is no longer Hermitian, e.g. with complex self-energies. Danielewicz has shown that when correlated initial conditions exist, the S-matrix is questionable at best [19].

The entire concept of the Keldysh contour is debatable. When electric fields or forces lead to currents in a semiconductor device, the system breaks time-reversal symmetry. If the device reaches a time-invariant steady state, this arises from a balance between the driving forces (derived from the applied potentials) and the dissipative forces (due to the scattering processes). These driving forces are not applied in an adiabatic manner but are switched on during a very short time span. Indeed, modern logic devices require switching times on the picosecond-to-subpicosecond scale [20]. When the device is switched on, there is a transient evolution to the stable state, if one exists. During this process, the device parameters evolve though a variety of intermediate conditions: e.g. homogeneous, inhomogeneous, linear, non-linear, etc. When the voltage is removed and the device relaxes toward its resting state, *the evolution is not the reverse of the excitation trajectory, its course may well pass through different phases.* This excitation/relaxation cycle almost certainly involves significant hysteresis. The adoption of a time-ordered contour is counter-intuitive and clearly contrary to this behavior.

Another important point related to the inappropriate use of the Keldysh path is the tendency for the normalization to be to the true ground state arising from the full Fermi sea at zero Kelvin, as is the case in true equilibrium at zero temperature. This assumption in the Keldysh path allows for the cancellation of disconnected diagrams in the scattering expansion, as illustrated in figure 5.5. But *this cancellation does not appear in nonequilibrium situations*, where the normalization is performed using the quantum distribution function and not the Fermi ground state [21]. Moreover, the expansion of the scattering from a unitary operator (such as that used in the Keldysh approach) into a perturbation series has never been shown to converge [22]. This is especially true when the terms have their own included commutators, which leads to nested commutators that create especially difficult diagrams. This is even more true for current–current correlation functions used to compute the conductivity [1]. The result of these issues is thought to cause a failure of Wick's theorem [23], and one then needs doubled diagrams (such as those used in the Bethe–Salpeter equation). Terms also appear that do not arise from normal Wick contractions, and this may lead to many complicated additional terms in the doubled diagrams [24].

From these considerations, one may conclude that basing NEGFs upon the Keldysh approach is quite a simplistic approach to quantum transport in real devices and may even risk the applicability of any results obtained. Moreover, the avoidance of two-particle Green's functions is a crude approximation that likely results in very serious deficiencies in the simulations. In processes that have a coulomb nature, such as impurity scattering and the polar optical phonon interactions, the avoidance of disconnected graphs can lead to significant phase errors that truly affect any quantum interferences in the conductance, as shown in figure 5.1. It was also shown that these phase interferences are responsible for the well-known conductance fluctuations in real nanoscale devices, in which only a small, limited number of impurities exist within the active regions.

Nevertheless, NEGFs have been used for quite some time for the simulation of semiconductor devices in total disregard of the above limitations. In the next section, our attention will turn to a discussion of some of these approaches. Yet it is important to keep the above considerations in mind when considering the usefulness of such approaches. Finally, it should be remarked that it is generally assumed, particularly in mesoscopic systems, that the optical phonon scattering is 'phase breaking.' That is, coherence is carried in the phase of the wave function, and it is exactly this phase that is modified by the relatively large phonon energy of either the emission process or the absorption process. Hence, the scattering event modifies the phase of the wave function through the energy change, and it is believed that this destroys any coherence of the wave function. Whether this is true or not has never been proven, but the argument gives a rationale for continued use of NEGFs for device simulation.

12.4 NEGFs in devices

The use of NEGFs in semiconductor devices has a relatively long history. These functions were used at low temperatures to evaluate the average electric field in an inversion layer as a function of the depletion and inversion charge densities [25]. Here, the conductance was determined using the full integral equation arising from the Bethe–Salpeter equation for coulomb scattering. This approach was then extended to Si MOSFETs at room temperature [26] and to InAlAs/InGaAs heterostructures at low temperature [27]. The approach was finally extended to the universal mobility in Si MOSFETs [28] and to the quantum effects on the gate capacitance [29].

One important aspect of the development of NEGFs for semiconductor devices is the recognition that the full energy-band structure is needed to evaluate the dynamics of carriers with high energies that arise in nanostructured devices. The coupling of band structure to Monte Carlo techniques was initiated by Shichijo and Hess [30]. Their approach utilized a local empirical pseudopotential band structure without spin–orbit coupling (see chapter 6). Subsequent development led to the well-known results from IBM for the DAMOCLES code [31, 32]. But this was still classical transport. The earliest extensions to quantum transport came from the semiconductor group in Rome, which developed the full band structure from first-principles tight-binding approaches and applied these to quantum transport in semiconductor nanostructures [33] and molecules [34]. This approach used exact exchange in the density functional to correct the band structure. A somewhat later development was the well-known Spanish Initiative for Electronic Simulations with Thousands of Atoms (SIESTA) program, which used first principles for tight binding and NEGFs for transport (leading to Trans-SIESTA) [35, 36]. The nano-electronic modeling (NEMO) approach used semiempirical tight-binding methods for the band structure and NEGF for the transport [37] and was first developed to study resonant-tunneling diodes. A more modern NEGF tool, which also uses an atomistic band structure from which the effective masses at various energies are extracted, has been developed in Glasgow [38]. Finally, another full-band NEGF simulator using empirical pseudopotentials has been developed in France [39]. However, the basics of the NEGF approach began to be applied to MOSFETs more than a decade ago [40–43].

12.4.1 MOSFETs

Some of the early work on MOSFETs extended the treatment reported earlier [41] to study the analog performance of double-gate devices with nanoscale gate lengths [44]. The authors of this paper observed very high transconductance at a 10 nm channel length, as the device had only limited scattering in the channel. They also observed cutoff frequencies which implied that the velocity was limited by the injection velocity at the source end of the channel. Although the authors did not discuss it, one expects to see significant drain-induced barrier lowering in devices with such limited scattering [45]. Somewhat later, the effect of remote coulomb scattering (scattering from impurities in the gate oxides) was studied in a full quantum NEGF approach to gate-all-around Si nanowire MOSFETs [46] using a modern high-dielectric constant oxide. These latter authors found that the performance of these devices was degraded by the inefficiency of the charge screening of these remote impurities. This led to a significant increase in the subthreshold slope and enhanced variability of the threshold voltage. It also led to a reduction in the impurity-limited mobility.

The role of fixed charges in Si tunnel field-effect transistors (FETs) was recently examined while taking full account of the atomistic positions of the charge in real space, which clearly showed how a few random dopants can greatly affect nanowire FETs [47]. Here, random dopants in the InAs drain region of a Si-channel tunnel FET were also examined. In other work, the role of phonon scattering in multi-material nanowire gate-all-around FETs (using Si, InAs, and GaAs) and FinFETs [48] was studied with NEGFs. Figure 12.5 shows the current (as a function of energy) overlaid on the potentials for two Si nanowire FETs of 2.2×2.2 nm^2 and

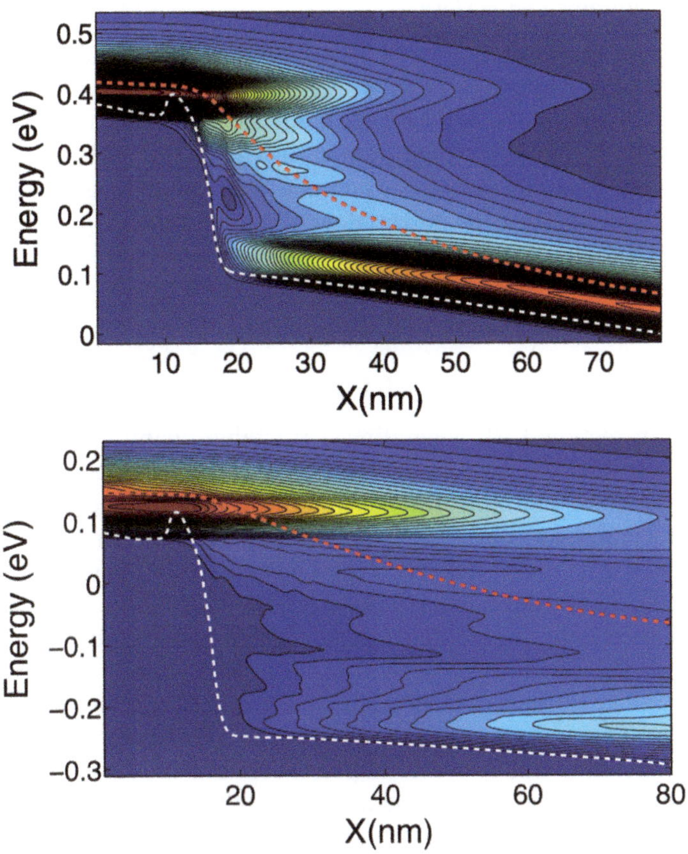

Figure 12.5. Current spectra obtained for Si nanowire FETs of (top) 2.2×2.2 nm^2 and (bottom) 36×3.6 nm^2. The drain region is 64 nm in length and $V_G = 0.9$ V. The dotted white line represents the first sub-band and the dashed red line represents the current–energy variation in the device (basically the integral over the cross-section of the product of current and kinetic energy). This dashed red line clearly shows that the hot carriers have not relaxed even after 65 nm of drain. Reproduced from [48]. CC BY 4.0.

3.6×3.6 nm^2 cross-sections. It may be seen that the device with the larger cross-section has less change in the current. It is remarkable that the hot electrons have not relaxed in the drain region of the larger device even beyond 65 nm into this region. Even the carriers that do relax are reaccelerated, even within the drain region. This behavior arises because the field is more than 8 kV cm^{-1} in the drain, clearly too high a field to allow equilibration of the carriers. The authors concluded from the comparison between the two devices that the larger density of states in the smaller wire led to a larger dissipative scattering than that of the slightly larger wire.

Others have also investigated the role of drain doping on nanowire tunnel FETs [49] and random doping in Si and GaAs FETs [50], evaluating the effect of the dopant atom on the overall energy structure using density-functional theory (DFT). Others have studied the role of a SiO_2 layer between the high-dielectric-constant gate insulator and the channel in FinFETs [51].

12.4.2 Other materials and devices

Studies of bilayer graphene tunnel FETs have been carried out using NEGFs for layers with armchair edges [52]. Later work found that the presence of strain varies the energy gap opened in the bilayer system and increases the effect of an interlayer electric field [53]. Thus, such structures are suitable for consideration in graphene logic devices [54]. Even later studies examined the role of vacancies [55], different substrates [56], and tailoring the potential with implantation [57]. NEGFs have even been used in the study of transport in single layers of phosphorene [58]. Atomistic band structure and NEGFs have also been used to study quantum ballistic transport in single sheets of MoS_2 [59].

12.5 The use of particles in NEGFs

As remarked in section 12.3, generating a tractable quantum transport equation using NEGFs has proven to be a difficult task. When one incorporates both the collisional broadening and the intracollisional field effect, it becomes more difficult adapt to the small space and time behaviors of modern semiconductor devices. However, at least part of this approach can be handled with particles using the Airy function transformations discussed in that section. Most of the problems in dealing with NEGFs arise from the coherence that is maintained throughout the transport region, even in the presence of scattering. This is especially true in the case of impurity scattering, which is coulombic in nature and therefore exists over relatively large distances. This was illustrated in figure 5.1. The failure of this scattering to destroy the coherence leads to the interferences that are observed in this latter figure as well as the conductance fluctuations exhibited in figure 3.8. It is easy to extend this caution to polar optical phonon scattering, as it too is coulombic in nature. Nevertheless, these concerns do not dissuade one from the use of particles in simulation. Rather, they merely confirm that care must be taken in such approaches.

The method by which particle simulations of NEGFs can proceed is based upon the results of section 12.3, in which the 'less-than' Green's function, or rather the corresponding particle distribution function, can be expressed as

$$f(s, \omega) = \frac{1}{\mathrm{Im}\{\Sigma_r(s, \omega)\}} \sum_{\nu=\pm 1} \left(N_q + \frac{\nu + 1}{2} \right) \\ \times \int ds' S(s, s', \omega + \nu\omega_0) As^2\left(\frac{s - s'}{3^{1/3} L} \right) f(s', \omega + \nu\omega_0) , \tag{12.63}$$

where

$$S(s, s', \omega + \nu\omega_0) = \frac{\sqrt{3} \mid M_q \mid^2 m^*}{\hbar} \left[\frac{\pi}{2} + a \tanh\left(\frac{\hbar\omega - E(\boldsymbol{k},s') - \mathrm{Re}\,\{\Sigma_r(,s', \omega + \nu\omega_0)\}}{\mathrm{Im}\,\{\Sigma_r(,s', \omega + \nu\omega_0)\}} \right) \right]. \tag{12.64}$$

This can now be compared to the equivalent integral equation for the Boltzmann transport equation. However, there is one significant difference, which is that the exponential in time in the latter equation is replaced by the square of the Airy function in position. Instead of the probabilistic setting of the drift time by the exponential, here, the position function sets the drift *length* probabilistically. That is, the distance s–s' determines the energy gained from the electric field as the particle accelerates from position s' to s. This is normalized by the characteristic length L, which itself is dependent upon the electric field [60].

12.5.1 The silicon model

In order to evaluate the usefulness of the particle approach, example simulations will be discussed in this section. For this purpose, Si high-field transport will be considered. Handling electrons in the conduction band is complicated by the multivalley nature of this band, since the conduction band has six equivalent ellipsoids centered in the {100} directions (see figure 6.9). The scattering processes themselves are very local; that is, the optical phonons are momentum independent and thus act like delta-function scatterers in real space. These do not require the more complicated higher-order corrections, and one only needs the simpler Fermi golden rule, with, of course, the modifications for the quantum case. The six equivalent ellipsoids in the conduction band are located about 85% of the way to the zone edge at X (figure 6.9). In order to maintain the balance among these valleys, the electric field will be taken in the (111) direction so that the field makes the same angle with each of the six valleys. The carrier scattering within each ellipsoid (*intra*valley scattering) arises from acoustic phonons and impurities, as the optical processes are forbidden. Acoustic-mode scattering is governed by the deformation potentials, whose values are determined by the two constants Ξ_u and Ξ_d. These are thought to have values of 9 eV and −6 eV, respectively [61]. The *effective* deformation potential is then about 3 eV. Nonpolar optical phonon scattering occurs between equivalent ellipsoids (*inter*valley scattering). Two types of phonons are involved in this scattering. One of these is the *g*-phonon, and this couples the two valleys that lie at opposite positions along the same (100) axis. This is an *umklapp* process, so that after reduction by a reciprocal lattice vector, it has a net phonon wave vector of $0.3\pi/a$. The symmetry properties of the crystal allow only the longitudinal optical (LO) mode to contribute to this scattering. At the same time, *f*-phonons couple the (100) valley to the (010) and (001) valleys, that is, they couple valleys on different {001} axes. The wave vector for this latter process has a magnitude of $2^{1/2}(0.85)\pi/a = 1.2\pi/a$, which again is an umklapp process. This wave vector lies in the square face of the second Brillouin zone along the extension of the (110) line into this second Brillouin zone [62]. The reduced phonons for this lie near the X-point in value but have a different symmetry. Again, only the upper branches are allowed.

The energies of the LO *g*-phonon and the longitudinal acoustic (LA) and transverse optical (TO) *f*-phonons nearly all have the same value, since, as remarked, the low-energy intervalley phonons are forbidden. However, Long [63] found that a weak low-energy intervalley phonon is required to fit the data for experimental mobility versus temperature. He treated the allowed high-energy phonons as a single equivalent intervalley phonon of 64.3 meV but introduced a low-energy phonon with an energy of 16.4 meV. The presence of these low-energy phonons was confirmed in studies of magneto-phonon resonance (in which the magneto-resistance shows a resonance when the phonon frequency is equal to a multiple of the cyclotron frequency) in Si inversion layers [64]. Evidence for this low-energy optical phonon was also found in simulations that aimed to fit the time-of-flight studies of velocity in Si [65]. Thus, it appears that the low-energy phonons, while weak scatterers, are necessary to understand the drift velocity.

The low-energy phonons are certainly forbidden by symmetry. Long treated them with a very weak coupling constant. It is more likely that this intervalley phonon may be treated by a first-order interaction that breaks the symmetry [66, 67]. This interaction is found to fit the data with a deformation potential of $\Xi_0 = 5.6$ eV. The allowed, higher-energy transition is treated with a interaction potential of $D = 9 \times 10^8$ eV cm^{-1}. This two-phonon model will be used in the following discussion of Si. The acoustic phonon scattering is also included. However, impurity scattering will not be included, as a focus on the phonons is considered appropriate to address the NEGFs.

The conduction bands away from the minima at Δ are known to be non-parabolic. This behavior arises more from repulsion between the normal conduction band and the upper (second) conduction in these directions than from a valence band interaction. Following the discussion in chapter 6, this non-parabolicity is handled with the normal hyperbolic bands with an effective 'energy gap' of 2.1 eV. This non-parabolicity will be important when the velocities are determined from the energy–momentum bands. The lattice temperature is uniformly assumed to be 300 K, and 10^5 particles are used in the simulation.

The retarded self-energies are easily calculated using the Fermi golden rule along with the inherent quantum broadening that arises from the properties of the spectral density following (12.48) and (12.50). The real and imaginary parts of the retarded self-energy are plotted in figure 12.6 for three different values of the electric field. Oscillations in the real part of the self-energy exhibit a quasi quantization in the density of states that is characteristic of two-dimensional sub-bands. The electric field tends to create a triangular potential well, known from introductory quantum mechanics, and this confinement provides the quantization. When the real part is positive, the energy is reduced. When it is negative, the energy is increased. This forms the step-like density of states, although here it is certainly not a true discretization of this effect. This tendency toward sub-band formation is also evident in the imaginary part of the self-energy.

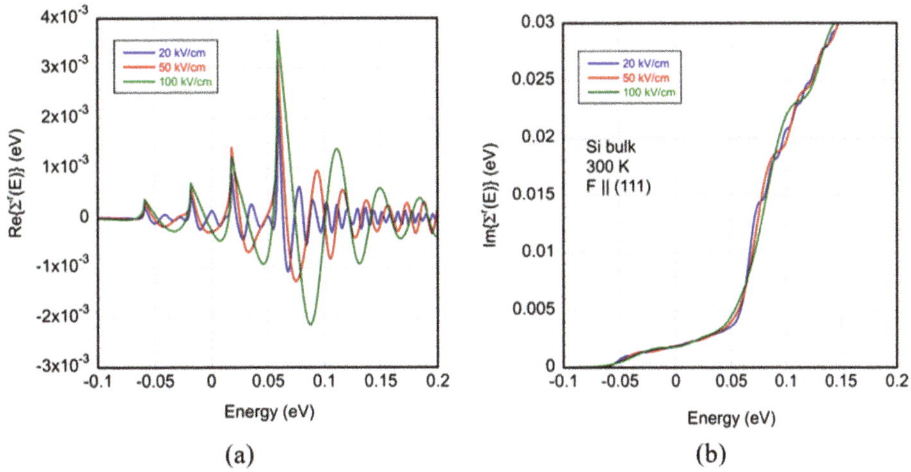

Figure 12.6. (a) The real part of the retarded self-energy at three different values of a high electric field for Si at room temperature. (b) The imaginary part of the retarded self-energy at three different values of a high electric field for Si at room temperature.

12.5.2 The Monte Carlo sequence

As may be seen from (12.63), the various steps followed in the Monte Carlo simulation follow a similar flow to that of the Boltzmann equation. In the present case, however, the exponential in time in (9.8) is replaced by the Airy function in space. That is, the temporal drift of the carriers is replaced by a spatial drift. Nevertheless, the sequence is drift first, then scattering in (12.64). The imaginary part of the retarded self-energy sits outside the integral and is only a final adjustment to the distribution function. In this process, the spatial motion of the particle arises from a random number r, which generates a *step* in space according to

$$r = \int_{-u_0}^{u} As^2(u')du' \Bigg/ \int_{-u_0}^{\infty} As^2(u')du', \tag{12.65}$$

so that

$$\Delta s = 3^{1/3} L u. \tag{12.66}$$

The Airy function has values over a semi-infinite range of values, and this creates a problem. However, since the scattering function in (12.64) seems to be taken over a single effective sub-band (the values of the bracketed term), a limit can be applied to $-u_0$, and this is taken to be the first zero of the Airy function. There is now a single maximum in the quantity integrated in (12.65). The range of values for Δs allows for the selection of a single value by the random number in (12.65). A different random number is used for each particle in the ensemble. Averaging over the ensemble and also over the number of iterations in the simulation may be used to determine an 'average' value of Δs. This resulting average is plotted as a function of the electric field in figure 12.7.

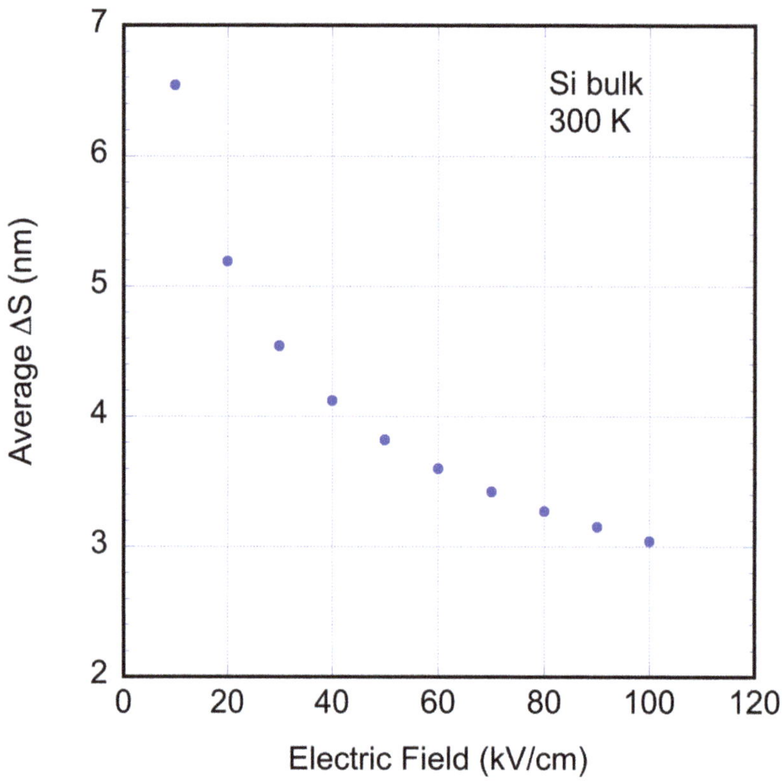

Figure 12.7. The acceleration distance Δs averaged over the ensemble and the number of iterations as a function of the electric field applied to the Si sample.

Once the value of Δs is determined for a particle, the energy of that particle is increased (or decreased depending upon the sign of Δs) as follows:

$$E = E + eF\Delta s. \tag{12.67}$$

In addition, the acceleration time may be determined from classical mechanics (but using the analytical hyperbolic band model to incorporate the non-parabolicity) to be

$$\Delta t = \sqrt{\frac{2m^* |\Delta s|}{eF}}\sqrt{1 + \frac{E}{E_G} + \frac{eF\Delta s}{2E_G}}, \tag{12.68}$$

where the average energy over the drift is used. The resulting change in momentum is

$$\Delta k = \frac{eF\Delta t}{\hbar}\text{sign}(\Delta s). \tag{12.69}$$

Finally, the new velocity of the particle is given, in the non-parabolic format, by

$$v_i = \frac{\hbar}{m_c}k_{z,\,i}\left(1 + \frac{E}{E_G}\right)^{-1}, \tag{12.70}$$

where m_c is the band-edge mass in the non-parabolic bands. (The mass correction is also inherent in (12.68).) Hence, the acceleration in the field gives the dynamical parameters that would be found in any semiclassical Monte Carlo procedure. From here, normal ensemble averages can be taken for any of these dynamical parameters.

In the above, it is important to note that Δs can be either positive or negative. In the latter case, the field is in the opposite direction to the momentum change, and the sign has to be maintained for (12.69) since Δt is always a positive quantity.

12.5.3 Results

As with the retarded functions, the scattering function that appears in (12.64) does not include the distribution function or the 'less-than' Green's function, so it can be evaluated outside the iterative loop. This scattering function does depend upon the electric field, however, and more so than the retarded functions. The total scattering function, which includes all processes plus emission and absorption, is shown in figure 12.8 for three values of the electric field. However, it is not this total rate that is used in the computation, but each individual scattering rate, just as in semi-classical Monte Carlo simulations. The particular rate for each particle (at each

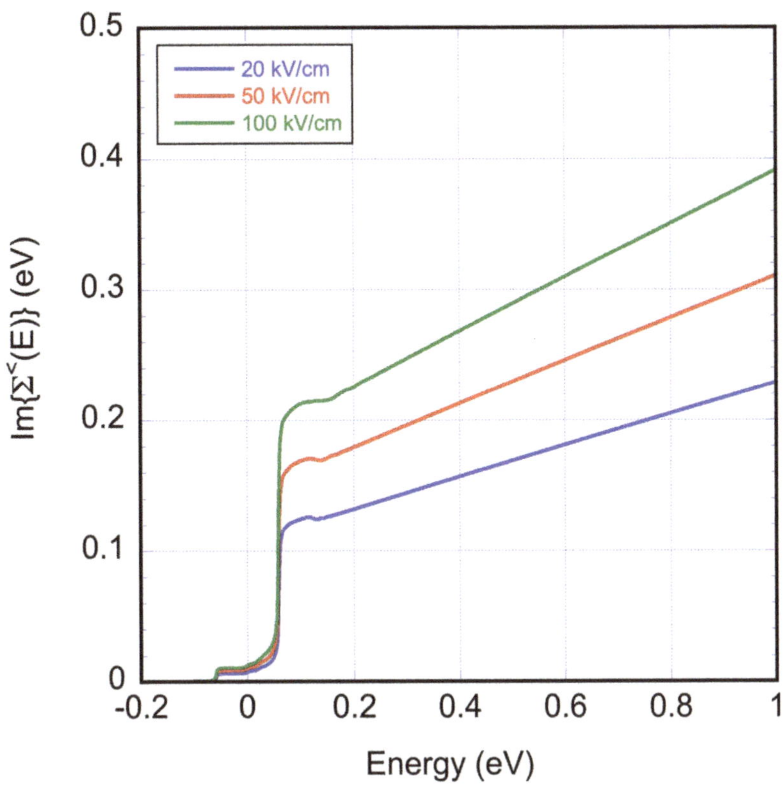

Figure 12.8. The total of all scattering rates that contribute to the 'less-than' self-energy. Each individual rate is utilized in the Monte Carlo simulation.

iteration) is determined randomly according to the relative size of each individual rate, just as is normally done.

In the simulation, each particle is followed for 200 iterations of drift and scattering. The parameters are then averaged over the particle ensemble and these 200 iterations. In figure 12.9, the average drift velocity is shown as a function of the electric field. While it appears that error bars are not shown, they are actually smaller than the dots used to depict the data points. The velocity is found to converge to a steady value within the first few iterations. Common wisdom suggests that velocity saturation takes place at around 10^7 cm s^{-1}. Obviously, a higher value is found here. Classically, the commonly expected value is found to arise from the rapid energy rise in scattering of the first-order low-energy intervalley phonon, as its more rapid rise with energy leads to its dominance under very high fields. In the quantum case, this first-order scattering is especially restricted by the intracollisional field effect [68]. Hence, it is not surprising that a larger 'saturated' velocity is found in quantum transport.

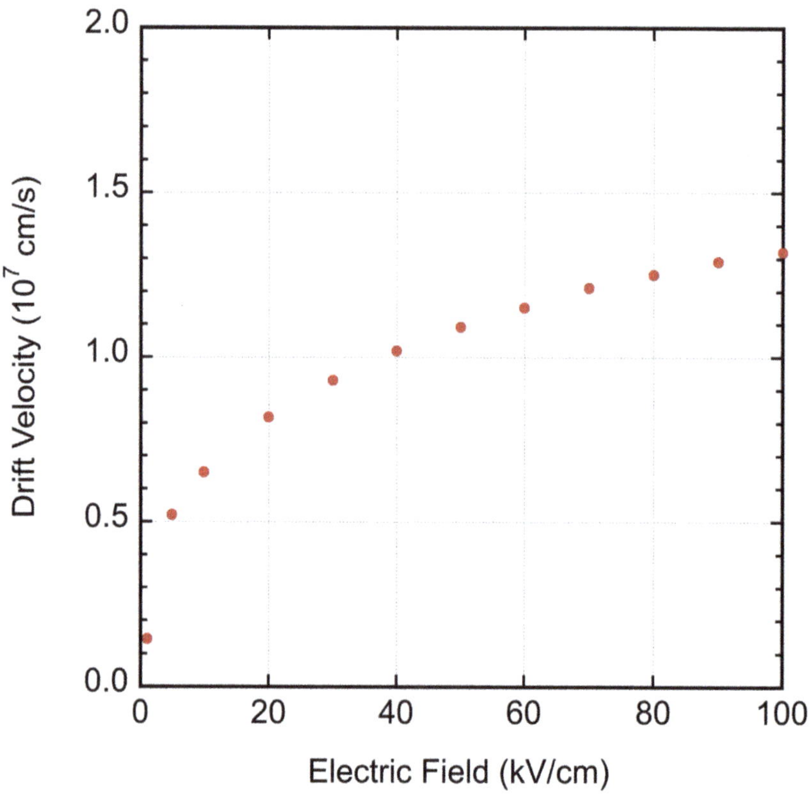

Figure 12.9. The average drift velocity as a function of the electric field. Saturation appears to occur at higher values of the velocity than in the semiclassical case.

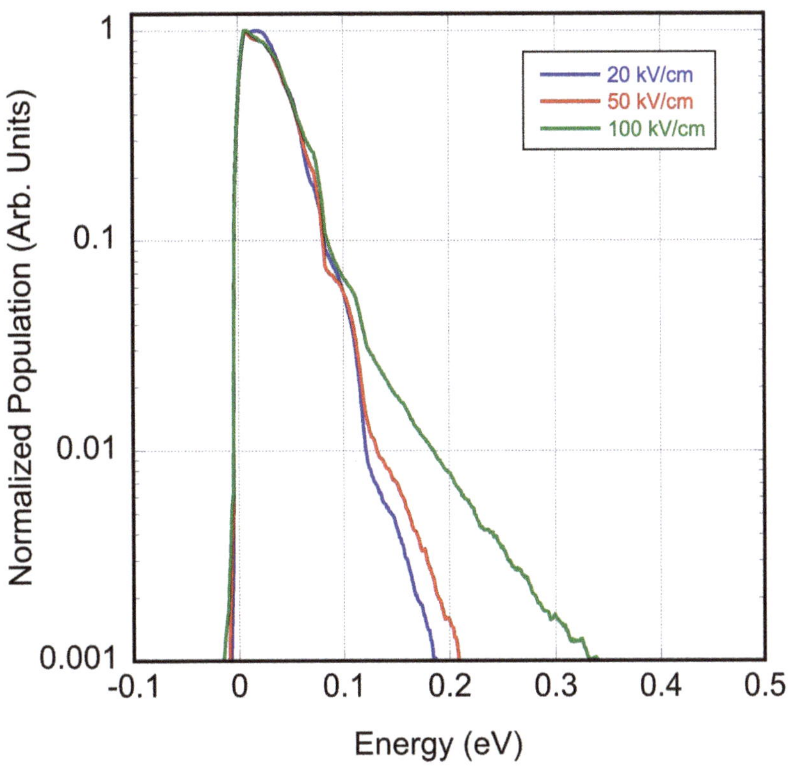

Figure 12.10. Relative particle populations as a function of energy for three different electric fields.

The particle distribution function is shown for these same three electric field values, used in early figures, in figure 12.10. This is not strictly the distribution function but the particle population as a function of energy. In classical situations, this would have to be corrected using the density of states at each energy. Also, the three curves are normalized with the peak at unity. The prime interest here is the spread of the energy distribution that increases with the electric field. Eventually, this spread will be sufficient to push some carriers to a high enough energy for impact ionization to begin, which is generally near 200 kV cm^{-1} in Si. It is interesting that kinks occur at about the same energy in all three curves, whereas it might be expected these would shift with the electric field if they were due to the tendency toward sub-band formation, as occurs in figure 12.6(b). Nor are these kinks at the energies that might be expected for phonon emission, as the energies do not correspond to any phonon energy. Further, the kinks in the slopes of figure 12.8 are at different energies. For example, at 50 kV cm^{-1}, the dip is about 7% of the total value and occurs near 138 meV, whereas at 100 kV cm^{-1}, the dip is an order of magnitude smaller and occurs near 118 meV. Again, these energies do not correspond to any phonon energy. These variations need to be examined further to determine their origins.

Finally, the speed of these particle-based computations is extraordinary. All computations were performed on a MacBook Pro with the Apple M1 Pro chip. A typical computation of all the functions at a single field value, with 200 iterations of the acceleration–scatter loop, required approximately 1.6–2.1 s, depending upon the field itself.

References

[1] Fetter A L and Walecka J D 1971 *Quantum Theory of Many-Particle Systems* (New York: McGraw-Hill)

[2] Ferry D K 2017 *An Introduction to Quantum Transport in Semiconductors* (Singapore: Jenny Stanford Publishing)

[3] Martin P C and Schwinger J 1959 *Phys. Rev.* **115** 1342

[4] Kadanoff L P and Baym G 1962 *Quantum Statistical Mechanics* (New York: W. A. Benjamin)

[5] Lipavský P, Ŝpiĉka V and Velický B 1986 *Phys. Rev.* B **34** 6933

[6] Feynman R P 1948 *Rev. Mod. Phys.* **20** 367

[7] Mahan G D and Hänsch W 1983 *J. Phys.* F **13** L47

[8] Mahan G D 1992 *Quantum Transport in Semiconductors* ed C Jacoboni, L Reggiani and D K Ferry (New York: Springer) pp 101–40

[9] Mahan G D 1984 *Phys. Rep.* **110** 321

[10] Keldysh L V 1965 *Sov. Phys. JETP* **20** 1018 http://jetp.ras.ru/cgi-bin/e/index/e/20/4/p1018?a=list

[11] Langreth D C 1976 *Linear and Nonlinear Electron Transport in Solids* ed J T Devreese (New York: Plenum) pp 3–32

[12] Aspnes D E 1966 *Phys. Rev.* **147** 554

[13] Abramowitz M and Stegun I 1964 *Handbook of Mathematical Functions with Formulas, Graphs, and Mathematical Tables* (Washington, DC: National Bureau of Standards)

[14] Ferry D K 2021 *Quantum Mechanics* 3rd edn (Boca Raton, FL: CRC Press) p 51

[15] Jauho A P and Wilkins J W 1984 *Phys. Rev.* B **29** 1919

[16] Bertoncini R, Kriman A M and Ferry D K 1989 *Phys. Rev.* B **40** 3371

[17] Bertoncini R, Kriman A M and Ferry D K 1990 *Phys. Rev.* B **41** 1390

[18] Bertoncini R, Kriman A M and Ferry D K 1990 *J. Phys.: Condens. Matter* **2** 5991

[19] Danielewicz P 1984 *Ann. Phys.* **152** 239

[20] Chau R *et al* 2005 *IEEE Trans. Nanotechnol.* **4** 153

[21] Dyson F 1949 *Phys. Rev.* **75** 1736

[22] Dyson F 1951 *Phys. Rev.* **82** 428

[23] Dyson F 1951 *Phys. Rev.* **83** 608

[24] Walhout T S 1999 *Phys. Rev.* D **59** 065009

[25] Vasileska D *et al* 1995 *J. Vac. Sci. Technol.* B **13** 1841

[26] Vasileska D *et al* 1996 *Physica* B **227** 333

[27] Vasileska D, Eldredge T and Ferry D K 1996 *J. Vac. Sci. Technol.* B **14** 2780

[28] Vasileska D and Ferry D K 1997 *IEEE Trans. Electron Devices* **44** 577

[29] Vasileska D, Schroder D K and Ferry D K 1997 *IEEE Trans. Electron Devices* **44** 584

[30] Shichijo H and Hess K 1981 *Phys. Rev.* B **23** 4197

[31] Laux S E and Fischetti M V 1988 *IEEE Electron Device Lett.* **9** 467

[32] Fischetti M V and Laux S E 1988 *Phys. Rev.* B **14** 9721

[33] Di Carlo A *et al* 1996 *Solid State Commun.* **98** 803
[34] Solomon G C *et al* 2006 *J. Chem. Phys.* **125** 184702
[35] Ordejón P *et al* 1996 *Phys. Rev.* B **53** 10441
[36] Branbyge M *et al* 2002 *Phys. Rev.* B **65** 165401
[37] Brown R C *et al* 1997 *J. Appl. Phys.* **81** 3207
[38] Berrada S *et al* 2020 *J. Comput. Electron.* **19** 1031
[39] M'foukh A, Pala M G and Esseni D 2020 *IEEE Trans. Electron Devices* **67** 5662
[40] Yu Z *et al* 2001 *Proc. SPIE* **4600** 117–22
[41] Ren Z *et al* 2001 *Proc. Int. Electron Devices Meeting* (New York: IEEE Press) p 107
[42] Ogawa M, Tsuchiya H and Miyoshi T 2002 *Int. Conf. on Simulation of Semiconductor Processes and Devices* 1034568 261–6
[43] Jovanovic D 2002 *Int. Conf. Modeling Simulation and Microsystems—MSM* 2002 pp 478
[44] Jiménez D *et al* 2004 *J. Appl. Phys.* **96** 5271
[45] Ferry D K *et al* 2020 *Entropy* **22** 1103
[46] Poli S and Pala M G 2009 *IEEE Trans. Electron Devices* **56** 1191
[47] Carrillo-Nuñez H *et al* 2018 *IEEE Electron Device Lett.* **39** 1473
[48] Martinez A *et al* 2016 *J. Comput. Electron* **15** 1130
[49] Mech B C, Koley K and Kumar J 2018 *IEEE Trans. Electron Devices* **65** 4694
[50] Wilson L S J, Barker J R and Martinez A E 2019 *J. Phys.: Condens. Matter* **31** 144003
[51] Bousari N B, Anvarifard N K and Haji-Nasiri S 2020 *Silicon* **12** 2221
[52] Fiori G and Iannaccone G 2009 *IEEE Electron Device Lett.* **30** 1096
[53] Khaliji K *et al* 2013 *IEEE Trans. Electron Devices* **60** 2464
[54] Agarwal T K *et al* 2014 *IEEE Electron Device Lett.* **35** 1308
[55] Nazari A, Faez R and Shamloo H 2016 *Superlattices Microstruct.* **97** 28
[56] Tiwari L D and Sivasankaran K 2018 *Superlattices Microstruct.* **113** 244
[57] Eshkalak M A and Anvarifard M A 2015 *ECS J. Solid State Sci. Technol* **5** M141
[58] Quhe R *et al* 2018 *Phys. Rev. Appl.* **10** 024022
[59] Song X J *et al* 2017 *J. Appl. Phys.* **121** 144505
[60] Ferry D K 2023 *Semicond. Sci. Technol.* **38** 055005
[61] Ridley B K 1982 *Quantum Processes in Semiconductors* 1st edn (Oxford: Clarendon)
[62] Ferry D K 2020 *Semiconductors: Bonds and Bands* 2nd edn (Bristol: IOP Publishing)
[63] Long D 1960 *Phys. Rev.* **120** 2024
[64] Eaves L *et al* 1975 *J. Phys.* C **8** 1034
[65] Canali C *et al* 1975 *Phys. Rev.* B **12** 2265
[66] Ferry D K 1976 *Phys. Rev.* **B14** 1605
[67] Siegel W, Heinrich A and Ziegler E 1976 *Phys. Status Solidi* A **35** 269
[68] Ferry D K, Barker J R and Grubin H L 1981 *IEEE Trans. Electron Devices* **28** 905

www.ingramcontent.com/pod-product-compliance
Ingram Content Group UK Ltd.
Pitfield, Milton Keynes, MK11 3LW, UK
UKHW051909100926
8556IPUK00005B/825